Evolutionary Algorithms
in Engineering Applications

Springer

Berlin
Heidelberg
New York
Barcelona
Budapest
Hong Kong
London
Milan
Paris
Santa Clara
Singapore
Tokyo

D. Dasgupta
Z. Michalewicz (Eds.)

Evolutionary Algorithms in Engineering Applications

**With 150 Figures
and 61 Tables**

 Springer

Editors:

Dipankar Dasgupta
University of Memphis
Mathematical Sciences Department
Memphis, TN 38152-6429, USA
dasgupta@zebra.msci.memphis.edu

Zbigniew Michalewicz
University of North Carolina
Department of Computer Science
9201 University City Boulevard
Charlotte, NC 28223-0001, USA
zbyszek@uncc.edu

Cataloging-in-Publication Data applied for

Die Deutsche Bibliothek - CIP-Einheitsaufnahme

Evolutionary algorithms in engineering applications / D.
Dasgupta ; Z. Michalewicz ed. - Berlin ; Heidelberg ; New
York ; Barcelona ; Budapest ; Hong Kong ; London ; Milan ;
Paris ; Santa Clara ; Singapore ; Tokyo : Springer, 1997
   ISBN 3-540-62021-4
NE: Dasgupta, Dipankar [Hrsg.]

ISBN 3-540-62021-4   Springer-Verlag Berlin Heidelberg New York

© Springer-Verlag Berlin Heidelberg 1997
Printed in Germany

The use of registered names, trademarks, etc. in this publication does not imply, even in the absence of a specific statement, that such names are exempt from the relevant protective laws and therefore free for general use.

Typesetting: Camera-ready by the editors
Cover design: *design & production* GmbH, Heidelberg
SPIN: 10538461   45/3142-5 4 3 2 1 0 – Printed on acid-free paper.

Preface

Evolutionary Algorithms (EAs) are general-purpose search procedures based on the mechanisms of natural selection and population genetics. These algorithms are appealing to many users in different areas of engineering, computer science, and operation research, to name a few, due to their simplicity, ease of interfacing, and extensibility. EAs have attracted wide attention and found a growing number of applications, especially in the last decade.

Evolutionary algorithms—genetic algorithms, evolution strategies, evolutionary programming, and genetic programming are increasingly used to great advantage in applications as diverse as architectural design, engineering optimization, factory job scheduling, electronic circuit design, signal processing, network configuration, robotic control, etc.

This book, entitled *Evolutionary Algorithms in Engineering Applications*, is concerned with applications of evolutionary algorithms and associated strategies in the domain of engineering. The book covers a wide range of topics from major fields of engineering in a systematic and coherent fashion and shows the elegance of their implementations. The book is intended to serve the needs of engineers, designers, developers, and researchers in any scientific discipline who are interested in applications of evolutionary algorithms. The book includes descriptions of many real-world problems together with their evolutionary solutions; we hope the reader will appreciate their simplicity and elegance.

Apart from the introductory part, this book consists of four other parts: Architecture and Civil Engineering, Computer Science and Engineering, Electrical and Control Engineering, and Mechanical and Industrial Engineering.

The introductory part consists of two chapters. The chapter by Dasgupta and Michalewicz deals with theoretical and computational aspects of evolutionary algorithms. It describes different variants of EAs such as genetic algorithms (GA), evolutionary strategies (ES), evolutionary programming (EP), and genetic programming (GP). It also discusses other (hybrid) methods and various constraint-handling techniques in evolutionary computation.

The next chapter, by Ronald, proposes a robust encoding which may be effective for a wide range of problem instances. In particular, the encoding technique is amenable to modification or extension to solve different problem types.

Part II deals with architecture and civil engineering problems. The chapter by Gero, Kazakov, and Schnier describes an extension of genetic algorithms which uses a set of ideas borrowed from the practice of genetic engineering of natural organisms. They show that genetic engineering techniques can be used in the solution of a spatial layout planning problem, a class of NP-complete problems which cannot be solved directly using traditional genetic algorithms.

Rosenman (next chapter) presents an evolutionary approach to architectural design using a hierarchical growth model. He argues that the evolutionary approach fits well to the generate-and-test approach in design and is especially suited to non-routine design situations where the (inter-)relationships between complex arrangements of elements and their behavior are not known. A bottom-up hierarchical model is used to avoid the combinatorial problems involved in linear models. In this approach, a genotype consists of chromosomes which comprise genes representing design grammar rules. Evaluation is carried out both through the use of a fitness function and through human interaction. Use of these concepts is demonstrated in the context of the design of house plans.

LeRiche and Haftka's chapter on optimization of composite structures discusses evolutionary techniques to optimize laminated composite structures. This chapter describes successful experiments with a Wing Box problem to minimize the thickness of the laminates in the skin.

The next chapter, by Louis, Zhao, and Zeng, describes a methodology based on genetic algorithms for the solution of inverse problems pertaining to the identification of flaw shapes and the reconstruction of boundary conditions in continua. They investigate the feasibility and utility of the approach through a series of numerical experiments involving the detection of rectangular flaws of various sizes and orientations.

King, Fahmy, and Wentzel, in their chapter, apply an evolutionary approach to the problem of optimizing the operation of a river/reservoir system for maximum economic return. They show the effectiveness of the genetic approach to a complex hydraulic/economic problem based on the Rio Grande Project in southern New Mexico.

In the following chapter, Roston explores some of the hazards associated with genetics-based design methodologies. This chapter highlights the two greatest dangers of biased artifact representations and unrepresentative evaluations. According to him, by focusing activities in these areas, researchers can facilitate the development of these key concepts which can then be applied to any number of specific methodologies.

Part III discusses some interesting problems from the area of computer science. The contribution by Pettey, White, Dowdy, and Burkhead deals with the identification and characterization of workload classes in computer systems. An evolutionary technique is used to search for the appropriate number of workload classes and is able to apply an evaluation function which is not based on Euclidean distances. The proposed technique appears to outperform K-means clustering, an accepted workload characterization technique.

Data compression (or source coding) is the process of creating binary representations of data which require less storage space than the original data. Finding the optimal way to compress data with respect to resource constraints remains one of the most challenging problems in the field of source coding. Ng, Choi, and Ravishankar use GAs for performing lossless and lossy compressions, respectively, on text data and Gaussian-Markov sources. They show that, in most of the test cases, the genetic approach yields better solutions than the conventional techniques.

Cedeño and Vemuri apply a Multi-Niche Crowding Genetic Algorithm (MNC GA) model to the File Design Problem (which is an NP-hard problem). They use a parallel MNC GA model to exploit the multimodality inherent in the File Design Problem and perform a more balanced search over the entire space.

The next chapter, by Yue and Lilja, describ a distributed genetic algorithm, running on a workstation and two multiprocessor systems exploiting current technologies in distributed computing, multithreading, and parallel processing to design new processor scheduling algorithms. It demonstrates how to automatically generate new scheduling algorithms in a relatively short period of time that produce good performance when subsequently executing similar types of parallel application programs on a shared-memory multiprocessor system.

Sen, Knight, and Legg propose a prototype learning system for constructing appropriate prototypes from classified training instances. After constructing a set of prototypes for each of the possible classes, the class of a new input instance is determined by the nearest prototype to this instance. Attributes are assumed to be ordinal in nature and prototypes are represented as sets of feature-value pairs. A genetic algorithm is used to evolve the number of prototypes per class and their positions on the input space.

The chapter by Baluja, Sukthankar, and Hancock investigates a reasoning system that combines high-level task goals with low-level sensor constraints to control simulated vehicles. Evolutionary algorithms are used to automatically configure a collected set of distributed reasoning modules for intelligent behavior in the tactical driving domain.

Iba, Iwata, and Higuchi describe the fundamental principle of the gate-level Evolvable Hardware (EHW) which modifies its own hardware structure according to changes environment in its. EHW is implemented on a programmable logic device (PLD), whose architecture can be altered by downloading a binary bit string, i.e., architecture bits. The architecture bits are adaptively acquired by genetic algorithms. The effectiveness of this approach is shown by comparative experiments and a successful application.

The task of VLSI physical design is to produce the layout of an integrated circuit. New performance requirements are becoming increasingly dominant in today's sub-micron regimes requiring new physical design algorithms. Lienig applies genetic algorithms in VLSI physical design to produce the layout of an integrated circuit. In his chapter, a parallel genetic algorithm for the routing problem in VLSI circuits is discussed.

Ieumwananonthachai and Wah present a method for generalizing performance-related heuristics learned by genetics-based learning for knowledge-lean applications. This is done by first applying genetics-based learning to learn new heuristics for some small subsets of test cases in a problem space, then by generalizing these heuristics to unlearned subdomains of test cases. Experimental results on generalizing heuristics learned for sequential circuit testing, VLSI cell placement and routing, branch-and-bound search, and blind equalization are reported.

Part IV deals with electrical, control, and signal processing problems. The optimal scheduling of power generation is a major problem in electricity generating industries. This scheduling problem is complex because of many constraints which cannot be violated while finding optimal or near-optimal scheduling. Dasgupta presents a GA-based scheduler for determining a near-optimal commitment order of thermal units in a small power system. He showes that GA-based methods can handle multi-level constraints efficiently to produce reduced cost unit commitment schedules.

The use of genetic algorithms and genetic programming in control engineering has started to expand because of the physical cost of implementing a known control algorithm and the difficulty of finding such an algorithm for complex plants. Dracopoulos reviews some of the most successful applications of genetic algorithms and genetic programming in control engineering and outlines some general principles behind such applications. In particular, he shows that some evolutionary hybrid methods are very powerful and successful for the control of complex nonlinear dynamic systems.

Li, Tan, and Gong develop a Boltzmann learning refined evolutionary method to perform model reduction for systems and control engineering applications. They show that the evolutionary approach is uniformly applicable to both continuous and discrete time systems in both the time and the frequency domains. Three examples involving process and aircraft model reductions verify that the approach not only offers higher quality and tractability than conventional methods, but also requires no *a priori* starting points.

Adaptive digital filters have been used for several decades to model systems whose properties are *a priori* unknown. Pole-zero modeling using an output error criterion involves finding an optimum point on a (potentially) multimodal error surface. The chapter by White and Flockton discusses the application of evolutionary algorithms as an alternative scheme for recursive adaptive filters when it is likely that the error surface to be explored is multimodal.

Edwards and Keane describe an evolutionary method that may be used to efficiently locate and track underwater sonar targets in the near-field, with both bearing and range estimation, for the case of very large passive sonar arrays. The method gives the precise location of targets with noise free data, and still performs well when the data are corrupted by noise. It is based on a simple power maximization approach but one that is corrected for target ranges as well as bearings and also allows for array shape changes. The array shapes are estimated using array compass data and further corrections are made by allowing for range and bearing shifts that occur while taking coherently overlapped Fast Fourier Transform averages in the time domain.

Radar imaging is an advanced remote sensing technique that maps the reflectivity of distant objects by transmitting modulated signals at radio frequencies and processing the detected echoes. By proper waveform selection, it is currently possible to image the surface of planets or asteroids from Earth with a relatively high degree of resolution, despite the astronomical distances to these objects. The chapter by Flores, Kreinovich, and Vasquez discusses the implementation of a genetic algorithm in the field of radar imaging.

Smith and Dunay apply a proprietary evolutionary algorithm to the task of discriminating stationary targets from clutter in both a radar and an imaging sensor. Three applications for radar target detection are considered. The EAs are used to derive an algorithm for radar target detection, to derive an algorithm for imaging sensor target detection, and to derive a fused detector, which combines the information from both sensors to make a better detection decision. A technique for real adaptation of target acquisition is also discussed.

In Part V of this volume, a few mechanical and industrial engineering problems are discussed. Parmee investigates the utility of evolutionary/adaptive search within the generic domain of the engineering design process as a whole. The objective is to develop co-operative frameworks involving a number of evolutionary/adaptive computing techniques and integrate them with each stage of the engineering design process.

The identification of mechanical inclusion, even in the linear elasticity framework, is a difficult problem, theoretically ill-posed: evolutionary algorithms are in that context a good tentative choice for a robust numerical method, as standard deterministic algorithms have proven inaccurate and unstable. Schoenauer, Jouve, and Kallel apply non-standard representations together with *ad hoc* genetic operators in EAs for non-destructive inclusion identification in structural mechanics.

In the next chapter, Deb uses a combination of binary GAs and real-coded GAs for solving nonlinear engineering design optimization problems involving mixed variables (zero-one, discrete, continuous). The approach is called *GeneAS* to abbreviate *Gene*tic *A*daptive *S*earch. The efficacy of GeneAS is demonstrated by solving three different mechanical component design problems, whose solutions were attempted using various traditional optimization algorithms. In all cases, it is observed that GeneAS has been able to find a better solution than the best previously available solutions.

The chapter by Ono and Watanabe describes two kinds of optimal material-cutting methods using genetic algorithms. The first is the optimal cutting of bars to satisfy two requirements on the minimum length of scrap produced and the balance among the number of finished products of each length to meet customer's order, when a series of raw bars are cut to various ordered lengths. The second is to determine the layout of various patterns to be cut on a sheet to make the required sheet length minimum. By combining layout determining algorithms (LDAs) with order-type genetic algorithms, the required calculation time is considerably reduced.

Scheduling problems occur wherever there is a need to perform a number of tasks with limited resources. This very wide definition naturally covers many different kinds of problem. Corne and Ross provide a brief survey to reflect the strengths and direction of EA-based scheduling with special emphasis on job-shop scheduling and related problems.

The successful manufacturing operation must focus outside of the walls of the manufacturing facility to accommodate a global supply chain, disaggregation of enterprise production resources, and customer focused product distribution. Fulkerson presents some examples of current and anticipated methods of operation and discusses the use of evolutionary computing to meet those challenges. According to him, decentralized control and emergent behaviors available from evolutionary computation offer the promise of adaptive solutions to organizational and systems design that can adapt quickly to customer driven requirements.

Editing a book that covers a wide range of engineering applications is a very difficult and time consuming task. The book could not have been successfully accomplished without support and constructive feedback from all the contributors. We would like to thank them for their effort, for reviewing each others' work, and for providing feedback on their own chapters.

Also, we express our gratitude to the executive editor of Springer-Verlag, Hans Wössner, for his help throughout the project, and to Frank Holzwarth (Springer-Verlag) for his help in formatting the final, camera-ready version of this book.

D. Dasgupta
Z. Michalewicz

March 1997

Table of Contents

Part IV Electrical, Control and Signal Processing

List of Contributors

Shumeet Baluja
School of Computer Science
Carnegie Mellon University
Pittsburgh
PA 15213, USA
baluja@cs.cmu.edu

Darrell Burkhead
Department of Computer
Science
Box 1679 Station B
Vanderbilt University
Nashville
TN 37064, USA
burkhead@vuse.vanderbilt.edu

Walter Cedeño
Independent Consultant
446 Creekside Drive
Downingtown
PA 19335, USA
wcedeno@llnl.gov

Sunghyun Choi
Electrical Engineering and
Computer Science Department
University of Michigan
Ann Arbor
MI 48109, USA
shchoi@eecs.umich.edu

Dave Corne
Department of Computer
Science
University of Reading
Whiteknights
PO Box 220 Reading
RG6 2AX, UK
D.W.Corne@reading.ac.uk

Dipankar Dasgupta
Mathematical Science Department
Room No. 378, Dunn Hall
The University of Memphis
Memphis, TN 38152, USA
dasgupta@zebra.msci.memphis.edu

Kalyan Deb
Department of Mechanical
Engineering
Indian Institute of Technology
Kanpur, UP 208 016
India
deb@iitk.ernet.in

Larry Dowdy
Department of Computer Science
Box 1679 Station B
Vanderbilt University
Nashville
TN 37064, USA
dowdy@vuse.vanderbilt.edu

Dimitris C. Dracopoulos
Department of Computer
Science
Brunel University
London Uxbridge Middlesex
UB8 3PH, UK
Dimitris.Dracopoulos@brunel.ac.uk

Bertrand D. Dunay
System Dynamics International
512 Rudder Road
St. Louis
MO 63026, USA
dunay@mo.net

D. J. Edwards
Department of Engineering
Science
University of Oxford
Park Road, Oxford
OX1 3PJ, UK
david.edwards@eng.ox.ac.uk

Hazem S. Fahmy
Department of Civil,
Agricultural and Geological
Engineering
New Mexico State University
Las Cruces
NM 88003, USA
hfahmy@nmsu.edu

Stuart J. Flockton
Royal Holloway
University of London
Egham Surrey
TW20 0EX, UK
S.Flockton@sun.rhbnc.ac.uk

Benjamin C. Flores
Department of Electrical and
Computer Engineering
University of Texas at El Paso
El Paso
TX 79968, USA
flores@ece.utep.edu

Bill Fulkerson
Deere & Company
Computer Information Systems
Technology Integration
John Deere Road
Moline
IL 61265, USA
wf28155@deere.com

John S. Gero
Key Centre of Design
Computing
Department of Architectural and
Design Science
The University of Sydney
NSW 2006
Australia
john@archsci.arch.su.edu.au

Mingrui Gong
Centre for Systems and Control
Department of Electronics and
Electrical Engineering
University of Glasgow
Rankine Building
Glasgow
G12 8LT, UK
M.Gong@elec.gla.ac.uk

Raphael T. Haftka
Department of Aerospace
Engineering
The University of Florida
Gainesville
FL 32611, USA
haftka@ufl.edu

John Hancock
School of Computer Science
Carnegie Mellon University
Pittsburgh
PA 15213, USA
jhancock@cs.cmu.edu

Tetsuya Higuchi
Machine Interface Section
Electrotechnical Laboratory
1-1-4 Umezono
Tsukuba
Ibaraki 305
Japan
higuchi@etl.go.jp

Hitoshi Iba
Machine Interface Section
Electrotechnical Laboratory
1-1-4 Umezono
Tsukuba
Ibaraki 305
Japan
iba@etl.go.jp

Arthur Ieumwananonthachai
Department of Electrical and
Computer Engineering
University of Illinois at
Urbana-Champaign
Urbana
IL 61801, USA
arthuri@manip.crhc.uiuc.edu

Masaya Iwata
Machine Interface Section
Electrotechnical Laboratory
1-1-4 Umezono
Tsukuba
Ibaraki 305
Japan
miwata@etl.go.jp

Francois Jouve
CMAP - URA CNRS 756
Ecole Polytechnique
F-91128 Palaiseau
France
Francois.Jouve@polytechnique.fr

Leila Kallel
CMAP - URA CNRS 756
Ecole Polytechnique
F-91128 Palaiseau
France
Leila.Kallel@polytechnique.fr

Vladimir A. Kazakov
Key Centre of Design
Computing
Department of Architectural and
Design Science
The University of Sydney
NSW 2006
Australia
kaz@arch.su.edu.au

Andy J. Keane
Department of Mechanical
Engineering
Southampton University
Highfield, Southampton
SO17 1BJ, UK
Andy.Keane@soton.ac.uk

Phillip King
Department of Civil,
Agricultural and Geological
Engineering
New Mexico State University
Las Cruces
NM 88003, USA
jpking@nmsu.edu

Leslie Knight
Department of Mathematics and
Computer Science
University of Tulsa
600 South College Avenue
Tulsa
OK 74104, USA
knight@euler.mcs.utulsa.edu

Vladik Kreinovich
Department of Computer
Science
University of Texas at El Paso
El Paso
TX 79968, USA
vladik@ece.utep.edu

Kevin Legg
Department of Mathematics and
Computer Science
University of Tulsa
600 South College Avenue
Tulsa
OK 74104, USA
legg@euler.mcs.utulsa.edu

Yun Li
Centre for Systems and Control
Department of Electronics and
Electrical Engineering
University of Glasgow
Rankine Building
Glasgow
G12 8LT, UK
Y.Li@elec.gla.ac.uk

Jens Lienig
Tanner Research
180 North Vinedo Avenue
Pasadena
CA 91107, USA
jens.lienig@tanner.com

David J. Lilja
Department of Electrical
Engineering
University of Minnesota
200 Union St. S.E.
Minneapolis
MN 55455, USA
lilja@ee.umn.edu

Sushil J. Louis
Department of Computer
Science
University of Nevada
Reno
NV 89557, USA
sushil@cs.unr.edu

Zbigniew Michalewicz
Department of Computer
Science
University of North Carolina
Charlotte
NC 28223, USA
zbyszek@uncc.edu

Wee K. Ng
Electrical Engineering and
Computer Science Department
University of Michigan
Ann Arbor
MI 48109, USA
wkn@eecs.umich.edu

Toshihiko Ono
Department of Communication
and Computer Engineering
Fukuoka Institute of Technology
3-30-1 Wajiro-higashi
Higashi-ku
Fukuoka 811-02
Japan
ono@cs.fit.ac.jp

Ian C. Parmee
Plymouth Engineering Design
Centre
University of Plymouth
Drakes Circus
Plymouth
PL4 8AA, UK
iparmee@soc.plym.ac.uk

Chrisila C. Pettey
Department of Computer
Science
Middle Tennessee State University
Murfreesboro
TN 37132, USA
cscbp@knuth.mtsu.edu

Chinya V. Ravishankar
Electrical Engineering and
Computer Science Department
University of Michigan
Ann Arbor
MI 48109, USA
ravi@eecs.umich.edu

Rodolphe G. Riche
Centre des Matériaux Ecole des Mines
de Paris
BP87
F-91003 Evry Cedex
France
leriche@mat.ensmp.fr

Simon Ronald
GPO Box 944
Adelaide
SA 5001
Australia
dna@createwin.com

Peter Ross
Department of Artificial
Intelligence
University of Edinburgh
80 South Bridge
Edinburgh
EH1 1HN, UK
peter@aisb.ed.ac.uk

Gerald P. Roston
Cybernet System Corporation
727 Airport Blvd
Ann Arbor
MI 48108, USA
gerry@cybernet.com

T. Schnier
Key Centre of Design
Computing
Department of Architectural and
Design Science
The University of Sydney
NSW 2006
Australia
thorsten@arch.su.edu.au

Marc Schoenauer
CMAP - URA CNRS 756
Ecole Polytechnique
F-91128 Palaiseau
France
Marc.Schoenauer@polytechnique.fr

Sandip Sen
Department of Mathematics and
Computer Science
University of Tulsa
600 South College Avenue
Tulsa
OK 74104, USA
sandip@euler.mcs.utulsa.edu

Steven P. Smith
System Dynamics International
512 Rudder Road
St. Louis
MO 63026, USA
sdi@mo.net

Rahul Sukthankar
School of Computer Science
Carnegie Mellon University
Pittsburgh
PA 15213, USA
rahuls@cs.cmu.edu

Kay Chen Tan
Centre for Systems and Control
Department of Electronics and
Electrical Engineering
University of Glasgow
Rankine Building
Glasgow
G12 8LT, UK
K.Tan@elec.gla.ac.uk

Roberto Vasquez
Department of Computer
Science
University of Texas at El Paso
El Paso
TX 79968, USA
robert@ece.utep.edu

V. Rao Vemuri
Department of Applied Science
University of California at Davis
Livermore
CA 94550, USA
vemuri@icdc.llnl.gov

Benjamin B. Wah
Department of Electrical and
Computer Engineering
University of Illinois at
Urbana-Champaign
Urbana
IL 61801, USA
wah@manip.crhc.uiuc.edu

Gen Watanabe
Department of Communication and
Computer Engineering
Fukuoka Institute of Technology
3-30-1 Wajiro-higashi
Higashi-ku
Fukuoka 811-02
Japan
mfm95016@ws.ipc.fit.ac.jp

Mark W. Wentzel
Department of Civil, Agricultural
and Geological Engineering
New Mexico State University
Las Cruces
NM 88003, USA
mwentzel@cage.nmsu.edu

Trish White
Department of Computer Science
Middle Tennessee State University
Murfreesboro
TN 37132, USA
cosc001c@knuth.mtsu.edu

Michael S. White
Royal Holloway University
of London
Egham Surrey
TW20 0EX, UK
uhap081@sun.rhbnc.ac.uk

Kelvin K. Yue
Department of Computer Science
University of Minnesota
200 Union St. S.E.
Minneapolis
MN 55455, USA
yue@cs.umn.edu

Xiaogang Zeng
Algor Inc.
Pittsburgh
PA 15238, USA

Fang Zhao
Department of Civil Engineering
Florida International University
Miami
FL 33199, USA
zhao@fiu.edu

Part I

Introduction

Evolutionary Algorithms — An Overview

Dipankar Dasgupta[1] and Zbigniew Michalewicz[2]

[1] Department of Mathematics and Computer Science, University of Missouri, St. Louis, MO 63121, USA

[2] Department of Computer Science, University of North Carolina, Charlotte, NC 28223, USA, and Institute of Computer Science, Polish Academy of Sciences, ul. Ordona 21, 01-237 Warsaw, Poland

Summary. Evolutionary algorithms (EAs), which are based on a powerful principle of evolution: survival of the fittest, and which model some natural phenomena: genetic inheritance and Darwinian strife for survival, constitute an interesting category of modern heuristic search. This introductory article presents the main paradigms of evolutionary algorithms (genetic algorithms, evolution strategies, evolutionary programming, genetic programming) and discusses other (hybrid) methods of evolutionary computation. Also, various constraint-handling techniques in connection with evolutionary algorithms are discussed, since most engineering problems includes some problem-specific constraints.

Evolutionary algorithms have been widely used in science and engineering for solving complex problems. An important goal of research on evolutionary algorithms is to understand the class of problems for which EAs are most suited, and, in particular, the class of problems on which they outperform other search algorithms.

1. Introduction

During the last two decades there has been a growing interest in algorithms which are based on the principle of evolution (survival of the fittest). A common term, accepted recently, refers to such techniques as *evolutionary algorithms* (EA) (or *evolutionary computation* methods). The best known algorithms in this class include genetic algorithms, evolutionary programming, evolution strategies, and genetic programming. There are also many hybrid systems which incorporate various features of the above paradigms, and consequently are hard to classify; anyway, we refer to them just as EC methods.

The field of evolutionary computation has reached a stage of some maturity. There are several, well established international conferences that attract hundreds of participants (International Conferences on Genetic Algorithms—ICGA [46, 47, 96, 10, 38, 26], Parallel Problem Solving from Nature—PPSN [102, 62, 16, 111], Annual Conferences on Evolutionary Programming—EP [33, 34, 103, 63, 35]); new annual conferences are getting started, e.g., IEEE International Conferences on Evolutionary Computation [84, 85, 86]. Also, there are many workshops, special sessions, and local conferences every year, all around the world. A journal, *Evolutionary Computation* (MIT Press) [21], is devoted entirely to evolutionary computation techniques; the first issue of a new journal *IEEE Transactions on Evolutionary Computation* will appear in May 1997. Other journals organized special issues on evolutionary computation (e.g., [30, 67]). Many excellent tutorial papers [8, 9, 90, 112, 31] and technical reports provide more-or-less complete bibliographies of the field [1, 44, 95, 77]. There is also *The Hitch-Hiker's Guide to Evolutionary Computation* prepared initially by Jörg Heitkötter and currently by David Beasley [49], available on comp.ai.genetic interest group (Internet), and a new text, *Handbook of Evolutionary Computation*, is in its final stages of preparation [5].

In this introductory article, we provide a general overview of the field. The next section provides a short introductory description of evolutionary algorithms. Section 3. discusses the paradigms of genetic algorithms, evolution strategies, evolutionary programming, and genetic programming, as well as some other evolutionary techniques. Section 4. provides with an overview of constraint-handling methods which have been developed in connection with evolutionary algorithms, and section 5. summarizes this chapter.

2. Evolutionary Algorithms

In general, any abstract task to be accomplished can be thought of as solving a problem, which, in turn, can be perceived as a search through a space of potential solutions. Since usually we are after "the best" solution, we can view this task as an optimization process. For small spaces, classical exhaustive methods usually suffice; for larger spaces special artificial intelligence techniques must be employed. Evolutionary algorithms are among such techniques; they are stochastic algorithms whose search methods model some natural phenomena: genetic inheritance and Darwinian strife for survival. As stated in [20]:

"... the metaphor underlying genetic algorithms[1] is that of natural evolution. In evolution, the problem each species faces is one of searching for beneficial adaptations to a complicated and changing environment. The 'knowledge' that each species has gained is embodied in the makeup of the chromosomes of its members."

As already mentioned in the Introduction, there are several variants of evolutionary algorithms, and there are also many hybrid systems which incorporate various features of these paradigms; however, the structure of any evolutionary method is very much the same; a sample structure is shown in Figure 2.1.

The evolutionary algorithm maintains a population of individuals, $P(t) = \{x_1^t, \ldots, x_n^t\}$ for iteration t. Each individual represents a potential solution to the problem at hand, and is implemented as some data structure S. Each

---

[1] The best known evolutionary computation methods are genetic algorithms; very often the terms *evolutionary computation* methods and *GA-based* methods are used interchangeably.

```
procedure evolutionary algorithm
begin
    t ← 0
    initialize P(t)
    evaluate P(t)
    while (not termination-condition) do
    begin
        t ← t + 1
        select P(t) from P(t − 1)
        alter P(t)
        evaluate P(t)
    end
end
```

Fig. 2.1. The structure of an evolutionary algorithm

solution x_i^t is evaluated to give some measure of its "fitness". Then, a new population (iteration $t + 1$) is formed by selecting the more fit individuals (select step). Some members of the new population undergo transformations (alter step) by means of "genetic" operators to form new solutions. There are unary transformations m_i (mutation type), which create new individuals by a small change in a single individual ($m_i : S \to S$), and higher order transformations c_j (crossover type), which create new individuals by combining parts from several (two or more) individuals ($c_j : S \times \ldots \times S \to S$).[2] After some number of generations the algorithm converges—it is hoped that the best individual represents a near-optimum (reasonable) solution.

Despite powerful similarities between various evolutionary algorithms, there are also many differences between them (often hidden on a lower level of abstraction). They use different data structures S for their chromosomal representations, consequently, the 'genetic' operators are different as well. They may or may not incorporate some other information (to control the search process) in their genes. There are also other differences; for example, the two lines of the Figure 2.1:

select $P(t)$ from $P(t − 1)$
alter $P(t)$

can appear in the reverse order: in evolution strategies first the population is altered and later a new population is formed by a selection process (see section 3.2). Moreover, even within a particular technique there are many flavors and twists. For example, there are many methods for selecting individuals for survival and reproduction. These methods include (1) proportional selection, where the probability of selection is proportional to the individual's fitness, (2) ranking methods, where all individuals in a population are sorted from

---

[2] In most cases, crossover involves just two parents, however, it need not be the case. In a recent study [25] the authors investigated the merits of 'orgies', where more than two parents are involved in the reproduction process. Also, scatter search techniques [39] proposed the use of multiple parents.

the best to the worst and probabilities of their selection are fixed for the whole evolution process,[3] and (3) tournament selection, where some number of individuals (usually two) compete for selection to the next generation: this competition (tournament) step is repeated population-size number of times. Within each of these categories there are further important details. Proportional selection may require the use of scaling windows or truncation methods, there are different ways for allocating probabilities in ranking methods (linear, nonlinear distributions), the size of a tournament plays a significant role in tournament selection methods. It is also important to decide on a generational policy. For example, it is possible to replace the whole population by a population of offspring, or it is possible to select the best individuals from two populations (population of parents and population of offspring)— this selection can be done in a deterministic or nondeterministic way. It is also possible to produce few (in particular, a single) offspring, which replace some (the worst?) individuals (systems based on such generational policy are called 'steady state'). Also, one can use an 'elitist' model which keeps the best individual from one generation to the next[4]; such model is very helpful for solving many kinds of optimization problems.

However, the representation used for a particular problem together with a set of 'genetic' operators constitute the most essential components of any evolutionary algorithm. These are the key elements which allow us to distinguish between various paradigms of evolutionary methods. We discuss this issue in detail in the following section.

3. Main Paradigms of Evolutionary Algorithms

As indicated earlier, there are a few main paradigms of evolutionary computation techniques. In the following subsections, we discuss them in turn; the discussion puts some emphasis on the data structures and genetic operators used by these techniques.

3.1 Genetic Algorithms

The beginnings of genetic algorithms can be traced back to the early 1950s when several biologists used computers for simulations of biological systems [40]. However, the work done in late 1960s and early 1970s at the University of Michigan under the direction of John Holland led to genetic algorithms as

---

[3] For example, the probability of selection of the best individual is always 0.15 regardless its precise evaluation; the probability of selection of the second best individual is always 0.14, etc. The only requirements are that better individuals have larger probabilities and the total of these probabilities equals to one.

[4] It means, that if the best individual from a current generation is lost due to selection or genetic operators, the system force it into next generation anyway.

they are known today. A GA performs a multi-directional search by maintaining a population of potential solutions and encourages information formation and exchange between these directions.

Genetic algorithms (GAs) were devised to model *adaptation processes*, mainly operated on binary strings and used a recombination operator with mutation as a background operator [50]. Mutation flips a bit in a chromosome and crossover exchanges genetic material between two parents: if the parents are represented by five-bits strings, say $(0, 0, 0, 0, 0)$ and $(1, 1, 1, 1, 1)$, crossing the vectors after the second component would produce the offspring $(0, 0, 1, 1, 1)$ and $(1, 1, 0, 0, 0)$.[5] Fitness of an individual is assigned proportionally to the value of the objective function for the individual; individuals are selected for next generation on the basis of their fitness.

The combined effect of selection, crossover, and mutation gives so-called the reproductive schema growth equation [50]:

$$\xi(S, t+1) \geq \xi(S, t) \cdot eval(S, t) / \overline{F(t)} \left[1 - p_c \cdot \frac{\delta(S)}{m-1} - o(S) \cdot p_m \right],$$

where S is a schema defined over the alphabet of 3 symbols ('0', '1', and '$\star$' of length m; each schema represents all strings which match it on all positions other than '$\star$'); $\xi(S, t)$ denoted the number of strings in a population at the time t, matched by schema S; $\delta(S)$ is the defining length of the schema S — the distance between the first and the last fixed string positions; $o(S)$ denotes the order of the schema S — the number of 0 and 1 positions present in the schema; Another property of a schema is its *fitness* at time t, $eval(S, t)$ is defined as the average fitness of all strings in the population matched by the schema S; and $F(t)$ is the total fitness of the whole population at time t. Parameters p_c and p_m denote probabilities of crossover and mutation, respectively.

The above equation tells us about the expected number of strings matching a schema S in the next generation as a function of the actual number of strings matching the schema, the relative fitness of the schema, and its defining length and order. Again, it is clear that above-average schemata with short defining length and low-order would still be sampled at exponentially increased rates.

The growth equation shows that selection increases the sampling rates of the above-average schemata, and that this change is exponential. The sampling itself does not introduce any new schemata (not represented in the initial $t = 0$ sampling). This is exactly why the crossover operator is introduced — to enable structured, yet random information exchange. Additionally, the mutation operator introduces greater variability into the population. The combined (disruptive) effect of these operators on a schema is not significant if the schema is short and low-order. The final result of the growth equation can be stated as:

---

[5] This is an example of so-called 1-point crossover. However, multi-point, uniform and other crossover operators are also being used.

Schema Theorem: Short, low-order, above-average schemata receive exponentially increasing trials in subsequent generations of a genetic algorithm.

An immediate result of this theorem is that GAs explore the search space by short, low-order schemata which, subsequently, are used for information exchange during crossover:

Building Block Hypothesis: A genetic algorithm seeks near-optimal performance through the juxtaposition of short, low-order, high-performance schemata, called the building blocks.

As stated in [40]:

"Just as a child creates magnificent fortresses through the arrangement of simple blocks of wood, so does a genetic algorithm seek near optimal performance through the juxtaposition of short, low-order, high performance schemata."

A population of *pop_size* individuals of length m processes at least 2^m and at most 2^{pop_size} schemata. Some of them are processed in a useful manner: these are sampled at the (desirable) exponentially increasing rate, and are not disrupted by crossover and mutation (which may happen for long defining length and high-order schemata).

Holland [50] showed, that at least pop_size^3 of them are processed usefully — he has called this property an *implicit parallelism*, as it is obtained without any extra memory/processing requirements. It is interesting to note that in a population of *pop_size* strings there are many more than *pop_size* schemata represented. This constitutes possibly the only known example of a combinatorial explosion working to our advantage instead of our disadvantage.

To apply a GA to a particular problem, it is necessary to design a mapping between a space of potential solutions for the problem and a space of binary strings of some length. Sometimes it is not trivial task and quite often the process involved some additional heuristics (decoders, problem-specific operators, etc). For additional material on applications of genetic algorithms, see, for example, [65].

3.2 Evolution Strategies

Evolution strategies (ESs) were developed as a method to solve parameter optimization problems [100]; consequently, a chromosome represents an individual as a pair of float-valued vectors,[6] i.e., $\mathbf{v} = (\mathbf{x}, \sigma)$.

The earliest evolution strategies were based on a population consisting of one individual only. There was also only one genetic operator used in the

---

[6] However, they started with integer variables as an experimental optimum-seeking method.

evolution process: a mutation. However, the interesting idea (not present in GAs) was to represent an individual as a pair of float–valued vectors, i.e., $\mathbf{v} = (\mathbf{x}, \boldsymbol{\sigma})$. Here, the first vector $\mathbf{x}$ represents a point in the search space; the second vector $\boldsymbol{\sigma}$ is a vector of standard deviations: mutations are realized by replacing $\mathbf{x}$ by

$$\mathbf{x}^{t+1} = \mathbf{x}^t + N(0, \boldsymbol{\sigma}),$$

where $N(0, \boldsymbol{\sigma})$ is a vector of independent random Gaussian numbers with a mean of zero and standard deviations $\boldsymbol{\sigma}$. (This is in accordance with the biological observation that smaller changes occur more often than larger ones.) The offspring (the mutated individual) is accepted as a new member of the population (it replaces its parent) iff it has better fitness and all constraints (if any) are satisfied. For example, if f is the objective function without constraints to be maximized, an offspring $(\mathbf{x}^{t+1}, \boldsymbol{\sigma})$ replaces its parent $(\mathbf{x}^t, \boldsymbol{\sigma})$ iff $f(\mathbf{x}^{t+1}) > f(\mathbf{x}^t)$. Otherwise, the offspring is eliminated and the population remain unchanged.

The vector of standard deviations $\boldsymbol{\sigma}$ remains unchanged during the evolution process. If all components of this vector are identical, i.e., $\boldsymbol{\sigma} = (\sigma, \dots, \sigma)$, and the optimization problem is *regular*[7], it is possible to prove the convergence theorem [6]:

> **Convergence Theorem**: For $\sigma > 0$ and a regular optimization problem with $f_{opt} > -\infty$ (minimalization) or $f_{opt} < \infty$ (maximization),
> $$p\{\lim_{t \to \infty} f(\mathbf{x}^t) = f_{opt}\} = 1$$
> holds.

The evolution strategies evolved further [100] to mature as

$$(\mu + \lambda)\text{–ESs and } (\mu, \lambda)\text{–ESs};$$

the main idea behind these strategies was to allow control parameters (like mutation variance) to self-adapt rather than changing their values by some deterministic algorithm.

In the $(\mu + \lambda)$–ES, μ individuals produce λ offspring. The new (temporary) population of $(\mu + \lambda)$ individuals is reduced by a selection process again to μ individuals. On the other hand, in the (μ, λ)–ES, the μ individuals produce λ offspring $(\lambda > \mu)$ and the selection process selects a new population of μ individuals from the set of λ offspring only. By doing this, the life of each individual is limited to one generation. This allows the (μ, λ)–ES to perform

---

[7] An optimization problem is regular if the objective function f is continuous, the domain of the function is a closed set, for all $\epsilon > 0$ the set of all internal points of the domain for which the function differs from the optimal value less than ϵ is non-empty, and for all $\mathbf{x}_0$ the set of all points for which the function has values less than or equal to $f(\mathbf{x}_0)$ (for minimalization problems; for maximization problems the relationship is opposite) is a closed set.

better on problems with an optimum moving over time, or on problems where
the objective function is noisy.

The operators used in the $(\mu + \lambda)$–ESs and (μ, λ)–ESs incorporate two-
level learning: their control parameter σ is no longer constant, nor it is
changed by some deterministic algorithm (like the $1/5$ success rule), but it is
incorporated in the structure of the individuals and undergoes the evolution
process. To produce an offspring, the system acts in several stages:

- select two individuals,
    $(\mathbf{x}^1, \sigma^1) = ((x_1^1, \ldots, x_n^1), (\sigma_1^1, \ldots, \sigma_n^1))$ and
    $(\mathbf{x}^2, \sigma^2) = ((x_1^2, \ldots, x_n^2), (\sigma_1^2, \ldots, \sigma_n^2))$,
and apply a recombination (crossover) operator. There are two types of
crossovers:
- discrete, where the new offspring is
    $(\mathbf{x}, \sigma) = ((x_1^{q_1}, \ldots, x_n^{q_n}), (\sigma_1^{q_1}, \ldots, \sigma_n^{q_n}))$,
    where $q_i = 1$ or $q_i = 2$ (so each component comes from the first or second
    preselected parent),
- intermediate, where the new offspring is
    $(\mathbf{x}, \sigma) = (((x_1^1 + x_1^2)/2, \ldots, (x_n^1 + x_n^2)/2), ((\sigma_1^1 + \sigma_1^2)/2, \ldots, (\sigma_n^1 + \sigma_n^2)/2))$.
Each of these operators can be applied also in a global mode, where the
new pair of parents is selected for *each* component of the offspring vector.
- apply mutation to the offspring $(\mathbf{x}, \sigma)$ obtained; the resulting new offspring
is $(\mathbf{x}', \sigma')$, where
    $\sigma' = \sigma \cdot e^{N(0, \Delta\sigma)}$, and
    $\mathbf{x}' = \mathbf{x} + N(0, \sigma')$,
where $\Delta\sigma$ is a parameter of the method.

The best source of complete information (including recent results) on
evolution strategies is recent Schwefel's text [101].

3.3 Evolutionary Programming

The original evolutionary programming (EP) techniques were developed by
Lawrence Fogel [36]. They aimed at evolution of artificial intelligence in the
sense of developing ability to predict changes in an environment. The environment was described as a sequence of symbols (from a finite alphabet) and
the evolving algorithm supposed to produce, as an output, a new symbol.
The output symbol should maximize the payoff function, which measures the
accuracy of the prediction.

For example, we may consider a series of events, marked by symbols
$a_1, a_2, \ldots$; an algorithm should predict the next (unknown) symbol, say a_{n+1}
on the basis of the previous (known) symbols, $a_1, a_2, \ldots, a_n$. The idea of
evolutionary programming was to evolve such an algorithm.

Finite state machines (FSM) were selected as a chromosomal representation of individuals; after all, finite state machines provide a meaningful

representation of behavior based on interpretation of symbols. Figure 3.1 provides an example of a transition diagram of a simple finite state machine for a parity check. Such transition diagrams are directed graphs that contain a node for each state and edges that indicate the transition from one state to another, input and output values (notation a/b next to an edge leading from state S_1 to the state S_2 indicates that the input value of a, while the machine is in state S_1, results in output b and the next state S_2.

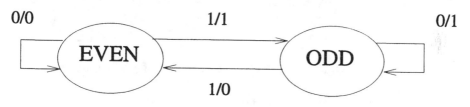

0/0   1/1   0/1

1/0

Fig. 3.1. A FSM for a parity check

There are two states 'EVEN' and 'ODD' (machine starts in state 'EVEN'); the machine recognizes a parity of a binary string.

So, evolutionary programming technique maintains a population of finite state machines; each such individual represents a potential solution to the problem (i.e., represents a particular behavior). As already mentioned, each FSM is evaluated to give some measure of its "fitness". This is done in the following way: each FSM is exposed to the environment in the sense that it examines all previously seen symbols. For each subsequence, say, $a_1, a_2, \ldots, a_i$ it produces an output a'_{i+1}, which is compared with the next observed symbol, a_{i+1}. For example, if n symbols were seen so far, a FSM makes n predictions (one for each of the substrings a_1, a_1, a_2, and so on, until $a_1, a_2, \ldots, a_n$); the fitness function takes into account the overall performance (e.g., some weighted average of accuracy of all n predictions).

Like in evolution strategies, evolutionary programming technique first creates offspring and later selects individuals for the next generation. Each parent produces a single offspring; hence the size of the intermediate population doubles (like in (pop_size, pop_size)-ES). Offspring (a new FSMs) are created by random mutations of parent population (see Figure 3.2). There are five possible mutation operators: change of an output symbol, change of a state transition, addition of a state, deletion of a state, and change of the initial state (there are some additional constraints on the minimum and maximum number of states). These mutations are chosen with respect to some probability distribution (which can change during the evolutionary process); also it is possible to apply more than one mutation to a single parent (a decision on the number of mutations for a particular individual is made with respect to some other probability distribution).

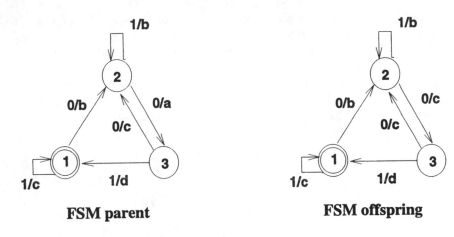

FSM parent   **FSM offspring**

Fig. 3.2. A FSM and its offspring. Machines start in state 1

The best *pop_size* individuals are retained for the next generation; i.e., to qualify for the next generation an individual should rank in the top 50% of the intermediate population. In original version [36] this process was iterated several times before the next output symbol was made available. Once a new symbol is available, it is added to the list of known symbols, and the whole process is repeated.

Of course, the above procedure can be extended in many way; as stated in [32]:

"The payoff function can be arbitrarily complex and can posses temporal components; there is no requirement for the classical squared error criterion or any other smooth function. Further, it is not required that the predictions be made with a one-step look ahead. Forecasting can be accomplished at an arbitrary length of time into the future. Multivariate environments can be handled, and the environmental process need not be stationary because the simulated evolution will adapt to changes in the transition statistics."

Recently evolutionary programming techniques were generalized to handle numerical optimization problems; for details see [28] or [32]. For other examples of evolutionary programming techniques, see also [36] (classification of a sequence of integers into primes and nonprimes), [29] (for application of EP technique to the interated prisoner's dilemma), as well as [33, 34, 103, 63] for many other applications.

3.4 Genetic Programming

Another interesting approach was developed relatively recently by Koza [58, 59]. Koza suggests that the desired program should evolve itself during the evolution process. In other words, instead of solving a problem, and instead of building an evolution program to solve the problem, we should rather search the space of possible computer programs for the best one (the most fit). Koza developed a new methodology, named Genetic Programming (GP), which provides a way to run such a search.

There are five major steps in using genetic programming for a particular problem. These are:

- selection of terminals,
- selection of a function,
- identification of the evaluation function,
- selection of parameters of the system, and
- selection of the termination condition.

It is important to note that the structure which undergoes evolution is a hierarchically structured computer program.[8] The search space is a hyperspace of valid programs, which can be viewed as a space of rooted trees. Each tree is composed of functions and terminals appropriate to the particular problem domain; the set of all functions and terminals is selected *a priori* in such a way that some of the composed trees yield a solution.

For example, two structures e_1 and e_2 (Figure 3.3) represent expressions $2x+2.11$ and $x \cdot \sin(3.28)$, respectively. A possible offspring e_3 (after crossover of e_1 and e_2) represents $x \cdot \sin(2x)$.

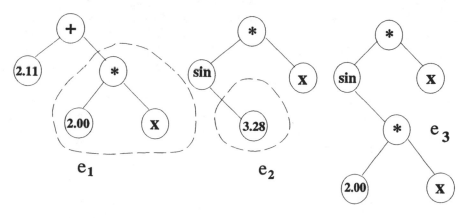

Fig. 3.3. Expression e_3: an offspring of e_1 and e_2. Broken line includes areas being exchanged during the crossover operation

The initial population is composed of such trees; construction of a (random) tree is straightforward. The evaluation function assigns a fitness value

---

[8] Actually, Koza has chosen LISP's S-expressions for all his experiments. Currently, however, there are implementations of GP in C and other programming languages.

which evaluates the performance of a tree (program). The evaluation is based on a preselected set of test cases; in general, the evaluation function returns the sum of distances between the correct and obtained results on all test cases. The selection is proportional; each tree has a probability of being selected to the next generation proportional to its fitness. The primary operator is a crossover that produces two offspring from two selected parents. The crossover creates offspring by exchanging subtrees between two parents. There are other operators as well: mutation, permutation, editing, and a define-building-block operation [58]. For example, a typical mutation selects a node in a tree and generates a new (random) subtree which originates in the selected node.

In addition to five major steps for building a genetic program for a particular problem, Koza [60] recently considered the advantages of adding an additional feature: a set of procedures. These procedures are called Automatically Defined Functions (ADF). It seems that this is an extremely useful concept for genetic programming techniques with its major contribution in the area of code reusability. ADFs discover and exploit the regularities, symmetries, similarities, patterns, and modularities of the problem at hand, and the final genetic program may call these procedures at different stages of its execution.

The fact that genetic programming operates on computer programs has a few interesting aspects. For example, the operators can be viewed also as programs, which can undergo a separate evolution during the run of the system. Additionally, a set of functions can consist of several programs which perform complex tasks; such functions can evolve further during the evolutionary run (e.g., ADF). In general, a GP tends to do a fairly good job at finding a general solution provided (a) it is given enough training cases to be able to deduce the general solution in the first place, and (b) that the function and terminal set have been chosen so as to bias the solution space appropriately for the given problem. Clearly, it is one of the most exiting areas of the current development in the evolutionary computation field with already a significant amount of experimental data (apart from [59] and [60], see also [57] and [2]).

3.5 Other techniques

Goldberg et al. [43] explored an alternative representation in a messy GA, where the variable-size encoding is used which allows overspecification with respect to the problem being solved. The conflict of redundancy in an overspecified string is generally handled by a specific dominance mechanism. Messy GAs solve problems by combining relatively short, well-tested substrings to form the solution space and have been mostly used for solving a certain class of problem known as deceptive problems.

Dasgupta and McGregor have investigated a more biologically motivated genetic model called the **Structured Genetic Algorithm** (sGA) [15, 12].

The model uses redundant genetic materials and a gene activation mechanism in a multi-layered chromosomal structure for developing an efficient search technique. This representation provides greater implicit genetic diversity and exhibits many advantages in search and optimization. The details of the model and its applications were described in [12, 13, 15, 14].

Other researchers modified further evolutionary algorithms by 'adding' the problem specific knowledge to the algorithm. Several papers have discussed initialization techniques, different representations, decoding techniques (mapping from genetic representations to 'phenotypic' representations), and the use of heuristics for genetic operators. Davis [19] wrote (in the context of classical, binary GAs):

> "It has seemed true to me for some time that we cannot handle most real-world problems with binary representations and an operator set consisting only of binary crossover and binary mutation. One reason for this is that nearly every real-world domain has associated domain knowledge that is of use when one is considering a transformation of a solution in the domain [...] I believe that genetic algorithms are the appropriate algorithms to use in a great many real-world applications. I also believe that one should incorporate real-world knowledge in one's algorithm by adding it to one's decoder or by expanding one's operator set."

Such hybrid/nonstandard systems enjoy a significant popularity in evolutionary computation community. Very often these systems, extended by the problem-specific knowledge, outperform other classical evolutionary methods as well as other standard techniques [64, 65]. For example, a system Genetic-2N [64] constructed for the nonlinear transportation problem used a matrix representation for its chromosomes, a problem-specific mutation (main operator, used with probability 0.4) and arithmetical crossover (background operator, used with probability 0.05). It is hard to classify this system: it is not really a genetic algorithm, since it can run with mutation operator only without any significant decrease of quality of results. Moreover, all matrix entries are floating point numbers. It is not an evolution strategy, since it did not encode any control parameters in its chromosomal structures. Clearly, it has nothing to do with genetic programming and very little (matrix representation) with evolutionary programming approaches. It is just an evolutionary computation technique aimed at particular problem.

There are a few heuristics to guide a user in selection of appropriate data structures and operators for a particular problem. It is a common knowledge that for numerical optimization problem one should use an evolutionary strategy or genetic algorithm with floating point representation, whereas some versions of genetic algorithm would be the best to handle combinatorial optimization problems. Genetic programs are great in discovery of rules given as a computer program, and evolutionary programming techniques can be used successfully to model a behavior of the system (e.g., prisoner dilemma problem). It seems also that neither of the evolutionary techniques is perfect (or even robust) across the problem spectrum; only the whole family of algorithms based on evolutionary computation concepts (i.e., evolutionary algorithms) have this property of robustness. But the main key to successful applications is in heuristics methods, which are mixed skillfully with evolutionary techniques.

4. Evolutionary Algorithms and Constrained Optimization

In this section, we discuss several methods for handling feasible and infeasible solutions in a population; most of these methods emerged quite recently. Only a few years ago Richardson et al. [93] claimed: "Attempts to apply GA's with constrained optimization problems follow two different paradigms (1) modification of the genetic operators; and (2) penalizing strings which fail to satisfy all the constraints." This is no longer the case as a variety of heuristics have been proposed. Even the category of penalty functions consists of several methods which differ in many important details on how the penalty function is designed and applied to infeasible solutions. Other methods maintain the feasibility of the individuals in the population by means of specialized operators or decoders, impose a restriction that any feasible solution is 'better' than any infeasible solution, consider constraints one at the time in a particular linear order, repair infeasible solutions, use multiobjective optimization techniques, are based on cultural algorithms (i.e., algorithms with an additional layer of beliefs which undergoes evolution as well [91]), or rate solutions using a particular co-evolutionary model (i.e., model with more than one population, where the fitness of an individual in one population depends on the current state of evolution in the other population [81]).

4.1 Rejection of infeasible individuals

This "death penalty" heuristic is a popular option in many evolutionary techniques (e.g., evolution strategies). Note that rejection of infeasible individuals offers a few simplifications of the algorithm: for example, there is no need to evaluate infeasible solutions and to compare them with feasible ones.

The method of eliminating infeasible solutions from a population may work reasonably well when the feasible search space is convex and it constitutes a reasonable part of the whole search space (e.g., evolution strategies do not allow equality constraints since with such constraints the ratio between the sizes of feasible and infeasible search spaces is zero). Otherwise such an approach has serious limitations. For example, for many search problems where the initial population consists of infeasible individuals only, it might be essential to improve them (as opposed to rejecting them). Moreover, quite often the system can reach the optimum solution easier if it is possible to "cross" an infeasible region (especially in non-convex feasible search spaces).

4.2 Penalizing infeasible individuals

This is the most common approach in the genetic algorithms community. The domain of the objective function f is extended; the approach assumes that

$$eval(p) = f(p) \pm Q(p),$$

where $Q(p)$ represents either a penalty for infeasible individual p, or a cost for repairing such an individual. The major question is, how should such a penalty function $Q(p)$ be designed? The intuition is simple: the penalty should be kept as low as possible, just above the limit below which infeasible solutions are optimal (so-called *minimal penalty rule*) [61]. However, it is difficult to implement this rule effectively.

The relationship between infeasible individual 'p' and the feasible part of the search space plays a significant role in penalizing such individuals: an individual might be penalized just for being infeasible, the 'amount' of its infeasibility is measured to determine the penalty value, or the effort of 'repairing' the individual might be taken into account.

Several researchers studied heuristics on design of penalty functions. Some hypotheses were formulated [93]:

- "penalties which are functions of the distance from feasibility are better performers than those which are merely functions of the number of violated constraints,
- for a problem having few constraints, and few full solutions, penalties which are solely functions of the number of violated constraints are not likely to find solutions,
- good penalty functions can be constructed from two quantities, the *maximum completion cost* and the *expected completion cost*,
- penalties should be close to the *expected completion cost*, but should not frequently fall below it. The more accurate the penalty, the better will be the solutions found. When penalty often underestimates the completion cost, then the search may not find a solution."

and in [105]:

- "the genetic algorithm with a variable penalty coefficient outperforms the fixed penalty factor algorithm,"

where a variability of penalty coefficient was determined by a heuristic rule.

This last observation was further investigated by Smith and Tate [106]. In their work they experimented with dynamic penalties, where the penalty measure depends on the number of violated constraints, the best feasible objective function found, and the best objective function value found.

For numerical optimization problems,

optimize $f(\overline{X})$, $\overline{X} = (x_1, \ldots, x_n) \in R^n$,

where

$g_j(\overline{X}) \leq 0$, for $j = 1, \ldots, q$, and $h_j(\overline{X}) = 0$, for $j = q+1, \ldots, m$,

penalties usually incorporate degrees of constraint violations. Most of these methods use constraint violation measures f_j (for the j-th constraint) for the construction of the *eval*; these functions are defined as

$$f_j(\overline{X}) = \begin{cases} \max\{0, g_j(\overline{X})\}, & if\ 1 \leq j \leq q \\ |h_j(\overline{X})|, & if\ q+1 \leq j \leq m \end{cases}$$

For example, Homaifar et al. [52] assume that for every constraint we establish a family of intervals that determines appropriate penalty values.

It is also possible (as suggested in [105]) to adjust penalties in a dynamic way, taking into account the current state of the search or the generation number. For example, Joines and Houck [53] assumed dynamic penalties; individuals are evaluated (at the iteration t) by the following formula:

$$eval(\overline{X}) = f(\overline{X}) + (C \times t)^\alpha \sum_{j=1}^{m} f_j^\beta(\overline{X}),$$

where C, α and β are constants.

Michalewicz and Attia [68] considered a method based on the idea of simulated annealing: the penalty coefficients are changed once in many generations (after the convergence of the algorithm to a local optima). At every iteration the algorithm considers active constraints only, the pressure on infeasible solutions is increased due to the decreasing values of the temperature of the system.

A method of adapting penalties was developed by Bean and Hadj-Alouane [7, 48]. As the previous method, it uses a penalty function, however, one component of the penalty function takes a feedback from the search process. Each individual is evaluated by the formula:

$$eval(\overline{X}) = f(\overline{X}) + \lambda(t) \sum_{j=1}^{m} f_j^2(\overline{X}),$$

where $\lambda(t)$ is updated every generation t in the following way:

$$\lambda(t+1) = \begin{cases} (1/\beta_1) \cdot \lambda(t), & if\ case\#1 \\ \beta_2 \cdot \lambda(t), & if\ case\#2 \\ \lambda(t), & otherwise, \end{cases}$$

where cases #1 and #2 denote situations where the best individual in the last k generation was always (case #1) or was never (case #2) feasible, $\beta_1, \beta_2 > 1$, and $\beta_1 \neq \beta_2$ (to avoid cycling).

Yet another approach was proposed recently by Le Riche et al. [61]. The authors designed a (segregated) genetic algorithm which uses two values of penalty parameters (for each constraint) instead of one; these two values aim at achieving a balance between heavy and moderate penalties by maintaining two subpopulations of individuals. The population is split into two cooperating groups, where individuals in each group are evaluated using either one of the two penalty parameters.

Some researchers [82, 73] reported good results of their evolutionary algorithms, which worked under the assumption that any feasible individual was better than any infeasible one. Powell and Skolnick [82] applied this heuristic rule for the numerical optimization problems: evaluations of feasible solutions were mapped into the interval $(-\infty, 1)$ and infeasible solutions—into the interval $(1, \infty)$ (for minimization problems). Michalewicz and Xiao [73] experimented with the path planning problem and used two separate evaluation functions for feasible and infeasible individuals. The values for infeasible solutions were increased (i.e., made less attractive) by adding such a constant, so that the best infeasible individual was worse that the worst feasible one.

It seems that the appropriate choice of the penalty method may depend on (1) the ratio between sizes of the feasible and the whole search space, (2) the topological properties of the feasible search space, (3) the type of the objective function, (4) the number of variables, (5) number of constraints, (6) types of constraints, and (7) number of active constraints at the optimum. Thus the use of penalty functions is not trivial and only some partial analysis of their properties is available. Also, a promising direction for applying penalty functions is the use of adaptive penalties: penalty factors can be incorporated in the chromosome structures in a similar way as some control parameters are represented in the structures of evolution strategies and evolutionary programming.

4.3 Maintaining feasible population by special representations and genetic operators

One reasonable heuristic for dealing with the issue of feasibility is to use specialized representation and operators to maintain the feasibility of individuals in the population. During the last decade several specialized systems were developed for particular optimization problems; these systems use a unique chromosomal representations and specialized 'genetic' operators which alter their composition. Some of such systems were described in [18]; other examples include Genocop for optimizing numerical functions with linear constraints and Genetic-2N [65] for nonlinear transportation problem. For example, Genocop assumes linear constraints only and a feasible starting point (or feasible initial population). A closed set of operators maintains feasibility of solutions. For example, when a particular component x_i of a solution vector $\overline{X}$ is mutated, the system determines its current domain $dom(x_i)$ (which is a function of linear constraints and remaining values of the solution vector $\overline{X}$) and the new value of x_i is taken from this domain (either with flat probability distribution for uniform mutation, or other probability distributions for non-uniform and boundary mutations). In any case the offspring solution vector is always feasible. Similarly, arithmetic crossover of two feasible solution vectors $\overline{X}$ and $\overline{Y}$ yields always a feasible solution (for $0 \le a \le 1$) in convex search spaces (the system assumes linear constraints only which imply convexity of the feasible search space).

Often such systems are much more reliable than any other evolutionary techniques based on penalty approach. This is a quite popular trend: many practitioners use problem-specific representations and specialized operators in building very successful evolutionary algorithms in many areas; these include numerical optimization, machine learning, optimal control, cognitive modeling, classic operation research problems (traveling salesman problem, knapsack problems, transportation problems, assignment problems, bin packing, scheduling, partitioning, etc.), engineering design, system integration, iterated games, robotics, signal processing, and many others.

4.4 Repair of infeasible individuals

Repair algorithms enjoy a particular popularity in the evolutionary computation community: for many combinatorial optimization problems (e.g., traveling salesman problem, knapsack problem, set covering problem, etc.) it is relatively easy to 'repair' an infeasible individual. Such a repaired version can be used either for evaluation only, or it can also replace (with some probability) the original individual in the population.

The weakness of these methods is in their problem dependence. For each particular problem a specific repair algorithm should be designed. Moreover, there are no standard heuristics on design of such algorithms: usually it is possible to use a greedy repair, random repair, or any other heuristic which would guide the repair process. Also, for some problems the process of repairing infeasible individuals might be as complex as solving the original problem. This is the case for the nonlinear transportation problem [65], most scheduling and timetable problems, and many others.

On the other hand, the recently completed Genocop III system [70] for constrained numerical optimization (nonlinear constraints) is based on repair algorithms. Genocop III incorporates the original Genocop system [65] (which handles linear constraints only), but also extends it by maintaining two separate populations, where a development in one population influences evaluations of individuals in the other population. The first population P_s consists of so-called search points which satisfy linear constraints of the problem; the feasibility (in the sense of linear constraints) of these points is maintained by specialized operators (as in Genocop). The second population, P_r, consists of fully feasible reference points. These reference points, being feasible, are evaluated directly by the objective function, whereas search points are "repaired" for evaluation. The first results are very promising [70].

4.5 Replacement of individuals by their repaired versions

The question of replacing repaired individuals is related to so-called *Lamarckian evolution*, which assumes that an individual improves during its lifetime and that the resulting improvements are coded back into the chromosome.

Recently Orvosh and Davis [78] reported a so-called *5%-rule*: this heuristic rule states that in many combinatorial optimization problems, an evolutionary computation technique with a repair algorithm provides the best results when 5% of repaired individuals replace their infeasible originals. In continuous domains, a new replacement rule is emerging. As mentioned earlier, the Genocop III system for constrained numerical optimization problems with nonlinear constraints is based on repair approach. The first experiments (based on 10 test cases which have various numbers of variables, constraints, types of constraints, numbers of active constraints at the optimum, etc.) indicate that the 15% replacement rule is a clear winner: the results of the system are much better than with either lower or higher values of the replacement rate.

At present, it seems that the 'optimal' probability of replacement is problem-dependent and it may change over the evolution process as well. Further research is required for comparing different heuristics for setting this parameter, which is of great importance for all repair-based methods.

4.6 Use of decoders

Decoders offer an interesting option for all practitioners of evolutionary techniques. In these techniques a chromosome "gives instructions" on how to build a feasible solution. For example, a sequence of items for the knapsack problem can be interpreted as: "take an item if possible"—such interpretation would lead always to feasible solutions. However, it is important to point out that several factors should be taken into account while using decoders. Each decoder imposes a relationship T between a feasible solution and decoded solution.

It is important that several conditions are satisfied: (1) for each feasible solution s there is a decoded solution d, (2) each decoded solution d corresponds to a feasible solution s, and (3) all feasible solutions should be represented by the same number of decodings d. Additionally, it is reasonable to request that (4) the transformation T is computationally fast and (5) it has locality feature in the sense that small changes in the decoded solution result in small changes in the solution itself. An interesting study on coding trees in genetic algorithm was reported by Palmer and Kershenbaum [79], where the above conditions were formulated.

4.7 Separation of individuals and constraints

This is a general and interesting heuristic. The first possibility would include utilization of multi-objective optimization methods, where the objective function f and constraint violation measures f_j (for m constraints) constitute a $(m+1)$-dimensional vector $\mathbf{v}$:

$$\mathbf{v} = (f, f_1, \ldots, f_m).$$

Using some multi-objective optimization method, we can attempt to minimize its components: an ideal solution x would have $f_j(x) = 0$ for $1 \le i \le m$ and $f(x) \le f(y)$ for all feasible y (minimization problems). A successful implementation of similar approach was presented recently in [109].

Another approach was recently reported by Paredis [81]. The method (described in the context of constraint satisfaction problems) is based on a co-evolutionary model, where a population of potential solutions co-evolves with a population of constraints: fitter solutions satisfy more constraints, whereas fitter constraints are violated by more solutions.

Yet another heuristic is based on the idea of handling constraints in a particular order; Schoenauer and Xanthakis [98] called this method a "behavioral memory" approach.

It is also possible to incorporate the knowledge of the constraints of the problem into the belief space of cultural algorithms [91, 92]. The general intuition behind belief spaces is to preserve those beliefs associated with "acceptable" behavior at the trait level (and, consequently, to prune away unacceptable beliefs). The acceptable beliefs serve as constraints that direct the population of traits. It seems that the cultural algorithms may serve as a very interesting tool for numerical optimization problems, where constraints influence the search in a direct way (consequently, the search in constrained spaces may be more efficient than in unconstrained ones!).

For more information on constraint-handling techniques in evolutionary methods, see [66] and [71]; a few real examples of applying these techniques for some engineering problems are presented in [69].

5. Summary

Evolutionary algorithms are general-purpose search procedures based on the mechanisms of natural selection and population genetics. Evolutionary algorithms have received a lot of attention both in academic and industries. The EA-based tools have started growing impact in companies — predicting financial market, in factories — job shop scheduling etc. with their power of search, optimization, adaptation and learning. One of the recent areas of applications for these techniques is in the field of industrial engineering; these include scheduling and sequencing in manufacturing systems, computer-aided design, facility layout and location problems, distribution and transportation problems, and many others.For the users of diversified fields, genetic algorithms are appealing because of their simplicity, easy to interface and ease to extensibility.

Although evolutionary algorithms have been successfully applied to many practical problems, as the application increases, there have been a number of failures as well. There is little understanding of what features of domains make them appropriate or inappropriate for these algorithms. Several modifications have been suggested to alleviate the difficulties both in the manipulation of encoded information and the ways of representing problem spaces. Three important claims have been made about why evolutionary algorithms perform well: (1) independent sampling is provided by populations of candidate solutions, (2) selection is a mechanism that preserves good solutions, and (3) partial solutions can be efficiently modified and combined through various 'genetic' operators.

Evolutionary algorithms have been widely used in science and engineering for solving complex problems. An important goal of research on evolutionary algorithms is to understand the class of problems for which EAs are most suited, and, in particular, the class of problems on which they outperform other search algorithms.

References

1. Alander, J.T., *An Indexed Bibliography of Genetic Algorithms: Years 1957–1993*, Department of Information Technology and Production Economics, University of Vaasa, Finland, Report Series No.94-1, 1994.
2. Angeline, P.J. and Kinnear, K.E. (Editors), *Advances in Genetic Programming II*, MIT Press, Cambridge, MA, 1996.
3. Arabas, J., Michalewicz, Z., and Mulawka, J., *GAVaPS — a Genetic Algorithm with Varying Population Size*, in [84].
4. Bäck, T., and Hoffmeister, F., *Extended Selection Mechanisms in Genetic Algorithms*, in [10], pp.92–99.
5. Bäck, T., Fogel, D., and Michalewicz, Z. (Editors), *Handbook of Evolutionary Computation*, Oxford University Press, New York, 1996.
6. Bäck, T., Hoffmeister, F., and Schwefel, H.-P., *A Survey of Evolution Strategies*, in [10], pp.2–9.
7. Bean, J.C. and Hadj-Alouane, A.B., *A Dual Genetic Algorithm for Bounded Integer Programs*, Department of Industrial and Operations Engineering, The University of Michigan, TR 92-53, 1992.
8. Beasley, D., Bull, D.R., and Martin, R.R., *An Overview of Genetic Algorithms: Part 1, Foundations*, University Computing, Vol.15, No.2, pp.58–69, 1993.
9. Beasley, D., Bull, D.R., and Martin, R.R., *An Overview of Genetic Algorithms: Part 2, Research Topics*, University Computing, Vol.15, No.4, pp.170–181, 1993.
10. Belew, R. and Booker, L. (Editors), Proceedings of the Fourth International Conference on Genetic Algorithms, Morgan Kaufmann Publishers, Los Altos, CA, 1991.
11. Brooke, A., Kendrick, D., and Meeraus, A., *GAMS: A User's Guide*, The Scientific Press, 1988.
12. Dasgupta, D. and McGregor, D R., *A more Biologically Motivated Genetic Algorithm: The Model and some Results*, Cybernatics and Systems: An International Journal, Vol.25, No.3, pp.447–469, May-June 1994.

13. Dasgupta, D. and McGregor, D R., *Designing Application-Specific Neural Networks using the Structured Genetic Algorithm*, Proceedings of the International Workshop on Combination on Genetic Algorithms and Neural Networks (COGANN-92), pages 87–96, IEEE Computer Society Press, June 6, U.S.A 1992.
14. Dasgupta, D. and McGregor, D R., *Genetically Designing Neuro-controllers for a Dynamic System*, Proceedings of the International Joint Conference on Neural Networks (IJCNN), pages 2951–2955, Nagoya, Japan, 25-29 October 1993.
15. Dasgupta, D. and McGregor, D R., *Nonstationary Function Optimization using the Structured Genetic Algorithm*, Proceedings of Parallel Problem Solving From Nature (PPSN-2), pages 145–154, Brussels, 28-30 September 1992.
16. Davidor, Y., Schwefel, H.-P., and Männer, R. (Editors), Proceedings of the Third International Conference on Parallel Problem Solving from Nature (PPSN), Springer-Verlag, New York, 1994.
17. Davis, L., (Editor), *Genetic Algorithms and Simulated Annealing*, Morgan Kaufmann Publishers, Los Altos, CA, 1987.
18. Davis, L., *Handbook of Genetic Algorithms*, New York, Van Nostrand Reinhold, 1991.
19. Davis, L., *Adapting Operator Probabilities in Genetic Algorithms*, in [96], pp.61–69.
20. Davis, L. and Steenstrup, M., *Genetic Algorithms and Simulated Annealing: An Overview*, in [17], pp.1–11.
21. De Jong, K.A., (Editor), *Evolutionary Computation*, MIT Press, 1993.
22. De Jong, K., *Genetic Algorithms: A 10 Year Perspective*, in [46], pp.169–177.
23. De Jong, K., *Genetic Algorithms: A 25 Year Perspective*, in [115], pp.125–134.
24. Dhar, V. and Ranganathan, N., *Integer Programming vs. Expert Systems: An Experimental Comparison*, Communications of ACM, Vol.33, No.3, pp.323–336, 1990.
25. Eiben, A.E., Raue, P.-E., and Ruttkay, Zs., *Genetic Algorithms with Multiparent Recombination*, in [16], pp.78–87.
26. Eshelman, L.J., (Editor), Proceedings of the Sixth International Conference on Genetic Algorithms, Morgan Kaufmann, San Mateo, CA, 1995.
27. Eshelman, L.J. and Schaffer, J.D., *Preventing Premature Convergence in Genetic Algorithms by Preventing Incest*, in [10], pp.115–122.
28. Fogel, D.B., *Evolving Artificial Intelligence*, Ph.D. Thesis, University of California, San Diego, 1992.
29. Fogel, D.B., *Evolving Behaviours in the Iterated Prisoner's Dilemma*, Evolutionary Computation, Vol.1, No.1, pp.77–97, 1993.
30. Fogel, D.B. (Editor), IEEE Transactions on Neural Networks, special issue on Evolutionary Computation, Vol.5, No.1, 1994.
31. Fogel, D.B., *An Introduction to Simulated Evolutionary Optimization*, IEEE Transactions on Neural Networks, special issue on Evolutionary Computation, Vol.5, No.1, 1994.
32. Fogel, D.B., *Evolutionary Computation: Toward a New Philosophy of Machine Intelligence*, IEEE Press, Piscataway, NJ, 1995.
33. Fogel, D.B. and Atmar, W., *Proceedings of the First Annual Conference on Evolutionary Programming*, La Jolla, CA, 1992, Evolutionary Programming Society.
34. Fogel, D.B. and Atmar, W., *Proceedings of the Second Annual Conference on Evolutionary Programming*, La Jolla, CA, 1993, Evolutionary Programming Society.
35. Fogel, L.J., Angeline, P.J., Bäck, T. (Editors), Proceedings of the Fifth Annual Conference on Evolutionary Programming, The MIT Press, 1996.

36. Fogel, L.J., Owens, A.J., and Walsh, M.J., *Artificial Intelligence Through Simulated Evolution*, John Wiley, Chichester, UK, 1966.
37. Fogel, L.J., *Evolutionary Programming in Perspective: The Top-Down View*, in [115], pp.135–146.
38. Forrest, S. (Editor), Proceedings of the Fifth International Conference on Genetic Algorithms, Morgan Kaufmann Publishers, Los Altos, CA, 1993.
39. Glover, F., *Heuristics for Integer Programming Using Surrogate Constraints*, Decision Sciences, Vol.8, No.1, pp.156–166, 1977.
40. Goldberg, D.E., *Genetic Algorithms in Search, Optimization and Machine Learning*, Addison-Wesley, Reading, MA, 1989.
41. Goldberg, D.E., *Simple Genetic Algorithms and the Minimal, Deceptive Problem*, in [17], pp.74–88.
42. Goldberg, D.E., Deb, K., and Korb, B., *Do not Worry, Be Messy*, in [10], pp.24–30.
43. Goldberg, D. E., and Korb, B. and Deb, D., *Messy Genetic Algorithms: Motivation, Analysis and First Results*, Complex Systems, Vol.3, pages 493–530, May 1989.
44. Goldberg, D.E., Milman, K., and Tidd, C., *Genetic Algorithms: A Bibliography*, IlliGAL Technical Report 92008, 1992.
45. Gorges-Schleuter, M., *ASPARAGOS An Asynchronous Parallel Genetic Optimization Strategy*, in [96], pp.422–427.
46. Grefenstette, J.J., (Editor), Proceedings of the First International Conference on Genetic Algorithms, Lawrence Erlbaum Associates, Hillsdale, NJ, 1985.
47. Grefenstette, J.J., (Editor), Proceedings of the Second International Conference on Genetic Algorithms, Lawrence Erlbaum Associates, Hillsdale, NJ, 1987.
48. Hadj-Alouane, A.B. and Bean, J.C., *A Genetic Algorithm for the Multiple-Choice Integer Program*, Department of Industrial and Operations Engineering, The University of Michigan, TR 92-50, 1992.
49. Heitkötter, J., (Editor), *The Hitch-Hiker's Guide to Evolutionary Computation*, FAQ in comp.ai.genetic, issue 1.10, 20 December 1993.
50. Holland, J.H., *Adaptation in Natural and Artificial Systems*, University of Michigan Press, Ann Arbor, 1975.
51. Holland, J.H., *Royal Road Functions*, Genetic Algorithm Digest, Vol.7, No.22, 12 August 1993.
52. Homaifar, A., Lai, S. H.-Y., Qi, X., *Constrained Optimization via Genetic Algorithms*, Simulation, Vol.62, No.4, 1994, pp.242–254.
53. Joines, J.A. and Houck, C.R., *On the Use of Non-Stationary Penalty Functions to Solve Nonlinear Constrained Optimization Problems With GAs*, Proceedings of the First IEEE ICEC 1994, pp.579–584.
54. Jones, T., *A Description of Holland's Royal Road Function*, Evolutionary Computation, Vol.2, No.4, 1994, pp.409–415.
55. Jones, T. and Forrest, S., *Fitness Distance Correlation as a Measure of Problem Difficulty for Genetic Algorithms*, in [26], pp.184–192.
56. Julstrom, B.A., *What Have You Done for Me Lately? Adapting Operator Probabilities in a Steady-State Genetic Algorithm*, in [26], pp.81–87.
57. Kinnear, K.E. (Editor), *Advances in Genetic Programming*, MIT Press, Cambridge, MA, 1994.
58. Koza, J.R., *Genetic Programming: A Paradigm for Genetically Breeding Populations of Computer Programs to Solve Problems*, Report No. STAN-CS-90-1314, Stanford University, 1990.
59. Koza, J.R., *Genetic Programming*, MIT Press, Cambridge, MA, 1992.
60. Koza, J.R., *Genetic Programming – 2*, MIT Press, Cambridge, MA, 1994.

61. Le Riche, R., Knopf-Lenoir, C., and Haftka, R.T., *A Segregated Genetic Algorithm for Constrained Structural Optimization*, in [26], pp.558–565.
62. Männer, R. and Manderick, B. (Editors), Proceedings of the Second International Conference on Parallel Problem Solving from Nature (PPSN), North-Holland, Elsevier Science Publishers, Amsterdam, 1992.
63. McDonnell, J.R., Reynolds, R.G., and Fogel, D.B. (Editors), Proceedings of the Fourth Annual Conference on Evolutionary Programming, The MIT Press, 1995.
64. Michalewicz, Z., *A Hierarchy of Evolution Programs: An Experimental Study*, Evolutionary Computation, Vol.1, No.1, 1993, pp.51–76.
65. Michalewicz, Z., *Genetic Algorithms + Data Structures = Evolution Programs*, Springer-Verlag, 3rd edition, 1996.
66. Michalewicz, Z., *Heuristic Methods for Evolutionary Computation Techniques*, Journal of Heuristics, Vol.1, No.2, 1995, pp.177-206.
67. Michalewicz, Z. (Editor), Statistics & Computing, special issue on evolutionary computation, Vol.4, No.2, 1994.
68. Michalewicz, Z., and Attia, N., *Evolutionary Optimization of Constrained Problems*, Proceedings of the 3rd Annual Conference on EP, World Scientific, 1994, pp.98-108.
69. Michalewicz, Z., Dasgupta, D., Le Riche, R.G., and Schoenauer, M., *Evolutionary Algorithms for Constrained Engineering Problems*, Computers & Industrial Engineering Journal, Vol.30, No.4, September 1996, pp.851–870.
70. Michalewicz, Z. and Nazhiyath, G., *Genocop III: A Co-evolutionary Algorithm for Numerical Optimization Problems with Nonlinear Constraints*, Proceedings of the 2nd IEEE International Conference on Evolutionary Computation, Vol.2, Perth, 29 November – 1 December 1995, pp.647–651.
71. Michalewicz, Z. and Schoenauer, M., *Evolutionary Algorithms for Constrained Parameter Optimization Problems*, Evolutionary Computation, Vol.4, No.1, 1996.
72. Michalewicz, Z., Vignaux, G.A., and Hobbs, M., *A Non-Standard Genetic Algorithm for the Nonlinear Transportation Problem*, ORSA Journal on Computing, Vol.3, No.4, 1991, pp.307–316.
73. Michalewicz, Z. and Xiao, J., *Evaluation of Paths in Evolutionary Planner/Navigator*, Proceedings of the 1995 International Workshop on Biologically Inspired Evolutionary Systems, Tokyo, Japan, May 30–31, 1995, pp.45–52.
74. Mühlenbein, H., *Parallel Genetic Algorithms, Population Genetics and Combinatorial Optimization*, in [96], pp.416-421.
75. Mühlenbein, H. and Schlierkamp-Vosen, D., *Predictive Models for the Breeder Genetic Algorithm*, Evolutionary Computation, Vol.1, No.1, pp.25–49, 1993.
76. Nadhamuni, P.V.R., *Application of Co-evolutionary Genetic Algorithm to a Game*, Master Thesis, Department of Computer Science, University of North Carolina, Charlotte, 1995.
77. Nissen, V., *Evolutionary Algorithms in Management Science: An Overview and List of References*, European Study Group for Evolutionary Economics, 1993.
78. Orvosh, D. and Davis, L., *Shall We Repair? Genetic Algorithms, Combinatorial Optimization, and Feasibility Constraints*, in [38], p.650.
79. Palmer, C.C. and Kershenbaum, A., *Representing Trees in Genetic Algorithms*, Proceedings of the IEEE International Conference on Evolutionary Computation, 27–29 June 1994, pp.379–384, 1994.
80. Paredis, J., *Genetic State-Space Search for Constrained Optimization Problems*, Proceedings of the Thirteen International Joint Conference on Artificial Intelligence, Morgan Kaufmann, San Mateo, CA, 1993.

81. Paredis, J., *Co-evolutionary Constraint Satisfaction*, Proceedings of the 3rd PPSN Conference, Springer-Verlag, pp.46–55, 1994.
82. Powell, D. and Skolnick, M.M., *Using Genetic Algorithms in Engineering Design Optimization with Non-linear Constraints*, Proceedings of the Fifth ICGA, Morgan Kaufmann, pp.424–430, 1993.
83. Potter, M. and De Jong, K., *A Cooperative Coevolutionary Approach to Function Optimization*, George Mason University, 1994.
84. Proceedings of the First IEEE International Conference on Evolutionary Computation, Orlando, 26 June – 2 July, 1994.
85. Proceedings of the Second IEEE International Conference on Evolutionary Computation, Perth, 29 November – 1 December, 1995.
86. Proceedings of the Third IEEE International Conference on Evolutionary Computation, Nagoya, 18–22 May, 1996.
87. Radcliffe, N.J., *Forma Analysis and Random Respectful Recombination*, in [10], pp.222–229.
88. Radcliffe, N.J., *Genetic Set Recombination*, in [114], pp.203 219.
89. Radcliffe, N.J., and George, F.A.W., *A Study in Set Recombination*, in [38], pp.23–30.
90. Reeves, C.R., *Modern Heuristic Techniques for Combinatorial Problems*, Blackwell Scientific Publications, London, 1993.
91. Reynolds, R.G., *An Introduction to Cultural Algorithms*, Proceedings of the Third Annual Conference on Evolutionary Programming, River Edge, NJ, World Scientific, pp.131–139, 1994.
92. Reynolds, R.G., Michalewicz, Z., and Cavaretta, M., *Using Cultural Algorithms for Constraint Handling in Genocop*, Proceedings of the 4th Annual Conference on Evolutionary Programming, San Diego, CA, pp.289–305, March 1–3, 1995.
93. Richardson, J.T., Palmer, M.R., Liepins, G., and Hilliard, M., *Some Guidelines for Genetic Algorithms with Penalty Functions*, in Proceedings of the Third ICGA, Morgan Kaufmann, pp.191–197, 1989.
94. Ronald, E., *When Selection Meets Seduction*, in [26], pp.167–173.
95. Saravanan, N. and Fogel, D.B., *A Bibliography of Evolutionary Computation & Applications*, Department of Mechanical Engineering, Florida Atlantic University, Technical Report No. FAU-ME-93-100, 1993.
96. Schaffer, J., (Editor), Proceedings of the Third International Conference on Genetic Algorithms, Morgan Kaufmann Publishers, Los Altos, CA, 1989.
97. Schaffer, J.D. and Morishima, A., *An Adaptive Crossover Distribution Mechanism for Genetic Algorithms*, in [47], pp.36–40.
98. Schoenauer, M., and Xanthakis, S., *Constrained GA Optimization*, Proceedings of the Fifth ICGA, Morgan Kaufmann, pp.573–580, 1993.
99. Schraudolph, N. and Belew, R., *Dynamic Parameter Encoding for Genetic Algorithms*, CSE Technical Report #CS90-175, University of San Diego, La Jolla, 1990.
100. Schwefel, H.-P., *On the Evolution of Evolutionary Computation*, in [115], pp.116–124.
101. Schwefel, H.-P., *Evolution and Optimum Seeking*, John Wiley, Chichester, UK, 1995.
102. Schwefel, H.-P. and Männer, R. (Editors), Proceedings of the First International Conference on Parallel Problem Solving from Nature (PPSN), Springer-Verlag, Lecture Notes in Computer Science, Vol.496, 1991.
103. Sebald, A.V. and Fogel, L.J., *Proceedings of the Third Annual Conference on Evolutionary Programming*, San Diego, CA, 1994, World Scientific.
104. Shaefer, C.G., *The ARGOT Strategy: Adaptive Representation Genetic Optimizer Technique*, in [47], pp.50–55.

105. Siedlecki, W. and Sklanski, J., *Constrained Genetic Optimization via Dynamic Reward–Penalty Balancing and Its Use in Pattern Recognition*, Proceedings of the Third International Conference on Genetic Algorithms, Los Altos, CA, Morgan Kaufmann Publishers, pp.141–150, 1989.
106. Smith, A. and Tate, D., *Genetic Optimization Using A Penalty Function*, Proceedings of the Fifth ICGA, Morgan Kaufmann, pp.499–503, 1993.
107. Spears, W.M., *Adapting Crossover in Evolutionary Algorithms*, in [63], pp.367–384.
108. Srinivas, M. and Patnaik, L.M., *Adaptive Probabilities of Crossover and Mutation in Genetic Algorithms*, IEEE Transactions on Systems, Man, and Cybernetics, Vol.24, No.4, 1994, pp.17–26.
109. Surry, P.D., N.J. Radcliffe, and I.D. Boyd, *A Multi-objective Approach to Constrained Optimization of Gas Supply Networks.* Presented at the AISB-95 Workshop on Evolutionary Computing, Sheffield, UK, April 3–4, 1995, pp.166–180.
110. Vignaux, G.A., and Michalewicz, Z., *A Genetic Algorithm for the Linear Transportation Problem*, IEEE Transactions on Systems, Man, and Cybernetics, Vol.21, No.2, 1991, pp.445–452.
111. Voigt, H.-M., Ebeling, W., Rechenberg, I., Schwefel, H.-P. (Editors), Proceedings of the Fourth International Conference on Parallel Problem Solving from Nature (PPSN), Springer-Verlag, New York, 1996.
112. Whitley, D., *Genetic Algorithms: A Tutorial*, in [67], pp.65–85.
113. Whitley, D., *GENITOR II: A Distributed Genetic Algorithm*, Journal of Experimental and Theoretical Artificial Intelligence, Vol.2, pp.189–214.
114. Whitley, D. (Editor), *Foundations of Genetic Algorithms-2*, Second Workshop on the Foundations of Genetic Algorithms and Classifier Systems, Morgan Kaufmann Publishers, San Mateo, CA, 1993.
115. Zurada, J., Marks, R., and Robinson, C. (Editors), *Computational Intelligence: Imitating Life*, IEEE Press, 1994.

Robust Encodings in Genetic Algorithms

Simon Ronald

University of South Australia, GPO Box 944, Adelaide, SA 5001

Summary. Problems of encoding brittleness have been observed in the Genetic Algorithm (GA) literature, where slightly different problems require completely different genetic encodings for good solutions to be found. As research continues into GA encoding schemes the idea of encoding robustness becomes more important. A robust encoding is one which will be effective for a wide range of problem instances that it was designed for. A robust encoding will also be amenable to modification or extension to solve different problem types. This chapter considers some of the practical and theoretical considerations vital to the construction of a more robust encoding which will allow the GA to solve a broader range of problem types.

1. Encoding Requirements

1.1 Introduction

Problems of *encoding brittleness* have been observed in the Genetic Algorithm (GA) literature, where slightly different problems require completely different genetic encodings for good solutions to be found. As research continues into GA encoding schemes the idea of encoding robustness becomes more important. A robust encoding is one which will effective for a wide range of problem instances that it was designed for. A robust encoding will also be amenable to modification or extension to solve different problem types. This chapter considers some of the practical and theoretical considerations vital to the construction of a more robust encoding which will allow the GA to solve a broader range of problem types.

The first issue considered in this chapter is how a particular encoding is chosen for a given GA. A few classic encodings such as binary, real valued, and permutation encodings are described next. A particular emphasis is placed on the physical representation of the encoding (the genotype) as compared to the qualities of the problem to be optimised by the GA (the phenotype). It is shown that for many complex problems there is a rich mapping between these two representations. Various issues in encodings are considered in detail. These include partial cover, when the genotype cannot fully describe the problem space, simulation-type mapping between the genotype and phenotype, genetic operators, building blocks, encoding isomorphisms, and mixed encodings. These are all important issues when designing GAs with flexible and robust encodings.

1.2 How encodings are chosen

Genetic algorithms are biologically-inspired computational models and much of the terminology has been borrowed from the field of genetics. The genotype is defined as all of the genes possessed by an individual. In genetics the genes form in long molecules called deoxyribonucleic acid (DNA) [Sta91]. In cells, the DNA is organised together with proteins into structures called genotypes. In GAs, a genotype is defined a string of genes. These genes reside at various positions or *loci* in the genotype, and have a value (an *allele*).

Currently, the literature indicates that encodings are chosen according to the following properties. The encoding:

1. embodies the fundamental building-blocks that are important for the problem type [Gol89];
2. is amenable to a set of genetic operators that can propagate these building blocks from parent genotypes though to the children genotypes during child generation[FM91];
3. is not subject to *epistasis* where the effect of one gene suppresses the action of one or more other genes [BBM93];
4. allows a tractable mapping to the phenotype, allowing fitness information to be calculated in the minimum number of steps [CRG93];
5. exploits a appropriate genotype-phenotype mapping process if a simple mapping to the phenotype is not possible (Section 4.4.1);
6. embodies feasible solutions if possible – penalty systems or repair strategies are options if illegal solutions are produced [MN95];
7. suppresses isomorphic forms, i.e. many genotypes that map to the one problem point [RAV95, Ron95a]
8. uses gene values taken from an alphabet of the smallest possible cardinality, with binary encodings considered the best if suited to the problem [Gol89];
9. represents the problem at the correct level of abstraction ranging from a completely specified point in the problem space or a set of aggregate qualities that describe a family of possible solutions [MP92, CRG93].

These are each ideal encoding features. Often, for typical engineering problems each of these encoding requirements are at play and the best compromise encoding must be found. This chapter will consider some of these compromises.

Each GA problem requires a particular GA encoding. As yet, there has not been a general methodology in which any problem may be encoded by one strategy. Even two similar problems may require a completely different encoding. For example, in [CP87], Cohoon and Paris used a matrix encoding to encode each of the VLSI modules and their position within the integrated circuit as represented by the genotype. A matrix encoding was possible because the problem related to regular modules, i.e. each of the modules were of identical size and shape as arranged on a grid. However, in [CHMR88],

Cohoon et al. presented an encoding for irregular shaped modules. The objectives are similar in both cases, conductor-length between modules was to be minimised. However, the second encoding was completely different as a Polish expression was used with operands represented by module identifiers and two operators for vertical and horizontal composition. The key difference between the two problems in [CP87] and [CHMR88] is whether the modules are a fixed, or variable shape. Although both scenarios had similar objectives, the encoding used was completely different. The choice of encoding also affected the genetic operators that acted on each encoding. The matrix encoding in [CP87] allowed a form of square-patch crossover and swap-based mutation, however, the unusual encoding in [CHMR88] required three specialised crossover operators which were capable of transferring meaningful building blocks from the parent Polish expressions to the child expressions. These examples show that there is not one encoding that is superior for all problem types, but there is a wealth of possible encoding types.

The ideal encoding for a complex problem is a difficult choice. Often many different encodings strategies are possible for one problem. In [BRE91], Bhuyan et al. identified three possible encoding strategies for the clustering problem: binary strings, an ordinal representation, and an ordered representation. The ordered representation was used as it allowed simple genetic operators to be developed (with a low time complexity), and it also allowed useful building blocks to propagate under the action of the genetic operators.

2. The Genotype

In genetics, the genotype is the abstract collection of genes possessed by an individual [Sta91]. The actual structure containing the genes is referred to as the *chromosome*. A gene has a value (*allele*) and a position in the genotype (*locus*). In nature there is a complex mapping between the genotype and properties of the organism (the *phenotype*). It has been observed in nature that a single gene can affect multiple qualities in the phenotype, and this is referred to as *pleiotropy*. The other mapping has been also observed where a single quality or character in the phenotype is controlled or affected by multiple genes (*polygeny*). As argued in [Fog95], naturally evolved systems make extensive use of polygeny and pleiotropy. Typical GAs abstract away much of the richness of natural evolution and do not exploit the complex encoding and crossover mechanisms that model those found in nature [Fog95]. Many examples can be found in the GA literature where genes have a one-to-one mapping to phenotypic properties. This type of mapping represents a simplification of natural evolution. The following discussion describes a number of key encoding mechanisms that have been used to solve various optimisation problems in the GA literature.

2.1 Binary encodings

Binary encodings are an excellent choice for problems in which a problem point naturally maps into a string of zeros and ones. A binary string is often used to represent such an encoding, e.g. $E = [10001010]$. In this example the encoding E has eight genes. The position or locus of the ith gene is simply the i bit in the bitstring, and the value or allele is given by the bitstring $E[i]$.

The boolean satisfiability problem [JS89] (SAT) maps well to a binary string as the solution to a SAT problem is an assignment of binary variables. However, for many problems a binary encoding is not appropriate because:

1. epistasis – the value of a bit may suppress the fitness contributions of other bits, B, in the genotype. This can result in the fitness of the genotype being insensitive to the values of the bits in B. This can have an adverse affect in the way the GA accumulates schemata;
2. natural representation – the problem to be solved requires a higher-order than binary symbol set, and these symbols can be arranged together to form building blocks that map to decomposable problem-specific properties – in these cases a higher-order alphabet can be more efficient;
3. illegal solutions – the genetic operators may produce illegal solutions with binary encodings as a binary encoding may not naturally describe a problem point.

In some situations problems that do not naturally exploit binary encodings may be mapped to problems that do. This is the approach taken in [JS89], to map the Hamiltonian circuit problem to that of boolean satisfiability, exploiting the fact that two NP-complete problems can be mapped to each other in polynomial time. This approach cannot usually be used for real-world engineering problems as such problems usually do not fit into the classic NP-complete mould.

2.2 Parameter encoding for real function optimisation

In real-function optimisation problems an n-dimensional function is given, e.g. $f(x, y, z)$. The optimisation objective is typically the requirement to locate the maximum (or minimum) value of the function in a given domain. In classical GAs, used in real-function optimisation problems, a potential solution to the problem is encoded into a bit-string. In a three dimensional example $f(x, y, z)$, a potential solution might be $f(x_1, y_1, z_1)$, and x_1, y_1, z_1 would be encoded as three binary substrings. These substrings would be concatenated to form a single bit-string genotype. Each parameter may be encoded as a single binary string, or as traditional floating point number (FPN). As a FPN each parameter can be encoded as three contiguous bitstrings; a sign bit, mantissa bits, and exponent bits.

An interesting encoding technique is to vary the encoding mechanism for a given parameter for each genotype in the GA at various times during evolution. A good example of this is the ARGOT strategy [Sha87]. Shaefer performs population measurement calculations at various times during evolution. These measurements determine what encoding changes should be applied. These properties include adding or removing bits and shifting the parameter boundaries (domain).

2.3 Ordered, permutation encoding

Permutation problems require the optimal arrangement of a set of symbols in a list. The TSP is such a problem where a symbol can be used to identify a city, and the arrangements of symbols in a list represent the order in which the salesperson visits each city to form a circuit of all cities. A permutation encoding can be represented by a list of distinct integer values, e.g. x = [4, 3, 0, 1, 2] [GGRG85, Gre87, WSF89, Ron95a, Ron94, RAV95]. Each integer value in the list directly encodes the relative ordering of some problem-specific object. This representation:

1. prohibits missing or duplicate allele values;
2. allows high-performance genetic operators (such as the edge recombination operator [WSF89]) to be used;
3. facilitates a simple decoding mechanism from the genotype to the phenotype for travelling salesperson problems.

A permutation encoding alone does not dictate what building blocks will be accumulated by a GA. In permutation-type problems the building-blocks of interest are jointly determined by the qualitative nature of the fitness function of the problem to be solved, the genetic operators, and the encoding. For example, in TSP two adjacent genes 4, 3 indicate that towns 4 and 3 are connected in the solution, adding a component of cost according to the distance between the two towns $d_{4,3}$. These adjacent genes contribute directly to the cost of the solution. If the genetic operators in a given GA transfer and recombine adjacent towns then the building blocks will be undirected edge relationships. However, other permutation problems lead to different building block types. For example, in some routing problems, the building blocks are defined by the location of an allele within the genotype. In [Ron95a], Chapter 4, five qualitatively different landscapes are identified which stem from the one permutation encoding. Knowing the nature of the problem landscape is important in choosing a robust encoding. The problem landscape, the genetic encoding, and the genetic operators must be matched in order for the useful problem-specific building blocks to be effectively accumulated.

Some applications require that one or more alleles in the genotype are repeated a number of times [CHMR88, GGRG85]. In [CHMR88] Cohoon et al. describe a VLSI integrated circuit that is composed of a number of variable sized micro-modules. These modules are composed together to form a complete integrated circuit floorplan. The genotype is a Polish expression which determines how each module is composed together. The two repeated symbols in this encoding are: *, the symbol for horizontal composition; and +, the symbol for vertical composition. Therefore, a genotype may contain multiple instances of either or both symbols. For example, the genotype [1, 4, 5, 6, *, +, +, 8, 7, *, 3, 2, * +, *] specifies how the modules identified by the symbols 1 to 8 are composed under the action of the two composition operators.

2.4 Other representations

Many other encoding strategies have been used in evolutionary algorithms. These include trees [Koz92], matrix encodings [CP87, BBM93, VM89], and structured heterogeneous encodings [Gib95] to name a few. The challenge for research is to examine ways in which these encoding schemes can be combined together to solve hybrid type problems (Section 7.).

3. The Phenotype

In genetics, a phenotype is defined as a distinctive trait or a measurable characteristic that an organism possesses [Sta91]. In some situations a phenotype can be easily observed, such as wing colouring in moths, and eye colour in mammals. In GAs, the phenotype means the defining characteristics or qualities of the entire genotype. In some situations, the phenotypic characteristics directly follow from the genotype. For example, if the genotype encodes a single binary parameter to a problem, then the problem point is described at a phenotypic level by decoding that binary parameter into a numeric quantity. This direct mapping between the genotype and phenotype is also evident in the Travelling Salesperson Problem (TSP) (Section 2.3) with a permutation encoding. The genotype string encodes each town identifier at consecutive loci. A TSP circuit is made directly from this list of town identifiers [Ron94, Ron95b]. However, in other problem types the phenotypic characteristics do not follow in a straight-forward manner. For example, in [CRG93] a job shop scheduling representation is described. The genotype encoded a set of job priority lists for each machine. When a number of jobs are waiting to be processed by a machine, that machine chooses the job with the highest priority as encoded in the genotype for that machine. In this way a schedule (a Gantt chart) was derived by a simulation of the genotype. The genotype was a set of machine-job priority lists and a complex mapping was required to map to the phenotype (a schedule).

4. Encoding Issues

4.1 Degree of encoding

A number of GA researchers have addressed the question of the degree of encoding. In many problems it is possible to encode various levels of detail. The classic example is the routing and scheduling problem. There are two approaches to such problems. The first is to perform routing and scheduling as two separate processes. In a GA implementation it is possible to encode only the routing details, and obtain the schedule by a simulation process. However, a given routing may map to many different schedules. One strategy is to perform a local search of all possible schedules. The other strategy is to simply make assumptions during the simulation process which prunes down the mapping to a single schedule as in [CRG93]. The other alternative is to encode all of the routing and scheduling information into the genotype [Bru93]. With a complete encoding, more detail is encoded into a genotype allowing the evolutionary process to pinpoint the exact solution. However, when a complete encoding is used the genetic operators may become more complex [Bru93], and the concept of a building block becomes more difficult to interpret relative to the schema theorem and the fundamental theorem of GAs [Hol75, Gol89].

4.2 Partial cover

A genetic encoding is often a compromise between many conflicting factors. One problem that arises, especially in indirect representations, is that the entire search space is not addressed by the encoding used [CRG93]. This can limit the GA search and preclude promising areas of the problem space. In the worst case, this may prevent the global maximum peak from being found. Direct representations address the entire search space, however they may have drawbacks. For example, in [Bru93], a direct representation of a production schedule is used. The genotype contains machine allocations, as well as start and stop times. This encoding addresses all of the search space, however, the design of the genetic operators becomes a complex task, and the initialisation of the population is more difficult. Additionally, direct representation GAs can use a considerable amount of problem-specific knowledge that detracts from the generality of the GA. When direct representations are used, it may become more difficult to determine whether meaningful building blocks are free to propagate in exponential numbers as time progresses [Hol75, Gol89].

4.3 Complex mappings from genotype to phenotype

In some applications it is a somewhat complex issue to determine the best encoding. In typical binary-encoded GA implementations the building blocks are low-order and short length schemata that have above-average fitness

[Gol89]. A building block is a solution component that can be arranged with other building blocks in a constructive way. Ideally, when a building block A is combined with another set of building blocks, A influences the final solution in an independent fashion. In practice this process can fail because:

1. The problem has no obvious building blocks (Section 5.).
2. The problem is highly polygenic, other genes also influence the primary characteristic that A is thought to be related to – this can result in a suppressive affect on the action of A on the property of interest (known as *epistasis*).
3. The problem is highly pleiotropy such that the interaction of many qualities gives a gene a defined meaning. This tends to confuse the precise meaning of a gene, and hence building blocks assembled from multiple genes amplify this confusion into the phenotype resulting in a non-linear mapping between a building block and a set of properties.
4. The building blocks are not adjacent or near-adjacent groups of genes but have large defining lengths [LR91, GL95].

In these situations it may be possible to map the encoding to a different problem that does not exhibit these problems to the same degree. This is discussed in detail.

4.4 Remapping the encoding

4.4.1 Why remap?. It is sometimes possible to remap an encoding to a second encoding in order to exploit a better encoding mechanism that can make use of better-defined building blocks. A case in point is the train scheduling problem in [MP92]. In this application a set of trains depart from a source arriving at a destination on a network. The network uses a single track for two-way traffic, and the trains must use a station crossing loop in order to pass each other in opposite directions, or for one train to overtake another. A GA was used to optimise a timetable of train departures in order to minimise late train arrivals at the destination points. The first devised encoding E_1 encoded the departure times for each train at each station. However it was found that merging a set of departure times from different parents resulted in very poor performance as completely different routing strategies would be introduced into newly created children. This was because a set of departure times was not a decomposable part of the overall problem as removing such a set removed the context that they worked in and rendered them ineffective. However, the second encoding that was used E_2 encoded a set of train priorities for each track segment. The train of the highest priority was allowed to enter the track segment with the lower-ranked train waiting in a crossing loop. The second encoding E_2 meant that a building block, a set of train priorities for different stations, had less dependent contextual information. Encoding E_2 resulted in high-quality timetables. Encoding E_2 represents the problem using a different mechanism, train priorities rather than individual times. The genotype encoded priorities and this was mapped to the phenotype that required a timetable. To do this Mills and Perkins used a complex mapping where the timetable was obtained through a simulation of the train network. A number of lessons can be derived from this example, i.e. it may be possible to:

1. modify the way the encoding describes the problem to exploit better building blocks;
2. map the problem to a genotype that encodes more general context-independent properties of the problem;
3. use a genotype-to-phenotype translation process that expresses the final solution in a form that is suitable for measuring fitness information.

It is important to recognise that the original problem has been modified in this process and that genotype-to-phenotype mapping process may not cover the entire problem space. This may result in the optimal-solution being unavailable to the GA [CRG93].

4.4.2 Simulation. A common practice in real-world GA problems is to encode the building-block information in the genotype, and to obtain the fitness of the phenotype through a simulation. For example, the GA in [DC87, CD87] was responsible for breeding a set of data-transfer speeds for a communication-link network of a given topology. The fitness of the genotype that encoded link speeds was determined by stochastic simulation that modelled the network with the speeds in the genotype, and which also modelled the effect of a number of stochastic constraints. The fitness value was the overall performance of such a simulation. Gibson in [Gib93] used a fitness function that invoked a stochastic simulation of a process control model. Each fitness evaluation required several minutes of CPU time. Simulations are sometimes necessary for the following reasons:

1. an indirect encoding can sometimes be used that has better building blocks – this indirect encoding may make simulation of the problem necessary to determine fitness information;
2. the problem itself is highly complex and requires a simulation to determine as single fitness sample [Gib93].

It is important to note that some simulations are stochastic and may result in a noisy approximation of the true fitness of the problem under test. Issues of noise are discussed in [Gib93].

5. Encodings and Genetic Operators

To solve a given problem a practitioner must choose an encoding and a suitable set of genetic operators. However, a naive choice can result in various problems which include;

1. the genetic operators do not transfer the most appropriate types of building blocks for the problem and the encoding used [Gol89];
2. the crossover operators do transfer the correct notion of building block but in the process of doing so introduce an unacceptable amount of disruption to the phenotype as a result of mutation around the crossover linkage sites [WSF89, Fog95];

In a typical GA implementations the genetic operators of crossover and mutation act in precise ways to transfer building blocks between parents and children. Different genetic operators are keyed to particular building blocks. For example, Whitley's genetic Edge Recombination Operator (ERO [WSF89]) acts on two permutation-encoded parents to produce two permutation-encoded children. This operator is designed to transfer and merge adjacent genes from the parents to the children while introducing as few new adjacency relationships not in either parent into the child. Goldberg's PMX operator [Gol89] acts on the same encoding types, permutations, but transfers unshifted contiguous blocks of genes from parents to children. In both cases the encoding is identical but the operators determine the type of building blocks that are accumulated as the GA is run.

Excessive building block disruption can be caused by inappropriate design of the genetic operators, the choice of the encoding strategy, or simply an artifact of the problem type. Mason in [Mas95], presented a crossover measure, the non-linearity ratio, for determining the degree in which crossover is likely to be effective for the problem space and the encoding that represents it. Mason's non-linearity ratio can determine whether crossover is effective in propagating building blocks from parents to children in binary domains. Mason primarily introduced such a measure to determine whether a given problem is suited to a genetic algorithm or to the many other computational models available [Fog95]. However, such a measure might be used when comparing two possible encoding strategies for a given problem. For example, whether Gray coding [Gol89] proves effective in reducing Hamming cliffs in a given domain for given operators.

6. Isomorphisms in the Encoding

6.1 Introduction

It has been observed in GA research on scheduling problems [NY91, FRC93] that some genetic encodings exhibit a high degree of redundancy. In the case of the job-shop-scheduling problem, in the two encoding schemes presented in [NY91, FRC93], a single point in the problem space could be represented by a large number of genotypes. In the case of [NY91], illegal schedules were possible, and a forcing step was used to modify the genotype to represent a legal point in the problem space. In [FRC93] the GA representations always encoded legal schedules, however, a high degree of redundancy was still present.

In these cases, a single problem point mapped to many different genotypes. Hence, a problem point defines an equivalence class P in which the members are the corresponding genotypic representations of that problem point. The genotypes in P represent isomorphisms of each other and serve to increase the size of the encoding space. It is not clear how all such isomorphisms in [NY91, FRC93] could be removed, and what the benefits would be when removed. However, the problem of isomorphisms in GA encodings is a relevant issue for robust encodings and should be considered by the practitioner.

6.2 The benefits of representational redundancy

It has been observed in the literature that some genetic encodings which allow representational redundancy result in better final solutions than encodings that preclude representational variants [CHMR88]. This has been observed in genetic programming experiments (GP) where normalising program trees down to the most simple structure leads to lower performance. In the GP examples the representational variants play a key role in diversity preservation due to the fact that crossover sees two representational variants as a greater source of information than a single simple representation when creating children from representationally variant parents.

It is conjectured that representation variants will be beneficial to a GA if the genetic operators see isomorphic encodings as a source of additional information. For example, if a crossover operator had to choose two parents from 100 identical genotypes, the created children would probably be identical to their parents. However, the situation is different if a crossover operator could choose from 100 different representationally-redundant genotypes that each encode a single point p in the problem space. In this case, it is feasible that the crossover operator might create children that do not encode p. It is suggested that these redundant encodings represent an additional source of diversity in an evolving population, and allows the process of implicit parallelism [Gol89] to occur for a greater period of time. It is expected, in such a scenario, that better final-population solutions would result.

However, representational redundancy is not always a beneficial phenomenon. If the genetic operators do not gain any additional information from representational variants, then representational redundancy represents the opposite situation, diversity loss [Ron95a]. In this case, the removal of isomorphic encodings results in a reduced encoding space and a higher degree of diversity throughout evolution. This allows the GA to solve a reduced problem and better results can be obtained.

6.3 Example of representational redundancy

In [Ron95a] it was found that isomorphisms in the encoding can be effectively removed with a two step process of normalisation and then duplicate removal.

It was found that this resulted in better GA performance in the various TSP benchmark problems that were examined. This process was achieved by:

1. normalisation – mapping all genotypes back to a canonical form – this effectively maps all isomorphisms contained in an equivalence class to a single representative genotype, and;
2. removal of subsequent identical genotypes that are in canonical form to enforce genotype uniqueness in the population.

The canonical representation chosen should not compromise the propagation of building blocks from parents through to the children under the action of the crossover operator. Secondly, this canonical representation should not introduce any positional or allelomorphic bias on the part of the genetic operators.

This two step process reduces the encoding space, making the problem simpler for the GA to solve.

In [Ron95a] it was observed through experimentation with the TSP problem that in the last phases of the GA run two types of genotype isomorphisms emerged in the population.

The first type of isomorphism encountered in the TSP was a cyclic-shift isomorphism. This isomorphism only applies to cyclic genotypes such as those genotypes used to encode the cyclic TSP problem. If the problem requires a search for Hamiltonian paths, then these isomorphisms would not be removed. In the cyclic TSP, the last town, in the list of towns, is considered adjacent to the first town. The following two genotypes $x_1 = [1\ 3\ 2\ 5\ 0\ 4]$ and $x_2 = [5\ 0\ 4\ 1\ 3\ 2]$ encode two isomorphisms of the TSP circuit $(0, 4, 1, 3, 2, 5)$.

The second kind of isomorphism encountered was an inverted-ordering isomorphism. The two genotypes $x_1 = [0\ 4\ 1\ 3\ 2\ 5]$ and $x_2 = [0\ 5\ 2\ 3\ 1\ 4]$ are examples of this. In this case x_1 and x_2 are an inverted orderings of each other.

Ronald in [Ron95a] used a 20 member population for the following four scenarios:

1. shift and inverted isomorphisms normalised;
2. shift isomorphisms normalised;
3. inverted isomorphisms normalised;
4. no isomorphisms normalised.

Each of these four scenarios was simulated for the two cases of 1) duplicates ignored and 2) duplicates removed (using an efficient hash tagging algorithm [Ron95a]) (Figure 6.1).

The results in Figure 6.1 and the discussion in [Ron95a] showed that:

– some problems, such as the TSP, have obvious isomorphic encodings which leads to simple normalisation routines;
– representational redundancy can be eliminated through normalisation followed by duplicate removal

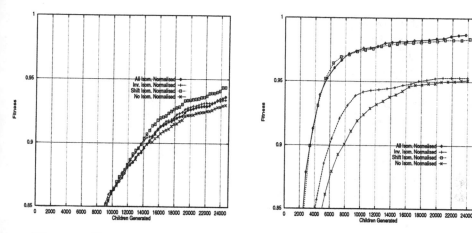

Fig. 6.1. The affect of encoding-normalisation with a regular GA for the Oliver 30-town TSP for population sizes of 20 (*left*) and the affect of isomorphism removal with, a regular GA, with all duplicates removed (*right*), Averaged over 50 runs, for the Oliver 30-town TSP

- representationally redundant encodings are unhelpful to a genetic algorithm if the genetic operators cannot benefit or gain additional information from isomorphic variations;
- the size of the encoding space is reduced by removing isomorphic encodings, this means a smaller problem for the GA;
- when representational redundancy is eliminated diversity loss is reduced allowing effective evolution (implicit parallelism) to proceed for a greater period of time

Representational redundancy occurs in many problem encodings. Hence, the discussion of normalisation for the TSP problem represents just one example of how the technique might be applied to other problems. This result is significant for GA practitioners as there is now strong evidence to suggest that encodings should be designed to naturally suppress redundant encoding forms, and genetic operators should be implemented in such a way that redundant forms do not benefit the operators. If it is not possible to suppress redundant forms in the encoding, then a normalisation process like the two presented in [Ron95a] may be warranted. If normalisation proves to be an intractable exercise then heuristic normalisation methods where most of the isomorphic forms are eliminated might be attractive.

7. Conclusion: Towards Mixed Encodings

An critical issue in the design of a flexible encoding is the way in which the encoding can be adapted to suit new requirements and constraints. The

GAME system [Fil94] provides a flexible encoding strategy. The encoding employs a tree structure where the leaves of the tree can be integers, characters, floating point values, or binary strings. Gibson, in his thesis [Gib95], designed a richer set of encoding primitives in the domain of process / factory modelling. Gibson defined general encoding components that could be assembled together to form a hybrid encoding. These components related to five distinct areas:

1. qualitative parameters, to model numerical quantities such as buffer sizes or processing times;
2. qualitative parameters to control enumerated categories in the model such as the type of machine to use;
3. structural choice parameters to allow control over which subsystems were used;
4. schedule parameters to allow sequential ordering and embedded looping;
5. and rulebase parameters – collections of if-then rules that can become regulating agents in the model.

Each of Gibson's encoding types are not new when considered individually, however his construction of composite encodings to form hybrid and extensible representations is of great practical interest. For each problem, Gibson built up a complex encoding out of these encoding components. The genetic operators understood the structure of the parent encodings and would apply themselves intelligently to the parents to produce children. Work such as this goes a long way towards the objective of being able to solve the type of complex problems that can be found in applications of all disciplines.

Many of the issues discussed in this chapter here must be considered before an encoding is deemed suitable for a GA. With new tools and techniques it should become easier in the future to design genetic encodings for new problems that comply with these theoretical and practical requirements. As more complex problems are tackled attention to such detail becomes critical to ensure that the selected encoding strategy is practical, efficient, and robust.

References

[BBM93] D. Beasley, D. Bull, and R. Martin. Reducing epistasis in combinatorial problems by expansive coding. In S. Forrest, editor, *Proceedings of the Fifth International Conference on Genetic Algorithms*, pages 400–407. Morgan Kaufmann Publishers, San Mateo, CA, 1993.

[BRE91] J. Bhuyan, V. Raghavan, and V. Elayavalli. Genetic algorithm for clustering and ordered representation. In R. Belew and L. Booker, editors, *Proceedings of the Fourth International Conference on Genetic Algorithms*, pages 408–415. Morgan Kaufmann Publishers, San Mateo, CA, 1991.

[Bru93] R. Bruns. Direct chromosome representation and advanced genetic operators for production scheduling. In S. Forrest, editor, *Proceedings of the Fifth International Conference on Genetic Algorithms*, pages 352–359. Morgan Kaufmann Publishers, San Mateo, CA, 1993.

[CD87] S. Coombs and L. Davis. Genetic algorithms and communication link speed design: Design, constraints and operators. In J. Grefenstette, editor, *Genetic Algorithms and their Applications. Proceedings of the Second International Conference on Genetic Algorithms*, pages 257–260. Lawrence Erlbaum Associates, Hilldale, New Jersey, 1987.

[CHMR88] J. Cohoon, S. Hegde, W. Martin, and D. Richards. Floorplan design using distributed genetic algorithms. In *IEEE International Conference on Computer-Aided Design*, pages 452–455. IEEE Computer Society Press, Washington, DC, USA, 1988.

[CP87] J. Cohoon and W. Paris. Genetic placement. *IEEE Transactions on Computer-Aided Design*, 6(6):1272–1277, 1987.

[CRG93] F. Croce, T. Roberto, and V. Giuseppe. A genetic algorithm for the job shop problem. Technical report, D.A.I Politecnico di Torino. To appear in *Computers and Operations Research*, Italy, 1993.

[DC87] L. Davis and S. Coombs. Genetic algorithms and communication link speed design: Theoretical considerations. *Genetic Algorithms and their Applications. Proceedings of the Second International Conference on Genetic Algorithms*, pages 252–256, 1987. Boston, MA.

[Fil94] R. Filho. The GAME system (genetic algorithms manipulation environment). In *IEE Colloquium on Applications of Genetic Algorithms*, pages 2/1–4. IEE, London, UK, 1994.

[FM91] B. Fox and M. McMahon. Genetic operators for sequencing problems. In G. Rawlins, editor, *Foundations of Genetic Algorithms*, pages 284–300. Morgan Kaufmann Publishers, San Mateo, CA, 1991.

[Fog95] D. Fogel. *Evolutionary computation. Towards a new philosophy of machine intelligence.* IEEE Press, Piscataway, NJ, 1995.

[FRC93] H. Fang, P. Ross, and D. Corne. A promising genetic algorithm approach to job-shop scheduling, rescheduling, and open-shop scheduling problems. In S. Forrest, editor, *Proceedings of the Fifth International Conference on Genetic Algorithms*, pages 360–367. Morgan Kaufmann Publishers, San Mateo, CA, 1993.

[GGRG85] J. Grefenstette, R. Gopal, B. Rosmaita, and D. Gucht. Genetic algorithms for the travelling salesman problem. In J. Grefenstette, editor, *Proceedings of an International Conf. on Genetic Algorithms and Their Applications*, pages 160–168, 1985.

[Gib93] G. Gibson. Application of genetic algorithms to mixed schedule and parameter optimisation for a visual interactive modeller. In *Fifth Workshop on Neural Networks: Academic/Industrial/NASA/Defence*, pages 119–124. SPIE Proceedings Series Volume 2204, 1993.

[Gib95] G. Gibson. *Application of Genetic Algorithms to Visual Interactive Simulation Optimisation.* PhD thesis, University of South Australia, 1995.

[GL95] J. Gero and S. Louis. Improving pareto optimo designs using an evolutionary approach. *Microcomputers in Civil Engineering*, 1995. To appear.

[Gol89] D. Goldberg. *Genetic Algorithms in Search, Optimization, and Machine Learning.* Addison-Wesley, 1989.

[Gre87] J. Grefenstette. Incorporating problem specific knowledge into genetic algorithms. In L. Davis, editor, *Genetic Algorithms and Simulated Annealing*, pages 42–60. Pitman Publishing, London, 1987.

[Hol75] J. Holland. *Adaption in natural and artificial systems*. The University of Michigan Press, 1975.

[JS89] K. De Jong and W. Spears. Using genetic algorithms to solve NP-complete problems. In J. Shaeffer, editor, *Proceedings of the Third International Conference on Genetic Algorithms and their Applications*, pages 133–139. Morgan Kaufmann Publishers, San Mateo, CA, 1989.

[Koz92] J. Koza. *Genetic programming: on programming computers by means of natural selection and genetics*. The MIT Press, Cambridge, MA, 1992.

[LR91] S. Louis and G. Rawlins. Designer genetic algorithms: genetic algorithms in structure design. In R. Belew and L. Booker, editors, *Proceedings of the Fourth International Conference on Genetic Algorithms*, pages 53–60. Morgan Kaufmann Publishers, San Mateo, CA, 1991.

[Mas95] A. Mason. A non-linearity measure of a problem's crossover suitability. In D. Fogel, editor, *Proceedings of the 1995 IEEE International Conference on Evolutionary Computation*, pages 68–73. IEEE Press, New York, 1995.

[MN95] Z. Michalewicz and G. Nazhiyata. Genocop III: A co-evolutionary algorithm for numerical optimization problems with nonlinear constraints. In D. Fogel, editor, *Proceedings of the 1995 IEEE International Conference on Evolutionary Computation*, pages 647–652. IEEE Press, New York, 1995.

[MP92] R. Mills and S. Perkins. Genetic algorithms applied to the scheduling of trains. the problem of string representation. Technical report, Scheduling and Control Group. The University of South Australia, South Australia, 1992.

[NY91] R. Nakano and T. Yamada. Conventional genetic algorithm for job shop scheduling. In *Proceedings of the Fourth International Conference of Genetic Algorithms and Their Application*. Morgan Kaufmann, San Mateo, CA, 1991.

[RAV95] S. Ronald, J. Asenstorfer, and M. Vincent. Representational redundancy in evolutionary algorithms. In D. Fogel, editor, *Proceedings of the 1995 IEEE International Conference on Evolutionary Computation*, pages 631–637. IEEE Press, New York, 1995. http://createwin.com/EC/pubs.html.

[Ron94] S. Ronald. Preventing diversity loss in a routing genetic algorithm with hash tagging. In R. Stonier and Xing Huo Yu, editors, *Complex Systems: Mechanism of Adaption*, pages 133–140. IOS Press, Amsterdam, 1994. http://createwin.com/EC/pubs.html.

[Ron95a] S. Ronald. *Genetic algorithms and permutation-encoded problems. Diversity Preservation and a Study of Multi-Modality*. PhD thesis, University South Australia. The Department of Computer and Information Science, E-mail s.ronald@unisa.edu.au, 1995. http://createwin.com/EC/pubs.html.

[Ron95b] S. Ronald. Genetic algorithms and scheduling problems. In L. Chambers, editor, *The Practical Handbook of Genetic Algorithms. Volume 1*, pages 367–430. CRC Press, Boca Raton, Florida, 1995. http://createwin.com/EC/pubs.html.

[Sha87] C. Shaefer. The ARGOT strategy: adaptive representation genetic optimizer technique. In J. Grefenstette, editor, *Genetic Algorithms and their Applications. Proceedings of the Second International Conference on Genetic Algorithms*, pages 50–58. Lawrence Erlbaum, Hilldale, New Jersey, 1987.

[Sta91] W. Stansfield. *Theory and Problems of Genetics*. McGrawHill, 1991.

[VM89] G. Vignaux and Z. Michalewicz. Genetic algorithms for the transportation problem. *Methodologies for Intelligent Systems*, 4:252–259, 1989.

[WSF89] D. Whitley, T. Starkweather, and D. Fuquay. Scheduling problems and traveling salesmen: The genetic edge recombination operator. In J. Schaffer, editor, *Proceedings of the Third International Conference on Genetic Algorithms and their Applications*, pages 133–139. Morgan Kaufmann Publishers, San Mateo, CA, 1989.

Part II

Architecture and Civil Engineering

Genetic Engineering and Design Problems

John S. Gero, Vladimir A. Kazakov, and Thorsten Schnier

Key Centre of Design Computing, Department of Architectural and Design Science, The University of Sydney, NSW 2006 Australia

Summary. In this chapter we describe an extension of genetic algorithms which utilizes a set of ideas borrowed from the practice of genetic engineering of natural organisms. We show how it can be constructed and implemented computationally and how it enhances the efficiency of the genetic algorithm's computational process. We apply it to the solution of a spatial layout planning problem as an example of an NP-complete problem and to case-based layout design as an example of a class of problem which cannot be solved directly using traditional genetic algorithms.

1. Genetic Engineering Extensions of GAs

Genetic engineering is a technology for direct intervention in the genetic structure of natural organisms. It is used for the rapid manufacture of improved organisms. Usually, its implementation includes two stages. The first stage is devoted to an analysis of the genetic material of highly successful (with respect to some criterion) organisms. The goal here is to identify genetic features (we shall call them 'evolved' genes) directly linked to their high level of success. The second stage includes attempts to manufacture improved organisms using these evolved genes. This includes some form of direct manipulation of genotypes of the trial organisms which results in the acquisition of these evolved genes by their genotypes. For example, severe combined immunodeficiency (SCID), an illness that occurs if a defective gene is inherited from both parents, can be treated by inserting normal copies of the gene into the person's white blood cells [1].

Genetic algorithms have proven to be both successful and popular in a wide range of engineering design problems [5, 8, 15]. It seems natural to extend genetic algorithm paradigms on the basis of this simple model of genetic engineering practice. Let us illustrate how it can be done using a simple problem where solutions are coded as strings of 19 different integer numbers from the range $[0, 18]$. Some fitness function is defined over the phenotype. Mutation here is defined as a pair-wise swap of two randomly picked genes. Assume that a standard GA is running for a predefined number of generations. Let us examine the genotypes of the evolved population. First, we classify each genotype as being either high fit or low fit, if the fitness of its phenotype is above or below some threshold. Then we try to identify features which distinguish the genotypes of the high fit phenotypes from the genotypes of the low fit phenotypes, Figure 1.1. Here it is easy to see that the fittest genotypes all contain the string of genes $A = \{0, 1, 2, 8, 4, 3\}$ and all that are less fit lack it.

48   John S. Gero et al.

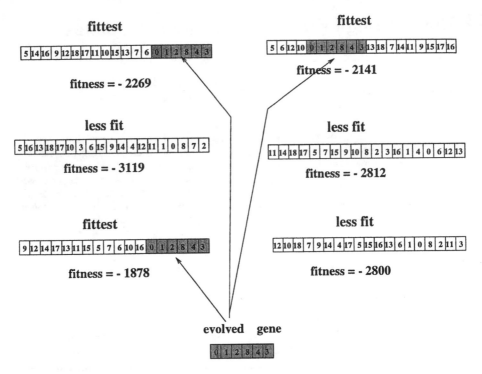

Fig. 1.1. The identification of the evolved gene $A = \{0, 1, 2, 8, 4, 3\}$ based on distinguishing between high and low fitness phenotypes

Hence, we can assume that A is an evolved gene and that its presence in a genotype leads to a high level of fitness. This hypothesis can be tested by generating two random sets of genotypes: one with and one without this evolved gene and using statistical tests to check if the former one is fitter than the latter one. If A withstands this testing we declare it our current evolved gene.

Now we operate under the assumption that the presence of high levels of A in the genetic pool leads to a fitter population. Hence, we want to increase the presence of A in the population. The first step is to encapsulate A into an evolved gene, and protect it from disturbance by genetic operations like cross-over and mutation. This can be done by marking the sections in the genotypes that contain the evolved gene, Figure 1.2, or by introducing a new symbol and replacing all or a high percentage of the occurrences of the evolved gene with the new symbol, Figure 1.3. In the first case, the length of the genotype does not change, and a fixed-length representation can be used. In the second case, a new symbol is added to the alphabet used in the genotypes and the genotypes may change length.

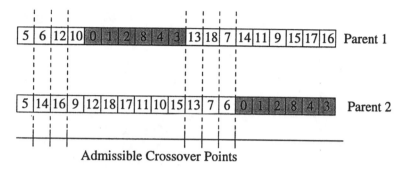

Fig. 1.2. Encapsulating evolved genes by restricting the choice of crossover points to regions not occupied by evolved genes (for a fixed length genotype)

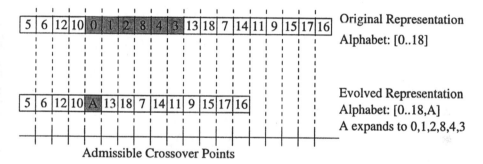

Fig. 1.3. Encapsulating evolved genes by replacing them with a newly introduced symbol

The encapsulated evolved genes function as fixed subassembled genetic blocks in the population. This is illustrated in Figure 1.4. Again, the problem is coded into strings of integer numbers in the range of [0, 18]. In the figure, the set of all possible genotypes of a certain length is represented by one of the concentric arcs. In Figure 1.4(a), only the basic alphabet of integer values is used. If analysis shows that individuals with the sequence 0, 1, 2 and the sequence 8, 4, 3 are found to be particularly successful (highlighted in the figure), then evolved genes can be created and added to the alphabet (represented by the symbols A and B). In Figure 1.4(b), the new situation is shown. The sequence 0, 1, 2, 8, 4, 3 can now be represented by a genotype of length two, instead of six as in Figure 1.4(a). If this sequence is found to be successful, a new evolved gene can be created for it in the next round of genetic engineering.

To further increase the proportion of the fit evolved genes in the population, the evolved genes can be artificially introduced into genotypes lacking it, as it is done in genetic engineering.

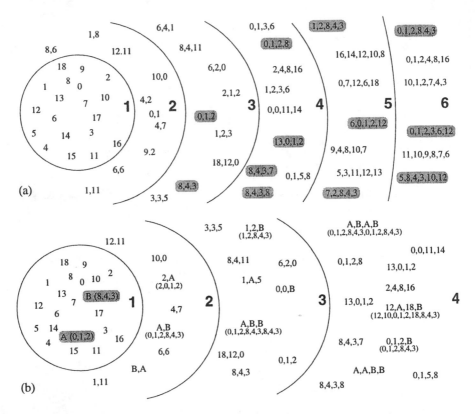

Fig. 1.4. Effect of evolved genes: (a) representation without evolved genes, genotypes consist of variable-length strings of integers between 0 and 18, individuals of high fitness are highlighted; (b) representation after evolved genes A and B are created, with some corresponding genotypes in original representation given in brackets

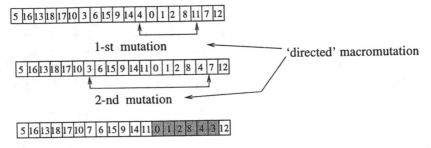

Fig. 1.5. The genetic 'operation' to create evolved gene $A = \{0, 1, 2, 8, 4, 3\}$ using directed macromutation

We try to make our trial population fitter before further reproduction by changing (the majority of) these genotypes which lack evolved gene A in such a way that they acquire it with minimal changes. It can be done using a 'directed' macromutation — a sequence of mutations which leads to the creation of A, Figure 1.5. Minimal changes to genetic material here mean that a number of elementary mutations in a macromutation should be minimal.

As a result of the introduction of evolved genes and the measures which ensure their survival at above the natural rate the evolutionary path of a standard GA is shortcut. If the genetic encoding is fixed-length and position-dependent then the evolved genes are just building blocks of a standard GA theory, and the genetic engineering extension of GA can be viewed as a way of explicitly handling such blocks. If the coding is position-independent and variable-length then it is clear that the effect of evolved genes usage instead of original genes is twofold: first, it allows the sampling of larger areas of a search space (the sample points are further apart from each other) subject to the same computational resources; and second, it allows the searching of a particular partition of the search space which contains the fittest points more thoroughly, again subject to the same computational resources.

Another advantage of the proposed approach is the possibility to use the evolved genes in a new environment. The evolved genes contain information about the problem that can be re-used to help solve similar problems.

After we have done this — identified newly evolved genes and made the population genetically healthier by 'operating' on those genotypes which lack the evolved genes using directed macromutations — we run a standard GA for the next predetermined number of generations. Since we do not want reproduction to damage the evolved genes which are now present in the genetic material we choose the crossover point only to be outside the genotype's sections occupied by this evolved gene in any of the parents, Figure 1.2, and set the mutation rate of the evolved gene's components at a much lower level than the standard mutation rate level.

Then we again analyze the genetic material of the evolved population, identify newly evolved genes, test them together with previously evolved genes to produce the current set of evolved genes and 'operate' on those genotypes which lack these genes. The cycle is then repeated.

In practice, the analysis of genetic material is not carried out 'manually' as we did in our illustrative example but automatically, by using one of a vast arsenal of string analysis techniques developed in genetic engineering and speech processing, [3, 10, 13, 16, 20]. In the general case the type of evolved genes naturally developing is problem and encoding dependent. It could be any arbitrary feature of a genotype, not necessarily a substring (fixed group of genes with arbitrary ordering, periodic pattern, etc.). If the type of evolved gene's characteristic for a particular type of problem is known then it is possible to design a much more efficient method for genetic analysis. Techniques employed during genetic 'operations' could also be designed

differently. It should produce a particular type of evolved gene in a particular encoded (position-dependent, position-independent, order-based, fixed
length, variable length, etc.) genotype and still lead to as little change of
genetic material as possible. These conditions still leave a significant freedom of choice. The simplest way of designing such an 'operation' technique
is to use macromutation (directed sequence of standard mutations) similar
to the example. Usually one of the techniques of genetic engineering of natural organisms (genetic surgery, genetic therapy, etc.) is used as a template
to design a particular operation technique. If one chooses genetic surgery
as a model then the genetic 'operation' changes only the genotype's pieces
which are similar enough (according to some measure) to the evolved genes
and not the whole trial genotype. If genetic therapy is chosen as a model of
genetic 'operation' then the evolved genes are inserted into trial genotypes
in variable-length genotypes or they replace pieces of equal length in fixed-
length genotypes, Figure 1.6. Usually, a random position for this insertion or
replacement is chosen.

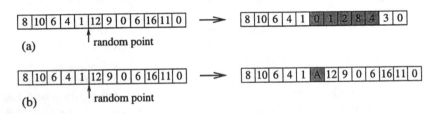

Fig. 1.6. Genetic therapy model: (a) evolved gene $A = 0, 1, 2, 8, 4, 3$ replaces randomly selected part of equal length in a fixed-length genotype; (b) evolved gene
$A = 0, 1, 2, 8, 4, 3$ is inserted at a random position into a variable-length genotype

Related ideas have been developed in genetic programming [2]. The main
differences to genetic engineering GAs include the way evolved genes are
identified (using the fitness function of partial genotype or by just random
selection of the parts of genotype), less with developed tools for manufacturing the population whose genotypes are reached using evolved genes, and the
concept of the re-use of evolved genes in related problems is absent.

In the remainder of this chapter we present two applications of genetic
engineering in GAs. The first deals with spatial layout planning problems
whilst the second deals with case-based layout design.

2. Spatial Layout Planning Problem

2.1 Introduction

Spatial layout problems have numerous applications in architectural design,
VLSI design, etc. We use its formalization as a capacitated quadratic assignment problem [11]. The number of activities to be placed, m, their areas

$\{a_0, \ldots, a_{m-1}\}$ (in terms of elementary square units) are given as well as the set of feasible locations, n, their areas, $\{l_0, \ldots, l_{n-1}\}$ and the matrix of distances between them, $|d_{k,n}|, k, n = 0, \ldots, n-1$. The 'forces' of interactions between activities, $|q_{i,j}|, i, j = 0, \ldots, m-1$, are also given. The objective is to find a feasible layout of activities such that its cost, I, is minimal

$$-I = \sum_{i,j} q_{i,j} d_{\rho(i),\rho(j)} \rightarrow \max$$

The mapping $\rho(i)$ of the activity i onto the feasible location is defined using an order-based genetic representation. In order to define it for this problem we choose a trajectory (path) along the possible locations. The path should be continuous within each of the closed locations, and adjacent non-overlapping elementary squares should cover all feasible locations.

2.2 Example of space layout planning problem

As the test example we use the problem of the placement of a set of office departments into a four-storey building. (See [11] and [7] for the details of the problem.) The areas of the $m = 19$ activities (office departments) are defined in terms of elementary square modules. There is one further activity (number 19) whose location is fixed. The interaction matrix $q_{ij}, i, j = 0, 19$ is given as data. There are $n = 18$ feasible zones or locations numbered from 0 to 17, Figure 2.1. The areas of these zones are derived from the geometry of the zones, Figure 2.1. The activity number 19 is an access area which has a fixed location — zones numbers 16 and 17, Figure 2.1. The travel matrix between these zones is similarly derivable. This yields an order-based genetic representation which is used to illustrate the principles of genetic engineering based GAs described in the previous section.

2.3 Evolved genes of the spatial layout problem

One feature of the example is the character of genetic regularities which naturally occur in spatial layout problems — there are clusters of genes, where the order of the genes within the cluster matters little. In other words in many spatial layout problems the cost of a layout whose genotype includes some cluster of genes is significantly better than the one without such a cluster (at least on average). This cluster of genes becomes an evolved gene. Since we employ a continuous genetic representation these clusters of genes (evolved genes) are mapped onto a spatially compact group of activities (although the actual configuration of these activities varies when their positions within genotypes or actual order of genes within clusters change). This is the equivalent of treating a group of activities as a new 'super-activity' which is never disaggregated. This new super-activity can be located anywhere in the layout in the same way as any other activity. The improvement of the cost that can

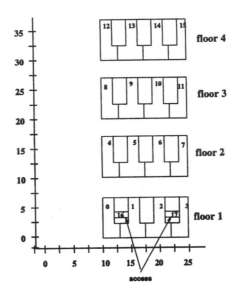

Fig. 2.1. Zone definition — graphical representation

be achieved by an optimal ordering of the genes within such a 'super-group' is much smaller. Correspondingly, a typical run of a standard GA proceeds in two stages. During the first stage, clusters are evolved and much of the cost improvement is achieved, while the second stage is devoted to optimal ordering of the genes within each cluster. Although the typical GA run actually follows this scenario it still tries to perform both of them simultaneously which leads to a waste of computational resources.

It seems natural to design an optimization procedure which operates according to this scenario, separating both searches. First, it identifies these gene groups and chooses their optimal positions with respect to each other (which is the major source of cost savings). This yields an optimization problem with a smaller number of parameters than the length of the initial genotype (instead of the positions of all genes in such group we have just the positions of the whole groups). Second, it chooses positions of the gene components (activities) within each of these groups (which improves the cost only marginally). Since the effective size of the search space in these two subsearches is smaller than the overall size of the search space of the original layout planning problem it can either be done faster or produce a better performing solution for the same computational expense. It is also clear that the local search in such a large-scale space of 'granulated' cells corresponds to the non-local search in the original state space.

Note that one does not have to establish beforehand the existence of the evolved genes in such a form — one can simply try an algorithm based on a corresponding notion of the 'group'-like evolved genes and a two level optimization process. Its success would be a sufficient condition for the existence of such genes. Otherwise the genetic engineering based GA degenerates into a standard GA.

2.4 Results

The result of the run of the standard GA which converge to the best solution (averaged over 10 such runs from different initial random seeds) is presented in Figure 2.2 (bold line). The population size was 200, probability of crossover 0.6 and the probability of mutation 0.01. We use elitist GAs with a generation gap of 3. 13 of the 100 GAs runs converged to the solution $\{7, 6, 13, 17, 14, 12, 15, 5, 18, 11, 10, 9, 0, 1, 2, 8, 3, 16, 4\}$ with a cost of the corresponding layout of 1834.

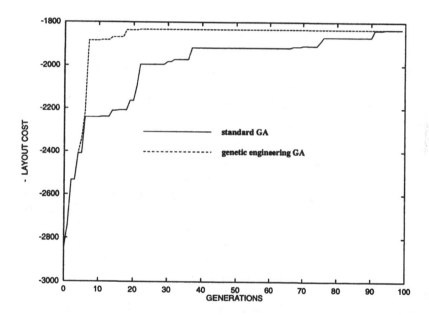

Fig. 2.2. The best fitness vs generation number for the standard GA and genetic engineering based GA. The results are averaged over 10 runs with different initial random seeds, which converge to the best solution found

The analysis of the evolutionary path of the GA shows that the search process consists of a first stage (5-10 generations), when the crossover serves as the major constructive tool and when non-local search takes place, and of a second stage, when pair-wise swapping of the genes is the driving force of the search. During the first stage the algorithm finds the point which belongs to the basin of attraction of one of the minima. During the second stage (about 90 generations), crossover does not make any contribution to the search, the actual improvement is due to mutations only and the algorithm actually searches locally within the basin of attraction previously found.

The results of the genetic engineering based GA are presented in Figure 2.2 with the dotted lines. Two evolved genes were identified after 5 generations: $\{0, 1, 2, 8, 3, 4\}$ and $\{12, 14\}$. After generation 10 two further evolved genes evolved, $\{6, 13\}$ and $\{15, 12, 14\}$, which include the previously evolved gene $\{12, 14\}$. On average just two runs were required with different initial seeds to find the layout with the cost 1834.

The computational savings in terms of generations are of the order of 90% (although some extra computations have to be done in order to facilitate the extra processing caused by additional gene analysis and processing).

2.5 Using previously evolved genes to solve families of layout planning problems

In many cases one has to solve a number of similar layout planning problems (for example, to place essentially the same set of activities into different plan shapes, etc). Usually, it is very difficult to use information about optimal layouts already designed to generate solutions for a similar problem.

One major advantage of the genetic engineering based GA is the possibility to re-use the evolved genes (information about optimal sublayouts) in a family of similar problems. This possibility is based on an intuitively obvious assumption that in many layout planning problems some activities 'gravitate' to each other much more strongly than they are attracted to the rest of the activities and therefore should be placed as a compact spatial group in any layout. If the possible locations (distance matrix) are changed then one still has to place such activities in a compact spatial group (although the actual physical placement could be quite different). Even the changes to the activity interaction matrix could preserve the attraction of some activities to each other compared to their attraction to the rest of activities, if the actual scale of attraction of the activities within the group with respect to the ones outside it do not change significantly.

In order to check this intuition we first modified the problem data in the example in Section 2.2, by changing the geometry of the system to be that shown in Figure 2.3. This changed the distance matrix. The activity interaction matrix remains the same. The results of the optimization are shown in Figure 2.4. We ran the standard GA and the genetic engineering based GA beginning with the empty set of evolved genes (dotted line) and from the previously evolved one (dotted line). We can see that the re-use of the evolved genes yields about a 20% savings in terms of the number of generations over the genetic engineering based GA.

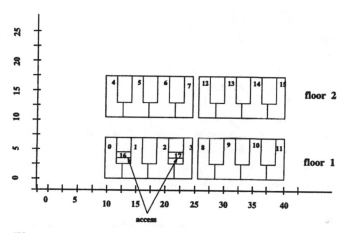

Fig. 2.3. Modified floor locations — graphical representation

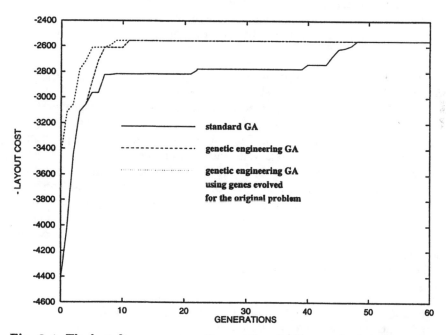

Fig. 2.4. The best fitness vs generation number for the standard GA and genetic engineering based GA in the modified example. The results are averaged over 10 runs with different initial random seeds, which converge to the best solution found

3. Case-Based Layout Design

3.1 Introduction

Case-based reasoning has been introduced into design to allow the re-use of knowledge from previous designs. It is based on the observation that designers very often re-use features of previous designs (their own or other designers) rather than having to start from first principles or compiled knowledge with every new design [12]. The word 'feature' is used here in a very general way, for a building, it could mean layouts of rooms, the general topology, materials and material combinations, structural details, form elements, etc. Case-based reasoning is different from knowledge-based design systems in that the expert knowledge is not compiled and stored, but is available only implicitly in a database of previous design cases. Since requirements and environmental conditions will usually vary between the retrieved case and the current design, aspects of the retrieved case have to be changed to fit new conditions. Design adaptation is therefore an important step in the application of case-based design.

> "Case adaptation can be simply stated as making changes to a re-called case so that it can be used in the current situation. Recognizing what needs to change and how these changes are made are the major considerations. Adapting design cases is more than the surface considerations of making changes to the previous design, it is a design process itself." [12]

Automatic adaptation of design cases has been studied by [4] and [17] in building design. These works show that the potential for adaptation of a case very much depends on its representation. Every adaptation operation first requires the transformation from a general case representation (e.g. a three-dimensional model, or a file from an architectural drawing program) into a special representation. The adaptation is applied only to this special representation, and the result is then transformed back into the general case representation. Usually, the transformation into a special representation also includes a strong parameter reduction. This reduces the size of the search space and is necessary to keep the computational complexity within bounds. However, it also often means that some, possibly more desirable, design solutions are excluded.

The adaptation process in case-based design can be seen as dealing with two problems:

1. deciding which features are characteristic for a design and are to be kept, and which features have to be changed; and
2. producing new designs, fulfilling the new requirements, re-using features that have been identified as characteristic, and adapting these features.

By using a specified representation for the case adaptation, the features of the design that can be changed and those that are left unmodified are defined. For example, if the adaptation is based on a representation that uses overall dimensions, then room areas can be adapted by stretching or shrinking parts of the design, but the topologies cannot be modified.

3.2 Basic and evolved representation

In order to use evolutionary systems in cased-based design, the designs have to be coded into a genotype suitable for genetic operations. As described in Section 3.1, the representation of the design case defines what adaptations are possible. Since our goal is to leave this decision to the evolutionary system, we have to start with a very simple representation that restricts as little as possible the designs that can be described by it. The representations used in the examples in this section are all based on a square grid, and consist of sequences of basic shapes (squares, vectors). To allow for shapes of different complexity, variable-length genotypes are used.

For the evolved genes, a number of basic shapes are composed into a more complex, evolved shape. A single evolved gene therefore can describe a complex shape and different evolved genes can describe shapes of different complexity.

3.3 Genetic engineering application

As mentioned above, the problem of adaptation in case-based design can be split into two parts. The genetic engineering approach described here is a two-stage process: one, evolving a representation; and two, using the evolved representation to create new designs.

In Figure3.1, Figure 1.4 is redrawn using vectors and shapes to describe this application. Genotypes represent outlines, they are constructed as sequences of vectors. The concentric circles of Figure 3.1 represent the search space for genotypes of different length, open arcs indicate that only part of the search space is shown. In Figure 3.1(a) the basic coding is used. Designs produced by genotypes of length one (the basic genes) are in the centre. The further away from the centre a design is, the larger is the space that has to be searched to find it, and the more complex it is. Every time an evolved gene is created the structure of the search space changes. The shape that is created by the evolved gene moves into the centre and all shapes that can be derived from the evolved gene move towards the centre with it. Figure 3.1 illustrates this: if the two closed shapes from the fourth circle in Figure 3.1(a) are identified as particularly successful and an evolved gene is created for them, the search space changes as shown in Figure 3.1(b). The two closed shapes are now basic building blocks and some shapes from the original fifth circle can now be found in the second circle, and the shape with four squares that previously was on the 14th circle (it requires a genotype of length 14, since the shape cannot be drawn without drawing two lines twice) is now on the fifth circle. Since the introduction of evolved genes increases the size of the alphabet, the size of the circles also grows.

The introduction of evolved genes obviously changes the probability that a part of a genotype maps onto a useful feature. While the number of different genes that can be used in the genotypes expands, the length of the genotype

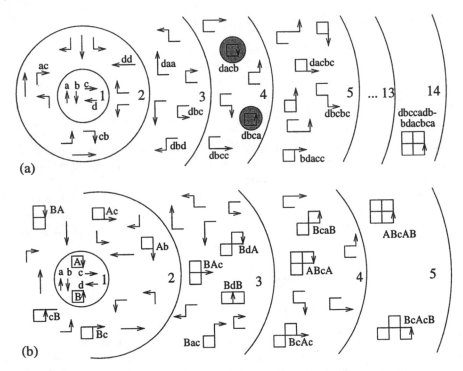

Fig. 3.1. Example of an evolving representation: (a) original representation and (b) representation with evolved genes. Some of the corresponding genotypes are given, capital letters denote evolved genes. The transformation from phenotype to genotype is not always unique, e.g. the genotypes 'ABc' and 'BAc' produce the same phenotype. Arc segments indicate that only part of the space is shown

that is required to describe a feature shrinks. The net effect is that a smaller space has to be searched. If, for example, we want to find the window-like shape composed of four squares that was used in the example above, and assume we already know how long the genotypes have to be to find it, then the original search space would have a size of about $4^{14} = 2^{28}$, while with the evolved representation, only a space of $6^5 \approx 2^{13}$ possible genotypes has to be searched.

In the adaptation step, it is important to note that the evolutionary system can make use of any or all of the features encoded in the new representation, or none of them. Solutions that are close to the example designs (in any respect) are easier to find than solutions that are less closely related, while any single feature can be adapted. In other words, the evolved transformation introduces a bias into the second-step search for results that share features with the examples, without restricting the space for possible results.

3.4 Example 1: Simple shapes

In this example, we show how the idea can be used in the design floor plans. (This example is described in more detail in [18].) As the basic coding, we use a turtle graphics-like coding with only four different basic genes, that either draw a line in the current direction, move the pen ahead, or change the current direction, Figure 3.2.

00: line forward   11: step forward   01: right turn   10: left turn

Fig. 3.2. Basic coding, arrows show pen position and current direction before and after the gene is drawn

3.4.1 Evolving the representation. The two floor plans shown in Figure 3.3 are used together as the designs to be represented. The fitness function compares individuals with this drawing, and rewards individuals depending on how much of the drawing they fit.

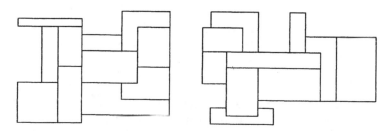

Fig. 3.3. Room plans used as example cases

To create evolved genes, successful combinations of genes in the population have to be identified. We have only to consider pairs of successive genes in the creation of evolved genes, using the following algorithm:

1. Create a table of all different pairs of successive genes that occur in the population.
2. For all individuals in the population divide the fitness by the length of the individual. In the table, add this value to all pairs occurring in the genotype of that individual.
3. Find the pair with the highest sum of fitnesses in the table.
4. Create a new evolved gene with a unique designator, and replace all occurrences of the pair in the population with the new evolved gene.

The number of evolved genes is kept to a certain percentage of the population (3% in the examples shown). Figure 3.4(a) shows a branch of the hierarchical composition of one of the evolved genes (no. 363) from lower-level evolved genes and basic genes (numbers in brackets). Figure 3.4(b) shows how an individual is composed from six evolved genes.

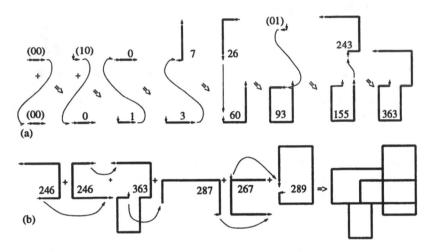

Fig. 3.4. Evolved representation: (a) part of the hierarchical composition of the evolved gene 363; and (b) composition of the individual with the genotype (289 267 287 363 246 246)

3.4.2 Using the representation. After a representation based on the examples in Figure 3.3 has been developed, it is used for new different fitness requirements. A standard evolutionary algorithm is used, where the fitness requirements are coded into the fitness function. The representation is not evolved further, instead the set of evolved genes learned from the examples is used, together with the original basic genes. As an example, the new requirement was to create a floor plan with minimal overall wall length, while at the same time fulfilling the following additional requirements:

1. no walls with 'open ends', that is, no walls that do not build a closed room;
2. 6 rooms; and
3. room sizes 300, 300, 200, 200, 100 and 100 units.

The additional requirements were given higher priority than the minimization of the wall length.

3.4.3 Results. Figure 3.5(a) shows the result of one run, after 150,000 crossovers were performed. The population size was 1,000 individuals. 320 evolved genes created from the examples in Figure 3.3 were introduced into the population by using them with equal probability in the generation of the initial random population, and by the mutation operator. Shown are the best individuals, with the exception that individuals that are rotated copies of already drawn individuals have been omitted.

To compare the performance with a standard genetic algorithm without evolving coding, Figure 3.5(b) shows the results of a run with only the basic coding, but identical fitness conditions. No rooms of more than unit size were produced (see Section 3.6).

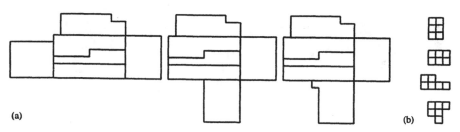

(a)   (b)

Fig. 3.5. New floor plans, using coding knowledge from the example cases (see text for fitness requirements): (a) using 320 evolved genes, (b) using only the basic genes

More than two thirds of the 15,400 genes used in the genotypes of the final population are evolved genes, encoding between 2 and 45 low-level genes. This shows that the new results indeed make use of the evolved representation to a very large degree, but at the same time use basic genes to 'fill in the holes' between evolved genes.

3.5 Example 2: Learning style features

In this second example the genetic engineering approach is used to learn style features from a set of example floor plans, in this case from floor plans of Frank Lloyd Wright houses [19] for a more detailed description of this application, see. An analysis of the style used shows that some of the style features are not simply products of outlines, but are linked to the functions of the rooms. To allow the system to pick up those style features as well, the basic coding is extended to allow it to represent lines of different colours, and the functions of the rooms are encoded in colours in the examples, Figure 3.6 given to the system.

Figure 3.7 shows examples of evolved genes created from the four example designs. Shown are some of the last evolved genes created from the examples. Clearly visible are the shapes of rooms, and the different line types, associated with the different functions. Two of the evolved genes shown (310 and 318) have the fireplace as part of the line drawing they code for (in this case from the Henderson house).

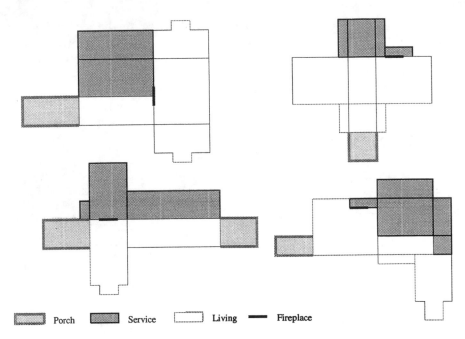

Fig. 3.6. Frank Lloyd Wright houses used to create the evolved coding: Henderson house (top left), Martin house (top right), Baker house (bottom left), Thomas house (bottom right)

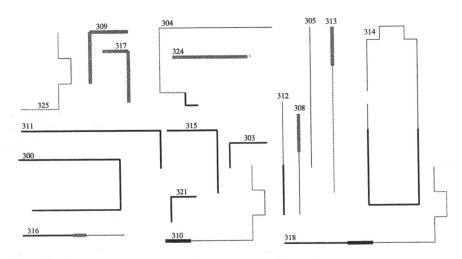

Fig. 3.7. Examples of evolved genes created from the example designs shown in Figure 3.6

3.5.1 Using the evolved representation. In the second phase, the representations evolved from the example cases are used to create new designs that show similarities in style to the example cases. For this, a standard evolutionary system is used, with the set of basic and evolved genes used in the coding. The evolved coding, as shown in Figure 3.7, captures information about shape and function of parts of the example designs. However, the way these parts are assembled to create new designs is only influenced by the fitness function that evaluates new designs. This means that any gene that codes for a room of a certain type, for example, has no influence on what other room is next to it; and there is nothing preventing a design from using two evolved genes that each include a fireplace. As a result, topological constraints are not automatically satisfied by using the evolved coding. Frank Lloyd Wright's prairie houses follow a number of topological constraints and they all have to be made part of the fitness. For the results presented here, fourteen different aspects influence the fitness. The following list shows some fitnesses:

- one porch, size between 9 and 12 units
- porch connected to living area, and not connected to service area
- two to four rooms in the service area, total size between 45 and 60 units
- two to four rooms in the living area, total size between 55 and 70 units
- only one service and one living area, i.e. all rooms of that type are connected
- one fireplace, two units length, between living and service area
- no 'dead ends', i.e. lines that do not enclose any room.

One way to handle a high number of individual fitnesses is to utilize 'Pareto optimization' [14]see, for example,, where only partial rankings between individuals are established. As an additional measure to prevent convergence, a 'niching' Pareto algorithm [9] is used. One of the effects of Pareto selection is that the population grows over time; for example, in one of the runs presented below, the population grew from 500 to 1,581 individuals.

3.5.2 Results. Two runs were executed using a set of 326 individuals created from the floor plans in Figure 3.6.

Run 1 ran for 60,677 loops, each loop creating two offspring individuals. The initial population was 1,000 individuals, the final population consisted of 1,652 individuals. From some 120,000 produced and tested individuals, 14,359 individuals were good enough to be introduced into the population.

Run 2 ran for 99,207 loops, the population grew from 500 to 1,581 individuals. Of the nearly 200,000 individuals produced, 12,502 made it into the population. Again, all 326 evolved genes where used.

The first result from Run 1, Figure 3.8(a), has a perfect fitness. The fitness function does not check if the fireplaces are straight, therefore a corner fireplace could result. Since none of the floor plans in the example drawing has a corner fireplace, this feature cannot have been part of the evolved coding. It therefore must be coded in basic genes. The second-best result from Run 1, Figure 3.8(b), has a penalty due to one segment of 'dead end' close to the fireplace, but fulfills all other fitness criteria.

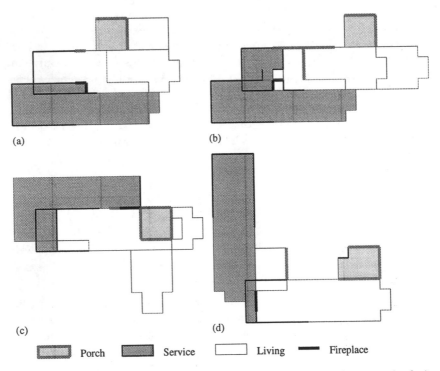

(a)   (b)

(c)   (d)

▨ Porch   ▨ Service   ☐ Living   ▬ Fireplace

Fig. 3.8. Floorplans created using the evolved genes from the example designs shown in Figure 3.6; (a) and (b) initial population 1,000 individuals, (c) and (d) initial population 500 individuals

Both results of Run 2, Figures 3.8(c) and (d), have perfect fitnesses. Again, the system has taken advantage of a weakness in the fitness function, that allowed it to put the porch inside the living space.

Since it was shown that a conventional GA without evolving coding was unable to create any designs that fulfill the simple criterions of Example 1 that also are part of the fitness criterion for this application (Figure 3.5(b)), no run without evolving coding was done for this application.

3.6 Comparison

The two examples described in this section again show how the genetic engineering approach can be used to extract information from one problem solution and re-use it in a different, but related, environment. It does this without the need for a specification of what features are to be extracted and without restricting the set of possible solutions when the new problem is solved. In the first example, the knowledge gained can be described as 'long straight lines, rectangular shapes and certain more prominent complex shapes'. How much this knowledge helps in creating new layouts can be seen in Figure 3.5. If only the basic coding is used without evolving coding to solve the problem used in the first example, the system still finds solutions with six rooms, but all of them have only unit size (see Figure 3.5(b)). Since any lines that do not totally enclose a room are penalized by the fitness function, the evolutionary system has to create the next larger room size in 'one step' from the small rooms. In other words, the first local maximum found in the fitness function is a layout with six unit-sized rooms, and the distance to the next local maximum is too far to be found easily by either mutation or cross-over. The 'granularity of possible solutions' seems to be too fine compared with the 'granularity of solutions rewarded by the fitness function'. It is important to see the difference between this fitness function and the one used to create the evolved coding. The latter is smooth, and any improvement in the code is rewarded appropriately. If the fitness that was used to create new layouts had been used to evolve a coding, the system would have been no more able to find larger rooms than without evolving coding. This highlights the point that the gain is not the evolving coding, but the evolved coding when used in similar applications. In the second example, an expanded basic coding was required to represent additional, semantic information in the layouts. This enabled the system to integrate more knowledge into the evolved representation.

Acknowledgments

This work is funded by the Australian Research Council. Computing resources have been provided by the Key Centre of Design Computing.

References

1. Anderson, W. F. (1995). Gene therapy, *Scientific American* **273**(3): 96–98B.
2. Angeline, P. (1994). Genetic programming and emergent intelligence, *in* K. Kinnear (ed.), *Advances in Genetic Programming*, MIT Press, Cambridge, MA, pp. 75–98.
3. Collins, J. and Coulson, A. (1987). Molecular sequence comparison and alignment, *Nucleic Acid and Protein Sequence Analysis: A Practical Approach*, IRL Press, Washington DC, pp. 323–358.
4. Dave, B., Schmitt, G., Faltings, B. and Smith, I. (1994). Case based design in architecture, *in* J. S. Gero and F. Sudweeks (eds), *Artificial Intelligence in Design '94*, Kluwer Academic Publishers, Dordrecht, pp. 145–162.

5. Gage, P. (1996). Variable-complexity evolution of shape grammars for engineering design, *in* J. S. Gero and F. Sudweeks (eds), *Artificial Intelligence in Design '96*, Kluwer Academic Publishers, Dordrecht, pp. 311–324.
6. Gero, J. and Kazakov, V. (1995). Evolving building blocks for genetic algorithms using genetic engineering, *Proceedings of the IEEE Conference on Evolutionary Computing*, pp. 340–345.
7. Gero, J. and Kazakov, V. (1996). Evolving design genes in space layout planning problems, *Artificial Intelligence in Engineering* (to appear).
8. Grierson, D. and Pak, W. (1993). Optimal sizing, geometrical and topological design using a genetic algorithm, *Structural Optimization* 6: 151–159.
9. Horn, J. and Nafpliotis, N. (1993). Multiobjective optimization using the niched pareto genetic algorithm, *Technical Report 93005*, Illinois Genetic Algorithms Laboratory (IlliGAL), University of Illinois at Urbana-Champaign.
10. Karlin, S., Dembo, A. and Kawabata, T. (1990). Methods for assessing the statistical significance of molecular sequence features by using general scoring scheme, *Proceedings of the National Academy of Science USA*, Vol. 87, pp. 5509–5513.
11. Liggett, R. (1985). Optimal spatial arrangement as a quadratic assignment problem, *in* J. S. Gero (ed.), *Design Optimization*, Academic Press, New York, pp. 1–40.
12. Maher, M. L., Balachandran, M. B. and Zhang, D. (1995). *Case-Based Reasoning in Design*, Lawrence Erlbaum, Hillsdale, NJ.
13. Needleman, S. and Wunsch, C. (1970). A general method applicable to the search for similarities in the amino acid sequence of two proteins, *Journal of Molecular Biology* 48: 443–453.
14. Radford, A. D. and Gero, J. S. (1988). *Design by Optimization in Architecture and Building*, Van Nostrand Reinhold, New York.
15. Rosenman, M. (1996). The generation of form using an evolutionary approach, *in* J. S. Gero and F. Sudweeks (eds), *Artificial Intelligence in Design '96*, Kluwer Academic Publishers, Dordrecht, pp. 643–662.
16. Sankoff, D. and Kruskal, J. (eds) (1983). *Time Warps, String and Macromolecules: The Theory and Practice of Sequence Comparison*, Addison-Wesley, Reading, MA.
17. Schmitt, G. N. (1993). Case-based reasoning in an integrated design and construction system, *in* K. Mathur, M. Betts and K. W. Tham (eds), *Management of Information Technology for Construction*, World Scientific Publishing, Singapore, pp. 453–465.
18. Schnier, T. and Gero, J. S. (1995). Learning representations for evolutionary computation, *Australian Joint Conference on Artificial Intelligence AI'95*, pp. 387–394.
19. Schnier, T. and Gero, J. S. (1996). Learning genetic representations as alternative to hand-coded shape grammars, *in* J. S. Gero and F. Sudweeks (eds), *Artificial Intelligence in Design '96*, Kluwer, Dordrecht, pp. 39–57.
20. Schuler, G., Altschul, S. F. and Lipman, D.J. (1991). A workbench for multiple alignment construction and analysis, *PROTEINS: Structure, Function, and Genetics* 9: 180–190.

The Generation of Form Using an Evolutionary Approach

M. A. Rosenman

Key Centre of Design Computing, Department of Architectural and Design Science, University of Sydney NSW 2006 Australia

Summary. This paper presents an evolutionary approach to design using a hierarchical growth model. It argues that the evolutionary approach fits well to the generate-and-test approach in design and is especially suited to non-routine design situations where the (inter)relationships between complex arrangements of elements and their behaviour are not known. The evolutionary approach is used as the computational method for the synthesis and evaluation stage of the design process. A bottom-up hierarchical model is used to avoid the combinatorial problems involved in linear models. The genotype consists of chromosomes which comprise genes representing design grammar rules. Evaluation is carried out both through the use of a fitness function and through human interaction. The concepts are exemplified in the context of the design of house plans.

1. Introduction

Design is a purposeful knowledge-based human activity whose aim is to create form which, when realized, satisfies the given intended purposes.1 Design may be categorized as routine or non-routine with the latter further categorized as innovative or creative. The lesser the knowledge about existing relationships between the requirements and the form to satisfy those requirements, the more a design problem tends towards creative design. Thus, for non-routine design, a knowledge-lean methodology is necessary. Natural evolution has produced a large variety of forms well-suited to their environment suggesting that the use of an evolutionary approach could provide meaningful design solutions in a non-routine design environment.

This work investigates the possibilities of using an evolutionary approach based on a genotype which represents design grammar rules for instructions on locating appropriate building blocks. A decomposition/aggregation hierarchical organization of the design object is used to overcome combinatorial problems and to maximize parallelism in implementation.

2. An Evolutionary Model Of Design

2.1 Genetic Algorithms

Genetic algorithms (GAs) are a class of algorithms, based on the adaptive process of natural evolution, employing a general uniform knowledge-lean methodology [2, 3]. While GAs have been used to solve mainly optimization, learning and control problems [4, 5], there has been very some research and applications in design [6]–[13] and in the generation of creative art forms [14]. The trend which has developed in GA applications is to encode the genotype as a string of, usually binary, characters representing a one-to-one correspondence to a property in the phenotype. This differs from nature where the genotype encodes instructions for locating building blocks of living form.

2.2 Evolution and design

The evolutionary approach is basically a generate and test approach which corresponds well to the procedures for design synthesis and evaluation in the design process. The specific characteristics of the approach are:

- a large pool or population of members (e.g. design solutions);
- members are selected for 'survival' using a biased random selection mechanism based on their 'fitness', i.e. relation to a fitness function;
- new members are generated from the existing ones using evolutionary mechanisms as crossover, inversion, 'gene splicing' and mutation.

The advantages of the evolutionary approach are:

- diverse sections of the state space can be investigated thus enhancing the possibility of discovering a variety of potential solutions;
- a probabilistic selection method directs the random generative process to produce meaningful and satisfactory solutions.

2.2.1 Growth of form. Whereas the process of generation of form in living systems involves the placement of different kinds of protein in particular locations, the process of generation of form in design involves the placement of units of different kinds of material in particular locations. We may describe an object, at a lower level of abstraction, as a composition of material units (cells or building blocks), where the type and scale of such units are chosen depending on the context and suitability for the level of abstraction required. An object can then be 'grown' by locating a required number of such units, one at a time in sequence. The form produced will depend on the form of the unit material and the location procedures, i.e. rules of growth, used, as in crystal growth.

The genotype for a homogeneous object is thus the sequence of coded instructions for selecting and locating material units, analogous to the DNA string in natural evolution. When this code is interpreted and executed, the phenotype, i.e. the object (or rather its representation), will be generated. A general model of form growth can be proposed as:

For given total units of material required
        SELECT a unit of material, M_m
        LOCATE unit of material, M_m relative to other units

A gene in such a model becomes $(Ot, M_m, L(M_m))$, where $L(M_m)$ is the instruction for locating the unit of material M_m relative to the generated object at each step, Ot. Initially, Ot is a single unit. The genotype is a sequence of such genes. Where a homogenous object is considered, the material identification is constant and the gene is basically a sequence of location operators. Obviously, such a model is computationally infeasible in general and a more computationally feasible approach is required.

2.2.2 Elements, components and assemblies. An object may be simple or complex. A simple object is termed an element and by definition is homogeneous otherwise it can be decomposed into separate homogeneous elements An element is thus a composition of material units. A complex object is termed an assembly and is composed of objects termed components. Recursively, components may be assemblies or elements. A formal description is as follows:

O :-	$A\|E$	*An object is an assembly or an element*
A :-	$(C, R(C))$	*An assembly is a set of components and*
C :-	$\{\cup(C_i)\}$	*a set of relationships among the components*
C_i :-	O	*i-th component is an object*
E :-	$(M_m, R(M_m))$	*An element is a set of material units of type*
M_m :-	$\{\cup(M_{mp})\}$	M_m *and a set of relationships among the*
		material units
M_{mi} :-	*i-th unit of*	
	material M_m	

2.2.3 Evaluation of design solutions. The evaluation of designs is carried out by interpreting the generated design solution, the phenotype, and determining its performance according to a set of behavioural requirements (formulated from the design requirements). The performance of the object may be determined using formulae or rules, etc. or by users exercising judgment in the case of qualitative behaviours. Such judgments are subjective and personal and the designer takes full responsibility for such evaluations. Such subjective evaluation coupled with user interaction is the approach taken in the generation of *Biomorphs* [15] and creative art forms [2].

Since an object exhibits more than one behaviour, the evaluation of the fitness of the object is a multiobjective problem and hence will involve evaluation using concepts similar to Pareto optimization [9, 16, 17]. Constraints can be implemented through the use of penalty functions.

2.2.4 Design through hierarchical decomposition/aggregation. Simon [18] points out, that even though organisms are very complex, it is only possible for them to evolve if their structure is organised hierarchically. The above formulation allows for the generation of objects through the recursive generation of its components until a level is reached where the generation becomes one of generating an element. Such an approach assumes a formulation of the decomposition structure of an object.

There are basically two approaches. The first is a top-down approach, used by Cramer [19], Dasgupta and McGregor [20], and also in Genetic Programming [21, 22]. The second is a bottom-up multi-level approach where, at each level, a component is generated from a combination of components from the level immediately below. At each level, an initial population is generated and then evolved over a number of generations until a satisfactory population of objects at that level is obtained. Members of that population are then selected as suitable components for generating the initial population at the next level. The process is repeated for all levels, Figure 2.1.

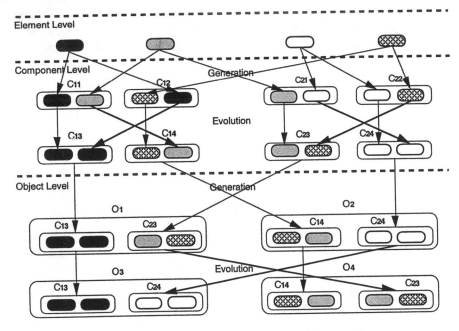

Fig. 2.1. Multi-Level Combination and Propagation

The advantages of a hierarchical approach are that only those factors relevant to the design of that component are considered and factors relevant to the relationships between components are treated at their assembly level. Instead of one long genotype consisting of a large number of low-level genes, the genotype is composed of a set of chromosomes relevant to their particular level. In addition to reducing the combinatorial problem substantially, parallelism is supported since all the different chromosomes (components) at a particular level can be generated in parallel. If the set of possible alternatives of component types is sufficiently large and varied, then many different combinations of members of different such sets are possible, at the next level, with a good chance of satisfying the criteria and constraints at that level. Only when no such possible combination satisfies such criteria is there a need for some generation of new alternatives at the lower level.

In a flat model of form generation, a genotype will consist of a string of a very large number of basic genes. In a hierarchical model, there are a number of component chromosomes, at different levels, consisting of much shorter strings of genes which are the chromosomes at the next lower level. All in all, the total number of basic genes will be the same in the flat and hierarchical models.

2.3 Design grammars

In order to generate a design solution a generative method, such as a design grammar, is required. A design grammar deals with a vocabulary of design elements and transformation on these elements and hence defines a design space [13, 23, 24]. While design grammars provide a syntactic generative capability, they lack the evaluative mechanisms for directing the generation towards meaningful solutions.

2.3.1 Recipes and blueprints – genotypes and phenotypes. The aim of the design process, in an evolutionary approach, is the attainment of a set of instructions, a genotype (recipe), that when executed, yields a design description of a product, a phenotype(blueprint), whose interpreted behaviours satisfy a set of required behaviours, the fitness function. In this approach, a grammar rule is a gene, the plan (sequence of rules) is the genotype and the design solution is the phenotype.

The advantage of the use of recipes rather than blueprints as the genetic information is in the simplicity of the information since the generative rules are fewer in number than solution parameters and generally less complex. Moreover, small changes in such rules or their combinations can lead to large and unexpected changes in the design solutions, a desirable property for creative design [13].

2.3.2 The evolution of new rules and plans. There are basically two approaches in the generation of genotypes of design grammar rules, analogous to the Michigan and Pitt approaches in classifier systems [5, 25]. The first approach, as taken by Gero et al. [7] attempts to 'learn' new grammar rules. The second approach, which is taken here, is based on the premise that the grammar rules are fundamental operators, which cannot be decomposed or recomposed, that the particular grammar contains all required rules and that the aim of the design process is to find satisfactory sequences of such rules.

2.3.3 A general model for an evolutionary approach to design. The general model of design using an evolutionary approach may be stated as follows:

    for all levels in the object hierarchy
        for all components at that level

> **GENERATE** initial population of members
> by synthesizing lower level units
> through random application of rules
> **EVOLVE** population until satisfactory

3. A House Design Example – Space Generation

The above concepts can be exemplified through the generation of 2-D plans for single-storey houses. Previous work demonstrated that a single-level approach was not able to converge towards satisfactory solutions mainly due to the interactions of the various factors of the fitness function required for the various elements [9, 26].

3.1 A house spatial hierarchy

A house can be considered to be composed of a number of zones, such as living zone, entertainment zone, bed zone, utility zone, etc. Each zone is composed of a number of rooms (or spaces), such as living room, dining room, bedroom, hall, bathroom, etc. Different houses are composed of different zones where each zone may be composed of different rooms. Each room is composed of a number of space units. Generally, in a design such as a house, the space unit will be constant. The scale (level of abstraction) of the space unit depends on the precision required in differences between various possible room sizes. The smaller the unit, the longer the genotype for a given size of room but the greater the shape alternatives.

3.2 Generation – The design grammar

In the above formulation, the generation of spaces, basically comes down to locating spatial component units for that level. At the room level, the component unit is a fundamental unit of space. At the zone level, the component unit is a room and at the house level the component unit is a zone.

The design grammar used here is based on the method for constructing polygonal shapes represented as closed loops of edge vectors [27]. The grammar is based on a single fundamental rule which states that any two polygons, P_i and P_j, may be joined through the conjunction of negative edge vectors, V_1 and V_2, (equal in magnitude and opposite in direction). The conjoining of these vectors results in an internal edge and a new polygon, P_k. The edge conjoining rule, $R1$ is:

$$V_m + V_n = V_{mn} = 0 \qquad \text{R1}$$

Rule $R1$ is commutative and applies for all values of vector direction. This rule ensure that new cells are always added at the perimeter of the new resultant shape. This rule is shown diagrammatically in Figure 3.1.

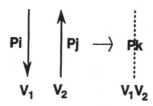

Fig. 3.1. Edge Vector Rule for the Construction of 2-D Shapes

The fundamental conjoining rule can be specialized for different types of geometries. Orthogonal geometries are based on the following four vectors of unit length: $W = (1, 90)$, $N = (1, 0)$, $E = (1, 270)$, $S = (1, 180)$, so that rule $R1$ becomes:

$$N + S = NS = 0 \qquad \qquad \text{R1a}$$
$$E + W = EW = 0 \qquad \qquad \text{R1b}$$

These two (sub)rules allow for the generation of all polyminoes. Orthogonal geometries will be used in this example without loss of generality. Other (sub)rules may be formed for other geometries [27].

3.2.1 Genotype and phenotype.
A polygon is described by its sequence of edge vectors. A suffix is used to identify individual edges of the same vector type. Thus the square cell of Figure 3.2 is described as $(W1, N1, E1, S1)$. The sequence of edge vectors for a shape is the phenotype providing the description of that shape's structure. The genotype for any generated polymino is the sequence of the two subshapes (polyminoes) used and the two edges joined. An example of the generation of a trimino is shown in Figure 3.2. Figure 3.2 shows a basic unit or cell, $P1$, which provides a starting point for the generation of polyminoes. Each generated shape is accompanied by its genotype and phenotype. The generation of these polyminoes occurs from a random selection of edges in the first shape conjoined with a random selection from equal and opposite edges in the second shape. At each step in the generation, the phenotype is reinterpreted to generate a new edge vector description and the conjoining (sub)rules applied. The genotype for the generated trimino is given as $(P2, P1, N2|S1)$. This can be expanded as $((P1, P1, E1|W1), P1, N2|S1)$. When the same units are used for generation, the unit can be omitted and the genotype represented as the sequence of edge vector conjoinings. That is $P3(g) = (E1|W1, N2|S1)$.

The length of the genotype (and phenotype) depends on the size of the polymino to be generated, that is on the area of the polymino. This corresponds to required room sizes.

Once a population of different rooms is generated for each room type in a given zone, the zone can be generated through the conjoining of rooms in a progressive fashion. Because of the cell-type structure of the polygons, the conjoining may occur at any appropriate pair of cell edges. Therefore, a large

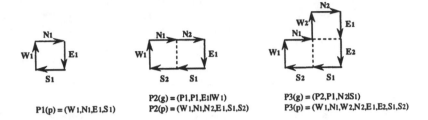

$$P1(p) = (W1,N1,E1,S1)$$

$$P2(g) = (P1,P1,E1|W1)$$
$$P2(p) = (W1,N1,N2,E1,S1,S2)$$

$$P3(g) = (P2,P1,N2|S1)$$
$$P3(p) = (W1,N1,W2,N2,E1,E2,S1,S2)$$

Fig. 3.2. Generation of a Trimino

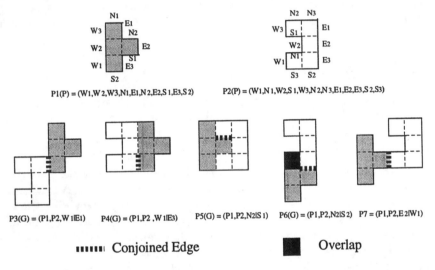

$$P1(P) = (W1,W2,W3,N1,E1,N2,E2,S1,E3,S2)$$ $$P2(P) = (W1,N1,W2,S1,W3,N2,N3,E1,E2,E3,S2,S3)$$

$$P3(G) = (P1,P2,W1|E1)$$ $$P4(G) = (P1,P2,W1|E3)$$ $$P5(G) = (P1,P2,N2|S1)$$ $$P6(G) = (P1,P2,N2|S2)$$ $$P7 = (P1,P2,E2|W1)$$

▮▮▮▮▮ Conjoined Edge   ■ Overlap

Fig. 3.3. Some Examples of Conjoining Two Polyminoes

number of possible zone forms can be generated from two rooms. An example of some possibilities arising from the conjoining of two polyminoes is given in Figure 3.3.

The two polyminoes, $P1$ and $P2$, represent instances of two different room types and the polyminoes resulting from the joining of the two rooms represent instances of a particular zone type. When one pair of edges are conjoined other edges may also be conjoined, e.g. $P4$, $P5$ and $P6$. In the case of overlap, as in $P6$, the resultant shape is discarded.

The same process used for generating zones is used to generate houses. The joining of different instances of different zone types generates different instances of houses.

3.2.2 Order of selection. At the zone and house level, the order of selection of the units to be joined may influence the solution and its performance.

However, there are problems with choosing a random order of room selections for every zone instance generation for the same population as crossover may lose some room types and include more than one of the same type. Thus for any population, the same order of room types must be kept. A given order may be chosen randomly or from an algorithm based on the number (and/or strength) of interconnections required in an interconnection matrix.

3.3 The evolution of house designs

The above grammar can be used to generate initial populations for each level in the spatial hierarchy. Each such initial population is then evolved, as necessary, so that solutions are 'adapted' to design requirements.

3.3.1 The evaluation criteria – fitness functions.

At each level, different fitness functions apply according to the requirements for that level. While the requirements for designs of houses involve many factors, many of which cannot be quantified or adequately formulated in a fitness function, some simple factors will be used initially to test the feasibility of the approach. For this example, the fitness function for rooms will consist of minimizing the perimeter to area ratio and the number of angles. This requirement tends to produce useful compact forms. For zones, the fitness function will consist of minimizing a sum of adjacency requirements between rooms reflecting functional requirements. At the house level, the fitness function will consist of minimizing a sum of adjacency requirements between rooms in one zone and rooms in other zones. This has the tendency to select those arrangement of zones where adjacency interrelations are required between rooms of different zones. In addition to these quantitative assessments, qualitative assessments will be made subjectively and interactively by a user/designer.

Although the above criteria have been described in terms of optimizing functions, the aim is not to produce global optimum solutions but rather to direct the evolutionary process to produce populations of good solutions either as components for higher levels or at the final level itself. So that, even though the global optimum solution for the shape of a room using the above criteria, may be known, this may not be the optimum solution at the zone and house levels. By selecting other non-optimal but good solutions, according to the given criteria, good unexpected results may be achieved for the overall design.

Other factors are required in more realistic design contexts. For example, a house needs to meet site requirements, both in terms of size and orientation for view or climate. Such factors can be formulated as constraints at various levels. These constraints can be handled explicitly as survival factors or as penalty functions. That is, a solution which does not meet a constraint is eliminated or, alternatively, a penalty can be added to the fitness of the solution according to the degree of violation. It is argued that with a sufficiently large population of room and zone alternatives such constraints can be met. If not, then redesign, i.e. new zone and/or room forms, must be produced.

3.3.2 Propagation – crossover. Simple crossover is used for the production of offspring during the evolution process. Looking first at the room level to see the effect of such a crossover process, crossover can occur at any of the four sites (see Figure 3.4) with two results as shown in Figure 3.5. Since we are always dealing with cells of the same space unit, the cell identification in the genotype representation has been omitted for simplicity.

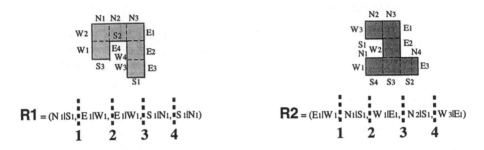

R1 = (N₁|S₁, E₁|W₁, E₁|W₁, S₁|N₁, S₁|N₁)

$$\text{R1} = (N_1|S_1, E_1|W_1, E_1|W_1, S_1|N_1, S_1|N_1)$$

$$\text{R2} = (E_1|W_1, N_1|S_1, W_1|E_1, N_2|S_1, W_3|E_1)$$

1   2   3   4

Fig. 3.4. Crossover at Room Level: initial rooms $R1$ and $R2$ generated from unit square cell $U1$

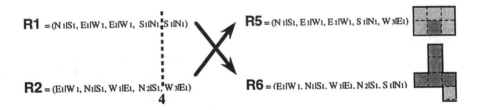

$$\text{R1} = (N_1|S_1, E_1|W_1, E_1|W_1, S_1|N_1, S_1|N_1)$$

$$\text{R5} = (N_1|S_1, E_1|W_1, E_1|W_1, S_1|N_1, W_3|E_1)$$

$$\text{R2} = (E_1|W_1, N_1|S_1, W_1|E_1, N_2|S_1, W_3|E_1)$$

4

$$\text{R6} = (E_1|W_1, N_1|S_1, W_1|E_1, N_2|S_1, S_1|N_1)$$

Fig. 3.5. Crossover at Room Level: crossover at site 4

At the zone level, crossover occurs as shown in Figures 3.6 and 3.7. Two initial instances of living zones, $Z1$ and $Z2$ are shown in Figure 3.6. Each zone has one instance of each of living room, dining room and entrance. Figure 3.7 shows crossover for one of the four possible sites. A similar process is followed at the house level.

4. Implementation and Results

A computer program written in C++ and Tcl-Tk under the Sun Solaris environment is under implementation. Currently, only the simple criteria described previously have been implemented. Each evolution run, for all levels, tends to converge fairly quickly to some dominant solution. Rather than use a mutation operator to break out of such convergence , it was found that

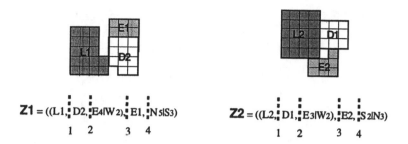

$$Z1 = ((L1, D2, E4|W2), E1, N5|S3)$$

$$1\quad 2\qquad 3\quad\ 4$$

$$Z2 = ((L2, D1, E3|W2), E2, S2|N3)$$

$$1\quad 2\qquad 3\quad\ 4$$

Fig. 3.6. Examples of Zone Crossover: rooms and initial zones, $Z1$ and $Z2$

SITE 2 - CROSSOVER

$$Z5 = ((L1,\ D2,\ E3|W2),\ E2,\ S2|N3)$$

$$Z6 = ((L2,\ D1,\ E4|W2),\ E1,\ N5|S3)$$

Fig. 3.7. Examples of Zone Crossover: crossover at site 2

a more efficient strategy was to generate multiple runs with different initial randomly generated populations. This produces a variety of gene pools thus covering a more diverse area of the possible design space. Moreover, such runs can be generated in parallel. Users can nominate the population size, number of generations for each run and select rooms, zones and houses from any generation in any run as suitable for final room, zone or house populations. These selections are made as solutions appear which find favour in users, based perhaps on factors not included in the fitness function. Such selections may therefore not be optimal according to the given fitness function.

Results are shown in the following figures, Figures 4.1 to 4.6 for room, zone and house solutions.

Figures 4.1 and 4.2 show the results of the 8th run of the generation of Living Rooms. Figure 4.1 shows 4 shapes selected from 7 previous runs together with an initial random population of 60. Figure 4.2 shows the 17th generation of the evolution of this population. A fifth shape was selected at the 14th generation and two more shapes are being selected. The upper line in the graph shows the evolution of the best solution while the lower line shows the evolution of the population average. All in all for this example, a total of 10 shapes were selected from 13 runs. The maximum number of generations before convergence for a run was 50 and the minimum 19 with an average of 32.

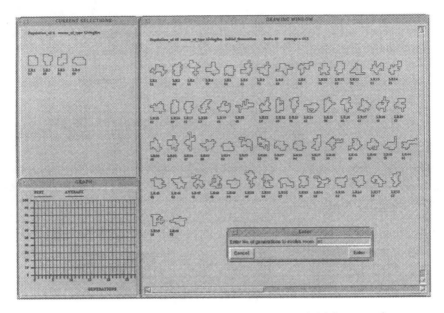

Fig. 4.1. Results of Living Room Generation: initial generation

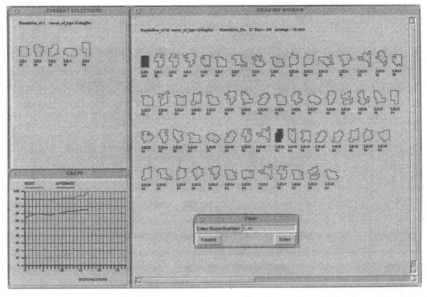

Fig. 4.2. Results of Living Room Generation: 17th generation

Other rooms were generated in a similar way. The room areas generated were: (a) Living Zone: Living Room 24; Dining Room 15; Kitchen 9; Entrance 4; (b) Bedroom Zone: Master Bedroom 15; Bedroom 12; Bathroom 6; Hall 3.

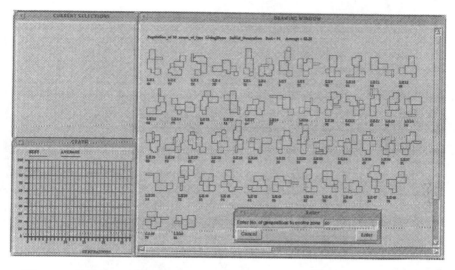

Fig. 4.3. Results of Living Zone Generation: initial Living Zone population

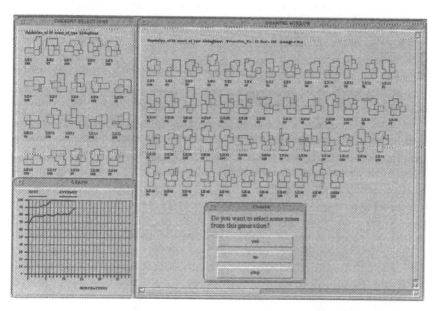

Fig. 4.4. Results of Living Zone Generation: evolved population

Figures 4.3 and 4.4 show the results of the Living Zone generation. Figure 4.3 shows an initial population of 50 Living Zones at run 1, randomly generated by selecting rooms from the final selections for the Living Room, Dining Room, Kitchen and Entrance. Figure 4.4 shows the 13th generation of the final run, run 7. Twenty Living Zones have been selected by the user.

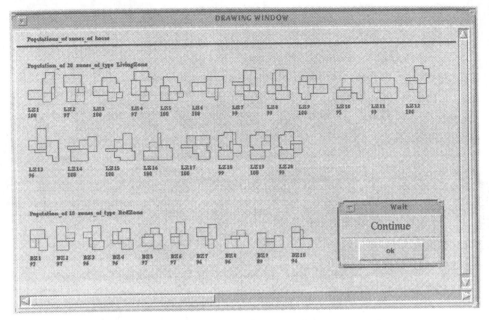

Fig. 4.5. Results of Bed and Living Zones Generation

Figures 4.5 shows the set of Bedroom and Living Zones selected. Figure 4.6 shows the final set of houses generated in this example. All in all 7 runs were carried out for a total of 12 suitable house plans.

The total area of this house type is 88 sq units, corresponding to a genotype of length 87 in a single genotype. The example of Jo [26] showed that no satisfactory convergence was obtained with a single genotype of this length. The size of the population and the number of generations and runs required to generate a satisfactory number of members depends on the genotype length. The longest genotype was of length 23, for the Living Room. The addition of more zones and rooms presents no problem for the hierarchical approach used here although it would present extra combinatorial problems for the single genotype approach.

5. Summary

This work has presented concepts for a general evolutionary approach to the generation of design solutions based on the growth of cells in a hierarchical organization. The main advantage of a hierarchical approach is that genotypes are shorter and the fitness functions relates only to the requirements for that component(element or assembly). In addition to reducing combinatorial problems, parallelism is supported. The critical element becomes the

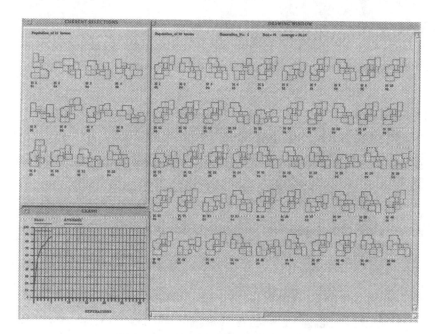

Fig. 4.6. Results of House Generation

genotype of the largest component genotype rather than the total length of a single-level genotype.

While the various fitness functions at the different levels of the component hierarchy have involved optimization criteria, the goal of the approach is not optimization of these factors per se but rather their use as a driving force in the generation of satisfactory form.

As an alternative to mutation in the evolution of populations, the use of multiple runs with new randomly generated initial populations was used. This also has the advantage of allowing parallel processing.

While the example presented is based on the generation of 2-D plans through the synthesis of a fundamental 2-D space unit, the approach can be generalized to the synthesis of any material cells. Although the example was based on orthogonal geometry, the method for growth is general for any polygonal geometry and may be extended to polyhedral geometry.

Further work will involve the inclusion of more realistic criteria and constraints as well as investigating the need for the recursive generation of new lower level components when no satisfactory assembly can be generated. Efficiency issues need to be investigated with respect to the type of crossover and selection mechanism.

Acknowledgments

This work is partially supported by the Australian Research Council.

References

1. Rosenman, M. A. and Gero, J. S. (1994). The what, the how, and the why in design, Applied Artificial Intelligence, 8(2):199-218.
2. Holland, J. H. (1975). Adaptation in Natural and Artificial Systems, The University of Michigan Press, Ann Arbor.
3. Beasley, D., Bull, D. R. and Martin R. R. (1993). An overview of genetic algorithms: Part 1, fundamentals, University Computing, 15(2):58-69.
4. Goldberg, D. E. (1989). Genetic Algorithms in Search, Optimization, and Machine Learning, Addison-Wesley Publishing Company, Reading, Mass.
5. Grefenstette, J. J. and Baker, J. E. (1989). How genetic algorithms work; a critical look at implicit parallelism, in J. D. Schaffer (ed.), Proc. of the Third Int. Conf. on Genetic Algorithms, Morgan Kaufmann, San Mateo, CA, pp.20-27.
6. Louis, S. and Rawlings, G. J. (1991). Designer genetic algorithms: genetic algorithms in structure design, in R. K. Belew and L. B. Booker (eds), Proc. Fourth Int. Conf. on Genetic Algorithms, Morgan Kaufmann, San Mateo, pp.53-60.
7. Woodbury, R. F. (1993). A genetic approach to creative design, in J. S. Gero and M. L. Maher (eds), Modeling Creativity and Knowledge-Based Creative Design, Lawrence Erlbaum, Hillsdale, NJ, pp.211-232.
8. Gero, J. S., Louis, S. J. and Kundu, S. (1994). Evolutionary learning of novel grammars for design improvement, AIEDAM, 8(3):83-94.
9. Bentley, P. J. and Wakefield, J. P. (1995). The table: an illustration of evolutionary design using genetic algorithms, 1st IEE/IEEEConference on Genetic Algorithms in Engineering Systems: Innovations and Applications (Galesin '95), pp.412-418.
10. Jo, J. H. and Gero, J. S. (1995). A genetic search approach to space layout planning,Architectural Science Review, 38(1):37-46.
11. Michalewicz, Z., Dasgupta, D., Le Riche, R. G. and Schoenaur, M. (1996). Evolutionary algorithms for constrained engineering problems, Computers and Industrial Engineering Journal, Special Issue on Genetic Algorithms and Industrial Engineering, 30(2):(to appear).
12. Schnier, T. and Gero, J. S. (1995). Learning representations for evolutionary computation, in X. Yao (ed.), AI'95 Eighth Australian Joint Conference on Artificial Intelligence, World Scientific, Singapore, pp.387-394.
13. Rosenman, M. A. (1996). A growth model for form generation using a hierarchical evolutionary approach, Microcomputers in Civil engineering, Special Issue on Evolutionary Systems in Design, 11:161-172.
14. Todd, S. and Latham, W. (1992). Evolutionary Art and Computers, Academic Press, London.
15. Dawkins, R. (1986). The Blind Watchmaker, Penguin Books.
16. Horn, J., Nafpliotis, N. and Goldberg, D. E. (1994). A niched Pareto genetic algorithm for multiobjective optimization, Proc. of the First IEEE Conf. on Evolutionary Computation (ICEC '94), Vol1, IEEE World Congress on Computational Intelligence, Piscataway, NJ: IEEE Service Center, pp.82-87.

17. Osyczka, A. and Kundu, S. (1995). A new method to solve generalized multicriteria optimization problems using the simple genetic algorithm, Structural Optimization, 10(2):94-99.
18. Simon, H.A. (1969). The Sciences of the Artificial, MIT Press, Cambridge, Mass.
19. Cramer, N. L. (1985). A representation for the adaptive generation of simple sequential programs, Proc. of an Int. Conf. on Genetic Algorithms and their Application, pp.183-187.
20. Dasgupta, D. and McGregor, D. R. (1993). sGA: A structured genetic algorithm, Technical Report No. IKBS-11-93, Dept. of Computer Science, University of Strathclyde, UK.
21. Koza, J. R. (1992). Genetic Programming: On the Programming of Computers by Means of Natural Selection, MIT Press, Cambridge, Mass.
22. Rosca, J.P. and Ballard, D.H. (1994). Hierarchical self-organization in genetic programming, Proc. of the Eleventh Int. Conf. on Machine Learning, Morgan-Kaufmann, San Mateo, CA, pp.252-258.
23. Stiny, G. and Mitchell, W. (1978). The Palladian Grammar, Environment and Planning B, 5:5-18.
24. Stiny, G. (1980). Introduction to shape and shape grammars, Environment and Planning B, 7:343-351.
25. Wilson, S. W. and Goldberg, D. E. (1989). A critical review of classifier systems, in J. D. Schaffer (ed.), Proc. of the Third Int. Conf. on Genetic Algorithms, Morgan Kaufmann, San Mateo, CA, pp.244-255.
26. Jo, J. H. (1993). A Computational Design Process Model using a Genetic Evolution Approach, PhD Thesis, Department of Architecural and Design Science, University of Sydney, (unpublished).
27. Rosenman, M. A. (1995). An edge vector representation for the construction of 2-dimensional shapes, Environment and Planning B:Planning and Design, 22:191-212.

Evolutionary Optimization of Composite Structures

Rodolphe Le Riche[1] and Raphael T. Haftka[2]

[1] Centre des Matériaux P-M. Fourt, Ecole des Mines de Paris, 91003 Evry, France
[2] Dept. of Aerospace Engineering, Mechanics and Engineering Science, University of Florida, Gainesville, FL 32611-6250, USA

Summary. Optimizing composite laminates is a combinatorial problem which has multiple solutions of similar performance. Evolutionary optimization techniques are particularly well suited for this kind of problems, because they can handle discrete variables, and yield many optimal or near-optimal solutions. The designer will then have a choice among various solutions. Nevertheless, evolutionary techniques are limited by their high computational cost. It is possible to reduce the cost of the optimization by specializing the evolutionary algorithm. An efficient evolutionary optimization method is first developed for computationally inexpensive plate problems. Those results are then extended to more complex composite structures design problems.

1. Introduction

Laminated composite structures are stacked from layers made up of fibers embedded in a matrix of isotropic material. Fibrous composites are usually manufactured in the form of layers of fixed thickness. Designing composite structures involves finding the number of layers and the fiber orientation of each layer inside each laminate that maximizes the performance of the structure under constraints such as failure, geometry, cost, etc. Compared to isotropic materials, the use of fibrous laminated composites permits higher stiffness to weight or strength to weight ratios, along with a finer tuning of the response of the structure.

Plates made of composite materials are the basic structural components used, for example, in the skin of wings of modern aircrafts. Because of manufacturing considerations and tradition, not only are the ply thicknesses fixed, but ply orientations are often limited to a set of angles such as 0°, 90°, $\pm 45^o$.

Much effort has been devoted to designing stacking sequences of composite plates. In response to the discrete nature of the problem, integer programming strategies based on the Branch and Bound algorithm [1] have been used: when the only failure mode considered is buckling, a linear formulation of the problem exists [2]. However, in general, designing a composite laminate is a non linear integer programming problem [5] [3]. Because continuous optimizers have a low computational cost and are widely available, stacking sequence problems have alternatively been treated using continuous optimization techniques [6], [7], [8].

All the above design procedures have pitfalls. First, problems involving the flexural and the in-plane response of laminates are nonlinear functions of the number of plies, the ply thicknesses, and the fiber orientations. Therefore, the design space contains local optima in which continuous optimization strategies may get trapped. Second, composite structures often exhibit many optimal or near optimal designs. The reason is that composite laminate performance is characterized by a number of parameters which is smaller than the number of design variables. Different sets of design variables can produce similar results, i.e., there are many optimal and near-optimal designs. Traditional design approaches not only have the drawback of sometimes converging to suboptimal designs, but also of yielding only one solution.

Probabilistic search methods offer a good potential for composite structures optimizations. They are global in scope and insensitive to problem nonconvexities and nonlinearities. Simulated annealing and evolutionary optimization have been applied to laminated plate optimization [13], [14], [15], [9], [10].

In spite of their theoretical advantages, evolutionary optimization strategies are often too expensive to use when the analysis of one candidate solution is computationally expensive. To alleviate this problem, it helps to specialize the algorithm to the problem by finding the best representation, operators and set of parameters for the problem considered.

This work is an illustration of how to specialize an evolutionary algorithm to a class of engineering optimization problems. In the first part of this paper, the laminate design problem is stated. In the second part, the specialized algorithm is developped on the computationally inexpensive laminate design problem. Because laminates are the building blocks of composite structures, it is hoped that the results of the specialization will apply, after some scaling, to more complex problems. This is the subject of the third part of the paper, where a composite wing box problem is considered.

2. Optimization of a Laminated Plate for Strength and Buckling

2.1 Analysis of a laminated plate

The simply supported laminate shown in Figure 2.1 is loaded in the x and y directions by N_x and N_y, respectively. The laminate is composed of N plies, each of thickness t, and is balanced and symmetric. A laminate is balanced when, for each ply at θ^o there is a ply at $-\theta^o$ somewhere in the laminate. Symmetry means that the laminate is symmetrical with respect to its laminate midplane (cf. example on Figure 2.1). The longitudinal and lateral dimensions of the plate are a and b, respectively. The behaviour of a symmetric laminate is defined by its extensional and flexural stiffness matrices, $\mathbf{A}$ and $\mathbf{D}$ respectively. $\mathbf{A}$ and $\mathbf{D}$ are given as,

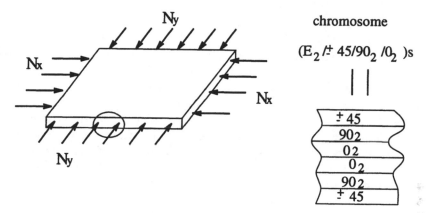

Fig. 2.1. Simply supported laminate and coding

$$\{A_{ij}, D_{ij}\} = \int_{-h/2}^{h/2} \bar{Q}_{ij}(\theta)\{1, z^2\}dz, \quad i,j = 1,6. \qquad (2.1)$$

where h is the laminate thickness, $\bar{Q}_{ij}(\theta)$ are the transformed reduced stiffnesses that depend on the material properties and the orientation of the fibers θ, and z is the vertical coordinate.

The laminate buckles into m and n half-waves in the x and y directions, respectively, when the loads reach the values $\lambda_b(m,n)\, N_x$ and $\lambda_b(m,n)\, N_y$. The buckling load factor $\lambda_b(m,n)$ is given in terms of the flexural stiffness coefficients D_{ij} as,

$$\lambda_b(m,n) = \pi^2 \frac{D_{11}(m/a)^4 + 2(D_{12}+2D_{66})(m/a)^2(n/b)^2 + D_{22}(n/b)^4}{(m/a)^2 N_x + (n/b)^2 N_y}.$$

$$(2.2)$$

The smallest value of λ_b over any m and n is the critical buckling load factor λ_{cb}. If λ_{cb} is larger than 1, the laminate can sustain the actual applied loads N_x and N_y without buckling.

The strength of the plate is predicted here by the maximum strain criterion. That is, the plate is assumed to fail if any of the ply strains exceeds its allowable value. The principal strains in the ith layer are calculated from the Classical Lamination Theory for symmetric and balanced laminates. The critical strength failure load factor λ_{cs} is defined as,

$$\lambda_{cs} = \min_i \left[\ min\left(\frac{\epsilon_1^{ua}}{f|\epsilon_1^i|}, \frac{\epsilon_2^{ua}}{f|\epsilon_2^i|}, \frac{\gamma_{12}^{ua}}{f|\gamma_{12}^i|}\right) \right], \qquad (2.3)$$

where ϵ_i^{ua} are the ultimate allowable strains, and a safety factor $f=1.5$ is used. If λ_{cs} is smaller than 1, the plate is assumed to fail. The critical load factor λ_{cr} is,

$$\lambda_{cr} = min(\lambda_{cb}, \lambda_{cs}). \qquad (2.4)$$

To alleviate matrix cracking problems, we also require that there are no more than four contiguous plies with the same fiber orientation.

The purpose of the optimization is to find the thinnest symmetric and balanced laminate satisfying the 4-ply contiguity constraint that will not fail because of buckling or excessive strains. The symmetry constraint is readily accommodated by optimizing only one half of the laminate, the other half being obtained by symmetry. The balance constraint is enforced by associating $+45°$ with $-45°$ into $\pm45°$ stack. Stacks of 0_2^o, $\pm45°$, and 90_2^o are our basic building blocks. We next formulate the objective function.

2.2 Optimization problem formulation

The objective is to minimize the total number of layers N and, at a given minimum thickness, maximize the failure load λ_{cr}. The buckling, strength, and 4-ply contiguity constraints are incorporated into the objective function through penalty functions. Our design variables are the number of plies in a laminate and the orientation of the fibers in each stack of two plies.

The 4-ply contiguity constraint is enforced by multiplying the objective by P_c^{nc}, where P_c is a penalty parameter for the ply contiguity constraint larger than 1 ($P_c = \sqrt{10/9}$ here), and nc is the number of stacks of two plies in excess of the constraint value of 2 stacks (i.e., 4 plies) in one half of the laminate. For example, a laminate in which there are 6 adjacent plies with fibers at $90°$ has its objective function multiplied by P_c.

The treatment of the failure and stability constraints is more complicated for two reasons. First, because of the discrete nature of the problem, there may be several designs of the same minimum thickness. Of these designs, we define the optimum to be the design with the largest failure load λ_{cr} by decreasing the objective in proportion to λ_{cr} when $\lambda_{cr} > 1$:

$$\text{if } \lambda_{cr} \geq 1, \ \Phi = P_c^{nc}(N + \delta(1 - \lambda_{cr})), \qquad (2.5)$$

where δ is a small parameter (=6 here). When the laminate fails ($\lambda_{cr} < 1$), the objective function Φ becomes,

$$\text{if } \lambda_{cr} < 1, \ \Phi = P_c^{nc}(\frac{N}{\lambda_{cr}^{P_l}}) + S. \qquad (2.6)$$

S and $1/\lambda_{cr}^{P_l}$ are two penalty functions. The use of a multiplying penalty function $1/\lambda_{cr}^{P_l}$ instead of an additive penalty term (N - penalty) is motivated by the scaling rule ([10]). Unlike S, $1/\lambda_{cr}^{P_l}$ is proportional to the distance to feasibility (λ_{cr}).

3. Specializing the Algorithm for Composite Structures

Even though we could apply a canonical evolutionary procedure to optimize simple laminates, such a strategy would become too costly for computationally expensive problems such as wing box design. Here, we specialize our algorithm by selecting operators and tuning their parameters. The algorithm in itself is a generational genetic algorithm with elitism. The tuning is made on a simple version of the problem, the laminate design problem, because massive testing of new ideas is possible. To discriminate between different possible implementations, a performance criterion is first defined.

3.1 Performance criterion

The effects of implementation choices will be discussed in terms of **price of the search** and **reliability**. The reliability is the probability of finding a **practical optimum** after n analyses. A practical optimum is defined here as a feasible optimal weight design with λ_{cr} within 0.1% of λ_{cr} of the global optimum. The price of the search is the number of analyses necessary to reach 80% reliability. The reliability is estimated by averaging the results of 200 searches of 6000 analyses, i.e., we perform 200 optimizations for each combination of operators and algorithm parameters. Our performance criterion is based on the assumption that we know the global optimum. This is a reasonable assumption, because many hundred of thousands searches have been performed on the laminate problems.

We consider a graphite-epoxy plate with longitudinal and lateral dimensions $a = 20in.$ and $b = 5in.$ respectively. The material properties are: $E_1 = 18.5 \times 10^6 psi.$; $E_2 = 1.89 \times 10^6 psi.$; $G_{12} = 0.93 \times 10^6 psi.$; $\nu_{12} = 0.3$; $t = 0.005in.$ (the basic ply thickness); $\epsilon_1^{ua} = 0.008$; $\epsilon_2^{ua} = 0.029$; $\gamma_{12}^{ua} = 0.015$. The maximum thickness for a laminate is assumed to be 64 plies. Four different loading cases are considered: load case 1, $N_x = 13000lb/in$ and $N_y = 1625lb/in$; load case 2, $N_x = 12500lb/in$ and $N_y = 3125lb/in$; load case 3, $N_x = 9800lb/in$ and $N_y = 4900lb/in$; and a multiple load case where the plate should simultaneously withstand $[N_x = 12000lb/in, N_y = 1500lb/in]$ and $[N_x = 10800lb/in, N_y = 2700lb/in]$ and $[N_x = 9000lb/in, N_y = 4500lb/in]$. The reliabilities, which are performance averages over 200 runs for each load case, are averaged over the four load cases to yield the reliability of an implementation of the algorithm independently of the load case.

3.2 Representation of the designs

An evolutionary algorithm works on a population of designs. Here, each design is coded in the form of a string that represents a laminate cross-section. The coding is the standard stacking sequence notation, with the character "E_2" added to denote two empty layers (see example in Figure 2.1). Non empty layers are restricted to stacks of 0_2^o, 90_2^o and $\pm45^o$ (cf. section 2.1), so that we use a four characters alphabet. By changing the number of E's in the string, one can change the laminate thickness. When an E_2 appears in the middle of the sequence, it is pushed outside so that there is no empty layer inside the laminate. Since the maximum number of plies is 64, and the symmetry and balance conditions each reduce the number of variables by 2, the string length is $64/4 = 16$.

A discussion of the laminate encoding based on schemata variance is given in Appendix A..

3.3 Failure load maximization: A first version of the algorithm

A first version of the algorithm was developped in [9] for maximizing the critical failure load λ_{cr} at a given fixed number of plies N. Since laminate thickness does not change, there is no E in the alphabet. The objective is to maximize λ_{cr} while satisfying the 4-ply contiguity constraint. The failure load maximization is formulated as: minimize $\Phi_l = P_c/\lambda_{cr}$. The operators of this first algorithm are standard genetic operators except for the permutation operator. A brief description of those operators is given below:

Selection Scheme: the selection was based on linear ranking of the designs. Recall that the algorithm is elitist, i.e., the best design found so far is kept unchanged from a generation to the next.

Crossover: the crossover used in the first version of the algorithm is a two-point crossover. Two-point crossover is the standard crossover in genetic algorithms because it minimizes the probability of separating digits in the string ([17]). Once two parents are selected, they have "probability of crossover" of creating a new design through crossover. Otherwise, one of the two parents is copied unchanged. An example of two-point crossover is given hereafter:

Parent 1: $(E_2/E_2/E_2/E_2/\pm45/ \overset{\vee}{\wedge} \pm45/90_2/ \overset{\vee}{\wedge} 90_2/0_2)_s$

Parent 2: $(E_2/90_2/\pm45/90_2/0_2/ \overset{\vee}{\wedge} 0_2/\pm45/ \overset{\vee}{\wedge} \pm45/\pm45)_s$

Child: $(E_2/E_2/E_2/E_2/\pm45/ \overset{\vee}{\wedge} 0_2/\pm45/ \overset{\vee}{\wedge} 90_2/0_2)_s$

Mutation: the design generated by crossover is next subjected to the mutation operator. Mutation performs random changes in an individual's string. Each bit has "probability of mutation" chances of changing. Mutation is illustrated below:

Before mutation: $(E_2/\underline{\pm45}/90_2/0_2/\underline{\pm45}/\pm45)_s$
After mutation: $(E_2/\underline{0_2}/90_2/0_2/\underline{E_2}/\pm45)_s$

Permutation: Permutation is a new operator created for laminate design in [9]. Permutation swaps the location of two pairs of stacks without changing the composition of the laminate. Therefore, the flexural properties of the laminate are modified while its in-plane characteristics are preserved. This operator has the property of moving in the design subspace of one of set of constraints (the constraint on flexural stiffness) while keeping a constant distance with respect to another set of constraints (the constraints on strength). Permutation is applied to new designs at a rate of "probability of permutation". During permutation, two points in the string are selected, creating three substrings in the chromosome. The two pairs of stacks at the two extremities of the central substring are then swapped (cf. example below).

Before permutation: $(\pm 45/ \overset{\vee}{\underset{\wedge}{}} \pm 45/90_2/ \pm 45/0_2/90_2/ \overset{\vee}{\underset{\wedge}{}} \pm 45)_s$

After permutation: $(\pm 45/ \overset{\vee}{\underset{\wedge}{}} 90_2/0_2/ \pm 45/90_2/ \pm 45/ \overset{\vee}{\underset{\wedge}{}} \pm 45)_s$

The algorithm was tuned in [9] by minimizing the price of the search on three load cases such that $N_y/N_x = 0.5, 0.25$, and 0.125, respectively (but no multiple load case). The total number of layers was kept constant at $N = 48$, i.e., there were 12 digits per chromosome. The following optimal parameters are obtained: population size $= 8$, probability of crossover $= 1.$, probability of mutation $= 0.01$, probability of permutation $= 1$, $P_c = 1.08$. This is a local optimum in the sense that moving any of those parameters away from the optimal setting results in an increase in price of the search (990 analyses at the optimum). The small population size (8 individuals) is explained by the intense use that is made of the random permutation operator.

3.4 An improved algorithm for minimum thickness design

The previous version of the algorithm can be applied on the laminate minimum thickness problem presented in Section 2.2 by adding empty ply symbols E in the alphabet. In that case, the price of the search went up from 990 to over 3300 analyses. For improving the performance of the basic algorithm on the varying, thickness problem, the following changes are proposed.

Objective Function: The key to the approach of equation 2.6 is to combine fixed and proportional-to-λ_{cr} penalties. The choice of the penalty parameters P_l and S is crucial to the performance of the algorithm.

The main problem of an objective function without S is that for some load cases, the critical load of infeasible designs that are one stack lighter than the optimum feasible design is very close to one. Then P_l needs to be very large for the objective function Φ of feasible optimum designs to be smaller than the objective function of good infeasible designs. But a large P_l impairs the efficiency of the genetic search ([9], [16]). On the other hand, it is found in [10] that the best performance is achieved for $S = 1$, $P_l = 0.5$.

No Identical Parents: In order to preserve population diversity, two identical designs are not allowed to mate, which would probably result in a clone.

Crossover for Varying Thickness Laminate: the two-point crossover has a problem for the varying thickness case in that the cutting points can fall in the empty part of the laminate, reducing the potential for crossover to truly recombine the properties of the parents. After testing eight alternatives, it was found in [10] that a single-point crossover, where the cutting point is restricted to fall in the non-empty part of the thicker laminate was the best. This crossover was dubbed "one-point thick crossover".

Differentiated Mutation: The traditional mutation operator, previously described, has a biased and uneven effect depending on the string length (i.e., maximal laminate thickness) and number of digits denoting full stacks (i.e., actual laminate thickness). For example, a laminate coded with a long string length (many E_2's) has a higher probability of increasing its thickness through mutation than the same laminate if it is coded using a shorter string. To correct this, the mutation operator is differentiated into thickness-changing mutation and fiber orientation changing mutation. Three separate probabilities of mutation are used: a probability of adding plies, a probability of deleting plies, and a probability of changing the fiber orientation per digit. The thickness mutations add or delete two stacks somewhere in the laminate if the corresponding probability test is passed. For change of fiber orientation, a test is implemented for each digit coding a full stack, and if it is passed, the fiber orientation is changed.

Single Stack-Swap Permutation: the original algorithm (cf. section 3.3) has been developped for a class of relatively easy problems for which only one load case was considered in the optimization problem. However, for more difficult problems where many loadings are considered, the basic version of permutation appears to shuffle the digits too much. Accordingly, a permutation operator that swaps a single pair of stacks instead of two is proposed.

The effect of each of the modification on the genetic search is described in [10]. Here, we will show the combined effect of all the modifications averaged over the four load cases. The parameters common to the basic and specialized algorithms are population size = 8, $P_c = \sqrt{10/9}$, and the probability of crossover and permutation equal to 1. The probability of mutation is equal to 0.01 for the basic algorithm, while the specialized algorithm uses this probability for stack change, and has a probability of 0.05 for stack addition or deletion. For penalty parameters, the basic algorithm has $P_l = 2$ and $S = 0$, while the modified algorithm has $P_l = 0.5$ and $S = 1$. The two versions of the algorithm are compared in Figure 3.1 in terms of the reliability as a function of the number of analyses. The plot results from 200 independent runs for each one of the 4 test cases described in section 3.1

The intersections of the curves with the 80% reliability line represent the price of the associated searches. The specialized evolutionary algorithm costs 1450 analyses, compared to 3300 for the basic algorithm, which represents a 56% decrease in the price of the search.

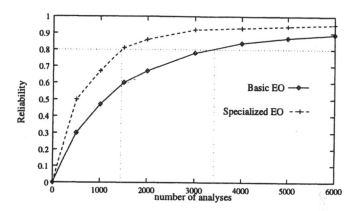

Fig. 3.1. Comparison of the basic and specialized algorithm, 200 runs

4. Application to a Wing Box Problem

4.1 Problem description and evolutionary implementation

The experience gained on the simple plate problems is now applied to a more complex problem, the optimization of a wing box skin [12]. We consider here an unswept, untapered wing box with four spars and three ribs. The wing box is clamped at the root and subjected at the tip to the applied load distribution shown in Figure 4.1.

The wing box is analyzed through a finite element model with 32 nodes and 72 degrees of freedom. The 18 laminates making up the skin, the ribs and the spars elements are 8 degrees of freedom rectangular elements.

The purpose of the optimization is to minimize the thickness of the laminates in the skin. The problem complexity is varied by changing L, the number of laminates that are independent in the wing box skin. L can vary from 1 (all the laminates in the upper and lower skin are the same) to 18 (all the laminates in the wing skin are different). In addition to the constraints used for the plate problem, we require that laminates in the upper skin should not buckle because of shear loads.

Optimizing a wing box skin is more difficult than optimizing an isolated laminate for three reasons: First, because the wing is modeled by finite elements, the analysis is computationally expensive. On an IBM AIX RISC/model 6000 version 3, it takes 0.1 sec CPU to calculate the response of one wing, which limits the number of experiments that can be made. It is thus necessary to extrapolate the results of the tests performed on the single plate. Second, the design space is larger for the wing than it is for the single laminate since many laminates can be considered simultaneously. The increase in size of the design space has also a price in terms of length of the evolutionary search. Third, the different laminates of the skin are mechanically coupled in the wing (by other laminates, and by spars and ribs). These

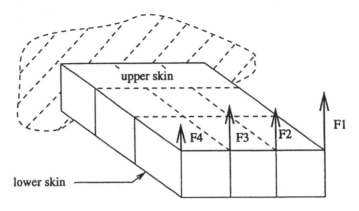

dimensions: 139.5 x 88.2 x 15 in.

F1=85467lbs, F2=F3=42239lbs, F4=20235lbs, ribs and spars are [(+/- 45)5]s.

Fig. 4.1. *Wing box*

couplings increase the interdependency of the plies, making the optimization problems more difficult ([12]).

The objective function formulation and the operators used for the wing are straightforward generalizations of those used for the plate. The total number of layers N is the sum of the layers in the laminates of the upper and lower skins. The coding of a wing box is the composition of the codings of its independent laminates. Crossover, mutation and stack swap are applied independently to each of the laminates. During crossover, the laminates that are recombined have the same relative positions in the parent designs. An example of crossover for Two-Laminate Wing Box problem is given below:

Parent 1: $(E_2/E_2/ \overset{\vee}{\underset{\wedge}{}} \pm 45/\pm 45/\pm 45/\pm 45/90_2/\pm 45/0_2)_s$ *upper skin*

$(E_2/E_2/E_2/\pm 45/90_2/0_2/90_2/ \overset{\vee}{\underset{\wedge}{}} 90_2/0_2)_s$ *lower skin*

Parent 2: $(E_2/90_2/ \overset{\vee}{\underset{\wedge}{}} \pm 45/90_2/\pm 45/0_2/0_2/\pm 45/\pm 45)_s$ *upper skin*

$(E_2/E_2/E_2/E_2/\pm 45/90_2/\pm 45/ \overset{\vee}{\underset{\wedge}{}} 0_2/90_2)_s$ *lower skin*

Child: $(E_2/E_2/ \overset{\vee}{\underset{\wedge}{}} \pm 45/90_2/\pm 45/0_2/0_2/\pm 45/\pm 45)_s$ *upper skin*

$(E_2/E_2/E_2/\pm 45/90_2/0_2/90_2/ \overset{\vee}{\underset{\wedge}{}} 0_2/90_2)_s$ *lower skin*

4.2 Results

4.2.1 Recommendations for scaling the problem parameters. Based on [12], recommendations on how to scale the algorithm parameters as the number of laminates L changes can be made.

Population size: The optimal population size for different number of laminates L in the structure is given in Table 4.1. It is seen that the optimal population size does not grow with the design space size (linked to L). If there is a limit on the total number of analyses, it can be detrimental to increase the population size as the complexity of the problem increases. Indeed, on the one hand, increasing the problem complexity (number of laminates for example) causes longer runs to achieve a given reliability, at a given population size. On the other hand, increasing the population size also has a price in terms of length of the search. The cumulative effect of an increased complexity and a larger population on the time to convergence can exceed computation capacities. This occurs for the 6 and 18-Laminate Wing Box problems. The optimal population size is 5 individuals, because larger populations do not even have time to start locating optimal regions of the design space in 30000 analyses. This contradicts most models for sizing population ([18],[20], [21]). The reason is that these models do not consider limits on the computer resources, which is an important constraint in engineering applications of evolutionary computation.

Table 4.1. Optimal population size vs. number of independent laminates L for various probabilities of stack swap pp. Population sizes of 5, 10, and 50 are tried. Maximum search length = 30000 analyses. Based on 50 independent runs. The performance criterion for $L=1$ is the price of the search. For all other L's, it is the weight of feasible final designs

L	1	2	6	18
pp = 0	5	10	5	5
pp = 1	5	5	5	5

Probability of stack swap: On the L-Laminate Wing Box problems, $L = 6$ and 18, a probability of stack swap $= 1/L$ performs better than 1 or 0, for population sizes equal to 5 and 10. The optimal rate of stack swap can thus be approximated by $1/L$, where L is the number of independent laminates .

Other operators: It is found in [12] that lowering the probability of mutating the fiber orientation from 0.01 to 0.005 is beneficial for $L = 6$ and 18. On the other hand, probabilities of adding and deleting stacks should not be reduced. Indeed, as the number of independent laminates increases, exploration of weight redistribution between laminates becomes increasingly critical. Unless inter-laminate stack swap is used [4], addition and deletion of stacks through mutation is the primary factor that creates new weight distributions. The probability of crossover is kept equal to 1 in all experiments.

4.2.2 Comparison between continuous and evolutionary optimizations. One-Laminate Wing Box Problem: The best design obtained by evolutionary optimization is $(\pm45^{\circ}_{20}/0^{\circ}_{2}/(90^{\circ}_{2}/0^{\circ}_{4})_{4}/90^{\circ}_{2}/0^{\circ}_{2})_{s}$, for a total of $18 \times 140 = 2520$ plies in the skin. This design is found in 80% of the runs after 3500 analyses. The same problem has been solved using a more traditional strategy: the thicknesses of the plies are taken as continuous variables, optimized by SQP [23], and finally rounded to multiples of the basic ply thickness t. The best design obtained is $(\pm45^{\circ}_{21}/90^{\circ}_{4}/0^{\circ}_{24})_{s}$. In terms of weights, the evolutionary and the SQP designs are the same, but the SQP design does not satisfy the ply contiguity constraint, because there is no easy way to implement it.

Two-Laminate Wing Box Problem: The best design found by the evolutionary algorithm is,

upper skin: $(\pm45^{\circ}_{1}6/0^{\circ}_{4}/(90^{\circ}_{2}/0^{\circ}_{4})_{s}/90^{\circ}_{2}/0^{\circ}_{2})_{s}$
lower skin: $(90^{\circ}_{2}/\pm45^{\circ}/90^{\circ}_{4}/0^{\circ}_{2}/(0^{\circ}_{4}/90^{\circ}_{2})_{2})_{s}$

This design corresponds to a total number of plies equal to $9\times(140+52) = 1728$. 80% of the evolutionary searches find it after 10000 analyses. Continuous optimization has yielded on this problem,

upper skin: $(\pm45^{\circ}_{19}/90^{\circ}_{2}/0^{\circ}_{30})_{s}$
lower skin: $(90^{\circ}_{12}/0^{\circ}_{10}/90^{\circ}_{6})_{s}$

The continuous design has $9\times(140+56) = 1764$ plies, so that it is 2% heavier than the evolutionary design and violates the contiguity constraints.

5. Conclusions

Evolutionary optimization is a powerful tool for designing composite structures but the cost of the search can be excessive. This paper describes an approach to reducing the cost and increasing the reliability of the optimization. With a specialized evolutionary algorithm, laminated plates can be reliably optimized in 1450 analyses in a design space containing about 4 billions points. The specialized algorithm can be applied to more complex composite structures, such as wing box beams.

Other approaches at improving evolutionary optimization of composite structures can be found in [24], [4], [25], [10], [11].

References

1. Garfinkel, R.S., and Nemhauser, G.L., *Integer Programming*, John Wiley and Sons, Inc., New York, 1972.
2. Haftka, R.T., and Walsh, J.L., *Stacking-Sequence Optimization for Buckling of Laminated Plates by Integer Programming*, AIAA Journal, 30(3), 1992, pp. 814-819.
3. Nagendra, S., Haftka, R.T., and Gürdal, Z., *Stacking Sequence Optimization of Simply Supported Laminates with Stability and Strain Constraints*, AIAA Journal, 30(8), 1992, pp. 2132-2137.
4. Nagendra, S., Jestin, D., Gürdal, Z., Haftka, R.T., and Watson, L.T., *Improved Genetic Algorithm for the Design of Stiffened Composite Panels*, Computers & Structures, 58(3), pp. 543-555, 1996. .
5. Mesquita, L., and Kamat, M.P., *Optimization of Stiffened Laminated Composite Plate with Frequency Constraints*, Engineering Optimization, 11, 1987, pp. 77-88.
6. Schmit, L.A., and Farshi, B., *Optimum Design of Laminated Composite Plates*, International Journal for Numerical Methods in Engineering, 11(4), 1979, pp. 623-640.
7. Pedersen, P., *On Sensitivity Analysis and Optimal Design of Specially Orthotropic Laminates*, Engineering Optimization, 11, 1987, pp. 305-316.
8. Gürdal, Z., and Haftka, R.T., *Optimization of Composite Laminates*, presented at the NATO Advanced Study Institute on Large Structural Systems, Berchtesgaden, Germany, Sept.23-Oct.4, 1991.
9. Le Riche, R., and Haftka, R.T., *Optimization of Laminate Stacking Sequence for Buckling Load Maximization by Genetic Algorithm*, AIAA Journal, 31(5), 1993, pp. 951-970.
10. Le Riche, R., and Haftka, R.T., *Improved Genetic Algorithm for Minimum Thickness Composite Laminate Design*, Composites Engineering, 3(1), 1995, pp. 121-139.
11. Le Riche, R., Knopf-Lenoir, C., and Haftka, R.T., *A Segregated Genetic Algorithm for Constrained Structural Optimization*, Proceedings of the Sixth ICGA, Morgan Kaufmann, 1995, pp. 558-565.
12. Le Riche, R., *Optimization of Composite Structures by Genetic Algorithms*, Ph.D. Dissertation, Virginia Polytechnic Institute and State University, Dept. of Aerospace Eng., Oct. 1994.
13. Lombardi, M., Haftka, R.T., and Cinquini, C., *Optimization of Laminate Stacking-Sequence for Buckling Load Maximization by Simulated Annealing*, Proc. of the 33rd AIAA/ASME/ASCE/SDM Conference, Dallas, Texas, 1992.
14. Ball, N.R., Sargent, P.M., and Ige, D.O., *Genetic Algorithm Representation for Laminate Layups*, Artificial Intelligence in Engineering, 8(2), 1993, pp. 99-108.
15. Callahan, K.J., and Weeks, G.E., *Optimum Design of Composite Laminates using Genetic Algorithms*, Composite Engineering, 2(3), 1994.
16. Richardson, J.T., Palmer, M.R., Liepins, G., and Hilliard, M., *Some Guidelines for Genetic Algorithms with Penalty Functions*, in Proceedings of the Third ICGA, Morgan Kaufmann, 1989, pp.191-197.
17. Spears, W.M., and De Jong, K.A., *An Analysis of Multi-point Crossover*, in Foundations of Genetic Algorithms, Rawlins G.J.E. (Eds), Morgan Kaufmann Publishers, San Mateo, CA., 1991, pp. 301-315.
18. Goldberg, D.E., Deb, K., and Clark, J.H., *Genetic Algorithms, Noise, and the Sizing of Populations*, Complex Systems, 6, 1992, pp. 333-362.

19. Goldberg, D.E., *Genetic Algorithms in Search, Optimization and Machine Learning*, Addison-Wesley, Reading, MA, 1989.
20. Thierens, D., and Goldberg, D.E., *Mixing in Genetic Algorithms*, Proc. of the Fifth International Conference on Genetic Algorithms, Morgan Kaufmann Publishers, Univ. of Illinois at Urbana-Champaign, July 17-21, 1993, pp. 38-47.
21. Alander, J.T., *On Optimal Population Size of Genetic Algorithms*, in Proc. of Compeuro 92, IEEE Computer Society Press, 1992, pp. 65-70.
22. Holland, J.H., *Adaptation in Natural and Artificial Systems*, The University of Michigan Press, Ann Arbor, MI, 1975.
23. Schittkowski, K., *NLPQL: a FORTRAN subroutine solving nonlinear programming problems*, Annals of Operation Research, 5, 1986, pp. 485-500.
24. Kogiso, N., Watson, L.T., Gürdal, Z., and Haftka, R.T., *Genetic Algorithms with Local Improvement for Composite Laminate Design*, Structural Optimization, 7(4), 1994, pp. 207-218.
25. Harrison, P.N., Le Riche, R., and Haftka, R.T., *Design of Stiffened Composite Panels by Genetic Algorithms and Response Surface Approximations*, Proceedings of the 36th AIAA/ASME/ASCE/AHS/ASC Structures, Structural Dynamics and Materials Conference, AIAA-95-1163-CP, New Orleans, LO, USA, 1995, pp. 58-68
26. Grefenstette, J.J., *Deception Considered Harmful*, in Foundations of Genetic Algorithms – 2, Whitley L.D. (Eds), Morgan Kaufmann Publishers, San Mateo, CA., 1993, pp. 75-92.

A. Laminates and the Static Building Block Hypothesis

In this paragraph, we discuss the laminate encoding in light of some theoretical results on what makes a problem difficult for a genetic algorithm. Historical studies [22], [19] explain that genetic algorithms work by juxtaposing building blocks. Within those theories, the rate of growth of a schemata H is approximated by $F(H,i)/\bar{F}(i)$ where $F(H,i)$ represents the average worth of the instances of H at generation i and $\bar{F}(i)$ is the average worth in population i. The previous relation, called the Static Building Block Hypothesis, tells that schemata H carying high performance building blocks should spread in the population. Yet, as noted in [26], schemata having a large variance can have a "dynamic" evaluation (accounting for the incomplete representation of a schema that occurs in a finite size population) very different from their real worth. Variance in the schemata evaluation is one of the reasons that makes a problem difficult for genetic algorithms.

We investigate here if the laminate problem and its associated coding satisfy the Static Building Block Hypothesis. Our attempt at answering this question is based on a simplified version of the minimum thickness design problem: the basic ply thickness is increased to $t = 0.02$ in. (instead of 0.005 in.), there is no empty layer (no E's) and no ply contiguity constraint ($P_c = 1$). As a result, optimal chromosomes are only four digits long. We can easily sample the $3^4 = 81$ points of the design space and calculate average objective functions and variances for the $4 \times 3 = 12$ first order schemata. Recall that in all of the experiments performed, load case 1 is the easiest problem, followed by load case 2, load case 3 and the multiple load case. The optima of each of the load cases and their associated objective functions are:

load case 1 $[\pm 45/0_2/0_2/0_2]_s$ $\Phi = 9.51$

load case 2 $[\pm 45/\pm 45/0_2/0_2]_s$ $\Phi = 11.36$

load case 3 $[90_2/\pm 45/0_2/0_2]_s$ $\Phi = 9.01$

multiple load case $[90_2/0_2/0_2/0_2]_s$ $\Phi = 10.35$

Table A.1. Average and variance of objective function of first order schemata, P_l = 0.5, $S = 1$, $P_c =$1.; * stands for any stack

schem- ata	ld. c. 1 average/ variance	ld. c. 2 average/ variance	ld. c. 3 average/ variance	mult. ld. c. average/ variance
$0_2 * **$	14.84/3.82	16.03/6.13	17.52/13.69	17.21/10.63
$\pm 45 * **$	16.32/9.34	15.67/7.03	13.78/1.82	16.04/5.64
$90_2 * **$	16.95/7.50	16.42/6.62	13.60/4.57	16.04/7.15
$*0_2 * *$	14.9/4.99	15.54/8.66	16.11/19.60	16.48/13.52
$* \pm 45 * *$	16.43/8.11	16.08/5.01	14.69/5.59	16.62/4.87
$*90_2 * *$	16.99/7.00	16.50/5.93	14.10/2.86	16.19/5.87
$* * 0_2*$	14.36/5.70	14.89/7.91	14.92/17.78	15.59/12.07
$* * \pm 45*$	16.65/6.80	16.45/5.51	15.04/8.72	16.91/6.83
$* * 90_2*$	17.09/6.11	16.79/4.51	14.93/3.75	16.80/4.36
$* * *0_2$	14.19/5.28	14.49/6.45	14.25/13.50	15.01/9.01
$* * * \pm 45$	16.71/6.84	16.51/5.34	15.05/8.93	17.03/6.97
$* * *90_2$	17.20/5.61	17.12/4.33	15.60/6.86	17.27/5.17

Average and variance of the objective functions are given in Table A.1. It is seen in the Table that in 75% of the cases, the higher performance (lower objective function) first order schemata contain the optimum design. This shows that laminates essentially satisfy the Static Building Block Hypothesis, which confirms the potential of evolutionary optimization for this type of problem. On the average in the Table, the third and the multiple load case are the load cases with the larger variances, and also the most difficult load cases for the optimization. This confirms the role of schema variance in what makes evolutionary optimization hard. Load case 3 clearly shows features of a deceptive problem: the schema $* \pm 45 * *$ contains the optimum but does not have the best average performance, while the schema $*90_2 * *$ does not contain the optimum but has the best average performance. In addition, both schemata have a low variance (5.59 and 2.86, respectively), which means that the misleading attraction towards $*90_2 * *$ is a phenomenon that occurs frequently in the populations. In contrast, for load case 2, the misleading first order schema $*0_2 * *$ has a larger variance than $* \pm 45 * *$ and $*90_2 * *$, which echoes the fact that load case 2 is easier than load case 3.

Even though schema average performances and variances are clearly related to the difficulty experienced during evolutionary optimization, it does not explain why, for example, load case 3 is easier than the multiple load case. Different explanations need to be sought in those cases. One could consider, for example, the number of active constraints as a factor of hardness.

Flaw Detection and Configuration with Genetic Algorithms

Sushil J. Louis[1], Fang Zhao[2], and Xiaogang Zeng[3]

[1] Dept. of Computer Science, University of Nevada, Reno, NV 89557
[2] Dept. of Civil Engineering, Florida International University, Miami, FL 33199
[3] Algor Inc., Pittsburgh, PA 15238

Summary. This paper describes a genetic algorithms for the identification of flaw shapes and the reconstruction of boundary conditions in a continua. Such problems often arise in scientific research and engineering practice. Using a genetic algorithm avoids some of the weaknesses of traditional gradient based analytical search methods including the difficulty in constructing well-defined mathematical models directly from practical inverse problems, and easily getting trapped or oscillating between local minima and thus failing to produce useful solutions. Numerical experiments for the detection of internal structure flaws for simple prototype problems suggest that good predictions of the flaw sizes and locations are possible. These predictions may also serve as the initial points for more traditional gradient methods. The experiments indicate that the proposed method has the potential to solve a wide range of inverse structural identification problems in a systematic and robust way.

1. Introduction

In practical engineering, most inverse problems in mechanics fall under three categories:

1. Reconstruction
2. Identification [2] and
3. A combination of the previous two.

Reconstruction problems refer to the determination of missing boundary conditions or some other externally applied unknown inputs, such as body forces and sources, temperatures, or material properties for a system with known geometry. Such inputs result in an output that is known or easy to obtain by experimental measurements. Identification problems usually deal with the determination of the unknown geometry of the object such as internal flaws, for which we have the boundary conditions of the system and the measured responses at selected points. Inverse problems that include both reconstruction and identification problems are also of interest. In such cases, both the boundary geometry and the associated boundary conditions are not appropriately or completely defined.

Numerical formulation and computational techniques for the solution of such problems may provide non-destructive evaluation and design tools for monitoring, diagnosing, and examining structural systems [8, 9], identifying contact interfaces, noise control or reduction in structural acoustics, as well as hybrid experimental and numerical methods for the analysis of structures [10, 1]. Although inverse problems have received increasing attention from the engineering community for the past few decades, solution methods are still in the initial stages of development and application. Most of the papers in the literature deal with problems that have simple geometric and structural configurations, and the missing conditions are usually specified by few unknown parameters or design variables.

Computational techniques for the solution of inverse problems usually consist of two parts:

– Numerical discretization methods for *ill-posed* structures or objects with *assumed* geometric and/or boundary conditions (for the missing geometric and/or boundary conditions) and
– Iterative procedures that are used to search for the actual structural and geometrical configuration and its solution from the boundary conditions and certain available observed values obtained from either internal or external boundary points.

The discretization schemes for structures or objects in the literature to date include finite difference methods, finite element methods (FEM), and boundary element methods (BEM). However, the difficult part in formulating and numerically solving realistic inverse problems lies in finding robust iterative procedures that can be applied to a large variety of complex problems and that will proceed towards the final configuration. A majority of the iterative procedures in the literature thus far are based upon gradient-based optimization and/or sensitivity analysis schemes in which the Jacobian and/or the Hessian matrices (or their approximations) of the system must be calculated iteratively. One of the advantages of the gradient-based solution methods is that for a well-defined problem, good initial starting points usually lead to fast convergence, especially for objects with simple structural and geometrical configurations. The difficulty with such methods lies in that it is not easy to construct well-defined mathematical models for practical inverse problems that are usually severely ill-posed due to the inaccessibility of crucial data. In fact, even for problems with simple structural and geometric configurations, cumbersome regularization procedures, various penalty constraint formulations, or other augmentation schemes are usually required to make the problems well-defined to ensure good convergence. In addition, these treatments are often problem dependent, requiring different constraints and regularization formulations for different problems. This results in the lack of generality or robustness of the gradient-based methods.

In this paper we describe a methodology based on genetic algorithms (GAs) [7, 6] for the solution of inverse problems pertaining to the identification of flaw shapes and the reconstruction of boundary conditions in a continua. Using this methodology, we express the inverse problem as a minimization problem with the objective function being the root mean square (RMS) of the differences between the measured quantities at each sensor location and the corresponding computed quantities from a candidate structural configuration. A parallel genetic algorithm uses this objective function to measure the fitness of individuals in a population of candidate structural configurations (solutions). Possible approximate solutions found by the genetic algorithm gradually lead to an increase in the average fitness of the population and thus better estimates of a flaw's configuration. The main advantage of GA methods is that they do not require any mathematical augmentation to the numerical solution methods used to represent the ill-posed structures. BEM, FEM or any other available numerical techniques are merely used as "response functions", the output of which provides general search guidance for the GA. This makes the GA method applicable for a wide range of practical problems and permits a variety of numerical discretization methods to be used. The GA-based methodology presented in this paper is not intended to replace but rather to complement the traditional gradient-based methods. When an inverse problem can be represented by a well-defined mathematical model, the traditional gradient-based methods are usually much more efficient than the GA methods. However, when it is too difficult or too expensive to construct a well-defined mathematical model or when it is impossible to obtain such a model due to practical conditions such as hazardous environment or inaccessibility of crucial data, the GA methods become more attractive. We believe that the best solution schemes may be the combination of GA and gradient-based methods. GA methods may first be used to narrow the search to promising regions. Objective functions defined on these regions may then be constructed and then used by the gradient-based methods or other traditional schemes to obtain the optimal point of the objective function.

The next section provides a short description of *our* genetic algorithm. Section 4. presents two numerical examples and conclusions are given in the last section.

2. Genetic Algorithms

Genetic algorithms are probabilistic parallel search algorithms based on natural selection [7]. GAs were designed to efficiently search large, non-linear, poorly-understood search spaces where expert knowledge is scarce or difficult to encode and where traditional optimization techniques fail. They work with a population of individuals (usually encoded as bit strings) that represent candidate solutions to a problem. The population is modified by the application of genetic operators from one iteration (generation) to the next until a stopping criterion is met. The algorithm evaluates, selects, and recombines members of the population to form succeeding populations. Evaluation of each string which encodes a candidate solution is based on a fitness function that is problem dependent. Since each candidate solution's evaluation is independent of others in the population we distribute the evaluations over available machines on our network using PVM [5]. For our problem, the RMS of the differences between the measured quantities at each sensor location and the corresponding computed quantities from the assumed structural configuration provides a measure of "fitness." This fitness measure is used by the selection operator to select relatively fitter individuals in the population for recombination. Crossover and mutation, the recombination operators, imitate sexual reproduction. Mutation insures against the permanent loss of genetic material. Crossover is the main recombination operator. It is a structured yet stochastic operator that allows information exchange between candidate solutions. Two point crossover is implemented by choosing two random points in the selected pair of strings and exchanging the substrings defined by the chosen points. Figure 2.1 shows how crossover mixes information from two parent strings, producing offspring made up of parts from both parents.

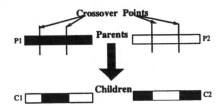

Fig. 2.1. Crossover of the two parents P1 and P2 produces the two children C1 and C2

3. Representation

In order to simplify the problem and investigate the feasibility of the method, we only consider inverse problems consisting of a known number of flaws with a known shape but unknown locations and sizes. In the current version of the genetic algorithm, a flaw is described by a set of parameters associated with a particular geometric shape and its location by a fixed point to which it is considered to be "attached." For instance, a rectangular flaw can be described by four parameters; the two lower left corner coordinates and the two upper right corner coordinates. We ensure that flaws are completely contained within the plate and do not overlap. The representation of a solution or an approximation for the true flaw configuration is a binary string of which each substring represents one flaw by describing its location and dimensions. Figure 3.1 gives an example of the GA representation of two rectangular flaws.

The fitness function is computed as the RMS of the differences between the given measurements and the calculated values using the BEM at fixed points on the plate boundary, i.e.

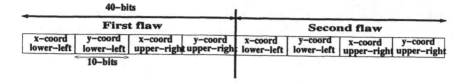

Fig. 3.1. Representing flaws

$$f_n \equiv \sqrt{\frac{1}{n} \sum_{i=1}^{n} \left(v_i^{(m)} - v_i^{(c)} \right)^2} \qquad (3.1)$$

where n is the number of sample points, $v^{(m)}$ and $v^{(c)}$ are, respectively, the measured and calculated values at the ith sample point. For a given or assumed structural and flaw configuration, the calculated values are obtained from the BEM.

We outline our algorithm below:

1. Initialize: randomly generate an initial population of solutions, each of which is an assumed flaw configuration;
2. Evaluation: use the BEM to compute the values of the structural behavior for the assumed configuration. This is done in parallel with m individuals being evaluated simultaneously where m is the number of machines available. Then compare the values computed from the BEM with measured values and compute the RMS error using the error expression defined in equation (1);
3. If the genetic algorithm has run for a pre-determined number of generations, stop;
4. Calculate population statistics, select, crossover and mutate the evaluated population to produce the next generation, and go to Step 2 for evaluating this new generation of individuals.

The method described above offers several advantages over traditional methods. First, unlike various traditional gradient based formulations, our method does not require the calculation of the Jacobian and/or the Hessian matrices or the various sensitivity parameters of the system. The corresponding numerical procedures are therefore much simpler and more stable than traditional methods. This is important because the Jacobian, Hessian matrices, and the various sensitivity parameters are usually sensitive to the ill-posed nature of the structure system. Second, unlike the traditional methods in which various penalty functions or Lagrangian multiplier procedures are used to impose the required constraints such as the internal structure flaws' sizes, location ranges etc., treatments for these constraints is usually much simpler in the GA. Third, because the GA starts the solution search from a population of random starting configurations, the problem of local minima in traditional gradient based analytical search methods is mitigated. Finally,

genetic algorithms are easily parallelizable and our parallel implementation
using PVM [5] achieved nearly linear speedup, allowing us to get results in
a reasonable amount of time (overnight). PVM or parallel virtual machine
is a function library that connects a network of workstations to simulate a
parallel machine.

4. Preliminary Numerical Experiments

We investigated the feasibility and utility of our approach through a series of
numerical experiments involving the detection of rectangular flaws of various
sizes and orientations. Figure 4.1 shows our original problem, which is a rectangular plate of 10×5 units with a rectangular hole/flaw centered at $(0,0)$
with sides of 2×1 units. The initial population consists of 40 random flaw
configurations, that is, the initial flaws are rectangular shapes with randomly
generated sides and center locations. We use a variant on the CHC elitist selection strategy [4]. Let the population size be N, selecting mating candidates
through the usual fitness proportional selection we double the population to
$2N$ with the offspring produced by crossover and mutation. From this $2N$
size population we then select the best N individuals to form the next generation. We use 0.85 probability of crossover, and a 0.055 mutation rate on
the problems described in this section.

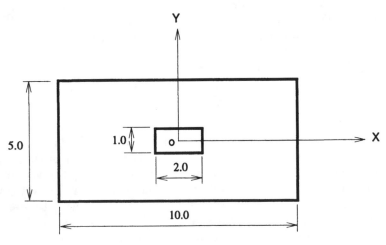

Fig. 4.1. One-Flaw Problem

We would like to have the RMS of the measured and computed values at
fixed boundary points be smaller than an acceptable value for terminating
the algorithm. However, for this preliminary investigation, we did not use this

condition and just ran the genetic algorithm for 40 generations. The results for eight different random seeds, to minimize the effect of a particular random seed, and the average values found for the variables are shown in Table 4.1.

Table 4.1. Results of Eight Runs for The One-Flaw Problem

	center coords		flaw dimensions		Error
	x	y	a	b	
1	-0.005	0.005	2.108	0.858	0.010
2	0.000	0.317	1.631	2.166	0.018
3	0.051	-0.107	2.414	0.341	0.031
4	-0.002	0.595	1.305	3.424	0.052
5	0.005	0.327	1.290	2.927	0.055
6	-0.002	0.532	1.217	3.532	0.066
7	-0.119	-0.132	2.327	0.468	0.076
8	0.061	-0.102	1.499	1.717	0.098
average	-0.002	0.179	1.724	1.929	0.051

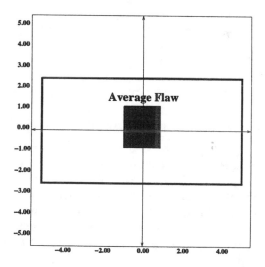

Fig. 4.2. Average flaw configuration found by the GA

The best result from the eight runs is the first one shown in the table, with the center being (-0.005, 0.005), rectangular flaw dimensions 2.108 × 0.858, and the RMS error 0.010. The center location as well as the X dimension of the flaw are very close to the actual values, while the Y dimension of the flaw has a 14.2% error. Figure 4.3 shows this best solution.

The average center location is (-0.002, 0.179) while the average flaw dimensions are 1.724 (13.8% error) by 1.929 (92% error). Average values are calculated by simply summing along a column in Table 4.1 and dividing by the number of rows, in this case, eight (8). Figure 4.2 shows the average location of the flaw. Looking closely at the second and last entries in the table, it may be observed that it is not obvious that the second solution is better than the last even though it has a smaller RMS error as a smaller RMS error may just indicate a local optimum. However the GA does manage to converge toward the observed flaw configuration. The flaw configuration corresponding to the highest RMS error, our worst result, is shown in figure 4.4.

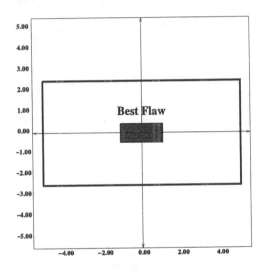

Fig. 4.3. Best flaw configuration found by the GA

The second example is also a rectangular plate but with two rectangular flaws as shown in Figure 4.5. Two-flaw problems, though similar to one-flaw problems, are significantly more complex for a gradient-based search method. The complexity arises from the difficulty of dealing with the topology and geometry of the flaws. In the present method, this problem is easily dealt with by ensuring that we only generate flaws (rectangles) that do not overlap and are within the plate.

For the second example, the initial population is also 40 and 40 generations are run. In addition to using our variant of the CHC algorithm we also linearly scale the fitnesses as this results in better performance. Figure 4.6 shows the average flaw locations over 11 runs with different random seeds at the 40th generation and Table 4.2 tabulates the results.

From the point of view of genetic algorithms, we find that scaling results in significant improvement in performance. Figures 4.6 and 4.7 depict average flaw configuration with and without scaling respectively. Scaling reduces average RMS errors and produces better flaw configurations as can be seen from the figures.

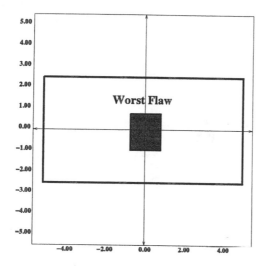

Fig. 4.4. Worst flaw configuration found by the GA

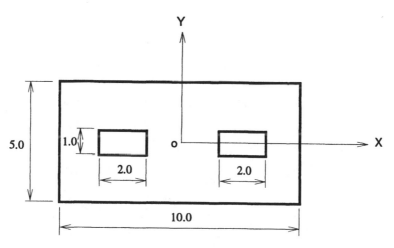

Fig. 4.5. Two-flaw Problem

We find that no single solution has the closest approximation of the actual flaw locations and dimensions. We also note that the best approximation for the left flaw is contained in one solution and that for the right flaw in another solution, the two solutions with the smallest RMS errors (The first two entries in the table). This is consistent with the solutions for the one-flaw problem in Figure 4.1 and suggests that better approximations have smaller RMS errors, which in turn suggests the use of the RMS error as a terminating condition for the GA. We may also do better by providing the genetic algorithm with more

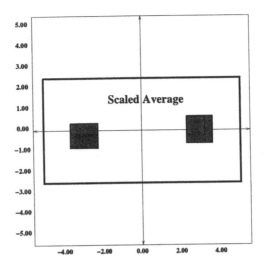

Fig. 4.6. Average of flaw configurations found by the scaled GA

Table 4.2. Best Two-flaw Problem Solutions of 11 Runs at Generation 40 in order of increasing error

	Flaw 1				Flaw 2				Error
	center coordinates		dimensions		center coordinates		dimensions		
			a	b			a	b	
1	-2.478	-0.007	2.078	1.061	2.546	-1.436	0.380	1.689	0.077
2	-3.180	-0.185	0.654	1.485	2.551	-0.226	1.658	1.081	0.104
3	-2.800	0.134	1.376	0.974	2.917	0.346	1.660	1.124	0.109
4	-2.654	-0.560	0.907	1.416	2.897	-0.514	0.771	1.528	0.110
5	-3.010	-0.187	1.151	1.334	2.839	0.419	1.551	0.969	0.123
6	-2.463	-0.044	2.654	1.183	2.688	-0.268	0.898	1.047	0.130
7	-2.668	-0.384	1.014	1.310	2.941	0.324	1.522	1.129	0.131
8	-2.444	-0.684	1.697	1.236	4.156	0.565	0.439	1.738	0.186
9	-2.819	0.905	2.293	0.988	2.619	0.937	3.005	0.720	0.223
10	-3.268	-0.088	0.888	1.037	3.341	0.219	1.093	1.514	0.241
11	-2.619	-0.394	1.639	1.416	-4.644	-1.353	0.283	0.492	0.321
Avg	-2.920	-0.237	1.386	1.161	2.950	0.036	1.298	1.254	0.160

information, perhaps about individual flaw errors rather than one number describing both flaws. For example, while a smaller RMS error gives a better solution in an overall sense than one that has a larger RMS error, it does not guarantee that the approximation for each of the flaws contained in the corresponding solution is better than that in the other. Figure 4.8 shows that for the first two solutions in Table 4.2 one flaw is well configured and the second is incorrect (way off), but the evaluation still produces a low RMS error. We also note that the genetic algorithm may in fact *crossover* these two partial solutions and with probability close to $1/l$, where l is the chromosome

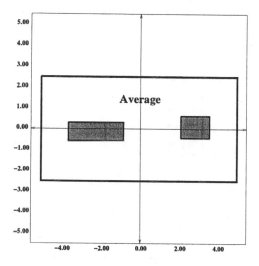

Fig. 4.7. Average of flaw configurations found by the GA without scaling

length (if using one-point crossover), produce a solution with both flaws well configured.

Figure 4.9 shows the worst flaw configuration produced by the scaled genetic algorithm. We believe that multiple runs for a problem and estimating the flaw configuration using averages will provide better, more consistent results than going with one run of the genetic algorithm.

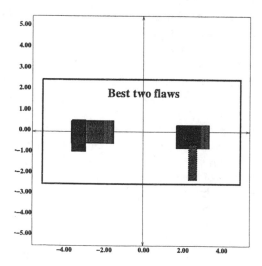

Fig. 4.8. The best two flaw configurations found by the scaled GA

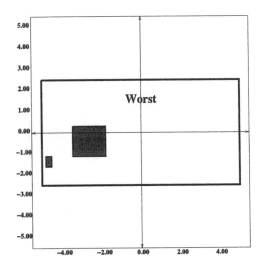

Fig. 4.9. The worst two flaw configurations found by the scaled GA

For complex problems we believe that genetic algorithm results will best
be used as good initial points for other gradient based numerical methods.
Genetic algorithms were not designed to be function optimizers, rather, they
tend to be very good at quickly finding promising *areas* of the search space [3].
It makes sense then, to use a two stage approach where the genetic algorithm
finds a population of good initial points and we use some other numerical
method to complete the search.

Although the mean values for all the parameters describing the flaw configuration are given for the first problem in Table 4.1 and for the two flaw
problem in Table 4.2, these averaged values are not obtained on an equal
basis because each solution has a different RMS error. The averaged values
may give better approximations if there is an upper bound on the RMS error
for the solutions being averaged and/or if we weight the averages based on
the RMS error.

These two examples as well as the results of other experiments indicate
that the proposed method is capable of finding the general locations and dimensions of the flaws and is useful for finding the solution of such inverse
problems. The accuracy of the solutions can probably be improved by changing the terminating condition and by using the generated solutions as an
initial point for other more traditional methods.

Note also that the genetic algorithm results are produced after 40 ×
40 = 1600 evaluations, which is population size multiplied by the number
of generations run. Enumerative and/or exhaustive search techniques would
need to search about half the search space before their chances of finding the

optimum became significant [1]. Since the search space size is 2^{40} for the first problem and 2^{80} for the second, we believe that using genetic algorithms may lead to significant savings in time and resources, even though the GA may not produce the optimum. Quantitatively, it is impressive that the genetic algorithm searched through a small $(1600/2^{40} = 0.00000000146)$ fraction of the search space to produce approximate but usable results on the first problem and through an even smaller fraction of the space on the second problem.

5. Conclusions

We presented a methodology based on GAs for solving inverse problems using two numerical examples. Preliminary results have demonstrated that our method is useful in predicting flaw sizes and locations and has the potential for either solving, or being a first step in solving, a range of inverse structural identification problems in a systematic and robust way. The most significant advantages of this methodology over the traditional search methods used for inverse problems are avoiding local optima, circumventing the mathematical difficulties encountered in the solution of inverse problems using traditional methods, and the difficulty in representing changing topologies.

Although the current experiments only involve rectangular flaw shapes, which require four parameters to describe the location and size of each flaw, more complex geometries may be represented by simply using more parameters in the form of polynomials. The key to successfully applying the methodology to practical inverse problems is computational efficiency. Each run of the parallel genetic algorithm, even with the small population sizes used, took many hours on four high-end workstations. We expect to be able to use larger populations and increase the number of iterations as more workstations become available. We are also planning to incorporate more domain knowledge both in generating the initial population and in the genetic operators. As noted earlier, using one point crossover may also improve performance. Evaluating fitness provides crucial feedback to the genetic algorithm and we will be experimenting with different evaluation functions. Finally, as delineated earlier, genetic algorithms are not function optimizers and we believe their best use may be in quickly generating a set of initial points for other, better understood, search and optimization methods.

Acknowledgments

The first author acknowledges support provided by the National Science Foundation under Grant IRI-9624130.

---

[1] This assumes a single optimum – a not unreasonable assumption, since in both problems we are searching for a single optimum flaw configuration

References

1. Balas, J., Sladek, J., and Drzik, M. (1983). Stress analysis by combination of holographic interferometry and boundary integral method. *BSSA*, 23:196–202.
2. Baumeister, J. (1981). *Stable solution of inverse problems.* Freidr, Braunschweig.
3. DeJong, K. A. (1993). Genetic algorithms are not function optimizers. In Whitley, D., editor, *Foundations of Genetic Algorithms-2*, pages 5–18. Morgan Kauffman.
4. Eshelman, L. J. (1991). The chc adaptive search algorithm: How to have safe search when engaging in nontraditional genetic recombination. In Rawlins, G. J. E., editor, *Foundations of Genetic Algorithms-1*, pages 265–283. Morgan Kauffman.
5. Geist, A., Beguelin, A., Dongarra, J., Jiang, W., Manchek, R., and Sunderam, V. (1993). Pvm 3 users's guide and reference manual. Technical Report ORNL/TM-12187, Oak Ridge National Laboratory, Oak Ridge, Tennessee.
6. Goldberg, D. E. (1989). *Genetic Algorithms in Search, Optimization, and Machine Learning.* Addison-Wesley, Reading, MA.
7. Holland, J. (1975). *Adaptation In Natural and Artificial Systems.* The University of Michigan Press, Ann Arbour.
8. Schnur, D. S. and Zabaras, N. (1992). An inverse method for determining elastic material properties and a material interface. *International Journal of Numerical Methods in Engineering*, 33:2029–2057.
9. Tanaka, M., Nakamura, M., and Nakano, T. (1988). Defect shape identification by means of elastodynamics boundary element analysis and optimization technique. *Advances in boundary elements*, 3:183–194.
10. Weathers, J. M., Foster, W. A., Swinson, W. F., and Turner, J. L. (1985). Integration of laser-speckle and finite element techniques of stress analysis. *Exper. Mech.*, 25:60–65.

A Genetic Algorithm Approach for River Management

J. P. King, H. S. Fahmy, and M. W. Wentzel

Department of Civil Agricultural & Geological Engineering,
New Mexico State University, Dept. 3CE, Las Cruces, NM 88003

Summary. A genetic algorithm can be effectively applied to the problem of optimizing the operation of a river/reservoir system for maximum economic return. Solution of such problems requires both a detailed model of the system and a powerful optimization technique. Standard approaches often represent a trade-off between model accuracy and optimization capability. When a large system is modeled in detail, optimization techniques such as dynamic programming may prove intractable. Alternately, techniques such as linear programming may not allow accurate system modelling. A genetic algorithm approach allows the use of an accurate system model while retaining powerful search capabilities. The effectiveness of this approach is demonstrated by the results of its application to a complex hydraulic/economic problem based on the Rio Grande Project in southern New Mexico.

1. Introduction

Optimizing economic benefits of river operation is a classic and persistent problem in water resource engineering and management. So many options exist for managing a river/reservoir system that a human operator cannot consider each possibility thoroughly. Attempts to use computers and mathematical programming techniques to find optimal management strategies that maximize economic benefits from water resources have been only marginally successful because of the complexity and wide range of possible operating policies. If a truly representative computer model of a river system is used for optimization, the number of options that must be considered is too large to be handled with traditional techniques. If the river model is simplified enough to be manageable, the model loses important representative characteristics. Therefore, most rivers are managed following protocols and policies developed through experience. While these methods may currently be adequate, they often are not optimum and do not adapt quickly to changing physical and economic conditions. As water demands increase and outstrip available resources, operating at non-optimal conditions becomes exceedingly expensive, making the development of improved optimization techniques even more crucial. This is a problem on nearly all major river systems of the world.

Improving techniques for optimization of economic benefits from river system operation is a continuing need in water resources management. Searching for optimal management strategies is complicated but crucial. River systems are increasingly multipurpose, supplying water for irrigation, hydropower generation, recreational, and municipal and industrial users. In addition, water quality improvement and fish and wildlife enhancement are gaining increasing importance in management decisions. In some systems, the requirements of navigation and flood control must also be considered. Constraints on the operation of the system may be quite complicated. Typical constraints include maximum and minimum storage, maximum and minimum releases, and equipment and facility limitations for each reservoir of the system. Additional constraints include obligations created by the various users of the system. The relationship between system response and management decisions is quite complicated. In addition, both inflow to the system and users' demands vary seasonally and with changing hydrological conditions. Identifying management strategies that optimize the benefits provided to users of multi reservoir river systems, subject to operating constraints, is difficult. Developing reliable techniques to aid in this task is an ongoing research effort in the field of operations research and water resources engineering.

1.1 Previous approaches

For over 20 years, engineers have been employing optimization techniques for the management of complex water resources systems. These efforts have been documented by Wurbs (1993), Simonovic (1992), Yeh (1985), and others. In his review of state-of-the-art methods for river and reservoir management, William Yeh (1985) discusses four broad classifications of optimization techniques. The first three, linear, dynamic, and nonlinear programming, are based on mathematical programming techniques.

Linear programming has been a popular technique because of ease of problem formulation and its ability to find truly global optimum points. Unfortunately, the problem of optimizing river basin management is nonlinear because of the nonlinear relationship between reservoir storage, surface area, and elevation. Benefit functions and operating constraints also are often nonlinear. Use of linear programming for these problems requires that the problem be "linearized." This results in simplifications that reduce the value of optimization results.

Dynamic programming is capable of handling nonlinear problems. This techniques works quite well for optimizing the operation of river systems with only a few reservoirs and control options. However, as noted by Fahmy, et al (1994) and others, for complex problems this approach becomes intractable. This so called "Curse of Dimensionality" has limited dynamic programming to optimizing operation of simplified, low dimension river basin management problems.

Nonlinear programming techniques, such as gradient descent methods, are also capable of handling nonlinear problems. Use of these techniques has been limited because they are computationally intensive and have slow rates of convergence. These techniques also require the calculation of derivatives for their search procedure, limiting their use to problems that are both smooth and continuous.

Each of the mathematical programming approaches requires a very specific formulation of the river management problem in order to apply the particular technique. All three approaches are proven optimizing techniques. In practice, however, successful application has been limited to relatively simple systems; either multi-user, single reservoir or single user and multi reservoir systems. When applied to multi-user, multi reservoir systems, these approaches are forced to operate on a simplified model of the system to avoid prohibitively large computing times. These simplifications have greatly devalued the results attainable with these techniques.

The final method for optimizing river management that Yeh reviewed was the use of computer simulations to model the behavior of river systems. These simulations provide a means to predict accurately the response of the system to specified inputs, including management decisions. In general, they model reservoir systems to a much greater level of detail than mathematical programming techniques. Unlike mathematical programming techniques, they do not directly optimize operation of the system. They have been used, however, to evaluate the merits of competing management alternatives. Recently, researchers have attempted to incorporate optimization methods within simulation models. This requires a means for selecting competing management strategies that can then be evaluated with the simulation model. Because of the greater detail retained by simulations, the number of possible strategies for these models is much greater than those for more simplified mathematical programming models. In addition, the problem is not formulated in a manner that readily allows the identification of infeasible strategies. Heuristic means of determining strategies have not proven capable of adequately searching the possible strategies. Mathematical methods of selecting strategies, such as gradient decent methods and mapping of the simulation response surfaces, have been unattractive because of huge computational requirements. River managers are still looking for an optimization technique that can adequately search for optimal strategies while retaining the level of detail of computer simulations. Water resources engineers are well aware that conventional approaches represent a trade-off between model accuracy and optimization capability. T.A. Austin (1986) found that this trade-off limited the usefulness of optimization techniques. He surveyed water resource engineers from state agencies and consulting firms regarding their use of computer models in planning, design, and operation of water resources systems. These engineers felt that one of the largest obstacles limiting the use of optimization models in their organizations was their inability to represent the "real world" situations they encountered. Considering this concern, it was not surprising that the survey also found simulation models were much more widely used than optimization models. Simulation models were considered to be more useful because they more accurately model water resource systems. These models could be used to predict the response of the system to individual management policies. They do not, however, provide a means to find optimal management policies. As a result, the management of most rivers is still very dependent on protocols and policies developed through experience.

1.2 Genetic Algorithms

Since their development in the late 1960s, genetic algorithms have been proven effective in searching large, complex solution spaces. They are capable of solving nonlinear problems. Because they do not require derivative information to direct their search, they are not limited to problems that are continuous. Instead of progressing from point to point, like other techniques, genetic algorithms search from a set of problem solutions to another. This feature allows them to escape local optimum, making their search more global in nature. They have proven to be very efficient in guiding searches within large solution spaces. Genetic algorithms have been effective in a broad range of optimization problems and may prove to be valuable for river management.

When applied to simple river management problems currently solvable with other techniques, genetic algorithms may not be competitive. They require numerous evaluations of the objective function, making them uncompetitive for simple problems. However, as problems become more complex, their computational requirements do not increase as quickly as those for other techniques. Fahmy, et al (1994) found that when a genetic algorithm was applied to river management problems of increasing complexity, running time increased at a significantly slower rate than that experienced by a dynamic programming approach. The attraction of genetic algorithms for river management lies in their ability to combine optimization capabilities with realistic simulation models of complex problems.

2. Genetic Algorithm Approach to River Management

A genetic algorithm approach to optimizing river management would parallel the approach taken by an expert manager using an accurate simulation model of the system. Provided that such a model of the physical response of the system and the associated economic benefits were available, a manager could improve operation by playing the "what if" game. The manager would select an operating policy for the system and provide this to the computer model. After analyzing the results of the computer simulation, he would use his experience and intuition to modify the operating policy in hopes of improving performance. The new policy would also be provided to the computer simulation and the results analyzed. Successive application of this procedure could lead to improved operating policies. Of course, this approach has some drawbacks. It's time consuming, labor intensive, and there's no guarantee of finding a globally optimal solution.

A genetic algorithm approach to the problem would provide some powerful advantages. Like the manager, the genetic algorithm could make use of an accurate and complex simulation model of the system in order to evaluate operating policies. Also like the manager, the genetic algorithm has a sophisticated procedure to develop new policies from old ones. This procedure is designed to effectively develop new strategies with improved economic performance. However, unlike the manager's procedure, that employed by the genetic algorithm is well defined and can be performed quickly by a computer. This allows the search for new strategies to be automated and carried out for many more iterations.

Unlike the manager, who searches from one operating policy to another (essentially from point to point), the genetic algorithm searches from one set of operating policies to another. Each of these sets, called a "generation," contains many operating policies. After the economic benefit of each individual operating policy in a generation has been evaluated, the genetic algorithm uses this information to develop a new set of operating strategies. Using information from a set of strategies results in a search method much more powerful than simply modifying a single good strategy. Searching from a set of many different strategies also allows the genetic algorithm to explore the entire realm of possible strategies much more effectively than point to point methods. As a result, globally optimal operating strategies can be found even when there are many locally optimal strategies. The use of a genetic algorithm provides water resource managers with a powerful tool for improving economic benefits of river basin management.

2.1 Basic elements of GA approach

Davis and Streenstrup (1987) described the basic elements necessary to solve a specific problem with a genetic algorithm. For the problem of optimizing river management, the components of a genetic algorithm can be described in the following way:

A. A chromosomal representation of problem solutions: The use of a genetic algorithm requires that operating policies be represented in binary strings. This can be accomplished by representing each management variable by a binary number. These numbers are then concatenated to form one long binary string that represents an operating policy for the system.

B. A method for creating an initial generation of problem solutions: To begin the search, a genetic algorithm must be supplied with an initial generation of operating policies. The initial generation may be seeded with operating policies that are known to work well or it may be generated entirely at random.

C. An evaluation function to determine the fitness of each string: For a river management problem, the evaluation function is a computer model of the river system. This model is used to compare the individual operating policies in terms of their economic benefits.

D. Genetic operators that produce the next generation of strings from the present generation: Many such operators have been developed and can be implemented with commercially available software packages.

E. Genetic algorithm parameter values: These parameters include the population size and values to describe the way that genetic operators are applied. These can be easily specified within the software package being used.

Once these five components are in place, a genetic algorithm based search for optimal operating policies can be initiated. The overall form of the genetic algorithm based search is shown in Figure 2.1. The simulation model/economic benefits estimator represents a "stand alone" component, it could be usefully operated individually without the genetic algorithm. A river manager could use the model to predict the system's response and the economic benefits generated by a specific operating policy. The genetic algorithm is used to generate policies which are evaluated by the computer model. Evaluation results are returned to the genetic algorithm and used to direct the search for new policies.

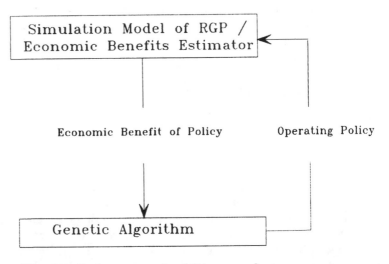

Fig. 2.1. Basic components of GA approach

2.2 Observations

Our experiences with using genetic algorithms and simulation models to improve river management have led us to make the following observations:

1. An accurate model is required to optimize a river system. As with all optimization techniques, the results of GA guided search are only as good as the model of the system used in the optimization process. The advantage of a GA based approach is that more detailed and complex models can be considered. But regardless of the capabilities of the GA, the results are only meaningful if the simulation model accurately reflects the behavior of the actual system. The ability of GA search to treat the fitness function like a "black box" makes it possible to incorporate sophisticated models of river systems. Even models developed within a computer river modeling software package can be utilized. Ideally, the simulation models that managers are already using to evaluate operating strategies for river systems should be used with a GA approach.

   Accurate simulation models are not always available, however. And developing such a model can be a difficult task. Development of the physical model may require extensive data collection and model calibration. Valuing the various uses of water in a river system can also be quite difficult. The value of any results are always limited by the accuracy of the model used during the optimization procedure. In the past, operations research techniques couldn't make use of truly accurate models anyway, making their development of less importance. The optimization capabilities of genetic algorithms, however, should justify the effort and cost to develop accurate and complex river system simulations.

2. Penalty functions should be carefully crafted to aid the search. The purpose of a penalty function is to remove the advantage of policies that violate a constraint. Technically, all policies that violate constraints are infeasible and some might argue that they should be ignored completely. However, it is seldom easy to identify which policies will be infeasible without first evaluating them with the computer model. If penalties can be made proportional to the degree of infeasibility, they can be used to guide the search toward feasible policies. We have found two types of penalty functions to be useful in doing this. For constraints related to specified system performance characteristics, the penalty can be set equal to a constant times the square of the constraint violation. For example, for a maximum reservoir elevation constraint, the penalty would be equal to a constant times the square of the reservoir water elevation in excess of the specified maximum. Penalties of this type would be calculated during each time increment of the simulation model and accumulate over the entire operating period.

   A second type of function can be used to penalize policies based on when they violate constraints. This form of penalty can be used to enforce a physical constraint whose violation results in termination of the computer model. For example, a policy that specifies continued releases from a reservoir that is already empty creates a violation of a physical constraint. This situation is usually enough to crash a computer model of the system.

In this case, the penalty can be set equal to a constant times the time remaining in the operating period when the constraint violation occurred. This function penalizes more severely policies that violate the constraint early on in the operating period.

Both of these types of penalties make use of constants that can be easily adjusted during preliminary runs. It is not necessary to find meaningful economic values for these constants. They are merely intended to prevent the genetic algorithm from favoring policies that violate constraints. A preliminary run of the genetic search can be carried out with arbitrary values for the constants. If the search is not quickly eliminating strategies that violate constraints, the value of the constants can be increased. Successive runs and increases in the value of the constants may be required to adequately enforce the constraints.

3. The choice of genetic algorithm search parameters is important, but not critical. The values of search parameters can have a great impact on the efficiency of genetic algorithm based search. However, their choice is not extremely critical as a wide range of search parameters will give relatively good search performance. We have experimented with many different values during preliminary runs. Our best results, however, were obtained with a set of parameters found by Grefenstette (1986) to have superior search performance when applied to a variety of functions. We have used these parameters for most of our work regarding river management.

4. Not all management variables are suitable for optimization with a genetic algorithm. One of the assumptions of genetic algorithm based search is that the variables being optimized are independent of one another. Additionally, small changes in solution strings (one or two bits) should not result in extremely large changes in the fitness of the strings. These conditions can sometimes be violated when attempting to optimize a river management system. For example, operation of a river system may include a one time decision at the beginning of the operating period regarding the size of the allotment to water right holders. The remaining management variables would be the reservoir releases to meet the demands of water right holders during the operating period.

An investigation by King, et al (1995) found that a genetic algorithm was not able to handle an optimization problem like this with both allotment and release variables. We believe this is due to the interdependence between the allotment and release variables. The value of a set of releases is very dependent on the allotment. A set of low releases would be unsuitable for a high allotment, and vice versa. If a good operating policy is found, the search tends to stall around that particular value of the allotment. Movement to a different allotment without a corresponding increase in every release variable would result in large constraint violation penalties. Such sweeping changes in the binary strings are improbable, slowing down the progress of the search considerably. Specifying the allotment, and then optimizing the releases for this allotment proved to be more efficient. In some situations, it may be inefficient to optimize all decision variables simultaneously.

5. Both initial populations generated at random and seeded with historical operating policies should be considered. An initial generation filled with randomly generated policies provides a thorough search of the range of possible solutions. Whether some or even most of these initial policies are infeasible is not of great concern. Like a manager faced with the task of improving operation, the genetic algorithm can gain valuable information by knowing what doesn't work, as well as what does. In general, a genetic algorithm operates to quickly reduce the number of infeasible solutions in successive generations. Beginning with an initial population generated entirely at random insures that the search is not biased by current operating procedures that may be good, but not optimal.

An initial generation can also be seeded with current operating policies. In this case, the search may neglect some regions of the operating space and no longer be truly global. However, it may also lead to a more efficient search with quicker improvement of operations. This may be adequate for the particular problem being considered.

6. Improvement of current operating policies is often adequate. One of the drawbacks of genetic algorithm based search is that the discovery of globally optimal solutions can't be guaranteed. For real world problems in river management, this is not necessarily a grave concern. These systems are already being managed with some sort of policy. If a genetic algorithm with a realistic computer model can find an improved operating policy, that represents an increase in real benefits. Even if the new policy can be improved in the future, its discovery is still of value. For many applications, finding near optimal solutions or solutions that represent improvements on current practice are sufficient.

7. It can be difficult to determine when a genetic algorithm search should be suspended. Even so, there are some steps that can be taken to gain some insight as to when further search is no longer profitable. As the search proceeds from generation to generation, the genetic algorithm can keep track of the policies with highest economic benefit found so far. Statistics regarding the progress of the search can also be stored for later reference. After a specified number of generations, the search can be suspended. By analyzing the search statistics provided by the algorithm, the manager can evaluate if further search is worthwhile. The search can be restarted from the final generation if continued improvement seems likely. If a large number of generations have passed without producing policies with improved benefit over the best policies found so far, further search may not prove useful. When search is suspended, the genetic algorithm provides the water resources manager with a set of operating policies with the highest fitness found so far. The manager can then evaluate these policies further and select one suitable for operating the system.

3. Example of GA Approach to River Management

The following case study demonstrates the application of a genetic algorithm to optimize the economic benefits of a complex river system. The problem is one of weekly operation of a multi-reservoir, multi-user river system with competing water demands. The problem is based on the Rio Grande Project (RGP) in southern New Mexico. Competing demands include hydroelectric production, agricultural, municipal and industrial, and recreational demands. The system's operation is constrained by required compliance to interstate and international treaties. These factors increase the number and complexity of operating policies. As a result, optimization of management policies for even this relatively simple system becomes very difficult for conventional methods. The major components of the system, as shown in Figure 3.1, are:

A. Elephant Butte Reservoir, with a storage capacity of about 2 million acre-feet, is operated to provide system storage and hydroelectric and recreational benefits.

B. Caballo Reservoir has a much smaller capacity, about 0.3 million acre-feet, and is operated predominantly as a regulatory reservoir with some recreational benefits. Releases from Elephant Butte can be scheduled for hydroelectric production and the water retained in Caballo until needed for irrigation.

C. Elephant Butte Irrigation District (EBID) provides water for about 94,000 acres of irrigated land from three diversion points on the Rio Grande: Percha, Leasburg, and Mesilla Dams. D. El Paso County Water Improvement District #1 (EPCWID#1) contains about 69,000 acres of water righted land. The system receives water from two diversions on the river as well as from canals originating in EBID. EPCWID#1's largest customer in terms of water use is the City of El Paso, which supplements groundwater with surface water for municipal and industrial usage. E. Mexico, the final user on the system, has water rights protected under a treaty with the United States.

The main system of the RGP is currently operated by the US Bureau of Reclamation. Yearly operation of this system requires setting the preseason allocation of water to the irrigation districts and determining the releases from the reservoirs during the year of operation. The allocation is 3.0 acre feet of water per water righted acre for a normal operating year. This value may be increased or decreased depending on the conditions from year to year. Releases from the reservoirs are scheduled to meet delivery requests from the irrigation districts. The amount of water that must be supplied to Mexico is also proportional to the preseason allocation.

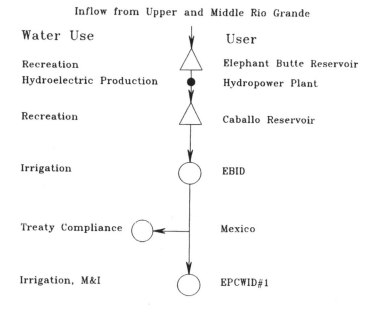

Fig. 3.1. Major components of Rio Grande project

3.1 Description of RGP physical simulation model

A computer model for the RGP system was developed to simulate the physical behavior and economic benefits of system operation. As shown in Figure 3.2, the model consisted of two main reservoirs (Elephant Butte and Caballo Reservoirs), four diversion dams (Percha, Leasburg, Mesilla, and American dams), and interconnecting reaches. Simple mass flow equations were used to model the behavior of the two main reservoirs. The historically recorded flow at San Marcial was used as the inflow for Elephant Butte Reservoir. Releases from Elephant Butte were used as the inflow for Caballo Reservoir. For each reservoir, equations were developed to estimate water surface elevation and surface area from storage volume. Evaporation losses were calculated using pan evaporation data available for both reservoirs. Our model of the RGP system included four nodes located at withdrawal points at Percha, Leasburg, Mesilla, and American Diversion dams. A Muskingum based routing procedure was used to route flows between nodes. Inflow to the system was determined from historical data for the year 1990. Withdrawals and outflow from the system were reduced or increased from the historical values in proportion to the preseason allocation.

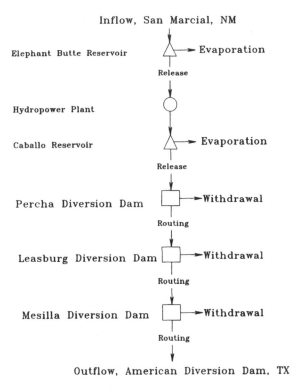

Fig. 3.2. Model layout for RGP system

3.2 Economic modeling of the RGP system

The total fitness of an operating strategy was determined by potential benefits minus any penalties for constraint violations. Economic benefits, measured in dollars, were determined by the sum of all agricultural, recreational, and hydroelectric benefits associated with a policy. Agricultural benefits for the model were based on benefits generated by five categories of crops: vegetables, pecans, field crops, forage, and cereals. Crop water production functions for southern New Mexico and historical data were used to estimate crop yields. Historical crop values were then used to determine the economic return from each crop. Recreational benefits for each time step were described as a function of reservoir volume. Hydroelectric power generation from Elephant Butte Reservoir was calculated as a function of the water surface elevation in the reservoir and the efficiency of the turbines. Turbine efficiency was a function of the release flow. An average value of $0.019 per kilowatt-hour was used to determine hydroelectric benefits.

Penalties for constraint violation were also given in units of dollars. Penalties were imposed on any operating strategies that failed to meet system operating constraints. These constraints included withdrawal requirements to satisfy agricultural users and interstate and international treaties, minimum and maximum volumes in the two storage reservoirs, and a minimum end of year storage volume in the main storage reservoir. Penalties for failing to meet withdrawal and outflow demands during the operating period and depleting the main storage reservoir were of the form of a constant times the square of the shortfall. Withdrawal and outflow demand penalties were imposed every week at each node where demand was not satisfied and accumulated throughout the model run. At the end of each model run, a one time penalty was imposed if the volume in the main storage reservoir, Elephant Butte, was below one million acre-feet. Operating strategies that resulted in reservoir volumes outside the region of accepted operation were assumed to be infeasible. Computer model runs were terminated at the point that such an infeasability occurred. The strategies were then given a penalty equal to a constant times the portion of the operating time remaining when the strategy failed. Penalty functions were adjusted during preliminary runs until the search no longer favored strategies with undesirable performance.

The behavior of our computer model was similar to that of the actual RGP system. A more detailed model of the system was not available. However, this model is sufficient to demonstrate the power of genetic algorithms to find improved operating strategies for water resource models of much greater detail than allowed by more conventional techniques.

3.3 Genetic Algorithm

For this case study, we made use of a commercially available genetic algorithm software package, GENEtic Search Implementation System (GENESIS, Version 5.0) developed by Grefenstette, et. al. (1991). The user must specify the options and parameters for genetic search and provide an evaluation function that determines the fitness of strings. In this case, the simulation model of the RGP and estimation of economic benefits were used to determine fitness. We made use of the genetic algorithm search parameters identified by Grefenstette (1986) as providing best performance for a variety of search problems.

Yearly operating policies for our model of the RGP specified the weekly releases from the two reservoirs in the system. These are the management decisions required for operation once the annual allocation to the irrigation districts is specified. A total of 104 decision variables were required for the model's year long operating period. Weekly reservoir releases were represented with a string segment of eight bits (256 possible states between 0 and 3200 cfs). This resulted in binary strings with 832 bits to represent operating policies.

3.4 Results

The results of applying genetic algorithm based search to this case study were very positive. The genetic algorithm proved very efficient at finding good results within the large search space. All runs were conducted on IBM compatible personal computers with 486 processors. Preliminary runs were conducted until penalties were adequately adjusted to eliminate infeasible policies. Final runs required approximately 14 hours and were carried out until roughly one million string evaluations had been made. This represented a very small portion of the 2832 possible operating strategies.

Several runs were made with the genetic algorithm guiding the search for optimal operating strategies. Various values for the annual allocation to the irrigation districts were specified. Initial populations were generated at random. Total benefits for the best strategies from these runs were $169.6 million, $185.0 million, and $195.8 million, for allotments of 2.5, 3.0, and 3.5 acre-feet per acre, respectively. Total benefits increased with the size of the allotment due to increased agricultural benefits. This trend would continue until allotments become too large to allow operating strategies to avoid demand and end-of-year penalties.

The best policy from the run with an allotment of 3.0 acre-feet per acre were compared to the historical operating policy, which also employed this allotment value. It is worth noting that when the allotment is fixed, agricultural benefits are fixed and the problem becomes one of optimizing the hydroelectric and recreational benefits. As mentioned previously, the best operating policy found by the genetic algorithm for this allotment had total benefits of $185 million. This represents a small but real increase in benefits over the historical operating policy, which had total benefits of $182.4 million. Both policies had agricultural benefits of $118.6 million and satisfied all system constraints. The genetic algorithm increased benefits by increasing hydroelectric and recreational benefits. Hydroelectric benefits increased from $1.92 million for the historical policy to $2.63 million for the policy found by the genetic algorithm. Recreation benefits increased from $61.9 million historically to $63.7 million for the policy found by the genetic algorithm.

The effect of seeding the initial population for a run with an allotment of 3.0 acre-feet per acre was also explored. The historical operating policy was included in the initial population. All other policies were generated at random. The search was again carried out for one million string evaluations. The best string from this run had total benefits of $186.8 million, slightly more than that obtained from the run with a randomly generated initial population. The best policies from the runs with and without the historical operating policy were very similar.

Figures 3.3 and 3.4 show the difference between historical operation and that suggested by the genetic algorithm. Figure 4 shows the behavior of Elephant Butte Reservoir for historical operation and the best strategies suggested by the genetic algorithm guided searches, with and without the historical policy in the initial population. From this figure, it can be seen that the policies found by the genetic algorithm release more water than the historical policy. The magnitude of releases for the genetic algorithm's policies also seemed to vary more from week to week. Both these factors may contribute to increased hydroelectric benefits. More release volume results in more hydroelectric generation. Making larger releases also results in greater turbine efficiency.

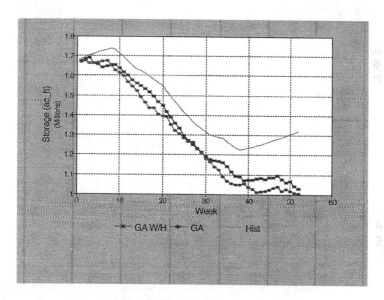

Fig. 3.3. Elephant Butte Reservoir storage volume

Figure 3.4 shows the behavior of Caballo Reservoir. The volume of Caballo Reservoir is maintained at a higher value for the strategies suggested by the genetic algorithm searches. This seems to be the driving force behind the increased recreational benefits. The historical operating policy has reduced evaporation losses because of the smaller volume stored in Caballo Reservoir. The genetic algorithm searches, however, seem to indicate that the value of water lost to evaporation is more than offset by increased recreation benefits.

This case study demonstrates how a genetic algorithm based search procedure can be applied to the economic optimization of a river management problem. The genetic algorithm proved capable of handling a complex model of the system. In this case, it was able to improve on the total benefits attainable with the historical operating policy.

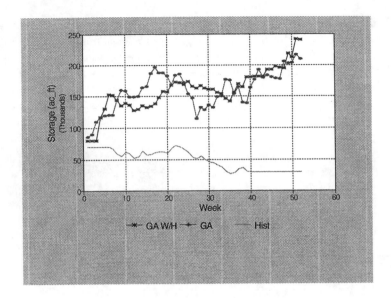

Fig. 3.4. Caballo Reservoir storage volume

4. Conclusions

We have reached the following conclusions regarding the application of genetic algorithms to the optimization of river management problems:

1. Genetic algorithms are capable of optimizing river system models of greater complexity than current techniques allow.
2. Problem formulations and operators designed to reduce the generation of infeasible solutions should be investigated.
3. Implementing parallel processing may improve the speed of genetic algorithm based search.
4. This approach used a genetic algorithm as a function optimizer. Using a genetic algorithm in a Learning Classifier System to discover operating rules may provide additional benefits for river management.

References

Austin, T. A., "Utilization of Models in Water Resources", Water Resources Bulletin, Vol. 22, No. 1, February 1986, pages 49-53.

Davis, L. and M. Streenstrup, "Genetic Algorithms and Simulated Annealing: An Overview", in Davis, L. (Editor), Genetic Algorithms and Simulated

Annealing, Morgan Kaufmann Publishers, Inc., Los Altos, CA, 1987, pages 1-11.

Fahmy, H., P. King, M. Wentzel, J. Seton, "Economic Optimization of River Management using Genetic Algorithms". Presented at the 1994 International Summer Meeting: Engineering for Sustainable Development, Paper No. 943034. ASAE, St. Joseph, MI, 1994.

Grefenstette, J., "Optimization of Control Parameters for Genetic Algorithms", IEEE Transactions of Systems, Man, and Cybernetics, Vol. SMC-16, No. 1, January/February 1986, pages 122-128.

Grefenstette, J., L. Davis, D. Cerys, "GENESIS and OOGA: Two Genetic Algorithm Systems", The Software Partnership, Melrose, MA, 1991.

King, P., F. Ward, H. Fahmy, M. Wentzel, "Economic Optimization of River Management Using Genetic Algorithms", WRRI Tech. Comp. Report No. 295, New Mexico Water Resources Research Institute, Las Cruces, NM, 1995.

Simonovic, S., "Reservoir Systems Analysis: Closing the Gap Between Theory and Practice", Journal of Water Resources Planning and Management, Vol. 118, No. 3, May/June 1992, pages 262-280.

Wurbs, R. A., "Reservoir-System Simulation and Optimization Models", Journal of Water Resources Planning and Management, Vol. 119, No. 4, July/August 1993, pages 455-472.

Yeh, W., "Reservoir Management and Operations Models: A State-of-the-Art Review", Water Resources Research, Vol. 21, No. 12, December 1985, pages 1797-1818.

Hazards in Genetic Design Methodologies

Gerald P. Roston

Cybernet System Corporation, 727 Airport Blvd., Ann Arbor, MI 48108

Summary. Genetics-based methodologies for the design of physical artifacts offer great promise but are also fraught with several hazards. This chapter serves to illuminate some of these hazards in the hope that researchers can use this as a starting point for further work in this burgeoning field. Although the hazards are illustrated with a particular genetics-based methodology, these hazards are present in most, if not all, computational design methodologies, of which genetics-based methodologies are a particularly interesting subset. This chapter briefly outlines the generic process employed by genetics-based design methodologies and the particular methodology employed for illustration. The remainder of the chapter explores some of the hazards associated with these methodologies.

1. Introduction

Although engineers have been designing artifacts[1] for millennia, there does not yet exist a systematic means to perform design tasks. To design an artifact, a rigorous evaluation of alternatives is typically not used, rather, the design is typically based on modifications of past experiences, prejudices and other constraints. This method is essentially *satisficing* — choosing an alternative that is satisfactory on all outcome attributes without computing utilities or comparing all alternatives [22].

However, to improve the design of an artifact, which in some cases have no precedent from which to work, a different method of design is required. This method is *optimizing*, and is based on choosing an alternative that optimizes certain criteria from the set of all alternatives. The discussion in this chapter is based on the optimizing approach to design because this definition subsumes the first. Specifically, optimization criteria can be defined that require satisfying a subset of the specified criteria while simultaneously optimizing a (possibly overlapping) subset of the specified criteria.

One recently popularized class of optimization methods for design is called computational design methodology (CDM). It encompasses the theory, methods and algorithms for assistive and automative design techniques via the computer. It has an active sense in that the act of design is considered (rather than, say, drafting).

The five following components are required to employ a CDM. The first two components essentially define the CDM. The last three components are problem specific and are provided by the artifact designer:

---

[1] In the context of this chapter, an artifact is a physical object that is intended to serve a specific purpose(s).

1. A method for artifact representation. To facilitate the employment of the CDM across a variety of problem domains, the representation methodology should be domain independent and amenable to computer manipulation.

2. An artifact representation manipulation scheme. The function of the manipulation scheme is to update the artifact representations in a prescribed manner in order to generate representations of (possibly) superior artifacts.

3. Artifact representations. For a given representation method, there are numerous ways to represent a specific artifact. The designer must chose a particular representation that both spans the design space and only generates syntactically correct artifacts. This representation must encompass the salient artifact characteristics necessary for solving the problem.

4. A numerable definition of the task requirements. This requirement subsumes two concepts. First, the designer must determine the essential task requirements for the problem to be solved. These requirements must be numerable, that is there must be fomulaic expressions for these requirements to permit automatic computer evaluation. Second, the designer must create a mapping between the artifacts' characteristics and the task requirements.

5. A method for evaluating the artifacts. This evaluation employs the previously described mapping to assess the performance of the artifact. For those cases with multiple mappings, this evaluation must combine multiple assessments into a single evaluation metric.

A synopsis of existing CDMs that satisfy some of these requirements are reviewed below. For a review of the state of mechanical design see [5, 6].

One of the most general techniques for representing systems and artifacts is the function logic method of value analysis [24]. With this technique, objects, and classes of object, are represented by a hierarchy of noun-verb pairs. Although this technique is capable of expressing a very wide range of concepts, computer implementation is difficult and it lacks some of the formality of other methods. A related methodology [2] presents the concept of designing from the basic, underlying principals.

Another approach used is bond-graphs [7]. In this paper, the authors state, "During the design process, a designer transforms an abstract functional description for a device into a physical description that satisfies the functional requirements." To achieve this goal, the authors propose using bond-graphs, a tool for describing generalized lumped-parameter dynamic systems, to achieve the required transformation. The bond-graph technique works well in the domain for which it is intended.

Another means of representation is the use of formal grammars. Some of the earliest work in this area was performed by Stiny [23] who developed the concept of shape grammars. Shape grammars, which are used to describe planar shapes, have been shown to be equivalent to other types of formal grammars [10]. More recently, Mullins and Rinderle [12] [16], Schmidt and Cagan [19], and Reddy and Cagan [14] have used grammars for the design of mechanical systems.

The introduction to the Mullins and Rinderle's paper [12] clearly states several reasons for using formal languages for mechanical design:

- Formal methods facilitate the characterization of the mechanical design problem more precisely.
- The absence of a fixed structure in mechanical decision-making (e.g. hierarchical decomposition) makes formal but less rigid methods attractive.
- Recent work in the representation of geometry as strings or graphs makes methods which operate on strings or graphs useful.
- A grammatical approach to design makes it possible to leverage related work in computer science.

The Mullins/Rinderle [12] paper presents two examples, one using a context-sensitive grammar and the other using a context-free grammar. Although the examples shown in this chapter use only CFGs, the Mullins/Rinderle paper shows that this is not a necessary restriction. In the concluding remarks of the Mullins/Rinderle paper, the authors state that there are several critical issues in applying grammatical formalism to mechanical designs. One of those issues is the generation of good designs. This paper highlights the difficulties in achieving this stated critical issue.

Schmidt and Cagan [19] propose an abstraction model for conceptual design. A conceptual design progresses along several levels of abstraction, from a high level, black-box description "convert electrical energy into mechanical energy" to a low level, component technology description "use an electric motor". The authors develop a grammar that can distinguish between different levels of abstraction while being compatible across the different levels. The strings formed by these grammars are manipulated by a recursive annealing process to produce results that optimize a designer-specified objective function.

Reddy and Cagan's [14] work uses shape grammars and simulated annealing to solve a variety of design problems. Reddy/Cagan's work is important because it shows how a grammatical representation of an artifact and a means of intelligent search can be used to generate optimal designs.

2. Genetic Design Methodology

The methodology used in this paper to highlight some of the hazards with genetics-based design methodologies is called Genetic Design (GD). This design methodology uses formal grammars as the method for artifact representation and genetic programming-like methods for representation manipulation.

The concept upon which GD is built is that physical systems to be designed are syntactically equivalent to computer programs. In a computer program, data are operated on by functions to produce results. The structure of a program is rigorously constrained by the definition of the grammar that describes the language. In a mechanical device,[2] physical objects are acted on by forces to produce results. With little loss of generality, the structure of certain classes of physical objects can be constrained by a grammar that describes its composition.

This analogy is not as farfetched as it might seem at first. A black-box description of a computer function and a mechanical device are essentially equivalent, see Figure 2.1. The left hand side of the figure shows a computer function operating on some data and producing a result. The right hand side shows mechanical device operating on some inputs and producing an output. As is clear from the diagram, a suitable mapping should allow representation of either of these systems in either domain, or perhaps a third, different domain. That this transformation is possible is evidenced by the fact that computer programs can be written to simulate mechanical systems and that mechanical computers can be built to run programs.

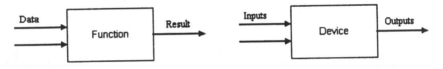

Fig. 2.1. Black-box description of a computer function and a mechanical device

The ability to map between these domains is important because all computer programs are based on formal grammars. Since sentences constructed from formal grammars can be represented as rooted trees, this suggests that an artifact representation that is constructed from a formal grammar can be manipulated in much the same manner as a computer program is manipulated using genetic programming. By employing these two proven methodologies, the first and second components for employing a CDM (Section 1.) are addressed.

The process of designing a system using GD is similar to the process of designing a program using more traditional genetics-based methodologies. GD is similar to genetic algorithms (GA) in the sense that the object directly manipulated by the computer, the artifact's genotype, is different from its "real-world" instantiation, its phenotype. GD is similar to genetic programming (GP) in the sense that the means of genotype representation and manipulation are quite similar. Figure 2.2, shows the process of design using the GD methodology.

---

[2] Other types of real-world systems also exhibit the same syntactic equivalence, e.g., electrical circuits.

```
process genetic design
begin
    initialize Population (t = 0)
    evaluate Individuals in Population(t)
    while termination Conditions not satisfied, do
    begin
        t = t + 1
        select Population(t) from Population(t − 1)
        recombine Individuals in Population(t)
        evaluate Individuals in Population(t)
    end
end
```

Fig. 2.2. Psuedo-code showing genetic design process

As compared to other design methodologies, genetic design:

- can explore a wide (grammar-defined) space of design alternatives. This is a marked advantage over those methodologies that do not incorporate grammars because with those methodologies, the designs considered are limited to only those conceived by the designer.
- is essentially domain independent. Although each design domain requires a grammar and a means of evaluation, the underlying genetic operations remain unchanged. Since GD does not require an integral evaluation routine, existing evaluators can be used. To use a non-integral evaluator, the designer is only required to write a program to translate between the GD representation and the representation required by the evaluator.
- does not require continuous functions, derivatives, etc. Traditional optimization functions are unsuitable to the task of artifact design because the evaluation functions are typically nonlinear, and possibly noncontinuous, making the application of traditional techniques very difficult.
- carries out synthesis and analysis automatically. Once the grammar and evaluation routines are developed GD will produce a number of viable alternatives for the designer to consider. Since the production takes place automatically, the designer is freed to carry out other tasks.
- performs in a "reasonable" time. Since GD is based on GA, the members of the population will successively improve with each generation. Thus, the "quality" of the answer generated by GD will be limited to the amount of computing resources available. Furthermore, since the calculation time is strongly dominated by the evaluation function (some tests indicate that the evaluation consumes in excess of 90% of the computing time), the time required for each generation can be estimated.
- uses a general grammar, a super set of other grammars, such as shape grammars, that is capable of defining higher level structures. This represents a potential improvement over other grammar based systems because they cannot handle these higher order structures, such as controllers. For-

mal grammars are not a panacea — the designer is still required to develop a proper artifact representation (see the Chapter by Simon Ronald).

However, despite these benefits, GD is subject to the same hazards that afflict most computational design methodologies. The remainder of this paper explores these hazards.

3. Pitfalls in Computational Design Methodologies

In Section 2., genetic design was introduced. This CDM is characterized by its use of formal grammars as the method for artifact representation and genetic programming-like methods for representation manipulation. This characterization addresses the first and second of the five components for the employment of CDMs were introduced in Section 1. This section focuses on the difficulties associated with the use of CDMs by exploring pit-falls in the third through fifth components for the employment of CDMs.

The key difficulty with the representation component is developing representations that are broad enough to cover the space of interest yet unbiased in their representation. The key difficulty with the task requirement component is developing evaluation schemes that yield meaningful results but also allow for population initialization. The key difficulty with the evaluation component is developing a single numeric value that fairly represents the quality of the design. Before exploring each of these difficulties, a typical design problem is introduced to provide an example of how this difficulties can be manifested.

3.1 Introduction to example problem—planar mechanism synthesis

A mechanism is a mechanical device that transfers motion/forces from a source to an output [19]. Elements of the subset of mechanisms whose members all lie on a single plane are called planar mechanisms. Mechanism design and analysis is one of the classical mechanical engineering problems, dating back to the mid-nineteenth century. Mechanisms are an important part of our daily lives — they appear in casement windows, windshield wipers, aircraft landing gear, etc.

Mechanisms are called upon to perform a variety of tasks. One of the more common tasks is called path generation. A typical path generation task is to convert a rotational input (from a motor, for example) into some specified path. The systematic method for carrying out this design is called mechanism synthesis. There are three aspects to mechanism synthesis: type (what type of mechanism is to be employed), number (how many elements are to be used) and dimension (what are the significant dimensions for the members in order to carry out the specified task). In this chapter, only linkages (mechanisms comprised solely of pin-jointed members) are considered.

Linkages are commonly referred to by the number of links in the mechanism. Figure 3.1 shows a six-bar linkage. The apparently missing sixth link is the ground link (not shown) and is notated as link 1. A typical path generation problem would be to specify the path for the coupler point to follow as the input link (for example, link 2) rotates. To be useful in a real-world application, the designer may require that the mechanism maintains closure. Closure means that the mechanism stays assembled over some specified range of motion of the input.

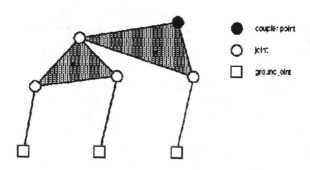

Fig. 3.1. Stephenson III six-bar mechanism

The classical method for designing mechanisms is to employ some form of Burmester theory. This theory, developed in the 1870s, shows that for a four bar mechanism, up to five *precision points* can be specified. Precision points are those points through which the mechanism must pass exactly. When five precision points are specified the order in which the mechanism passes through the points cannot be specified.

There are several algorithms for performing the kinematic analysis of planar mechanisms. The algorithm employed for this example is a general solution for planar mechanism analysis ([21], pages 181-187). The algorithm is based on the Newton-Raphson numerical method for iteratively finding the zeros of equations. To implement the method outlined, the user need only supply the vector-loop equation(s) that describes the mechanism being analyzed. The solution is found by reducing the error in the loop closure equation by using the Jacobian matrix of the mechanism. Although an iterative approach may seem to be inefficient, in practice, the equation is found to converge within five iterations — if a solution exists.

3.2 Representation

There are at least three levels of indirection that exist for representing artifacts in a manner amenable to computer manipulation:

1. Same: The representation (genotype) and the actualization (phenotype) are the same. This applies to the design of computer programs using genetic programming since the program is its own representation.
2. Direct: The representation (genotype) and the actualization (phenotype) are related by a one-to-one mapping. The applies to most common applications of genetic algorithms, genetics design, etc. In these cases, a specific part of the genome maps to a specific part of the object.
3. Indirect: The representation (genotype) is related to an intermediate representation by a one-to-one mapping which is in turn related to the actualization (phenotype). This is similar to the method used by nature — DNA does not map directly to the body but rather specifies a set of instructions for constructing the body. This method of representation allows multiple representations of the same physical instantiation.

One of the difficult problems in representing a planar mechanism is that they are connected graphs, and connected graphs are difficult to represent with context-free grammars. To generate a tree (a graphical representation of a context-free grammar) that represents the structure of a connected graph directly is not feasible because at some point in the tree, a link is required to a previously defined node. Because of this link, traditional GP techniques cannot be directly applied to this type of a representation. (However, see [14], a recent work that might offer a means to create a direct representation.)

To represent mechanisms as rooted trees (that is, representable by a context-free grammar), a different means of representation is required. This new representation can be envisioned not as the mechanism itself, but rather as a program that builds the linkage. By using this scheme, the mechanism is representable with a context-free grammar and can be manipulated using GD. The approach used will define commands, which require parameters, for building a linkage. The definition of a mechanism then becomes an instantiation of a program that calls these commands with certain supplied parameters.

The following two procedures are sufficient for designing any planar mechanism that is comprised of binary revolute joints and links. These procedures can be extended to other types of binary joints without changing the underlying theory. Non-binary joints can be modeled as a set of binary joints joined by a zero-length link:

$$\boxed{\begin{array}{l} \text{link } (n) \ [n = 2, 3, ..., N] \\ \text{joint } (L_i, L_{ij}, L_k, L_{kl}) \end{array}}$$

where

- n is the number of incident joints at the newly created link,
- N is the maximum number of incident joints,
- L_i is the first link to be joined, where i ranges from 1 to the number of links in the mechanism,

- L_{ij} is the j-th joint on link i,
- L_k is the second link to be joined, where j ranges from 1 to the number of links in the mechanism and $j \neq i$,
- L_{kl} is the l-th joint on link k.

The grammar created to embody these ideas is context free since these grammars are typically unambiguous and are easily translated into rooted tree representations. It is important to note that the grammar need not embody the ability to create the mechanism nor must it only be able to represent legal mechanisms. Rather it must embody the program that was developed to describe the mechanism. Such a grammar is shown in Table 3.1. (Productions are shown in *italics*, tokens in **bold** and terminals in *courier font*.)

Table 3.1. Planar mechanism context-free grammar

mechanism →	**L** *links* **J** *joints* **G** *ground*
links →	*link_type links* — *link_type*
link_type →	**b** *binary* — **t** *ternary* — **q** *quaternary*
binary →	*real_1*
ternary →	*real_1 real_2*
quaternary →	*real_1 real_2 real_2*
joints →	*joint_description joints* — *joint_description*
joint_description →	*(integer_4)*
ground →	*real_3*
real_1 →	*real_number*
real_2 →	*real_number, real_number*
real_3 →	*real_number, real_number, real_number*
integer_4 →	*int_number, int_number, int_number, int_number*

Although this method of representation seems to be relatively straightforward, there are two major difficulties that tend to cause the means of representation to create biases in the population distribution:

1. More complex (greater number of links) mechanisms are more difficult to construct than simpler ones. When designing a mechanism, it is usually desired that the mechanism maintains closure throughout a complete revolution (or some large fraction thereof) of the input crank. Mechanisms with a greater number of links are more likely to be evaluated poorly since it is less likely that they will maintain closure throughout the range of motion of the input link. Consider the case of a Watt II six-bar linkage, which is essentially two concatenated four-bar linkages. Let the probability for closure of a randomly generated four-bar linkage be equal to n, where $n < 1$. The probability for closure of a randomly generated Watt II mechanism cannot then exceed n^2, where $n^2 < n$.

   To demonstrate this, an experiment was run to determine how many randomly generated mechanisms need to be examined in order to find 20 that exhibit closure. The results are shown in Table 3.2.

Table 3.2. Number of mechanisms to find 20 mechanisms with closure

four-bar	six-bar	eight-bar	ten-bar
29	163	277	693

Thus, in a randomly generated population of 400 individuals, the expected distribution would be 302 four-bars, 54 six-bars, 32 eight-bars and 12 ten-bars.

2. The effects of genetic operations on more complex mechanisms are less likely to improve the mechanism than they are for simpler mechanisms. The reason for this is related to the indirect method of object representation. For example, with this scheme, identical physical representations can be represented differently. If a *link* command is swapped from one representation to the other, both representations need to be updated because the *joint* commands will likely no longer be valid. If these are complex mechanisms, it is likely that these new mechanisms will be inferior to their predecessors. This is shown in which shows the number of mechanisms by type over the course of a typical experiment.

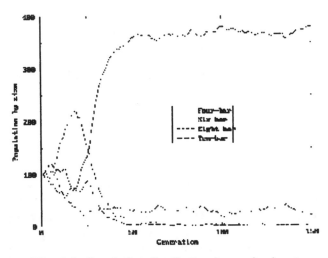

Fig. 3.2. Population distribution by mechanism type

The indirect method of artifact representation offers the greatest freedom of expression for the designer. This method is also potentially difficult to employ. Without employing any "tricks", the planar synthesis program will tend to produce four bar mechanisms as the solution to all problems, regardless of the quality of the solution, because of the greater probability that four-bar mechanisms will satisfy the closure requirements. The planar synthesis program had to be coerced to generate more complex mechanisms. This coercion, even though an attempt to remove bias, is itself a strong bias and may tend to produce results that are not those intended by the designer. Further work in the area of representations will hopefully yield results that fairly represent the design space without introducing biases.

3.3 Evaluation schemes

Three criteria must be satisfied when developing an evaluation scheme:

1. Provide sufficient distinction between closely related individuals. Without sufficient discrimination, non-identical solutions will appear to be identical.
2. Do not over emphasize local maxima. If the scheme over-awards partial solutions, it might be difficult for individuals to leave local maxima and find more global maxima.
3. Allow the initial population to be generated in a reasonable period of time. To minimize run-time for problems that employ satisficing requirements, the initial population should be comprised only of individuals that satisfice all of the task requirements.

The challenge is to satisfy both of these difficulties. Consider again the planar mechanism synthesis problem.

The strength of using a genetic algorithm method for solving planar mechanism synthesis problems lies in the ability to fashion an evaluation function to the needs of the designer. For example, Burmester theory shows that there may exist four-bar mechanisms that pass through five specified points, though the order in which the mechanism passes through these points cannot be guaranteed. By using an appropriate evaluation function, it is likely that mechanisms can be synthesized that *almost* pass through the five points in the specified order. The word *almost* is used because these synthesized solutions cannot violate Burmester theory and for certain cases (such as five precision points with a prescribed order) an exact solution probably does not exist.

But how is *almost* defined? If each of the nine independent parameters needed to define a four-bar mechanism were defined using 16 bits of precision, the design space would include in excess of 2.2×10^{43} possible mechanisms. Even if there are several billion possible mechanisms that are *almost* good enough, a random search of the design space, for the purpose of initializing the population, is impossible. One method for avoiding the search problem while still synthesizing mechanisms that are *almost* good enough, is to initially enlarge the space of *almost* good enough and to contract it as the synthesis process proceeds. To implement this scheme, a computational method for evaluating mechanisms must first be developed.

One such method is to compare the points along the path swept out by the coupler point of the candidate mechanisms with the desired points. The closer the points on the swept path come to the specified points, the higher the fitness of the mechanism. There are two potential problems with this method. First, the points along the swept path must necessarily be discrete points. This leads to the possibility of having the desired points fall between the path points, causing a good configuration to be evaluated poorly. This problem is somewhat alleviated by selecting a finer granularity for the swept points although this will increase the evaluation time. The second problem is that standard genetic methodologies are designed to maximize fitness evaluation. However, with this method, point pairs that are closer, that is are better approximations, would receive numerically smaller fitness values. Three schemes for changing these small numerical values into higher fitnesses are presented below:

- Sum the difference of the position error and a user specified constant:
$$f = \sum_{i=0}^{pts} K - e_i, \tag{1}$$
where f is the fitness, K is the constant, pts is the number of specified points, and e_i is the smallest distance between specified point i and a point on the mechanism's swept path. If $e_i > K$, the right hand side of the equation (for point i) is set equal to zero. The problem with this scheme is that if K is large enough so most points evaluate to a non-zero value, then K is also probably large enough that the function cannot clearly distinguish between swept points that are close to the specified point. (Fails to satisfy the first criteria.)
- Take the reciprocal of the sum of the position errors:
$$f = 1/\sum_{i=0}^{pts} e_i. \tag{2}$$
This scheme eliminates the arbitrary constant and does correctly reward a mechanism for minimizing the error at each of the specified points. However, in practice, this scheme failed to produce good results because once several points were satisfied, the system would fail to satisfy the remaining points because moving to alternate solutions requires a significant decrease in the evaluation function. (Fails the second criteria.)
- Take the sum of the reciprocals of the errors:
$$f = \sum_{i=0}^{pts} 1/e_i. \tag{3}$$
At first glance, this scheme seems to offer little advantage over the previous scheme, in fact, it can lead to significant problems as a single close point will result in very large values for f. However, by reformulating Equation (3) as

$$f = \sum_{i=0}^{pts} \begin{cases} 1/C, & if \ e_i \leq C \\ 1/e_i, & if \ e_i > C \end{cases} \tag{4}$$

and changing the value of C dynamically, all three criteria are satisfied. With this function, C defines the neighborhood of almost. A large initial value of C allows the creation of an initial population of valid solutions without exhaustively searching the design space. By maximizing the value for a close point, the system can explore other solutions because moving away from an existing solution cannot greatly reduce the fitness function. Finally, by continuing to reduce C, the different solution can be easily distinguished. One method for updating C that proved to be successful was to update C using $C \leftarrow C/K$, where k is an user specified constant once a user specified number of mechanisms satisfy $f = pts/C$.

Figure 3.3 shows the solution to a four-bar synthesis problem with 24 specified points using this evaluation function. In this figure, the diamonds indicate the specified points and the dashed line shows the coupler path of the generated mechanism.

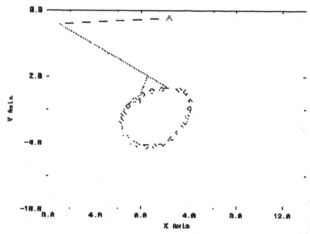

Fig. 3.3. Four-bar mechanism synthesis example

The key idea is to dynamically change the evaluation function during the course of the design evolution. Although this is an important technique, it too has a major drawback. If the change in the evaluation function is based on the quality of the design then there may be few problems, but the design quality may not improve over time. If the change in the evaluation function is based on the number of generations, surviving designs will necessarily improve if and only if they can survive the more challenging evaluation function.

This latter method tends to limit population diversity. After several generations, the members of the population typically seem to cluster in certain parts of the design space.[3] As the evaluation function becomes more restrictive, some of these clusters are eliminated from further evaluation as a whole because the new requirements cause the entire cluster to be evaluated poorly. (The phenomenon appears to be similar to mass extinctions — large portions

---

[3] This is the reason that multiple runs of any computational design methodology are strongly recommended.

of the existing genetic material is lost and succeeding generation are derived from a limited set of survivors.) Thus, the surviving population tends to be composed of genetic material from fewer clusters, which, by definition, implies limited diversity.

3.4 Evaluation

To design a complex artifact, a rigorous evaluation of alternative concepts is required. Unlike simple textbook examples, real-world artifacts must satisfy a large, and usually self-contradictory, set of requirements. Consider the following example which ranks two candidate mechanisms on two ten-point scales:

	Fitness Eq. (4) (larger is better)	Cost (smaller is better)
mechanism 1	8	7
mechanism 2	5	3

Which mechanism should be chosen? Mechanism 1 is better, but Mechanism 2 is cheaper. How is the designer to chose between these alternatives? This example illustrates the central core difficulty with the evaluation component of a computational design methodology — how to assign a representative fitness evaluation to an artifact when there are multiple objectives.

This issue is not new — in the past, the designer would employ past experiences and intelligence to manually determine a set of likely alternative designs. Modern computational design methodologies automatically produce sets of likely alternative designs and must therefore automatically evaluate these alternatives.

In both cases, it is necessary that the designer have an evaluation metric by which the alternative designs can be evaluated. For the first method, the metric is applied manually and the need for rigorous definition of the evaluation metric is diminished. For the second case, however, there is great need for a rigorous definition of the evaluation metric since the computational method cannot have the designer's understanding of the problem. Thus, the results produced by the computational method will be exactly those specified by the evaluation metric whereas by performing the analyses manually, the designer can modify the evaluation of the alternatives based on his knowledge and expertise in the design domain.

Combining multiple objectives into a single value is typically done by assigning a weighting factor to each of the objectives and summing the result:

$$O = \sum w_i o_i, \tag{5}$$

where O is the overall fitness, o_i is the fitness of objective i and w_i is the weight assigned to this objective. This fitness obtained in Equation (5) is sometimes referred to as an arithmetic weighted mean (AWM) method.

There are several problems with using an AWM, including:

- Widely different characteristics, from easily quantifiable parameters like power consumption to subjective assessments of risk, are combined into a single weighted number.
- Assumes the same degree of confidence for all characteristics, which is typically not true. For example, the power utilization for hexapod walking robot can be readily estimated, but the complexity of the planning problem may be hard to estimate *a priori*.
- Choice of scores and weighting factors can be subjective.
- The weightings are the same regardless of the values of the fitness of objective i.

What these points really say is that this method is rather subjective and limited in its ability to express the designer's true intent. Despite these difficulties, this method is frequently used. One readily accessible source is the journal *Info World*. In each issue, the editors rank certain products using a AWM. Within the context of a journal, this is a valid approach since the editors provide the scores and weightings for each product and the products are closely related. Readers interested in the product can reweigh the scores according to their own needs.

This approach does not work well for an automated design tool for two reasons [18]. First, the designer must decide a priori how to assign the weights. Second, for certain design problems, the artifacts being evaluated might be quite different from each other, thus raising the issue of the validity of the weighting. For an extraterrestrial vehicle, both wheeled and legged machines would typically be considered. Since these machines are quite different from each other, a simple weighting scheme may do nothing more than reflect the prejudices of the designer. Of course, the first objection can be overcome by having the designer interact with the system regularly to update the values, but there is no reason to believe that this would yield objectively superior results.

A second method is to use a rank-based fitness assessment for multiple objectives. This is done by creating, for each individual, a vector of the fitness measures to be considered. With this method, there is no longer a best individual, but rather a set of best individuals. This set is known as the Pareto-optimal set. First defining [4]:

Definition 1 (inferiority)
A vector $\mathbf{u} = (u_1, \ldots, u_n)$ is said to be inferior to $\mathbf{v} = (v_1, \ldots, v_n)$, iff $\mathbf{v}$ is partially less than $\mathbf{u}$, i.e., $\forall i = 1, \ldots, n : v_i \leq u_i$ and $\exists i = 1, \ldots, n : v_i < u_i$.

Definition 2 (superiority) A vector $\mathbf{u} = (u_1, \ldots, u_n)$ is said to be superior to $\mathbf{v} = (v_1, \ldots, v_n)$, iff $\mathbf{v}$ is inferior to $\mathbf{u}$.

Definition 3 (non-inferiority) Vectors $\mathbf{u} = (u_1, \ldots, u_n)$ and $\mathbf{v} = (v_1, \ldots, v_n)$ are said to be non-inferior to one another, iff $\mathbf{v}$ is neither inferior nor superior to $\mathbf{u}$.

Each point in the Pareto-optimal set is a non-inferior solution to the multi-objective problem. For a two objective problem, an example of a Pareto-optimal set is shown in Figure 3.4. In essence, by creating such a set, the designer is claiming that there is no way to combine the multiple objective functions into a single function and that all points in the Pareto-optimal set are equally acceptable. Some research has been done using of Pareto-optimal sets in conjunction with GA [5]. This method orders the solutions by rank, based on their relative superiority to other solutions. For limited-dimension objective space, this method works well. However, if there are a large number of objectives to be met, and the space of artifacts is sufficiently rich, it is not hard to envision the case in which a large percentage of the population falls on the portion of the convex-hull formed by the Pareto-optimal set. Such results would prove of limited value to a designer.

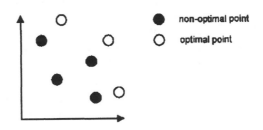

Fig. 3.4. Two-objective function showing Pareto-optimal set

Another possibility mentioned in the literature is to use a "distance" in Pareto-space as a single valued functional valuation [4]. Subjectivity is introduced with this method because the point from which the "distances" are measured must be specified by the designer. In many cases, however, the origin can be specified easily with little prejudice. To design a small part like a sewing machine bobbin for purposes of manufacturing, two "natural" axes are manufacturing cost and mean time between failures (MTBF). For the cost axis, the "optimal" point would be zero cost and for the MTBF axis, an infinite life is optimal. While this seems to be a good solution, there are two drawbacks. The first is that the different axes of the Pareto-space will typically need to be scaled because without scaling the values one axis may tend to dominate the other. Consider again the sewing machine bobbin. If the units of the cost axis measures are dollars and the unit of the MTBF axis are hours, the affect of cost would not be properly reflected in the design since these parts are inexpensive and last a long time. Second, even if the axes are scaled "properly", such that the range of all axes is equivalent, the use of a Euclidean distance metric can yield results that are far from optimal. Consider the case of a Pareto-space with two axes using a Euclidean distance metric. If the value for one of the axes is at its maximum value and the other is at zero, the fitness will be 71% of the maximum possible fitness. As more axes are added, this effect is magnified — with 3 axes, with two maximized and one at zero, the resulting fitness is 82% of the maximum. In certain instances, this might be a valid evaluation, but for other designs, minimum values along certain axes may be required, thus the use of an unmodified distance metric may not be acceptable.

A third solution is to extend the concept of the AWM to make it more general. Instead of assigning a simple proportional constant to each of the multiple objectives, a more general form can be used:

$$O = f_1(o_1, \ldots, o_n)w_1 + \ldots + f_n(o_1, \ldots, o_n)w_n. \tag{6}$$

Since GD does not require that the evaluation function be continuous, the weighting functions in Equation 11(6) can be used to express a wide range of possibilities. However, despite the generality of Equation (6), it is still rather subjective in nature.

A fourth solution that is unique to genetic methods is to apply niching techniques [8]. These techniques apply different weights to solutions that lie in different regions of the design space. Although these methods have not been widely used, they seem to offer great promise.

There do not exist techniques for generating a single valued evaluation metric that is based on multiple objectives without introducing subjectivity. To minimize the effects of this subjectivity, designers can perform the same design task several times with employing different weights to see the range of results. In all cases, the weights used must be carefully documented to permit duplication of results.

4. Summary

Computational design methodologies offer great promise in their ability to generate novel design concepts. However, these methods have some inherent pitfalls that must be addressed before they become widely accepted and employed. This chapter highlights the two greatest dangers — biased artifact representations and unrepresentative evaluations. By focusing their activities in these areas, researchers can facilitate the development of these key concepts which can then be applied to any number of specific methodologies.

References

1. M. Buckley. Multicriteria evaluation: measures, manipulation, and meaning. Environment and Planning B, 15(1):55–64, 1988.
2. Jonathan Cagan and Alice M. Agogino. Innovative design of mechanical structures from first principles. AI EDAM, 1(3):169–189, 1987.
3. C. L. Dym. Representation and problem-solving: the foundations of engineering design. Environment and Planning B, 19(1):97–105, 1992.

4. H. Eschanauer, J. Koski, and A. Osyczka, editors. Multicriteria Design Optimization. Springer-Verlag, 1990.
5. Susan Finger and John R. Dixon. A review of research in mechanical engineering design. part I: Descriptive, prescriptive and computer-based models of design processes. Research in Engineering Design, 1:51–67, 1989.
6. Susan Finger and John R. Dixon. A review of research in mechanical engineering design. part II: Representations, analysis and design for the life cycle. Research in Engineering Design, 1:121–137, 1989.
7. S. Finger and J. R. Rinderle. A transformational approach to mechanical design using a bond graph grammar. In Design Theory and Methodology, 1989.
8. Carlos M. Fonseca and Peter J. Fleming. Genetic algorithms for multiobjective optimization: Formulation, discussion and generalization. In Proceedings of the Fifth International Conference on Genetic Algorithms, pages 416–423, 1993.
9. John Gero, ed. Microcomputers in Civil Engineering: Special Issue: Evolutionary Systems in Design, May 1996.
10. J. Gips and G. Stiny. Production Systems and Grammars. Environment and Planning B. 7:399–408, 1980.
11. Ralph L. Keeney and Howard Raiffa. Decisions with Multiple Objectives: Preferences and Value Tradeoffs. Cambridge University Press, New York, 1993.
12. Scott Mullins and James R. Rinderle. Grammatical approaches to engineering design, part 1: An introduction and commentary. Research in Engineering Design, 2:121–135, 1991.
13. G. Pahl and W. Beitz. Engineering Design. Springer-Verlag, Berlin, 1984.
14. Poli, Riccardo. Discovery of Symbolic, Neuro-Symbolic and Neural Networks with Parallel Distributed Genetic Programming. University of Birmingham, School of Computer Science, CSRP-96-12, August, 1996.
15. G. Reddy and J. Cagan. An improved shape annealing method for truss topology generation. Accepted in ASME Journal of Mechanical Design, 1994.
16. James R. Rinderle. Grammatical approaches to engineering design, part II: Melding configuration and parametric design using attribute grammars. Research in Engineering Design, 2:137–146, 1991.
17. Gerald P. Roston. A Genetic Methodology for Configuration Design. Ph.D. Thesis (also available as a technical report — CMU-RI-TR-94-42), Carnegie Mellon University, December 1994.
18. T.L. Saaty. Decision Making for Leaders. Lifetime Learning Publications, Belmont, CA, 1982.
19. George N. Sandor and Arthur G. Erdman. Advanced Mechanism Design: Analysis and Synthesis, Volume 2. Prentice-Hall, Inc., Englewood Cliffs, NJ, 1984.
20. L.C. Schmidt and J. Cagan. Recursive annealing: A computational model for machine design. Accepted in Research in Engineering Design, 1994.
21. Joseph E. Shigley and John J. Uicker. Theory of Machines and Mechanisms. McGraw Hill Book Company, 1980.
22. Simon, H. A. Models of man: Social and rational. New York: Wiley, 1957.
23. G. Stiny. Introduction to Shape and Shape Grammars. Environment and Planning B. 7:343–351, 1980.
24. R. H. Sturges, K. O'Shaughnessy and R. G. Reed. A systematic approach to conceptual design based on function logic. International Journal of Concurrent Engineering, 1(2):93–106, 1992.
25. D. G. Ullman, T. G. Dietterich, and L. A. Stauffer. A model of the mechanical design process based on empirical data. AI EDAM, 2(1):33–52, 1988.
26. H. Voogd. Multicriteria evaluation: measures, manipulation, and meaning - a reply. Environment and Planning B, 15(1):65–72, 1988.

Part III

Computer Science and Engineering

The Identification and Characterization of Workload Classes

Chrisila C. Pettey[1], Patricia White[1], Larry Dowdy[2], and Darrell Burkhead[2]

[1] P.O. Box 48, Computer Science Department, Middle Tennessee State University, Murfreesboro, TN 37132
[2] Department of Computer Science, Vanderbilt University, Nashville, TN 37235

Summary. In all areas of engineering, it is important to be able to accurately forecast how a system would react to some change. For instance, for a proposed new bridge design, what would be the effect of 10 large trucks simultaneously crossing the bridge? Or, for an existing parallel computer system, what would be the expected response time if three processors failed simultaneously? To be able to answer such performance prediction questions, an engineer needs a correct characterization of the current system.

In this chapter, we apply genetic algorithms (GAs) to the problem of correctly characterizing the workload of computer systems. The GA identifies both the number of workload classes and the class centroids. The proposed new technique appears to outperform K-means clustering, an accepted workload characterization technique.

1. Motivation

The ability to predict the performance of a new design is central to all areas of engineering. In chemical engineering, it is important to understand underlying chemical processes in order to predict the efficiency of a proposed new wastewater treatment plant. In civil engineering, understanding the structural properties of steel and concrete are required prior to predicting the allowable weight limit imposed on a proposed new bridge. In electrical engineering, knowledge of the behavior of transistors is needed in order to predict the speed of a suggested new integrated circuit. In mechanical engineering, principles of statics and dynamics of particles and rigid bodies must be understood to be able to predict the performance of a new aircraft design. To be able to predict some proposed new feature accurately, a basic comprehension of the underlying components and their interactions is essential.

The same is true in computer engineering. For example, given a computer system, if the number of users were to double over the next six months, would the average response time for a particular database query remain under five seconds? For a given budget, would it be better to purchase extra memory or to upgrade the processor by adding a new floating point unit? Which files should be stored on a local, but slower, private disk, as opposed to a remote, but faster, shared disk? When a parallel program arrives at a multiprocessor system, how many processors should be allocated to the program and how many (if any) should be reserved for anticipated future arrivals? Is a first-come-first-served scheduling strategy better than a priority based scheduling strategy? The answer to such questions involves predicting the performance of various alternatives. For example, in order to choose between adding memory or upgrading the processor, a prediction "model" is constructed of each alternative. Each model is evaluated, and the resulting predictions are used to determine the best option. Thus, the art of modeling is important, and it is crucial that the model accurately captures those key underlying components of the computer system and their interactions.

Accurate models are necessary in order to obtain accurate predictions. The only completely accurate model of a computer system is the system itself. For instance, the best way to know if it is better to purchase extra memory or to purchase a floating point processor is to configure the system under each alternative, operate under each option, observe the resulting performance, and select the better alternative. However, one rarely has the luxury, capability, or time to conduct such experimentation. This is particularly true when evaluating several, quite different, "paper" design alternatives. Thus, many decisions are based on some type of model — an approximation at best. At issue is the model accuracy.

There are various types of models. With respect to computer system modeling, four common types include intuition driven models, analytic models, simulation models, and hybrid models. Intuition driven models are based on the experience, knowledge, and insight of the modeller. "I recall a similar situation where the workload was even less memory bound than this situation and in that case adding the extra memory was the right choice." Analytic models (e.g., queuing theoretic models) attempt to capture the interaction between the workload and the system components using a set of mathematical formulae (e.g., the global balance equations of an appropriate Markov state space model). By simply changing the value of some parameters (e.g., the parameters representing the memory latency time), the model can be used for prediction purposes. Simulation models seek to mimic the behavior of the system by taking an actual (or generated, or proposed) trace of workload events and hypothesizing what the system state would be at each point in time. By running the simulation long enough, appropriate confidence intervals of performance indices such as processor utilization, system throughput, and job response time can be obtained. By simulating a system with extra memory and by simulating a system with a new floating point processor, a prediction of which is the better option can be obtained. A hybrid model is a combination of other modeling techniques. For example, an analytic model may be used to model the overall interaction between the processor and a disk, while a simulation model is used to model the internal SCAN scheduling policy used by the disk. There are various tradeoffs between and within the various modeling approaches, including cost, accuracy, and flexibility. Such tradeoffs are not the focus of this chapter.

One aspect is common across all types of system models: the workload characterization. Workload characterization provides the parameters required by the model. For instance, to be able to predict accurately the performance of a computer system, it is necessary to know accurately the workload parameters of the predicted system. At a minimum, the workload is characterized by the number and type of the submitted jobs and the amount of time required by each job type at each hardware component. For instance, the workload might be characterized by eight CPU-bound jobs and five I/O-bound jobs, where, on average, each CPU-bound job requires 15 seconds of CPU time and 3 seconds of disk I/O, while each I/O-bound job requires 5 seconds of CPU time and 10 seconds of disk I/O. Each workload type is typically referred to as a workload (or customer) class. The problem with characterizing the workload is that usually the number of classes and the amount of time required at each device by each class (the class centroids) are unknown. Typically, all that is known is monitor data from each device. This monitor data typically contains only a list of the burst times spent by jobs at that device. To characterize the workload, it is necessary to extract the number of classes and the class centroids from such a collection of monitor data for each device. Accurately characterizing the workload of a computer system is the focus of this chapter.

The specific workload characterization problem addressed here can be summarized as follows. Given the composite raw measurement data from the system devices, the goal is to categorize the data into a set of customer classes (i.e., workload job types) and to determine the load that each customer class places on each of the system devices. This information can then be used to directly parameterize a system model. The system model can then be used to answer performance prediction questions. Thus, there are two subproblems: 1) determine the number of customer classes and 2) determine the load placed on the system devices by each customer class.

A traditional and commonly accepted approach for extracting the customer classes from composite measurement data is clustering [Ande73] [JD88] [Spat80]. A clustering algorithm (e.g., K-means) accepts the measurement data as input, and attempts to find the best set of data clusters. A set of K cluster centroids is guessed and a pass through the measurement data is made, assigning each data point to its nearest (e.g., with respect to, say, its Euclidean distance) centroid. The centroid means are then recalculated based on the assigned data points. The algorithm iterates until no data points shift between clusters. If cluster centroids get too close to each other, the clusters are merged, and if the intra-cluster distances get too large, the cluster splits. The resulting clusters represent the customer classes and the cluster centroids are used to determine the device loadings for each cluster.

There are two primary problems with clustering algorithms such as K-means. First, the determination of the appropriate number of clusters (i.e., the splitting and merging strategies) is quite ad hoc. Typically, the desired number of customer classes is specified by the analyst or by other external factors. Second, such clustering algorithms are typically based on Euclidean distance criteria. That is, two data points are placed in the same cluster if the distance between the two points is small. Said differently, one goal of such algorithms is to minimize the variance within each cluster. However, many system models allow customer classes whose data points are much more widely distributed. For example, queuing network models assume classes are exponentially distributed. Thus, if a queuing network model is to be constructed, K-means is inappropriate since it is not designed to extract exponentially distributed classes from the measurement data. Constructing a system model which assumes exponentially distributed classes and then using K-means to identify appropriate customer classes leads to internal modeling inconsistencies and the resulting model may not be accurate and may not be able to predict future performance effectively.

An alternative to the K-means workload characterization technique is described in this chapter. The iterative K-means clustering algorithm is replaced by a genetic algorithm (GA) search procedure. Effectively, the GA applies an "exponential sieve" to the measurement data in an attempt to extract exponentially distributed customer classes. The resulting classes are then used to parameterize a system model which assumes exponentially distributed classes. Thus, by using the GA technique, the resulting model is internally consistent and it is believed that the resulting model will be able to predict future behavior more accurately.

To summarize, accurate predictions are important in all engineering disciplines. Since direct experimentation is often not feasible, prediction models are used. These models require accurate parameters, based on previously obtained measurement data. A GA-based technique is described which is useful in extracting the required model parameters from the measurement data.

The outline for the rest of this chapter is as follows. In Section 2, the problem is more clearly specified using a motivating example. In Section 3, the GA-based technique is described. The technique is experimentally applied to a specific example and the results are evaluated in Section 4. Section 5 summarizes the quality of the proposed technique and indicates future research opportunities.

2. Specification of the Problem

Consider the following motivating example. Suppose the actual system were modeled as shown in Figure 2.1.

In this figure, the three circles represent servers (e.g., CPU, disk 1, and disk 2), while the boxes represent device queues (to hold customers while waiting for the requested device to become available). There are three customer classes, each with a single customer. Thus, the overall multiprogramming level (MPL, the number of circulating customers) is three, with each customer in its own unique customer class. The first customer (i.e., class 1) requires on average 15 time units of service from the CPU, 60 time units of service from disk 1, and

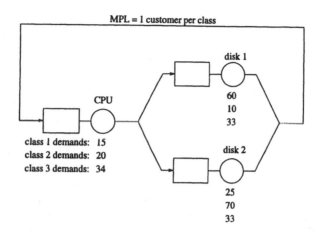

Fig. 2.1. An example system

25 time units of service from disk 2. Thus, disk 1 is the bottleneck for class 1. Class 1 represents that part of the workload that is "disk 1 bound". Similarly, the second customer places demands of 20, 10, and 70 on the three devices, respectively. Thus, disk 2 is the bottleneck for class 2. Class 2 represents that part of the workload that is "disk 2 bound". The third customer places demands of 34, 33, and 33 on the three devices, respectively. Class 3 represents that part of the workload that is "balanced", since it places nearly equal demands on each of the system devices. [Note: In this example, the class demands for each customer happen to sum to 100. This is done (1) to indicate the percentage of the demand that each customer places on each device and (2) to ensure that each customer contributes roughly equally to the total load. This restriction is for convenience only and the techniques described here are not dependent on, nor affected by, this restriction.]

Figure 2.1 represents a typical queuing network model of some system. Each customer circulates between the system devices, obtaining service from each device. As soon as a customer completes service from the system, it exits the system, only to be replaced by another job from the same customer class. This models a closed system where the number of customers in the system remains fixed. Closed models are a reasonable way to limit the number of simultaneously executing jobs due to memory or buffer constraints.

As customers circulate, they obtain service from each visited device. For instance, when the customer in class 1 visits the CPU, it requires an average of 15 time units. However, this 15 time units is the mean of a random variable with an associated distribution. That is, class 1 customers do not always require exactly 15 time units from the CPU, just as actual jobs take more or less time at each device, depending upon factors such as the job input variables, the current state of the memory cache, and the amount of allocated memory. From experimental data and for mathematical tractability reasons, the service demands are typically assumed to be exponentially distributed. This allows the model to be analytically solved (e.g., using mean value analysis (MVA), convolution, or direct Markovian analysis) to obtain performance measures such as system throughput, device utilizations, job response times, and device queue lengths [LZGS84].

Assume for the moment that the actual system behaves exactly as the Figure 2.1 model (e.g., closed, 3 customers, 3 classes, exponentially distributed demands). Thus, the actual throughput, found via a standard MVA algorithm [RL80], would be .006331 customers per unit time for class 1, .006328 customers per unit time for class 2, and .006012 customers per unit time for class 3. The total throughput would be .018672 customers per unit time.

Suppose that monitor data were collected on each system device. For instance, consider disk 1. Since the actual class throughputs are .006331, .006328, and .006012, respectively, 33.91% of the observed burst times would have a mean of 60 time units (exponentially distributed), 33.89% of the observed burst times would have a mean of 10 time units (exponentially distributed), and 32.20% of the observed burst times would have a mean of 33 time units (exponentially distributed). Since the monitor cannot distinguish between the customer classes, only the composite data is collected. Figure 2.2 illustrates this situation.

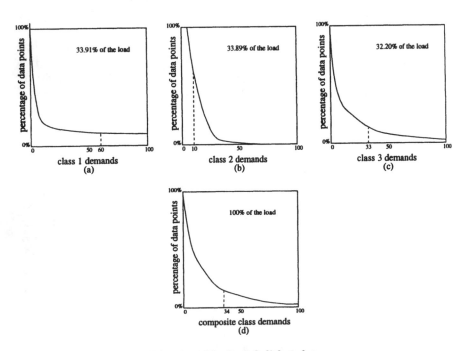

Fig. 2.2. Monitored disk 1 data

The composite data (i.e., Figure 2.2(d)) is obtained from the weighted addition of the individual class demands. The composite data has an observed mean demand of 34 time units, and although it "looks" exponentially distributed, it is not. (Actually, since it is a mixture of exponentials, the composite data has a hyper-exponential distribution.) Similar composite data is collected for the CPU and disk 2. Thus, the workload characterization problem is to take the raw composite data for each device (e.g., Figure 2.2(d) and extract both the number of customer classes and the means of each class. That is, effectively, the data in Figure 2.2(d) must be passed through a "sieve" to extract the data in Figures 2.2(a)-(c).

The traditional approach to extracting the underlying class data from the composite data is to apply a standard clustering technique such as K-means. An initial problem is that such techniques often require that the number of desired clusters (i.e., classes) be known a priori and specified as an input parameter to the clustering technique. From the monitored composite data, one cannot glean such information. However, suppose that by some oracle the correct number of classes were known a priori. For example, suppose that it were known that three classes formed the composite data shown in Figure 2.2(d). K-means will partition the data into three clusters in such a way so as to minimize the sum of the squared Euclidean distances from the cluster centroids to the data points assigned to each cluster. The cluster means (i.e., their centroids) are taken as estimates of the actual customer class means. To illustrate, by applying the K-means clustering algorithm to the composite data in Figure 2.2(d), the clusters found would have means of 183, 14, and 76. (See Figure 2.3.) These means are quite different from the actual means of 60, 10, and 33 from Figures 2.2(a), 2.2(b), and 2.2(c).

Likewise, if K-means were applied to the composite data for the CPU and disk 2, the results would be as indicated in Table 2.1. From this table, it is clear that the classes found by using K-means do not closely match the actual classes. [Note: There is another problem when using this approach. Even though three classes are found for each device, it is not clear which class from the CPU data corresponds to which class from the disk 1 and disk 2 data. This is a difficult problem. The corresponding classes shown in Table 2.1 are those found from making an exhaustive search of all possibilities and reporting the best.] By using the customer classes found using K-means (assuming that it is known that there are three classes and by exhaustively searching for the best correspondence between the customer classes), the resulting model gives a throughput estimate of .008004. The actual throughput is .018672. Thus, using the K-means, the model is in error by 57.1% with respect to total system throughput. (See Table 2.2.) Thus, even though the actual system is idealized to the point where an exact theoretical model exists, such an exact model cannot be accurately parameterized using traditional workload characterization techniques. A better technique is needed.

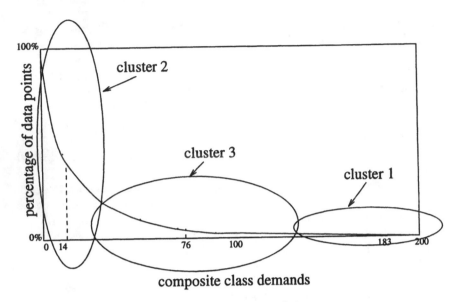

Fig. 2.3. Disk 1 clustered data

Table 2.1. Actual versus K-Means Generated Class Demands

Device	Class	Class Demands	
		Actual	K-means
CPU	1	15	10
	2	20	44
	3	34	107
disk 1	1	60	183
	2	10	14
	3	33	76
disk 2	1	25	21
	2	70	237
	3	33	94

Table 2.2. Actual versus K-Means Generated Class Throughputs

	Throughputs		% Error
	Actual	K-means	
class 1	.006331	.003417	46.0
class 2	.006328	.002373	62.5
class 3	.006012	.002214	63.2
Total	.018672	.008004	57.1

3. A GA-Based Technique for Workload Characterization

In order to determine the usefulness of GAs in extracting the number of workload classes and their mean demands from the monitored device data, a parallel diffusion model GA is constructed and implemented. Previous attempts [PWD94, PWD95] have been limited due to prespecifying the number of customer classes, normalizing the customer demands, considering only two server systems, and executing the GA on a serial workstation. Although these earlier GAs outperformed K-means clustering on a given task, the goal here is to extract both the number of workload classes and their centroids from larger, unnormalized, systems. Since the earlier GAs took several hours to execute, it was apparent that in order to solve larger systems efficiently, a parallel GA would be needed. The GA described here is a parallel diffusion model GA which can be used to characterize the workload of larger systems much more rapidly. For an in depth discussion of the diffusion model see [HEC97]. The remainder of this chapter will be devoted to describing the GA, the experiments, and the results.

In order to describe the implementation, it is helpful to review the pseudocode for a diffusion model GA.

```
Process(i):
    1. t ← 0;
    2. initialize individual i;
    3. evaluate individual i;
    4. while (t ≤ t_max) do
    5.        individual i ← select individual from neighborhood
    6.        choose parent 1 from neighborhood
    7.        choose parent 2 from neighborhood
    8.        individual i ← recombine(parent1,parent2)
    9.        individual i ← mutate(individual i)
   10.        evaluate individual i;
   11.        t ← t + 1;
          od
```

In the diffusion model, each individual of the population resides on its own processor in a massively parallel environment. The process task running on each processor is responsible for initializing its individual, evaluating its individual, and evolving its individual. Operations such as selection and recombination are performed by a process task with only the knowledge of the fitnesses of the neighboring individuals. It is possible, however, to maintain elitism by doing a global search for the best individual. Elitism is used in the described experiments.

The GA described in this chapter is implemented on a MasPar MP-1 with 2048 processors resulting in a population of 2048 individuals, one per processor. The underlying interconnection network of the MP-1 is a 2-D torroidal mesh, with each processor connected to a north, south, east, and west neighboring processor. There are internal commands that allow each processor to easily access the northwest, northeast, southwest, and southeast neighbor-

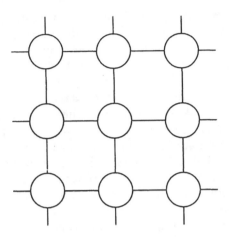

Fig. 3.1. An Individual's Neighborhood

ing processors. The neighborhood of an individual in this implementation consists of the individual and its eight neighbors (see Figure 3.1).

The GA is implemented with binary chromosomes. Each individual represents a possible workload characterization. Thus, an individual in the population consists of a guess for the number of classes followed by a list of guesses for the demands for each class for each device. For example, if an individual were to correctly characterize the assumed actual system shown in Figure 2.1, then it would appear as in Figure 3.2.

The first 3 bits in the binary representation of an individual represent the number of classes in the individual and is allowed to range from 1 to 8.

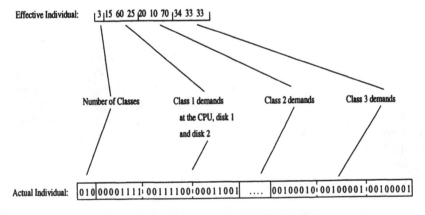

Fig. 3.2. Representation of an Individual

Since 3 bits result in binary numbers between 0 and 7, the binary value is incremented by 1 to determine the guessed number of classes. The demands, which are allowed to range from 0 to 255, are each stored in 8 bits. For simplicity, each individual is implemented as a vector of size

$$3\_bits + maximum\_number\_of\_devices \times maximum\_number\_of\_classes$$
$$\times 8\_bits\_per\_demand.$$

Since there were 3 devices in the system, an individual with the maximum of 8 classes would contain 24 demands and be 195 bits long. Notice that if the number of classes is less than 8, then 195 bits are not needed. When an individual is evaluated, any excess bits are simply ignored. It is interesting to note that, technically, this representation results in variable length individuals.

The selection strategy used is proportional selection in the neighborhood. A process task chooses exactly one individual for the next generation from the 9 individuals in its neighborhood based on the probability distribution created by the 9 individuals' fitnesses. For example, if the nine fitnesses were 1, 2, 1, 0, 2, 1, 0, 1, and 1, with 2 being a better fitness than 1, then those individuals with a value of 2 would be twice as likely to be chosen as those with a value of 1. Elitism ensures that the globally best individual will survive into the next generation.

Mating occurs in the diffusion model by selecting two parents from the neighborhood. Each process task determines if it will perform crossover with a probability of 0.6. If a crossover is to be performed, the individual is always chosen as the first parent. The second parent is chosen at random from the neighborhood. In the diffusion model, each process task can only maintain one individual. Therefore, one of the two children resulting from crossover is selected at random. In order to perform crossover, a type of "sexual" one point crossover is used. In an individual, the entry for the number of classes represents that individual's "sex". When two individuals of different sex are chosen to mate, two children result – one of each sex. The crossover point never occurs within the entry for the number of classes. Since the number of classes in an individual determines the number of bits in the individual that are "worthwhile" bits, it is important that no extraneous bits at the end of the individual be used in the crossover operation. If the individual is supposed to have 4 classes, then it cannot contribute a demand for class 5 to the crossover operation. The crossover point is always between the entry for the number of classes and the end of the shortest parent (see Figure 3.3).

Each process performs mutation on its individual with a probability of 0.001. In addition, a form of local hill-climbing allows the GA to explore additional combinations of the centroid guesses. Every tenth generation, the demands for devices 2 and 3 (i.e., the two disks) are systematically swapped and the individual's fitness re-evaluated. The changes are kept only if a better fitness is found.

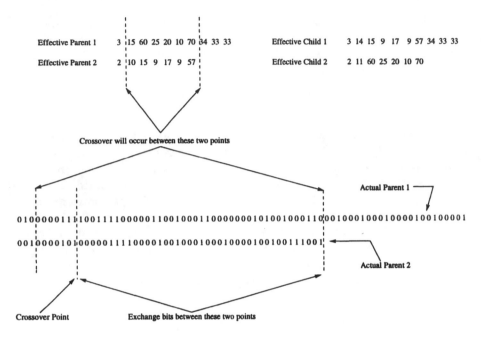

Fig. 3.3. Crossover of Two Individuals

4. Experiments and Results

The system described in Section 2 is evaluated. This system consists of three devices (i.e., a CPU and two disks) and an MPL of three customers, each in a separate customer class. By solving the system theoretically and finding the system throughput per class it is found that 33.91% of the load is generated by class 1, 33.89% is generated by class 2, and 32.20% is generated by class 3. A data set of 10,000 customer demands for each device is generated. These data points come from the three classes, weighted appropriately. That is, for the CPU, 3391 data points are generated for class 1 with a mean value of 15 that is exponentially distributed. For class 2 at the CPU, 3389 data points are generated with a mean value of 20. For class 3 at the CPU, 3220 data points are generated with a mean value of 34. Similarly, for disk 1, 3391 data points are generated for class 1 with a mean value of 60, 3389 data points are generated for class 2 with a mean value of 10, and 3220 data points are generated for class 3 with a mean value of 33. For disk 2, 3391 data points are generated for class 1 with a mean value of 25, 3389 data points are generated for class 2 with a mean value of 70, and 3220 data points are generated for class 3 with a mean value of 33. The composite set of 10,000 data points for each device is taken as the monitored measurement data. That is, if the actual system is as shown in Figure 2.1, and a system monitor is used to measure 10,000 data points at each device, the generated data set is accurate and would have been measured. The generated data set is sorted, grouped into bins, and presented as its composite probability mass function, pmf, similar to that displayed in Figure 2.2(d). This pmf for each device is used as input to the GA. From this composite "measurement" data for each device, the GA is used to try to identify the original number of classes (i.e., three) and the individual class demands.

The evaluation of an individual in the GA is nontrivial. Given an individual (i.e., the 195 bits representing the guessed number of classes and the guessed demands for each class), a system model is generated, similar to that shown in Figure 2.1. Remember, each individual may have a different number of devices and a different set of device demands. This guessed model is then theoretically solved using MVA to generate the throughputs for each class. These class throughputs are then used to derive the associated monitored data sets (i.e., the hyper-exponentially distributed probability mass functions, pmfs), one per device, that *would* have resulted had the actual system been the guessed system. These guessed pmfs are compared to the actual observed pmfs which are used as input to the GA. An individual's fitness evaluation is the simple sum of absolute differences between the resulting guessed pmfs and the observed input pmfs. Thus, a nontrivial amount of computation is involved in determining each individual's fitness. By assigning each individual to a separate MasPar processor, the overall computational effort is easily parallelized.

The GA was run 10 times, each with a different random number seed which provided a different initial population. Each experimental run executed for 100 generations. Each run required about 20 minutes to execute on the MasPar. Of the 10 runs, 1 run suggested that the workload consists of 2 customer classes, 4 runs suggested 3 classes, 2 runs suggested 4 classes, 2 runs suggested 5 classes, and 1 run suggested 6 classes. If we choose the answer that was generated most frequently, then the GA predicted 3 classes. If we average the predictions, the GA predicted 3.8 classes. Based on this, it is concluded that the GA can find the correct number of customer classes, relatively accurately.

Of the 4 GA runs that suggest 3 classes, the errors that result are all smaller than the error found using K-means. The error is measured by taking the difference between the actual overall measured throughput, and the throughput found by using the customer demands guessed by the GA and by K-means. As seen in Table 2.1, K-means generates an error in total throughput of 57.1%. The errors generated by the GA range from 1.1% to 11.7%, with an average error of 5.9%. A typical GA run (i.e., one that just happened to have the average error of 5.9%) is shown, along with the actual and K-means results, in Tables 4.1 and 4.2. These tables indicate the effectiveness of the GA approach. Figure 4.1 summarizes the actual model, the K-means model, and the GA model.

Table 4.1. Actual, K-Means, and GA Generated Class Demands

| Device | Class | Class Demands | | |
		Actual	K-means	GA
CPU	1	15	10	15
	2	20	44	18
	3	34	107	31
disk 1	1	60	183	73
	2	10	14	9
	3	33	76	32
disk 2	1	25	21	22
	2	70	237	89
	3	33	94	35

Table 4.2. Actual, K-Means, and GA Generated Class Throughputs

| | Throughputs | | | % Error | |
	Actual	K-means	GA	K-means	GA
class 1	.006331	.003417	.005944	46.0	6.1
class 2	.006328	.002373	.005565	62.5	12.1
class 3	.006012	.002214	.006066	63.2	0.9
Total	.018672	.008004	.017575	57.1	5.9

The primary use of system models is for predictive purposes. The results of two prediction studies are reported here. In the original Figure 2.1 system, the overall utilization of the CPU is 42.6%, the utilization of disk 1 is 64.1%, and the utilization of disk 2 is 79.9%. Thus, disk 2 is the overall system bottleneck. The first prediction study is to predict the performance if disk 2 were replaced by a disk which is twice as fast. The actual performance of such an upgrade can be found by solving the actual system model shown in Figure 2.1 with each of the disk 2 demands reduced by 50%. That is, each customer will require only half as much time at the new, faster disk. Similarly, to predict the performance using the K-means (GA) model, the K-means (GA) demands for disk 2 are halved and the model solved. These models are illustrated in Figure 4.2. The error using the K-means prediction model is 57.7%, while the error using the GA prediction model is 4.1%. These prediction studies indicate that the magnitude of an error in the original model propagates to the prediction model. Thus, it is important to obtain as accurate an original model as possible. Using a GA model appears to reduce the error, in comparison to a K-means model, by a factor of 10.

The second prediction study is to predict the performance if the workload were to double. That is, instead of three customers, one customer in each customer class, the models are used to predict the performance of six customers, two in each class. It is only necessary to increase the MPL to 2 in each customer class. These models are illustrated in Figure 4.3. The error using the K-means prediction model is 58.8%, while the error using the GA prediction model is 7.0%.

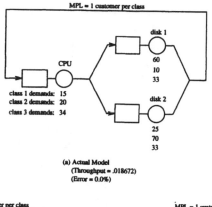

(a) Actual Model
(Throughput = .018672)
(Error = 0.0%)

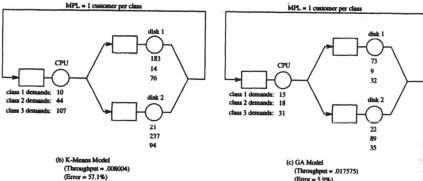

(b) K-Means Model
(Throughput = .008004)
(Error = 57.1%)

(c) GA Model
(Throughput = .017575)
(Error = 5.9%)

Fig. 4.1. Model comparison: actual, K-means, GA

5. Summary and Future Directions

Performance prediction is a recurring issue throughout all engineering disciplines. For this purpose, some type of system performance prediction model is useful. Such models seek to capture the primary interactions between the various system components. Accurate model parameterization is key.

In a computer system prediction model, characterizing the workload accurately is the most difficult subproblem. One common type of prediction model is an analytic (queuing network) model. In such models, the workload is characterized by specifying the number of distinct workload classes and the demands that each class places on each system device. An accepted method to extract the various workload classes from measurement data is the K-means clustering technique. However, the primary difficulties with K-means are the determination of the number of distinct classes and the ability to extract workload classes which are not based on strict Euclidean distances.

In this chapter a GA based technique is described. The technique is able to search for the appropriate number of workload classes and is able to apply an

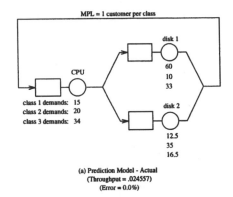

(a) Prediction Model - Actual
(Throughput = .024557)
(Error = 0.0%)

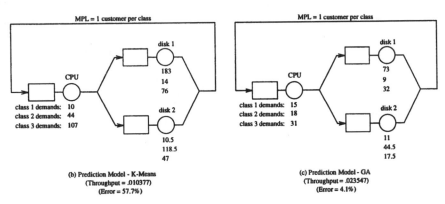

(b) Prediction Model - K-Means
(Throughput = .010377)
(Error = 57.7%)

(c) Prediction Model - GA
(Throughput = .023547)
(Error = 4.1%)

Fig. 4.2. Prediction study 1: upgrade disk 2

evaluation function which is not based on Euclidean distances. The technique is easily adapted to run in a parallel environment.

A small case study is provided which compares the accuracy of a K-means based model and a GA based model. When applied to an actual (albeit, theoretical) system, the GA model outperforms the K-means model by a factor of 10. This improvement applies to both the baseline models and to the prediction models. It appears that GAs can be used effectively to help solve workload characterization problems.

Further study is needed. It may be possible to develop better theoretical underpinnings for this work. That is, it may be possible to prove that GAs always outperform K-means techniques. Or, it may be possible to determine those systems and sets of assumptions where one technique is superior to the other. Performance bounding techniques would also be quite helpful to the analyst, to know the maximum possible error that might be caused by using a GA model. Much experimental work remains. To date, only relatively small theoretical systems have been studied. It would be interesting to apply these techniques to large production systems with hundreds of devices

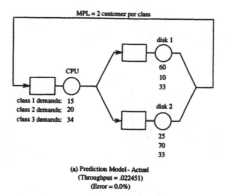

(a) Prediction Model - Actual
(Throughput = .022451)
(Error = 0.0%)

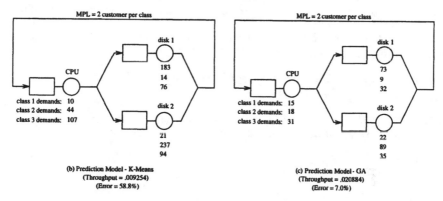

(b) Prediction Model - K-Means
(Throughput = .009254)
(Error = 58.8%)

(c) Prediction Model - GA
(Throughput = .020884)
(Error = 7.0%)

Fig. 4.3. Prediction study 2: doubling the workload

and tens of customer classes. However, preliminary studies such as the one described in this chapter indicate the wide applicability GA techniques to diverse engineering applications.

Acknowledgements

We are grateful for the earlier work [Wagn93] and helpful discussions with Professor Thomas Wagner, U.S. Military Academy, West Point, New York.

References

[Ande73] M. R. Anderberg, *Cluster Analysis for Applications*, New York: Academic Press, 1973.

[HEC97] T. Bäck, D. Fogel, Z. Michalewicz, editors, *Handbook of Evolutionary Computation*, New York: Oxford University Press, 1997.

[JD88] A. K. Jain and R. C. Dubes, *Algorithms for Clustering Data*, Englewood Cliffs, New Jersey: Prentice-Hall, 1988.

[LZGS84] E. D. Lazowska, J. Zahorjan, G. S. Graham, and K. C. Sevcik, *Quantitative System Performance. Computer System Analysis Using Queueing Network Models*, Englewood Cliffs, New Jersey: Prentice-Hall, 1984.

[PWD94] C. Pettey, T. Wagner, and L. Dowdy, "Applying genetic algorithms to extract workload classes," *CMG'94 International Conference on Management and Performance Evaluation of Computer Systems* (December 1994), 880-887.

[PWD95] C. Pettey, T. Wagner, and L. Dowdy, "Using GAs to characterize workloads," *6th International Conference on Genetic Algorithms* (July 1995).

[RL80] M. Reiser and S. S. Lavenberg, "Mean-value analysis of closed multichain queuing networks," *Journal ACM* **27** (1980), 313-322.

[Spat80] H. Spath, *Cluster Analysis Algorithms for Data Reduction*, United Kingdom: Ellis Horwood, 1980.

[Wagn93] T. Wagner, New Directions in Workload Characterization, Ph.D. Dissertation, Vanderbilt University, 1993.

Lossless and Lossy Data Compression

Wee Keong Ng, Sunghyun Choi, and Chinya Ravishankar

Department of Electrical Engineering and Computer Science, The University of Michigan, Ann Arbor, MI 48109–2122

Summary. Data compression (or source coding) is the process of creating binary representations of data which require less storage space than the original data [7, 14, 15]. Lossless compression is used where perfect reproduction is required while lossy compression is used where perfect reproduction is not possible or requires too many bits. Achieving optimal compression with respect to resource constraints is a difficult problem. For instance, in lossless compression, it has been shown to be NP-complete [13]. In this paper, we present genetic algorithms for performing lossless and lossy compressions respectively on text data and Gaussian-Markov sources.

1. Introduction

Finding the optimal way to compress data with respect to resource constraints remains one of the most challenging problems in the field of source coding. As genetic algorithms [5, 8] are becoming a widely-used and accepted method for very difficult problems [2], we present a variety of genetic algorithms for performing both lossless and lossy data compressions in this paper.

Our presentation is divided into two parts. Part I, consisting of Sections 2.–5., and part II, consisting of Sections 6.–9., present and evaluate genetic algorithms for performing lossless and lossy data compressions respectively. Within each part, we give a precise statement of the problem, describe and contrast conventional and genetic algorithmic approaches, and implement and evaluate the performance of the proposed genetic algorithms. We conclude this paper with a summary of our work and directions for future research.

2. Lossless Data Compression

Lossless compression arises frequently in the context of textual compression [1]. Due to its lossless requirement, the primary paradigm for lossless data compression is string substitution. Frequently occurring strings in the text are detected and substituted by shorter codes to achieve compression. These codes permit perfect reproduction of the original text. The strings and their corresponding codes are stored in a *dictionary*. Hence, these techniques are commonly referred to as *dictionary techniques*.

A dictionary technique usually performs two passes through the given text. During the first pass, it segments the text looking for strings to construct the dictionary. It then performs the actual coding using the dictionary during the second pass. The dictionary affects the compression efficiency because it determine not only which strings in the text to substitute but their lengths as well. Since the construction of an optimal dictionary for a given text is NP-complete [13], optimal text compression is hard.

Let A be a finite set of symbols or *alphabet*. A^+ is the set of all non-empty, finite strings $w = a_1 a_2 \cdots a_n$, $n \geq 0$, $a_i \in A$, composed of symbols from A. A text file (or simply text) T is a string from A^+; i.e., $T \in A^+$. Let $|T|$ denote the number of characters or bytes in the text. Note that $|T| \times 8$ gives the number of binary bits required to store T, assuming that each symbol is a byte of 8 bits.

A dictionary $D = \langle W, C \rangle$ consists of a finite, ordered set W of $m > 0$ strings and its corresponding ordered set C of codes, i.e., for each string w_i in W, there is a corresponding code c_i in C. The size of D is the number of strings in W; i.e., $\|D\| = m$. During coding, every occurrence of w_i in T is mapped into a code c_i for each $w_i \in W$, $c_i \in C$, $1 \leq i \leq m$.

A dictionary D must satisfy the *completeness* and *prefix* properties. D is *complete* with respect to text T if and only if T can be *parsed* entirely into strings of D. If D is complete with respect to T, we say that D *encodes* T. A dictionary D satisfies the prefix property if no string in W is a prefix of another string in W.

Example 2.1.: Let $A = \{a, b, d, e, f, h, i, l, n, o\}$ be an alphabet. The following text is a string of A^+:

> hinnfbondnfoiaedeoiaoianhoianhndileeililoianhilileeoiaedendee
> aedeedendaboaedendedehinilhinhineehinhinnfboeeiloiabonfedendo
> iandhinoiaedenhhinhineeboedeilndeeeendaedenfoiailnheendbohin

The dictionary for the text is shown in Table 2.1(a). It can be verified that the dictionary satisfies the completeness and prefix properties. The text can be unambiguously parsed using W into words that can be replaced by codes in C. ∎

The text in the previous example illustrates the complexity of dictionary construction. Unlike English text where the space character is often a delimiter of words, there may be no clear demarcations in the text. That is why choosing the right words (of the right lengths) for an arbitrary text in order to obtain an optimal dictionary is hard.

Example 2.2.: An alternative dictionary D' for the text in Example 2.1 is shown in Table 2.1(b). Let us compare the compression efficiencies of D and D'. A common measure for efficiency is the rate of coding, which is the average number of bits produced by each code per input symbol. In Example 2.1, the total number of bits in the encoded text is equal to the size (in bits) of each code (column C) multiplied by its frequency of occurrence in the text; i.e., $4 \times 4 + 5 \times 4 + \cdots + 10 \times 3 = 254$. The total number of text symbols is

Table 2.1. Two dictionaries for encoding a piece of text

Word (W)	Frequency of Occurrence	Probability	Huffman code (C)	Word (W')	Frequency of Occurrence	Probability	Huffman code (C')
a	4	0.0513	1111	f	5	0.0275	11111
nh	5	0.0641	1110	b	6	0.0330	11110
nf	5	0.0641	1101	l	9	0.0495	1110
bo	6	0.0769	1100	a	14	0.0769	1101
ee	9	0.1154	101	h	15	0.0824	1100
il	9	0.1154	100	o	16	0.0879	011
hin	10	0.1282	011	d	20	0.1099	010
nd	10	0.1282	010	i	29	0.1593	101
oia	10	0.1282	001	n	30	0.1648	100
ede	10	0.1282	000	e	38	0.2088	00

$$\text{(a) } D = \langle W, C \rangle \qquad\qquad \text{(b) } D' = \langle W', C' \rangle$$

computed to be 182 characters. This gives a coding rate of 254/182 = 1.40 bits per symbol for dictionary D. Considering that each symbol is an ASCII character of one byte (8 bits), this is an improvement of 82.5%. The coding rate for D' is 538/182 = 2.96 bits per symbol. Since it is higher than the coding rate of D, D is more desirable as the dictionary for encoding T. ∎

Given text T, there may exist more than one dictionaries capable of encoding T. The problem of finding the optimal dictionary may be defined as follows:

> Given a non-empty text T, find a dictionary D that minimizes the size of T_D, the compressed version of T using D.

To facilitate a genetic algorithmic implementation, we fix the size of the dictionary as m and define an upper limit n on the length of strings in the dictionary.

3. Conventional Algorithm

The conventional algorithm for computing the dictionary for a given text works as follows: The text is parsed at the character level and the frequency of occurrence for each distinct character in the text is computed. Suppose character a occurs n_a times in the text, then its frequency of occurrence is $p_a = n_a/N$, where N is the total number of characters in the text.

Although the text may also be parsed at the string level, it is usually not done so unless it is justified by a priori information about the text. For instance, if it is known that the text consists mostly of words of length 5, then parsing may be performed at the string level (of length 5). Unfortunately, there are no generally accepted word lengths for texts in English or other languages.

With this set of frequencies for all characters, the text is coded in such a way that the size of a code assigned to a character is inversely proportional to its frequency of occurrence; i.e., more frequently occurring characters are

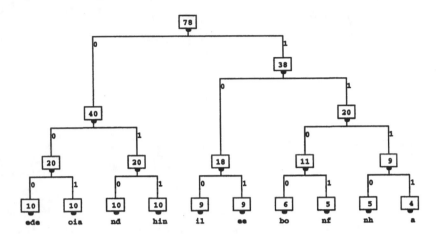

Fig. 3.1. Huffman tree for the dictionary in Example 1

assigned shorter codes. Examples of such coding techniques include Huffman coding [9] and *arithmetic coding* [16]. The Huffman tree for dictionary D in Example 2.1 is shown in Figure 3.1. We shall refer to this type of conventional algorithm by the abbreviation CO.

4. Genetic Algorithm Design

This section describes our implementation of a genetic algorithm that approximates a solution to the optimal dictionary construction problem. We shall refer to this implementation by the abbreviation GA.

4.1 Chromosome representation

Each chromosome in our genetic algorithm represents a dictionary. For example, $\langle w_1 w_2 \cdots w_m \rangle$ represents a chromosome where each gene $w_i \in A^+$ is a string whose length is bounded; i.e., $1 \leq |w_i| \leq n$. The alphabet A is the set of *printable* ASCII characters consisting of the 26 letters of the English alphabet in both lower and upper cases, the numerals and the special symbols. The population consists of a set of chromosomes whose genes are randomly initialized with characters from A; i.e., each gene has a length of one initially.

Our implementation incorporates the notion of gene *ranking*; an idiosyncratic reflection of the optimal dictionary construction problem. As each gene is a string whose occurrence in the text is to be substituted, we define its rank as its frequency of occurrence in the text. Given a set W of strings comprising the dictionary, the set of codes for encoding the text is usually constructed such that the size of each code is inversely proportional to the frequency of occurrence of the string it replaces. Therefore, higher ranking genes are more desirable. We refer to high and low ranking genes as *strong* and *weak* genes respectively.

4.2 Fitness value

The fitness of each chromosome is its *coding rate R* with respect to a text (see Example 2.2). It is defined as the ratio b/s where b is the number of bits in the text encoded via the chromosome and s is the number of symbols in the uncoded text. The coding rate for any chromosome is always strictly positive. Strong chromosomes are those that yield lower coding rates.

4.3 Reproduction

Selection:. We have three strategies for choosing parent chromosomes to undergo genetic operations: (1) parent chromosomes are randomly selected from the population; (2) parent chromosomes of relatively the same fitnesses are selected; i.e., strong chromosomes are matched with strong chromosomes and vice versa; and (3) parent chromosomes of widely differing fitnesses are selected; i.e., strong chromosomes are matched with weak chromosomes.

Exchange:. The exchange operator exchanges genes between two parent chromosomes. Two strategies are available for selecting genes for exchange: (1) genes from each parent chromosomes are randomly selected, and (2) the weak gene of one parent chromosome is replaced by the strong gene of the other parent chromosome; this is *one-way* exchange.

Fusion:. The fusion operator *fuses* two genes into a single gene. For example, strings ab and cd may be fused into $abcd$. This operator provides the means to lengthen the strings of the dictionary. Again two strategies are available for selecting genes for fusion: (1) genes from each parent chromosomes are randomly selected, and (2) the weak gene of one parent chromosome is fused with the weak gene of the other parent chromosome. Each fusion operation between two parent chromosomes results in four possible offspring chromosomes. For example, the following results when the d and n genes of two parent chromosomes are fused:

$$
\begin{array}{c}
\langle a \quad b \quad c \quad \boxed{d} \quad e \rangle \\
+ \\
\langle m \quad \boxed{n} \quad o \quad p \quad q \rangle
\end{array}
\quad \longrightarrow \quad
\begin{array}{l}
\langle a \quad b \quad c \quad dn \quad e \rangle \\
\langle a \quad b \quad c \quad nd \quad e \rangle \\
\langle m \quad dn \quad o \quad p \quad q \rangle \\
\langle m \quad nd \quad o \quad p \quad q \rangle
\end{array}
$$

Diffusion:. The diffusion operator is the reverse of fusion; it splits genes into smaller genes. This operator shortens strings of the dictionary. Two strategies are available for selecting genes for diffusion: (1) genes are randomly selected, and (2) a weak gene is diffused into two genes. Depending on the word length n, each diffusion results in $n-1$ possible pairs of offspring chromosomes. For example, with a word length of 4, the diffusion of $wxyz$ in chromosome $\langle a \ b \ c \ wxyz \ e \rangle$ results in 3 pairs of offspring chromosomes:

$$\langle a \ b \ c \ \ w \ \ e\rangle \quad \langle a \ b \ c \ wx \ e\rangle \quad \langle a \ b \ c \ wxy \ e\rangle$$
$$\langle a \ b \ c \ xyz \ e\rangle \quad \langle a \ b \ c \ yz \ e\rangle \quad \langle a \ b \ c \ \ z \ \ e\rangle$$

Operator selection:. When two parent chromosomes are selected to undergo reproduction, the choice of applying any of the three operators is made probabilistically. In our implementation, the exchange, fusion and diffusion operators are employed with probabilities p_1, p_2, and p_3 respectively where $p_1 + p_2 + p_3 = 1$.

Validity checking:. Since a chromosome represents a dictionary, each new offspring chromosome must satisfy the prefix property. After the application of a reproduction operator, we accept only those offspring chromosomes that are valid. We waive the completeness property because the restrictions on the size of the dictionary and the word lengths imply that the dictionary may not necessarily encodes a text completely. This issue is elaborated in Section 5.1

5. Performance Evaluation

In this section, we perform simulations to compare the performance of GA dictionaries (dictionaries discovered by the genetic algorithm) and CO dictionaries (dictionaries computed by the conventional algorithm) when encoding texts. Our primary interest is in determining the capability of GA and CO to derive a set of words to encode a text given a restriction on the size of the set.

5.1 Measures of interest

Several measurements are made in the simulations. The first one is the coding rate. Given a dictionary $D = \langle W, C\rangle$, let U be the set of strings in the text that remain uncoded by D. (The existence of U will be explained later on.) Let $f : A^+ \rightarrow Z$ be a function that returns the frequency of occurrence of a string in a text. Then, the coding rate (or fitness) of D is $R = (S_1 + S_2)/|T|$ where $S_1 = 8\sum_{u \in U} f(u)|u|$ is the size of the uncoded text in bits, assuming that each symbol is a byte (8 bits); and $S_2 = \sum_{w \in W} f(w)g(w)$ is the size of the coded text in bits. Function $g : A^+ \rightarrow Z$, defined as $g(w) = \phi\left(\log_2(V/f(w))\right)$ gives the number of bits in the binary representation of code $c \in C$ corresponding to $w \in W$, where $V = \sum_{u \in W} f(u)$ and ϕ is the rounding function.

The set of uncoded strings U exists due to the restriction on dictionary size, which implies that the words in the dictionary may not always completely encode the text. Thus, the dictionary is incomplete. This possibility arises in both GA and CO. This restriction implies that the sizes of both GA and CO dictionaries at the same word length will be approximately equal. Thus, we omit the dictionary size in the computation of the coding rating.

The second measure is the degree of *incompleteness* of a dictionary. It is given by the ratio $C = S_1/S_2$ where S_1 and S_2 are as defined previously. A *complete* dictionary has $C = 0$. Dictionary with a higher C value is more incomplete; there are more strings in the text that remain uncoded.

The third measure is the number of iterations it takes for GA to converge or equivalently for the code rating to stabilize. The measure is taken of multiple GAs executed at different word lengths (see below).

The fourth measure is the number of iterations needed for the completeness value of chromosomes to stabilize. Again, the measure is taken of multiple GAs executed at different word lengths.

The final measure is the word length, which is the upper limit on the size of words in the dictionary. In all the simulations, the above measurements are taken of multiple GAs and COs executed at different word lengths.

5.2 Simulation parameters

We performed a total of 9 sets of 5 simulations with both GA and CO. Each set of simulation has the following configuration of parameters: (1) the alphabet size $|A| = 80$, (2) the dictionary size $m = 20$, (3) the text size $|T| = 1000$ characters (or bytes), (4) the population size is 30, (5) the parent selection strategy is random; (6) the exchange strategy is such that weak genes of one parent chromosome are replaced by strong genes of another parent chromosome; and the (7) fusion and (8) diffusion strategies both select weak genes for fusion and diffusion respectively. (9) The probabilities of invoking exchange, fusion and diffusion are $0.75, 0.15, 0.10$ respectively.

Each set of simulation is performed at different word lengths of $n = 1, 2, \ldots, 9$. For each word length, 5 simulations are repeated and the measurements averaged.

At word length n, the CO parses the text n characters at a time while the GA generates dictionaries of words whose lengths are at most n. Once a dictionary is obtained from either algorithms, the frequencies of words in the dictionary is computed with respect to the text. These frequencies are used to compute S_1 and S_2 as described in the previous section.

5.3 Simulation results

Figures 5.1 and 5.2 show the results of the simulations. The following observations may be made:

1. From Graphs 1 and 2, the average coding rate of CO dictionaries increases with longer word lengths while that of GA dictionaries decreases up to a saturation point. Beyond the point, the average coding rate of GA dictionaries also increases with longer word lengths.
2. The average coding rate of CO dictionaries is consistently higher than that of GA dictionaries. In fact, some of the coding rates are higher than 8 bits per symbol.

3. From Graphs 3 and 4, the average incompleteness of CO dictionaries increases with longer word lengths while that of GA dictionaries decreases up to a saturation point beyond which, the average incompleteness also increases with longer word lengths. This behavior is similar to that of coding rates.

4. From Graph 3, the lowest average incompleteness value of GA dictionaries is 0.001 at word length 4. In reality, some individual GA dictionaries attain full completeness, i.e., they have 0 incompleteness value. These individual completeness values which are used to compute the average are not shown in the graphs.

5. From Graph 5, the average number of iterations it takes for the genetic algorithm to converge increases with longer word lengths. On the other hand, the average number of iterations for the completeness of GA dictionaries to stabilize decreases up to a saturation point beyond which it increases with longer word lengths (Graph 6).

6. Graphs 7 and 8 do not exhibit much differences among the two types of convergence schedules for GAs at word lengths of $1, 3, 6, 9$.

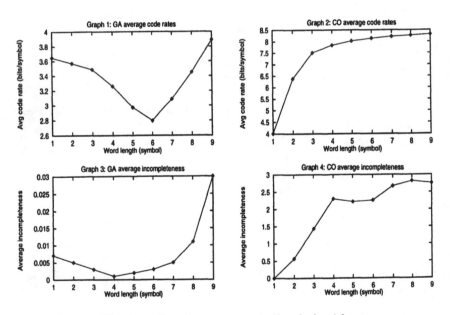

Fig. 5.1. Genetic versus conventional algorithms

5.4 Discussion

As the word length increases, CO dictionaries parse the text with larger segments. In general, when words in a dictionary are short, there is a higher

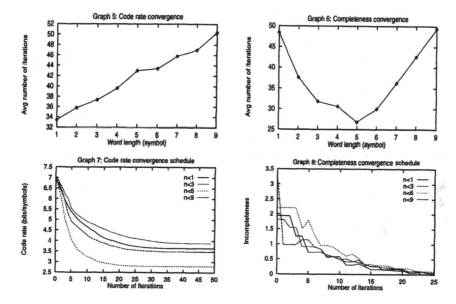

Fig. 5.2. Genetic versus conventional algorithms

probability of them occurring more frequently in the text than when the words are long. Thus, a larger amount of the text will be encoded by the dictionary, resulting in lower coding rates. The opposite is true when the words are long. Hence, we observe the average coding rates of CO dictionaries increasing with longer word lengths (Graph 2).

GA dictionaries behave differently. At longer word lengths, they have more flexibility to mix and match symbols to produce words of different lengths that best encode the text. Compared with their counterparts in CO at the same word length n, some of the words in the GA dictionaries will be shorter than n when the GA discover that having all words of length n does not yield good dictionaries. This flexibility allows the GA to yield dictionaries with lower average coding rates (Graph 1).

This same flexibility also allows both the average incompleteness values (Graph 3) and the average number of iterations for completeness stability (Graph 6) to decrease with longer word lengths because the GA is able to discover the best mix of words (of different lengths) to *cover* more of the text, resulting in less amount of uncoded text. Thus, the dictionaries are more complete and they also stabilize more quickly.

The reduction in average coding rate of GA dictionaries continues up to a saturation point due to the restriction in dictionary size (Graph 1). As the word length continue to increase, GA looses its earlier flexibility of obtaining a good mix of words. As a result, the average lengths of words in the dictionary increase. GA begins to resemble CO now because the increase in the average word lengths results in higher coding rates. Hence, we observe average coding rates increasing as the word length extends beyond the saturation point.

These effects are also observed in the average incompleteness values (Graph 3) and the average number of iterations for completeness stability (Graph 6). In particular, the increase in incompleteness is now observed by both the GA and the CO, which has increasing incompleteness all along (Graph 4).

The average number of iterations for convergence shows a steady increase with longer word lengths (Graph 5). This increase has to do with the increasingly amount of combinations to mix and match words that best encode the text. Because there are now more possible combinations, there is a higher probability of generating an invalid chromosome, resulting in more rejections. Hence, more iterations are needed.

One observation that was made above is that some of the dictionaries produced by GA are complete. This illustrates an additional bonus for GA because complete dictionaries are clearly more useful and *correct* than incomplete dictionaries.

6. Lossy Data Compression

Lossy data compression is studied in the field of source coding where Vector Quantization (VQ) is often the technique used for compression. VQ is a lossy source coding technique that maps a sequence of continuous or discrete k-dimensional vectors into a digital sequence suitable for communication over or storage in a digital channel [3, 6]. The goal is data compression: to reduce the bit rate so as to minimize communication channel capacity or digital memory requirements while maintaining the necessary fidelity of the data.

Given an input sequence vectors, VQ attempts to construct a set of representative vectors (*codewords*) called the *codebook* such that each input vector may be replaced by a vector from the codebook. As the input vector and its representative vector are not often identical, a *distortion* is introduced. The problem of VQ design is to find a codebook that has the least overall distortion for a given set of input vectors. The problem may be formally defined as follows:

> Given a set X of m k-vectors, find a codebook Y of n k-vectors such that the total distortion $\sum_{i=1}^{m} d(\bar{x}_i, Q(\bar{x}_i))$, $Q(\bar{x}_i) \in Y$, is minimized, where the mean-square distortion $d(\bar{x}_i, Q(\bar{x}_i)) = \|\bar{x}_i - Q(\bar{x}_i)\|^2$.

This problem is known to be difficult and there are no known closed-form solutions [4]. The conventional technique for designing a codebook works through a process of *iterative refinements* of an initial codebook [10]. A brief description is given in Section 7. This technique does not guarantee optimality, it sometimes yield locally optimal codebooks. In Sections 7. to 9., we present a genetic algorithm to approximate a solution to the optimal codebook design problem and evaluate its performance.

7. Generalized Lloyd Algorithm

A widely used technique for codebook design is the Generalized Lloyd Algorithm (GLA) [10] as shown in Figure 7.1. The algorithm begins with a set of input vectors and an initial codebook. For each input vector, a codeword from the codebook is chosen that yields the minimum distortion. If the sum of distortions from all input vectors does not improve beyond some threshold, the algorithm stops. Otherwise the codebook is modified as follows: Each codeword is replaced by the *centroid* of all input vectors that have previously chosen it as their output vector. This completes one iteration. The GLA iteratively refines a codebook.

Step 0   Given: An input sequence X and an initial codebook Y.
Step 1   Map each vector $\bar{x} \in X$ into a vector in Y that yields the minimum distortion.
Step 2   Compute total distortion for all $\bar{x} \in X$.
Step 3   If total distortion does not improve beyond a threshold, quit.
Step 4   Update the codebook Y as follows: Replace each codeword $\bar{y} \in Y$ by the centroid of all input vectors that are mapped into it in Step 1.
Step 5   Go to Step 1.

Fig. 7.1. Generalized Lloyd Algorithm

GLA is a *descent* algorithm; each iteration reduces the average distortion and corresponds to a *local* change in the codebook, i.e., the new codebook is not drastically different from the old codebook. Given an initial codebook, the algorithm leads to the *nearest local* codebook, which may not be optimal. As codebook design is a complex optimization problem, it has many local minima. Thus, the chance of GLA pinpointing the optimal codebook is slim.

8. Genetic Algorithm Design

We implemented three versions of genetic algorithms for the codebook design problem based on GLA. They are collectively called the Genetic Generalized Lloyd Algorithms (GGLA) and the three versions are referred to as GGLA-I, GGLA-II and GGLA-III. We first describe features that are common to all three; namely chromosome representation, genetic operators and fitness value definition:

– **Chromosome representation:** Each chromosome in our genetic algo-rithm represents a codebook. The set of genes in a chromosome corresponds to the codewords. A population consists of a set of codebooks.
– **Fitness value:** The fitness value of a chromosome is the *signal-to-noise* ratio (SNR) defined as follows: Given a set X of m k-vectors and a codebook Y of n k-vectors, the average distortion per symbol $D = 1/(km)\sum_{i=1}^{m}\|\bar{x}_i - Q(\bar{x}_i)\|^2$ where $Q(\bar{x}_i) \in Y$ is the codeword for $\bar{x}_i$. Then, SNR $= 10\log_{10}(\sigma^2/D)$ where variance $\sigma^2 = 1/(km)\sum_{i=1}^{m}\sum_{j=1}^{k}(x_{i,j} - \mu)^2$ and mean $\mu = 1/(km)\sum_{i=1}^{m}\sum_{j=1}^{k}x_{i,j}$.
– **Genetic operators:** The *mutation* operator modifies a gene (vector) by multiplying its elements by a random number between 0.8 to 1.25 with probability 0.001. The *crossover* operator exchanges segments of two chro-mosomes. The size of the segment is determined randomly. We also adopt a chromosome replacement strategy by replacing low fitting chromosomes with high fitting chromosomes. If the SNR of the chromosome with the lowest SNR is lower than the average SNR of all chromosomes less one, it will be replaced by the chromosome with the highest SNR. This process is repeated with subsequent pairs of chromosomes containing the next lowest and highest SNRs. The number of replacements is limited to no more than one third the size of the population.

The three versions of GGLA and their differences are described below. All three versions perform chromosome replacements at the beginning of each iteration.

– **GGLA-I:** All chromosomes are randomly paired up. If two chromosomes in any pair show no further SNR improvements, they are operated with the crossover operator. Otherwise, the GLA is used to refine the chromosomes.
– **GGLA-II:** Any two chromosomes that show no further improvements are operated with the crossover operator. Otherwise, they are refined by GLA. This differs from GGLA-I in that the chromosome are not paired up ini-tially.
– **GGLA-III:** All chromosomes are randomly paired up. Each pair of chro-mosome is either operated with the crossover operator or simply refined by the GLA with probabilities p and $(1 - p)$ respectively.

In all three versions, the mutation operator is always applied at the end of the iteration.

9. Performance Evaluation

We perform numerous simulations on the three versions of GGLAs using sam-ples generated by first-order Gaussian-Markov processes. The SNRs obtained are compared with those achieved by the GLA.

9.1 Gaussian-Markov source

Each first-order Gaussian-Markov sample x_i is generated using the equation: $x_i = \alpha x_{i-1} + w_i$ where α is a correlation coefficient and w_i is a sample of independent white Gaussian noise. We used three α values of 0.0, 0.5 and 0.9. The generated samples are used to initialize the population of chromosomes as follows: The sequence of samples are first partitioned into blocks of k samples to form k-vectors. Then, n of these vectors are grouped to form a chromosome. The elements of the vectors are multiplied by a random number between 0.8 and 1.25 so that the vectors obtained are different. We generated two sets of samples. In the first set, there are 4096 samples blocked into 1024 4-dimensional input vectors. In the second set, there are 8192 samples blocked into 4096 2-dimensional input vectors.

Table 9.1. GLA performance for first-order Gaussian-Markov source. k and n are the vector dimension and codebook size respectively

α	k	n	Max SNR (dB)	α	k	n	Max SNR (dB)	α	k	n	Max SNR (dB)
	4	4	2.079		4	4	3.209		4	4	6.946
		8	3.558			8	4.609			8	8.836
		16	4.967			16	6.057			16	10.645
		32	6.531			32	7.591			32	12.330
0		64	8.314	0.5		64	9.298	0.9		64	14.010
	2	16	9.845		2	16	10.442		2	16	13.625
		32	12.691			32	13.245			32	16.290
		64	15.729			64	16.196			64	19.184
		128	18.824			128	19.362			128	22.167
		256	22.050			256	22.726			256	25.507

9.2 Simulation parameters

All simulations are performed with population size of 10 and 200 iterations. For the first sample set, the codebook sizes are 4, 8, 16, 32, 64. In terms of coding rate, these correspond to rates of $R = 0.5 \sim 1.5$. ($R = \log_2 n/k$ where n is the codebook size and k is the vector dimension.) For the second sample set, the codebook sizes are 16, 21, 64, 128, 256 which yield rates of $R = 2.0 \sim 4.0$. For each combination of codebook size, vector dimension and correlation coefficient α, GLA and the three GGLAs are executed.

9.3 Simulation results

Table 9.1 shows the performance of GLA. The performances of the three GGLAs are shown in Figure 9.1. The maximum SNR figures in Table 9.1

are used as a reference for comparison with the GGLA performances in Figure 9.1. We observe that in almost all cases, the GGLAs outperformed the GLA and the improvements are larger for higher rates. In particular, GGLA-III with $p = 0.7$ made a remarkable 0.6 dB improvement for $\alpha = 0.9$, $k = 2$ and $R = 4.0$.

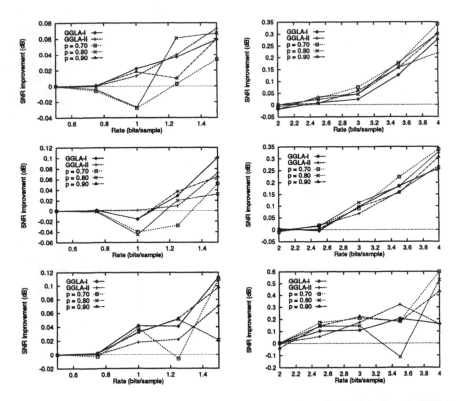

Fig. 9.1. The SNR improvements obtained with GGLA-I, GGLA-II and GGLA-III. The simulation parameters are: (a) $\alpha = 0$, $k = 4$, (b) $\alpha = 0$, $k = 2$, (c) $\alpha = 0.5$, $k = 4$, (d) $\alpha = 0.5$, $k = 2$, (e) $\alpha = 0.9$, $k = 4$ and (f) $\alpha = 0.9$, $k = 2$. The graphs on the left and right correspond to the two sets of Gaussian-Markov samples

Figure 9.2 shows another interpretation of the differences between GLA and the GGLAs. The parameters used are: $\alpha = 0.9$, $k = 4$, $n = 64$ and $R = 1.5$. Compared with the GGLAs which handle multiple codebooks at a time, the conventional algorithm operates on a single codebook only. To present a better comparison of the SNRs achieved, we made a modification to GLA so that when the SNR improvement of the codebook is less than 0.0001 (dB), GLA discards the codebook and starts with a new one. This explains the spikiness of the solid graph in Figure 9.2. In addition, the total number

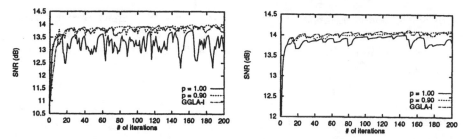

Fig. 9.2. Average and maximum SNR versus numbers of iterations. The parameters are: $\alpha = 0.9$, $k = 4$, $n = 64$ and $R = 1.5$

of initial codebooks used by GLA is approximately equal to the number of new chromosomes produced by each GGLA.

We observe that (1) the average SNR of GLA fluctuates even though GLA is a decent algorithm; (2) there are many occasions in GLA where the maximum SNR saturates and then drops abruptly only to rise again; and (3) the average and maximum SNRs achieved by GLA is less than those of GGLAs.

9.4 Discussion

A brief conclusion we can make is that GGLA outperforms GLA in most cases. Let us examine intuitively why this is so. As noted earlier in Section 7, GLA makes small local changes at each iteration. Depending on the initial codebook, it may reach a local minimum where some of its codewords are not chosen by any of the input vectors for output. The introduction of various genetic operators permits the elimination of these codewords and introduces new codewords from another codebook. In other words, these genetic operators effect a global change that is impossible with GLA.

10. Conclusions

We have developed genetic algorithms for approximating optimal solutions to lossless and lossy data compressions. Our study showed that in most cases, the genetic approach yields better solutions than the conventional algorithms.

With respect to lossless compression, the technique we have examined belongs to a class of techniques called the *semi-adaptive* compression techniques [1]. As part of our future work, we plan to extend the genetic algorithmic approach to *completely adaptive* compression techniques; these are techniques that build a dictionary incrementally at the same time as coding progresses in one single pass.

Many design alternatives remain to be explored. These include other possible reproduction operators; probabilistic selection of genes; probabilities with which operators are applied; other strategies for parent chromosome selection and removal; combination of alternatives not already investigated, etc. We would like to conduct a thorough investigation of these issues in our future work.

References

1. T. C. BELL, J. G. CLEARY, I. H. WITTEN. *Text Compression*. Prentice Hall Inc., Englewood Cliffs, New Jersey, 1990.
2. K. A. DEJONG, W. M. SPEARS. Using Genetic Algorithms to Solve NP-Complete Problems. *Proceedings of the 3rd International Conference on Genetic Algorithms*, Morgan Kaufmann, San Mateo, California, 1989.
3. A. GERSHO. On the Structure of Vector Quantizers. *IEEE Transactions on Information Theory*, Vol. 28, No. 2, pp. 157–166, March 1982.
4. A. GERSHO, R. M. GRAY. *Vector Quantization and Signal Compression*. Kluwer Academic Publishers, Boston, Massachusetts, 1992.
5. D. E. GOLDBERG. *Genetic Algorithms in Search, Optimization and Machine Learning*. Addison-Wesley, Reading, Massachusetts, 1989.
6. R. M. GRAY. Vector Quantization. *IEEE ASSP Magazine*, Vol. 1, pp. 4–29, April 1984.
7. R. M. GRAY. *Source Coding Theory*. Kluwer Academic Publishers, Boston, Massachusetts, 1990.
8. J. H. HOLLAND. *Adaptation in Natural and Artificial Systems: An Introductory Analysis with Applications to Biology, Control and Artificial Intelligence*. MIT Press, Cambridge, Massachusetts, 1992.
9. D. A. HUFFMAN. A Method for the Construction of Minimum-Redundancy Codes. *Proceedings of the I.R.E.*, Vol. 40, No. 9, pp. 1098–1101, 1952.
10. S. P. LLOYD. Least Squares Quantization in PCM. *IEEE Transactions on Information Theory*, Vol. 28, No. 2, pp. 127–135, March 1982.
11. W. K. NG, C. V. RAVISHANKAR. A Preliminary Study of Genetic Data Compression. *Proceedings of the 6th International Conference on Genetic Algorithms*, L. Eshelman (ed.), Morgan Kaufmann, San Francisco, California, 1995.
12. W. K. NG, S. CHOI, C. V. RAVISHANKAR. An Evolutionary Approach to Vector Quantizer Design. *Proceedings of the 2nd IEEE International Conference on Evolutionary Computing*, Perth, Western Australia, November 29–December 1, 1995.
13. J. A. STORER, T. G. SZYMANSKI. Data Compression via Textual Substitution. *Journal of the ACM*, Vol. 29, No. 4, pp. 928–951, 1982.
14. R. VELDHUIS. *An Introduction to Source Coding*. Prentice-Hall, New York, New York, 1993.
15. R. N. WILLIAMS. *Adaptive Data Compression*. Kluwer Academic Publishers, Boston, Massachusetts, 1991.
16. I. H. WITTEN, R. M. NEAL, J. G. CLEARY. Arithmetic Coding for Data Compression. *Communications of the ACM*, Vol. 30, No. 6, pp. 520–540, June 1987.

Database Design with Genetic Algorithms

Walter Cedeño[1] and V. Rao Vemuri[2]

[1] Research & Development, Hewlett-Packard Company, 2850 Centerville Road, Wilmington, DE 19808
[2] Department of Applied Science, University of California, Davis & Lawrence Livermore National Lab., Livermore, CA 94550

Summary. The design of distributed databases requires a configuration of the data such that queries are satisfied by accessing a minimum number of locations and the system load is equitably distributed among all locations. This problem is called the File Design Problem. This problem is NP Hard and requires the optimization over conflicting objectives. A genetic algorithm based on multi-niche crowding combines heuristics with parallel processing to provide a suitable approach to solve this problem. Performance of the algorithm is tested using multiple data sets on different system platforms. The new method holds promise in providing suitable solutions for this problem.

1. Introduction

Consider a distributed database system with multiple nodes, as shown in Figure 1.1, that contains all the data for a company-wide database comprising millions of records. In order to use the resources equitably and efficiently, the data in the database must be organized so that queries for database records are balanced among all nodes. That is, all nodes must handle about the same number of queries on average. Additionally, the information in the database must be organized so that a query can be satisfied by accessing only a small number of nodes. Minimizing the number of nodes accessed on a single query reduces greatly the communication overhead on the network. This requirement asserts that records with the same attributes should be placed together, if possible, at one location. These two criteria must be balanced to obtain an optimal database configuration. This problem is known as the File Design Problem (FDP).

The File Design Problem is known to be NP-hard. The goal is to find an assignment of database records to files that minimizes the average number of files examined over all single attribute queries. Techniques using Artificial Neural Networks [17] have been applied to this problem in the past. In this work we describe a solution to the File Design Problem using a Genetic Algorithm (GA)[16]. In particular we describe the application of the Multi-Niche Crowding Genetic Algorithm (MNC GA) [1] to this problem. Our implementation of the MNC GA is written in SISAL (Streams and Iterations in a Single Assignment Language) [18], a functional language that takes advantage of parallel architectures. Using the portability and architecture independence inherent in SISAL a parallel model of the MNC GA is

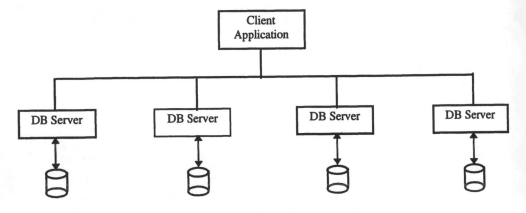

Fig. 1.1. Example of a distributed database system

defined that provides increased performance without losing the convergence properties of the algorithm.

Additionally we introduce the use of heuristics in the crossover and mutation operators used during the mating step. These operators prove to be essential in finding optimal solutions to the FDP. The use of heuristics and the ability of the MNC GA to search for multiple conflicting database configurations provides us a promising hybrid approach for solving complex combinatorial problems. Results with various test cases are shown. Performance of the algorithm is shown for different computer platforms.

This paper is organized as follows. Section 2 presents an overview of the MNC GA model used to solve this problem. Section 3 describes the File Design Problem in detail and presents examples. Section 4 describes the heuristic based genetic operators used for this problem. Section 5 defines the experimental setup used to test the performance of the MNC GA. Section 6 describes the results and the performance of the approach. Finally, Section 7 contains some comments and discussion about the applicability of the method.

2. The Multi-Niche Crowding GA

DNA, the building block of every living creature provides organisms a way to evolve and adapt to changing environments. Only organisms well adapted to their environment can survive from one generation to the next, transferring on the traits, that made them successful, to their offspring. Competition for resources and the ever changing environment drives some species to extinction and at the same time others evolve to maintain the delicate balance in nature.

The ability of organisms to evolve and adapt to their environment by means of natural selection has provided mother nature with a diverse set of species. This foundation, which is part of modern evolutionary thinking, was laid by Charles Darwin after the publication of his work "On the Origin of Species by Means of Natural Selection". Only organisms well adapted to their environment can survive from one generation to the next, transferring on the traits that made them successful to their offspring. Competition for resources between organisms and the ever changing environment drives some species to extinction and at the same time others evolve to maintain the delicate balance in nature. It is through this interaction between nature and organisms, that species containing favorable traits for a given environment emerge. In this work we apply the same principles present in nature to create a genetic algorithm that evolves a population of mathematical solutions containing different categories of solutions adapted to niches in a multimodal environment.

In this Section we describe the MNC GA, a computational metaphor to the survival of species in ecological niches in the face of competition. The MNC GA maintains stable subpopulations of solutions in multiple niches in multimodal landscapes. The algorithm introduces the concept of *crowding selection* to promote mating among members with similar traits while allowing many members of the population to participate in mating. The algorithm uses *worst among most similar replacement* (WAMS) policy to promote competition among members with similar traits while allowing competition among members of different niches as well.

The benefits of an approach that can locate multiple optima and maintain them throughout the search are many. Consider, for example, a dynamic environment where the optima are constantly changing, a technique that can locate and maintain multiple optima can inform the user when the current configuration is no longer the best based on the parameters in the environment. More details can be found in Cedeno [1], Cobb & Grefenstette [4], Dasgupta & McGregor [6], Goldberg & Smith [12], and Ng & Wong [21]. In other cases abnormal situations may cause changes in the current configuration; having viable alternatives at hand can allow for a more smoother transition to the new configuration. An approach that can use a set of solutions to locate multiple optima is more practical for these types of environments. Additionally, there exist many problems where the location of the best K optima are needed in order to compare different answers and point out further experimentation. The benefits of the MNC GA have been already shown in applications to problems in DNA mapping [2] and aquifer management [3]

Figure 2.1 shows an overview of the MNC GA. Initially, all the individuals in the population (size n) are created at random and evaluated in parallel. Once the initial population is created, the operations of selection, mating and mutation, and replacement are applied for a given number of generations. In each generation all individuals in the population are selected for mating and their mates are chosen in parallel using crowding selection. Then in parallel, all n pairs participate in mating producing $2n$ offspring. The $2n$

```
Generate initial population of size n at random.
Evaluate initial population.
For gen = 1 to MAX_GENERATIONS
   Use crowding selection to find mate for all individuals
   Mate and mutate all pairs
   Insert offspring in population using WAMS replacement
```

Fig. 2.1. Overview of parallel application of the Multi-Niche Crowding GA

offspring undergo mutation and those different than their parent are allowed
to participate in replacement. The offspring left are then inserted, one at a
time, into the population using WAMS replacement. These steps are repeated
for the specified number of generations.

In the MNC GA both the selection and replacement steps are modified
with some type of crowding [8]. The idea is to eliminate the selection pressure
caused by fitness proportionate reproduction (FPR) and allow the population
to maintain diversity throughout the search. This objective is achieved in
part by encouraging mating and replacement within members of the same
niche while allowing some competition for the population slots among the
niches. The result is an algorithm that (a) maintains stable subpopulations
within different niches, (b) maintains diversity throughout the search, and (c)
converges to different local optima. No prior knowledge of the search space
is needed and no restrictions are imposed during selection and replacement
thus allowing exploration of other areas of the search space while converging
to the best solutions in the different niches.

2.1 Crowding selection

In MNC, the FPR selection is replaced by what we call *crowding selection*. In
the parallel application of crowding selection used in this work each individual
in the population gets a chance for mating in every generation. Application
of this selection rule is done in two steps. First, each individual I_i from
the population is selected as a parent for mating. Second, its mate I_j is
selected, not from the entire population, but from a group of individuals of
size C_s (crowding selection group size), picked uniformly at random (with
replacement) from the population. The mate I_j thus chosen must be the
one who is the most "similar" to I_i. The similarity metric used here is not
a genotypic metric such as the Hamming distance, but a suitably defined
phenotypic distance metric.

Crowding selection promotes mating among members having similar traits
and allows all the members of the population to participate in mating. This
allows members of the same niche to participate in mating more often and preserve those traits that define their species. At the same time mating between
different species may occur giving rise to new species. Unlike mating restric-

tion [9] that only allows individuals from the same niche to mate, crowding selection allows some amount of exploration to occur while at the same time looking for the best individual in each niche.

2.2 Worst among most similar replacement

During the replacement step, MNC uses a replacement policy called *worst among most similar* (WAMS). The goal of this step is to pick an individual from the population for replacement by an offspring. Implementation of this policy follows these steps. First, C_f "crowding factor groups" are created by picking uniformly at random (with replacement) s (crowding group size) individuals per group from the population. Second, one individual from each group that is most similar to the offspring is identified. This gives C_f individuals that are candidates for replacement by virtue of their similarity to the offspring. The offspring will replace one of them. From this group of most similar candidates, we pick the one with the lowest fitness to die and be replaced by the offspring. Figure 2.2 shows a pictorial view of this replacement policy.

After the offspring becomes part of the population it competes for survival with other individuals when the next offspring is inserted in the population. In WAMS replacement offspring are likely to replace low fitness individuals from the same niche. It can also happen that it replaces a high fitness individual from the same niche or an individual from another niche. This allows a more diverse population to exist throughout the search. At the same time it promotes competition between members of the same niche and between members belonging to different niches. A similar technique was used by Goldberg [13] in classifier systems but he replaced the most similar individual out of a group of low fitness candidates.

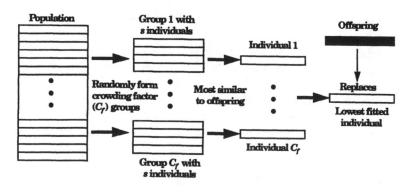

Fig. 2.2. Worst among most similar (WAMS) replacement policy

Worst among most similar replacement promotes competition among members with similar traits belonging to the same niche while allowing competition among members of different niches as well. This replacement technique accomplishes two things. First, by promoting competition among members of the same species in a niche it applies the *survival of the fittest* rule that is so prevalent in nature. Only those that are fit to their environment can survive for many generations, thus allowing the species to evolve to their best potential within their niche. Second, by allowing competition between different species as well, those species that are a better fit for their environment tend to occupy more slots in the overall population.

Both the selection and replacement steps in the MNC are primarily based on a similarity metric. Fitness is also considered during replacement to promote competition among members of the same niche. Competition among members of different niches occurs as well.

3. The File Design Problem

The File Design Problem is an NP-hard problem; that is, the number of possible solutions increases exponentially as the problem size increases linearly. It arises in the context of database design for a distributed system.

The problem is defined as follows. Given a set of N records, and each is characterized by a single attribute A that takes h different values $\{a_1, a_2, ..., a_h\}$. There are n_i records corresponding to attribute a_i, i.e., $n_1 + n_2 + ... + n_h = N$. The assumption is made that queries for records of any given attribute are equally likely. The query distribution is used only during the calculation of the fitness function, as we will show in Section 4.4. Other, more practical, distributions can be applied easily by modifying the fitness function accordingly, without affecting the behavior of the MNC GA. We also have K files of size b such that $K * b = N$. Different file sizes can also be accomodated to represent realistic configurations. The approach presented here is not limited by the simplifications made to the problem. The constants K, b, n_i, N, and h are all positive integers.

The problem is to find an assignment of the N records to the K files such that the average number of files (ANF) accessed over all possible single-attribute queries is minimized. In other words, an assignment of the records to the files must be found such that (on average) queries for the records with the same attribute can be satisfied by reading from as few files as possible.

For example, consider a database with $N = 12$ employee records characterized by their last name (here the last name refers to attribute A). Moreover, assume that all records have a last name in the set $A = \{Blattner, Cede\tilde{n}o, Feo, Vemuri\}$ with $n = \{2, 7, 1, 2\}$ records respectively. Here we have a total of $h = 4$ possible last names (attribute values). These records will be placed in a database consisting of $K = 2$ files of size $b = 6$. The problem now is to save the employee records in the database such that queries for records

File 1	File 2	fex(B)	fex(C)	fex(F)	fex(V)	ANF
C C C C C C	C B B V V F	1	2	1	1	1.25
C C C B V V	C C C C B F	2	2	1	1	1.50
C C C C B V	C C C B V F	2	2	1	2	1.75
C C C B B F	C C C C V V	1	2	1	1	1.25

Fig. 3.1. Possible configurations for 12 records in 2 files of size 6. The attributes values are {B, C, F, V} and have {2, 7, 1, 2} records respectively

with a given last name (attribute value) access the minimum number of files (on average). Using the first letter of each last name the first two columns in Figure 3.1 shows some sample configurations for this example.

The ANF for a configuration is given by the formula

$$\frac{\sum_{i=1}^{h} fex(a_i)}{h},$$

where the function called $fex(a_i)$ returns the number of files that must be accessed to retrieve all the records with attribute a_i. From Figure 3.1, the second configuration has a value of 2 for $fex(B)$ since both files contain a record with attribute value B. The first and last configurations in the table are examples of optimal solutions for this problem. Even though the ANF values are the same, in some contexts the last solution is better because it has a more balanced configuration. If requests for the attributes are distributed uniformly, file 1 and file 2 will be accessed 25% and 100% of the time respectively in the first solution, where as the last solution will be accessed 50% and 75% of the time respectively. This idea is incorporated when evaluating solutions generated by the MNC GA.

GAs have been successfully applied to a variety of optimization problems, such as the Traveling Salesman Problem [25], Scheduling [23] [19], and the Bin Packing Problem [10]. In some cases better results were obtained when the mating operator was designed to capture the essential information in the problem. With this in mind, the mating operator for the File Design Problem was designed using the "first fit" and "best fit" heuristics (to be described later). Such heuristics, group records with the same attribute together. The multimodal search space in the problem is explored in many directions by using selection and replacement operators in the MNC GA that encourages mating and replacement between solutions from the same extrema.

In this work we apply a parallel version of the generational MNC GA. The intent is two fold. First, we want to show that the generational version of the MNC GA exhibits the same properties as its steady state counterpart. Second, we want to show the advantages of the parallel version of the generational MNC GA, namely, the straight forward implementation on parallel architectures. SISAL was selected as the language for the parallel implementation because it is portable and easy to learn. The application can be ported to multiple platforms, including SGIs, Crays, and SUNs. Performance can then be evaluated using different number of processors. Additionally, SISAL is a deterministic functional language which guarantees the same solutions on different platforms.

There are basically three parallel GA models [14] exhibiting different degrees of parallelism; fine grain, distributed, and direct. In a fine grain model [7] [11] [22], each solution in the population is mapped to a processor with genetic operators applied between nearest neighbors. In a distributed model [20] [24], processors are assigned subpopulations, which converge locally and exchange genetic material among them at fixed intervals. Direct models [15], exploit the parallelism inherent in the GA operators and the GA structure while having the same properties of a sequential GA. Our SISAL implementation of the MNC GA follows the direct model while having the localized convergence exhibited in the other models.

The parallelism inherent in the generational MNC GA, and in the operators, is easily exploited. Performance is enhanced while maintaining the necessary computation for solving the problem. The best solution was found in all test cases with a speedup of at least 2.2 with four processsors.

4. The GA Model for the File Design Problem

The SISAL version of the MNC GA was designed to capture the parallelism in the model while maintaining the search for multiple solutions. In this model, multimodality is exploited by encouraging mating and replacement between solutions from the same peak. Improved performance is obtained by creating the offspring in parallel. The offspring are then inserted into the population sequentially to preserve replacement between members of the same peak.

The solutions in the initial population are created in parallel by assigning records to files at random. There are K files with b slots each for a total of N slots. The slots are uniquely numbered with a value between 1 and N. Each record is then assigned a slot number corresponding to a unique position in a file. The constraints of the problem are easily maintained without the need for counters for each of the files. The *fitness*, a measure of "goodness" of a solution, is then calculated for each member of the population.

The algorithm is executed for a fixed number of generations. Each generation consists of creating all the offspring and inserting them into the population. Three steps are involved to create two offspring: select the parents, apply the mating operator to the parents, and calculate the fitness to the offspring. Mutation is applied by the mating operator as part of the mating process. Each offspring is inserted sequentially in the population by selecting an existing solution to die.

To create the offspring each solution in the population is selected as a parent. This allows every individual in the population to mate at least once in every generation. All the mates for the parents are selected in parallel using crowding selection. After selection, mating produces two offspring and their fitness are computed. The number of offspring created can be up to two times the number of solutions in the population. We create a total of n (population size) mating pairs and each pair produces 2 offspring with probability c (crossover probability). All offspring are created in parallel with a given crossover and mutation probability.

The offspring are inserted one at a time in the population using the worst among most similar (WAMS) replacement policy. Replacement is applied sequentially. After an offspring is inserted in the population it immediately becomes a candidate for replacement and must compete with the other solutions in the population to survive. Some offspring are indeed replaced in the same generation before getting a chance to reproduce. As in selection, the replacement operator is biased toward solutions within the same extrema. Convergence is improved by the replacement operator which eliminates solutions with lower fitness.

The following sections describe the encoding and genetic operators for the File Design Problem. They were designed to preserve the constraints of the problem and take full advantage of the implementation of SISAL arrays.

Record number:	1	2	3	4	5	6	7	8	9	10	11	12
Record attribute:	B	B	C	C	C	C	C	C	C	F	V	V
Chromosome 1:	0	0	0	0	1	1	1	1	2	2	2	2
Chromosome 2:	0	0	1	1	2	2	2	0	0	1	1	2

Fig. 4.1. Encoding for the file design problem

4.1 Chromosome encoding

A *chromosome* represents a valid solution to the problem. It consists of an array of N *alleles* corresponding to each of the records in the problem. Each allele may assume a value between 0 and $K - 1$ inclusive, indicating the file containing the record. A valid encoding is a N digit number in base K where all digits appear exactly b times. An example is shown in Figure 4.1 for the records defined in Figure 3.1. To make clear that the chromosomes are not binary we selected in this case $K = 3$ files of size $b = 4$.

4.2 Similarity metric

Similarity between two solutions is measured from the number of records assigned to the same file. An example is shown in Figure 4.2 using the data

Records:	B	B	C	C	C	C	C	C	C	F	V	V
Chromosome 1:	1	0	2	2	0	1	0	1	1	0	2	2
Chromosome 2:	2	0	1	2	0	1	0	1	2	0	1	2
Similar assignments:		1		2	3	4	5	6		7		8

Fig. 4.2. Using similarity to select a mate during crowding selection

from Figure 3.1. As in the previous section we have 3 files of size 4. Each digit indicates the file where a record is located. In this example we have a similarity value of 8, that is 8 records have been assigned to the same file.

4.3 Mating operator

The crossover operator for the File Design Problem creates two offspring and was designed with two goals in mind. First, the characteristics expressed in both parents will be expressed in the offspring, thus preserving the schemata in both solutions. Second, fitness should be improved when combining two similar solutions. "Best fit" and "first fit" heuristics (described later) are used for this. Incorporating these features in the mating operator improves convergence of solutions from the same extrema. When two solutions from different extrema mate, offspring from other extrema can be created. This way the operator is not restricted to small areas in the search space.

```
Offspring inherits similar alleles from parents:
    Record Attribute: B B C C C C C C F V V
            Parent 1: 0 0 1 2 2 2 0 1 0 1 2 1
            Parent 2: 0 1 0 1 2 2 1 2 0 0 1 2
            Offspring: 0 - - - 2 2 - - 0 - - -
Unassigned records by attribute: B:1, C:4, F:1, V:2
```

```
Assignment of records in sorted order to both offspring:
         Offspring 1              Offspring 2
        Best Fit Method          First Fit Method
C:4  0 - 2 2 2 0 0 0 - - -    0 - 1 1 2 2 1 1 0 - - -
V:2  0 - 2 2 2 0 0 0 - 1 1    0 - 1 1 2 2 1 1 0 - 0 0
B:1  0 1 2 2 2 0 0 0 - 1 1    0 2 1 1 2 2 1 1 0 - 0 0
F:1  0 1 2 2 2 0 0 0 1 1 1    0 2 1 1 2 2 1 1 0 2 0 0
```

Fig. 4.3. Mating operator for the File Design Problem

The first step in the mating operator is to transfer similar characteristics from the parents to the offspring. This is done by transferring the records assigned to the same file in both parents to the same file in the offspring. Those records not assigned are counted for each attribute and sorted in decreasing order. One offspring is created using a best fit method based on the contents of files. In this approach unassigned records are located into files where records with the same attribute reside. The main idea is to group files with the same attribute in the same file as much as possible. The second offspring is created using a first fit method based on the empty space in the files. Here the unassigned records will be located where file space is available for records with the same attribute. Using the configuration in Figure 3.1 an example is shown in Figure 4.3.

Given the parents in Figure 4.3, the offspring inherits only four alleles; 3 records with attribute C and 1 record with attribute B. Using the best fit method the other 4 records with attribute C are assigned to file 2 and file 0 because those files contain records with the same attribute. Using the first fit method the 4 records are assigned to file 1 because that file is the most empty and all records can be placed together. If not all records fit in one file then the remaining records are placed in the next file having the most available space. The next attribute having the highest number of unassigned records is selected and its records are assigned in a similar manner.

Mutation is applied with a fixed probability for each allele. When an allele is selected for mutation another position in the chromosome is selected at random and the two values are interchanged. Such mutations may introduce a new configuration in succeeding generations.

4.4 Fitness function

The fitness function captures three important characteristics of an optimal solution: low ANF, records with the same attribute are grouped together, and records with the same attribute are spread equally among the minimum number of files needed to store them. The last two points are captured in a grouping term (GT) and balancing term (BT) respectively. The two terms are contradictory in the sense that GT wants to group records together, while BT wants to spread records equally across files. These three terms are added together, with different weight values, to get the fitness of a solution. The GT value is given a higher weight over the BT value because it promotes lower ANF values in the solution.

Recall from Section 3 that the ANF value is given by the formula:

$$ANF = \frac{\sum_{i=1}^{h} fex(a_i)}{h},$$

where $fex(a_i)$ returns the number of files containing attribute a_i. From this formula we can compute an upper and lower bound to the ANF term. The

lower bound represents a configuration where the records for all attributes
are assigned to the least number of files needed to contain them. The upper
bound can be calculated from the configuration containing the records for all
attributes spread among the maximum number of files possible. The lower
and upper bound are called min_anf and max_anf respectively and are:

$$min\_anf = \frac{\sum_{i=1}^{h}\lceil n_i/b \rceil}{h} \leq ANF \leq max\_anf = \frac{\sum_{i=1}^{h} min(n_i, K)}{h}.$$

Recall that n_i denotes the number of records with attribute i, h denotes
the number of attributes, b denotes the size of the files, and K denotes the
number of files.

To compute the GT value we need to know how the records of a given
attribute are spread in the files. Since we want as many records as possible
of the same attribute grouped together, we came up with an equation that
looks at the ratio of records with the same attribute in each file. The GT
value is computed by adding the normalized number of records squared for
each attribute in every file. The higher the number of records of the same
attribute in a file the higher the GT value. The formula for the GT value is:

$$GT = \sum_{i=1}^{h} \sum_{j=1}^{K} (attr(a_i, j)/n_i)^2,$$

where $attr(a_i, j)$ returns the number of records of attribute a_i in file j.

On the other hand the BT value wants to spread the records with the
same attributes equally among the minimum number of files needed to fit the
records. The BT value is then computed by adding the absolute value of the
difference between the number of records for each attribute and a balance
configuration for the attribute. Only files containing records for the given
attribute are included in the summation. The formula for this term is:

$$BT = \sum_{i=1}^{h} \sum_{j=1}^{K} \lfloor |attr(a_i, j) - \frac{n_i}{\lceil n_i/b \rceil}| \rfloor,$$

for i and j where $attr(a_i, j) \neq 0$.

Here $\lceil n_i/b \rceil$ returns the number of files needed to store the records of
attribute a_i. Values of BT closest to zero represent more balanced configurations.

The three terms ANF, BT, and GT are used to define the fitness value for
a solution. Since higher positive values are used to indicate a better solution,
the terms are normalized to return values between 0.0 and 1.0. A percentage
of each term is then added to form the final fitness value as indicated by the
following formula:

$$fitness = 0.70\frac{GT}{h} + 0.25\frac{max_anf - ANF}{max_anf - min_anf} + 0.05\frac{1.0}{1.0 + BT}.$$

The fitness value for any solution is a number between 0.0 and 1.0. Solutions where the fitness value is 1.0 represent configurations where the min_anf value is achievable and all the records for any attribute can fit in the minimum number of files. Having the property of fitting records with the same attribute in one file eliminates the conflict between BT and GT while obtaining a maximum value of h for GT.

5. Experimental Data

To evaluate the behavior of the algorithm six test cases, having different properties, were created. Some of the test cases contain solutions achieving the min_anf lower bound. In other test cases we have attributes with the number of records exceeding the file size (therefore the $fitness < 1.0$ and a min_anf configuration may not exist). In all test cases, multiple attributes per file were mixed to create the different configurations. For all configurations 100 records were used. Figure 5.1 summarizes all configurations created.

Case Num.	Num. Files	File Size	Num. Attr.	Number of Records per Attribute $n_1\ n_2\ n_3\ n_4\ n_5\ \ldots\ n_h$	min_anf exist
1	5	20	10	7, 2, 3, 1, 5, 17, 18, 13, 15, 19	Yes
2	10	10	10	7, 2, 3, 1, 5, 17, 18, 13, 15, 19	Yes
3	5	20	10	7, 4, 3, 8, 6, 11, 18, 15, 10, 18	No
4	10	10	10	7, 4, 3, 8, 6, 11, 18, 15, 10, 18	No
5	5	20	21	7, 4, 3, 8, 6, 1, 8, 5, 10, 8, 1, 2, 4, 9, 5, 1, 6, 2, 3, 3, 4	Yes
6	5	20	15	7, 4, 9, 7, 7, 4, 9, 5, 9, 7, 9, 5, 6, 7, 5	No

Fig. 5.1. Configuration for all test cases

To evaluate the performance of the implementation three different platforms were used: the SGI Iris 4D, Cray Y-MP, and Cray C90. The execution time from one to four processors was collected for the algorithm using case 1 in Figure 5.1. The MNC GA parameters used for each run are:

Population size:	100
Number of generations:	50
Mating probability:	0.95
Mutation probability:	0.01
C_s for selection:	4

C_f for replacement: 3
s for replacement: 5

These parameters were chosen after a trial and error period. They represent a good set of choices for the test data shown in Figure 5.1.

6. Results

The generational MNC GA was very successful for the test data in Figure 5.1. For all test cases, multiple optimal solutions were found and retained for many generations. In four of the six test cases at least one optimal solution was found prior to generation 6. More generations were needed for the test cases 3 and 4. These test cases have the property that the min_anf is not achievable and there are attribute values where the number of records is higher than the file size. In those cases, the solutions were competing between themselves for a very small improvement in fitness.

Figure 6.1 shows solutions for all test cases and the generation number on which they were obtained. Each solution is represented by the file number to which each record is assign. The records with the same attribute value are separated by commas to verify how many files are used to save them. For example, the first solution has the records for attribute 1 assigned to file 4, records with attribute 2 assigned to file 1, and so on.

Case	Gen	Best Solution
1	3	4444444, 11, 333, 0, 22222, 33333333333333333, 111111111111111111, 4444444444444, 22222222222222, 0000000000000000000
2	4	9999999, 77, 111, 6, 88888, 33333333113311111, 222222222277777777, 5555555555999, 444444444488888, 00000000006666666666
3	25	2222222, 0000, 222, 44444444, 333333, 44444444444, 0000000001000001000, 333333433333333, 2222222222, 111111111111111111
4	15	8888888, 9999, 888, 55555555, 999999, 22222222227, 000000000066666666, 111111111155667, 3333333333, 444444444477777777
5	5	0000000, 3333, 222, 33333333, 222222, 0, 00000000, 11111, 2222222222, 44444444, 2, 44, 4444, 11111111, 3333, 4, 111111, 44, 333, 444, 0000
6	5	0000000, 2222, 222222222, 4444444, 222222, 3333, 333333333, 44444, 000040000, 3333333, 111111111, 11111, 111111, 4444444, 00000

Fig. 6.1. Best solution and the number of generations needed for all test cases

By examining test case 4 more closely we can observe that the optimal configuration required the records for the eighth attribute (111111111155667) assigned to 4 different files. All other attribute values were assigned to 1 or

2 files only. In the same run other configurations were found were the ANF was the same and all the records for the eight attribute were stored in 2 or 3 files. In such cases other attribute values were assign to 3 or 4 files.

In general, the use of heuristics improved the convergence of the MNC GA. We tried other crossover operators, but they required many more generations to achieve similar results. At the same time, the MNC GA did not allow the population to converge prematurely to a local optimum. Mixing heuristics with the GA allowed us to obtain results which are better than using the heuristics alone. Heuristics alone tend to locate local optima frequently, whereas the MNC GA allows different solutions to converge at the same time giving a higher likelihood to obtain the global optimum, as defined by the fitness function, in search spaces with multiple optima.

Platform	1 Proc.	2 Proc.	3 Proc.	4 Proc.
Y-MP C90	12.9799	8.6748	6.4425	5.4930
Y-MP	18.6106	10.4642	8.9153	6.3648
SGI Iris	25.8900	16.5300	13.4200	11.9000

Fig. 6.2. MNC GA execution time in seconds for 50 generations

We also observed the increase in speed that can be obtained using a parallel implementation of the MNC GA. A speedup between 2.2 and 2.9 was achieved with four processors in the three different platforms. Figure 6.2 summarizes the performance from one to four processors in the different platforms.

The best speedup was obtained for the Cray Y-MP platform and the worst speedup on the SGI Iris platform. The speed of the Cray is not much faster than that of the SGI when we take into account that the SGI does not have vector calculations and has worse cache locality. Better speedup times can be obtained for more complex (in terms of evaluation time) fitness functions. This is because selection and mating are done in parallel whereas replacement is done sequentially. Since the fitness of a new offspring is calculated at the end of the mating step a more complex fitness function will benefit from the parallelism.

7. Summary

The results obtained with the parallel version of the generational MNC GA model are encouraging. Diversity was maintained during the run, just as the steady state algorithm did, and the last generation contained multiple optima. Exploiting the multimodality inherent in the File Design Problem resulted in a more balanced search over the entire space.

Creating genetic operators that use heuristics enhanced the convergence of the algorithm while at the same time allowing multiple solutions to coexist. In this work we developed a model from the problem's point of view. We enhanced the MNC GA with problem specific operators to provide a better way to search for optimal configurations. The convergence to optimal solutions was achieve in all cases while improving the performance using SISAL.

A better speedup can be achieved using more complex fitness functions or by introducing a parallel version of the WAMS replacement operator. The parallel version must retain the important properties of WAMS. Competition among solutions within the same peak is encouraged while allowing competition among the multiple peaks as well.

Like the steady state MNC GA, the algorithm located and maintained multiple solutions throughout the run while maintaining diversity in the population. Creating genetic operators that use heuristics enhanced the convergence of the algorithm while at the same time allowing multiple solutions to coexist. These operators enhanced the ability of the MNC GA to search for optimal configurations. The convergence to optimal solutions was achieve in all cases while improving the speed with a parallel version developed using SISAL. In the future we want to investigate in more detail the use of heuristics for genetic operators.

References

1. Cedeño, W. (1995). The multi-niche crowding genetic algorithm: analysis and applications. *UMI Dissertation Services*, 9617947.
2. Cedeño, W., Vemuri, V., and Slezak, T. (1995). Multi-Niche crowding in genetic algorithms and its application to the assembly of DNA restriction-fragments. *Evolutionary Computation*, 2:4, 321-345.
3. Cedeño, W. and Vemuri, V. (1996). Genetic algorithms in aquifer management. *Journal of Network and Computer Applications*, 19, 171-187, Academic Press.
4. Cobb, H. J. and Grefenstette, J. J. (1993). Genetic algorithms for tracking changing environments. In S. Forrest (ed.) *Proceedings of the Fifth International Conference on Genetic Algorithms*, Morgan Kaufmann Publishers San Mateo, California, 523-530.
5. Darwin, C. (1859). *On the Origin of Species by Means of Natural Selection*.
6. Dasgupta, D. & McGregor, D. R. (1992). Non-stationary function optimization using the structured genetic algorithm. In R. Manner and B. Manderick (eds.), *Parallel Problem Solving from Nature 2*, Amsterdam: North Holland, 145-154.
7. Davidor, Y. (1991). A naturally occurring niche & species phenomenon: The model and first results. In R. K. Belew and L. B. Booker, (Eds.), *Proceedings of the Fourth International Conference on Genetic Algorithms*, San Mateo, CA: Morgan Kaufmann, 257-263.

8. De Jong, K. A. (1975). An analysis of the behaviour of a class of genetic adaptive systems. Doctoral dissertation, University of Michigan. *Dissertation Abstracts International* 36(0), 5140B. (University Microfilms No. 76-9381).

9. Deb, K. and Goldberg, D. E. (1989). An investigation of niche and species formation in genetic function optimization. In J. D. Schaffer (Ed.), *Proceedings of the Third International Conference on Genetic Algorithms*, San Mateo, CA: Morgan Kaufmann, 42-50.

10. Falkenauer, E. and Delchambre, A. (1992). A genetic algorithm for bin packing and line balancing. *Proceedings of the 1992 IEEE International Conference on Robotics and Automation.*

11. Gorges-Schleuter , M. (1989). ASPARAGOS an asynchronous parallel optimization strategy. In J. D. Schaffer (Ed.), *Proceedings of the Third International Conference on Genetic Algorithms*, San Mateo, CA: Morgan Kaufmann, 422-427.

12. Goldberg D. E. & Smith R. E. (1987). Non-stationary function optimization using genetic algorithms with dominance and diploidy. In J. J. Grefenstette (Ed.), *Proceedings of the Second International Conference on Genetic Algorithms*, Hillsdale, NJ: Lawrence Erlbaum Associates

13. Goldberg, D. E. (1989). *Genetic Algorithms in Search, Optimization & Machine Learning.* Reading MA: Addison-Wesley.

14. Gordon, V.S., Whitley, D., and Böhm, A.P.W. (1992). Dataflow parallelism in genetic algorithms. In R. Manner and B. Manderick (eds.), *Parallel Problem Solving from Nature 2*, Elsevier Science Publishers.

15. Grefenstette, J.J. (1981). Parallel adaptive algorithms for function optimization. Technical Report No. CS-81-19, Vanderbilt University, Computer Science Department.

16. Holland, J. H. (1975). *Adaptation in natural and artificial systems*, Ann Arbor MI: The University of Michigan Press.

17. Liang, J., Chang, C. C., Lee, R. C. T., and Wang, J. S. (1991). Solving the file design problem with neural networks. *Tenth Annual International Phoenix Conference on Computers and Communications.*

18. McGraw, J., et. al. (1985). SISAL - Streams and iterations in a single-assignment language. Language reference manual, version 1.2. Lawrence Livermore National Laboratory manual M-146 (Rev. 1), Livermore, CA.

19. Michalewicz, Z. (1992). *Genetic Algorithms + Data Structures = Evolution Programs.* New York, NY: Springer-Verlag.

20. Mühlenbein, H. (1989). Parallel genetic algorithms, population genetics and combinatorial optimization. In J. D. Schaffer (Ed.), *Proceedings of the Third International Conference on Genetic Algorithms*, San Mateo, CA: Morgan Kaufmann, 416-421.

21. Ng, K. P. & Wong, K. C. (1995). A new diploid scheme and dominance change mechanism for non-stationary function optimization. In L. J. Eshelman (ed.), *Proceedings of the Sixth International Conference on Genetic Algorithms*, San Mateo, CA:Morgan Kaufmann, 159-166.

22. Spiessens, P. and Manderick, B. (1991). A massively parallel genetic algorithm - implementation and first analysis. In R. K. Belew and L. B. Booker, (Eds.), *Proceedings Fourth International Conference on Genetic Algorithms*, San Mateo, CA: Morgan Kaufmann, 279-287.

23. Syswerda, G. and Palmucci, J. (1991). The application of genetic algorithms to resource scheduling. In R. K. Belew and L. B. Booker, (Eds.), *Proceedings of the Fourth International Conference on Genetic Algorithms*, San Mateo, CA: Morgan Kaufmann, 502-508.

24. Tanese, R. (1989). Distributed genetic algorithms. In J. D. Schaffer (Ed.), *Proceedings of the Third International Conference on Genetic Algorithms*, San Mateo, CA: Morgan Kaufmann, 434-440.
25. Whitley, D., Starkweather, T., & Fugway, D. (1989). Scheduling problems and traveling salesmen: the genetic edge recombination operator. In J. D. Schaffer (Ed.), *Proceedings of the Third International Conference on Genetic Algorithms*, San Mateo, CA: Morgan Kaufmann, 133-140.

Designing Multiprocessor Scheduling Algorithms Using a Distributed Genetic Algorithm System

Kelvin K. Yue[1] and David J. Lilja[2]

[1] Dept. of Computer Science, University of Minnesota, 200 Union Street S.E., Minneapolis, MN 55455
[2] Dept. of Electrical Engineering, University of Minnesota, 200 Union Street S.E., Minneapolis, MN 55455

Summary. Processor scheduling algorithms for parallel computing systems are often designed using analytical modeling, simulations, and/or trial-and-error experiments. These approaches are usually quite time-consuming and produce only satisfactory solutions. To automate this process, we utilize current techniques in distributed computing, multithreading, and parallel processing to develop a distributed genetic algorithm system that designs new processor scheduling algorithms. A simple genetic algorithm running on a local workstation is used to generate possible scheduling algorithms. The performance of these algorithms is then evaluated in real-time on two shared-memory multiprocessor systems. The resulting performance data are returned to the local workstation for incorporation into the genetic algorithm. Using this approach, we are able to automatically generate new scheduling algorithms in a relatively short period of time that produce good performance when subsequently executing similar types of parallel application programs on a shared-memory multiprocessor system.

1. Introduction

One of the many ways to reduce the execution time of an application program is to partition the iterations of a loop in the application program into independent tasks and then execute these tasks concurrently on multiple processors. Exploiting this *loop-level parallelism* has proven to be an effective means of increasing the performance of multiprocessor systems [6]. However, when scheduling these iterations on to the processors for execution, one needs to tradeoff between processor load balancing and the scheduling overhead.

Several parallel loop scheduling algorithms have been proposed [4, 5, 7, 10] to schedule the iterations onto the processors for execution such that the workload among the processors is approximately balanced. However, none of these algorithms perform well for all types of loops [11]. In this chapter, we extend our previous work [12] to implement a distributed genetic algorithm system to systematically evaluate the performance of trial scheduling algorithms in real-time and thereby match an algorithm to the loop characteristics. Compared to using a simulator to evaluate each potential algorithm, this approach provides a more realistic evaluation of the trial scheduling algorithms generated by the Genetic Algorithm. The main goal of this work is to correlate loop characteristics with the scheduling algorithms by using the Genetic Algorithm to *discover* new generic scheduling algorithms that can be implemented in existing machines. This chapter describes the implementation of this distributed genetic algorithm system and how it is applied to the parallel loop scheduling problem.

This chapter is organized as follows: Section 2. provides background information on existing parallel loop scheduling algorithms. Section 3. presents our methodology for finding scheduling parameters using the Genetic Algorithm, while Section 4. discusses the results of applying this strategy to loop-level parallelism. Section 5. concludes the chapter.

2. Current Scheduling Algorithms

Shared-memory multiprocessor systems have been widely commercialized and are often used as a high-performance general purpose-machine. Figure 2.1 shows a diagram of a typical shared-memory multiprocessor system in which a number of processors are connected together with the main memory through a high-speed network. In this system, all processors run at the same speed and each of the processors executes its instructions independently of the others.

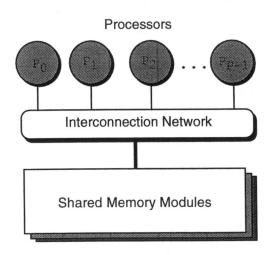

Fig. 2.1. Shared memory multiprocessor architecture

Loop-level parallelism is one of the most common approaches to utilize the computational power of shared-memory multiprocessors. With this approach, a programmer or a compiler parallelizes a sequential program by recognizing its parallelizable loops. Each processor then executes iterations of these loops concurrently, which thereby reduces the execution time of the application. The simplest form of a parallelizable loop is a Doall loop, as shown in

Figure 2.2. In this type of loop, each iteration is independent of the other iterations, and, therefore, multiple iterations can be executed concurrently by multiple processors.

```
DOALL i = 1, N
   a(i) = b(i) + c(i)
ENDDO
```

Fig. 2.2. Example of a parallel loop

When a Doall loop is executed on a shared-memory multiprocessor system, the iterations of the loop are distributed to the processors using some loop scheduling algorithm. There are two main categories of scheduling algorithms: static and dynamic [6]. *Static scheduling*, or *prescheduling*, assigns iterations to the processors at compile-time. For instance, the compiler could assign iterations to the processors based on the processor number so that processor 0 executes iterations $1, P+1, 2P+1, \cdots$, processor 1 executes iterations $2, P + 2, 2P + 2, \cdots$, and so on, where P is the number of processors. The main advantage of this approach is that there is no extra scheduling overhead at run-time since the iteration assignment is done entirely at compile-time. However, one of the problems with static scheduling is *processor load imbalance* [6]. The different iterations of a loop often require different execution times, but since static scheduling assigns the iterations to the processors at compile-time, the assignment cannot be adjusted to compensate for the dynamically varying workload of the processors.

Dynamic scheduling, on the other hand, assigns iterations to processors at run-time and can, therefore, adjust the schedule to the individual processors' workloads. Self-scheduling (SS) [2], chunk scheduling (CS) [5], guided self-scheduling (GSS) [7], factoring (FS) [4], and trapezoid self-scheduling (TSS) [10] are some of the common algorithms for scheduling parallel loop iterations dynamically.

Self-scheduling is the simplest form of dynamic scheduling. With self-scheduling, each idle processor obtains the index of the next iteration it should execute by accessing a shared work queue. By taking one iteration at-a-time, this algorithm balances the workload very well, but the scheduling overhead is large since the shared work queue must be accessed once for each iteration.

To reduce the scheduling overhead, *chunk scheduling* assigns groups of iterations as a single unit to the processors during run-time. The most commonly used *chunk size*, which is the number of iterations in a chunk, is N/P, where N is the total number of iterations in the loop, and P is the number of processors. This chunk size often produces an imbalanced workload, however. Kruskal and Weiss [5] analyzed load imbalances with this strategy and proposed the optimal chunk size to be $[(\sqrt{2}Nh)/(\sigma P \sqrt{\log P})]^{2/3}$, which takes the standard deviation of the distribution of iteration execution times, σ, and the scheduling overhead, h, into consideration. However, they assumed that the central-limit theorem holds for the iteration execution times, which is valid only when N is large.

To reduce load imbalance while maintaining low scheduling overhead, some algorithms use a decreasing chunk size. There are two strategies for decreasing the chunk size: *linear* decreases and *nonlinear* decreases. *Guided self-scheduling* (GSS) [7] decreases the chunk size nonlinearly by allocating iterations with a chunk size equal to $\lceil R_i/P \rceil$, where R_i is the number of iterations remaining to be executed at scheduling step i and $R_1 = N$. This algorithm allocates large chunk sizes at the beginning of a loop's execution to reduce the scheduling overhead. As the number of iterations remaining to be executed decreases, smaller chunks are allocated to balance the load.

A similar algorithm called *factoring* (FS) [4] also uses a nonlinearly decreasing chunk size. It allocates iterations in batches of P equal-sized chunks. After a batch is scheduled, the new chunk size is calculated to be $\lceil R_i/(xP) \rceil$, where R_i is the number of iterations remaining at step i, and x typically is chosen to be 2. The initial chunk size for FS is smaller than GSS. As a result, it has more iterations remaining at the end of the loop's execution to balance the load. However, FS requires many more scheduling steps than GSS.

Trapezoid self-scheduling (TSS) [10] decreases the chunk size linearly to achieve a better tradeoff between the scheduling overhead and the distribution of the processors' workload compared to the nonlinear strategies. The number of chunks, C, is equal to $\lceil 2N/(f + l) \rceil$ and the chunk size is decreased by a factor of $(f - l)/(C - 1)$ at each scheduling step, where typically $f = N/(2P)$ and $l = 1$. TSS does not allocate chunks as large as GSS in the beginning, and it does not require as many scheduling steps as FS. However, the linearly decrementing chunk size may create large load imbalances if the execution time differences between the last few chunks are large.

To summarize, one-iteration-at-a-time self-scheduling balances the workload to be within a single iteration time, but it requires a large scheduling overhead. Dynamic chunk scheduling, on the other hand, produces less overhead, but it also produces greater load imbalances. To tradeoff load imbalances with scheduling overhead, some scheduling algorithms, such as guided self-scheduling, factoring, and trapezoid self-scheduling, use a decreasing chunk size. A large chunk size is used at the beginning of a loop's execution to reduce the scheduling overhead. Small chunk sizes are then used near the end to balance the workload. However, the performance of these algorithms is sensitive to the characteristics of the loop and the system environment so that no single algorithm performs best in all cases [11].

It has been shown that, if the variance in iteration execution times is large, GSS may not balance the workload well since it does not save enough single-iteration chunks for the end [4, 10]. Factoring saves enough single-iteration chunks to balance the load, but these chunks may cause extra scheduling overhead for some types of loops [6]. Trapezoid self-scheduling assigns small initial chunks, as does factoring, and it requires fewer scheduling steps than FS [11], but the difference in execution time between the last few chunks might be large due to the linear decrement in the chunk size. This large difference may create correspondingly large load imbalances [4]. It can be concluded that the performance of the loop scheduling algorithms depends on the characteristics of the loops and that it can be quite difficult to design a scheduling algorithm tailored to an individual type of loop.

3. A New Scheduling Methodology

In the previous section, we reviewed the current dynamic scheduling algorithms and concluded that no single algorithm produces the best performance in all cases. We have previously described a generalization of these scheduling algorithms in which the algorithms' characteristics are parameterized and the genetic algorithm is used in conjunction with a multiprocessor simulator to search for a set of parameters to thereby construct a new scheduling algorithm that best matches the loop characteristics [12].

We now extend this generalized parallel loop scheduling methodology by developing a distributed genetic algorithm system (DGAS) that evaluates the performance of the trial scheduling algorithms in real-time on a targeted system instead of simply simulating the performance. In this section, we describe the necessary hardware components and the software implementation of our distributed genetic algorithm system. We then discuss how to use this system to automatically design parallel loop scheduling algorithms for different classes of parallel loops.

3.1 Experimental hardware

The distributed genetic algorithm system was implemented using one Silicon Graphics Indy workstation and two Silicon Graphics Challenge L [8] multiprocessor systems interconnected with a 10 Mbyte/sec Ethernet local area network. The Indy workstation has a MIPS R4600 CPU and a MIPS R4610 floating point processor running at 132 MHz. It includes a 16 Kbyte data cache and a 16 Kbyte instruction cache with 64 Mbytes of main memory. The two Challenge L systems are identical single-bus based shared-memory multiprocessors. Each of these systems is equipped with four MIPS R4400 processors along with four MIPS R4010 floating point units all running at a clock rate of 200 MHz. Each processing unit has 16 Kbytes of on-chip data cache and 16 Kbytes of on-chip instruction cache. The secondary cache is a 4 Mbyte unified instruction and data cache. Each system has 512 Mbytes of four-way interleaved physical memory. All of the systems run version 5.3 of the IRIX operating system with the XFS file system.

3.2 Software architecture

The software to utilize this distributed architecture is developed using a modular approach with the arrangement of the various modules depicted in Figure 3.1.

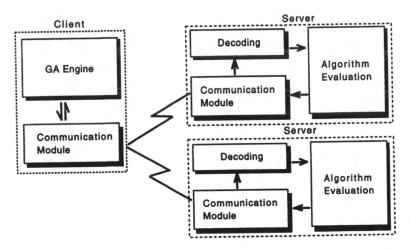

Fig. 3.1. Organization of the software modules in the distributed genetic algorithm system

The Indy workstation, which we call the client, executes the GA Engine and a communication module. The GA Engine, shown in Figure 3.2, is an implementation of the simple genetic algorithm [3]. It generates new *genetic structures* or *chromosomes* based on the "fitness" of the chromosomes from the previous generation. To generate a new chromosome, the *crossover* and *mutation* operations are used.

To evaluate the performance of the newly generated scheduling algorithm represented by a given chromosome, the chromosome is sent to one of the SGI Challenge multiprocessors, which we call the *servers*, through the communication modules. These modules communicate using standard Unix sockets with the TCP communication protocol [9].

Running on the Challenge multiprocessors are three main modules: the communication module, the decoder, and the algorithm evaluation module. The communication module waits for the connection from the client and accepts the chromosome of the candidate scheduling algorithm. It then passes the chromosome to the decoder where it is decoded to construct the scheduling algorithm. In addition to constructing the scheduling algorithms, the decoder also prescreens and filters out the unrealistic algorithms using some predefined rules by assigning the filtered-out algorithms a fitness value of zero. Since the system will not waste its time evaluating unrealistic algo-

```
/* Initialization */
Randomly generate the initial population

Send a chromosome to the communication module
Wait for the performance results

DO until (population converges)
    or (no. of generations > predefined value)

  /* Selection Phase */
  Select the chromosomes with the highest
     fitness values

  /* Reproduction Phase */
  Generate new chromosomes using the
     crossover operator and the mutation operator.

  /* Evaluation Phase */
  Send the chromosome to the communication module
     and wait for the performance results

ENDDO
```

Fig. 3.2. GA engine module running on the SGI Indy client

rithms, this procedure can reduce the total time required by the system to evaluate the scheduling algorithms. The specific filtering rules are discussed in the next subsection.

Once the scheduling algorithm is constructed, a benchmark application that uses this scheduling algorithm is started and its resultant performance is measured. This performance result is then returned to the client through the communication module. The basic operation of the server is shown in Figure 3.3.

Once the communication module on the client side receives the performance data, it passes this data to the GA Engine. Since the overall goal is to reduce the execution time produced by the scheduling algorithm represented by the given chromosome, the smaller the value of this evaluation function (i.e. the execution time), the higher is the "fitness" of the corresponding chromosome. Two identical multiprocessor systems are run in parallel so that two different chromosomes can be evaluated concurrently. To utilize this arrangement, part of the GA Engine is multithreaded concurrently executing with two independent processes. These two processes interact with the communication module simultaneously. This multithreaded approach reduced the time to evaluate all of the chromosomes by almost a factor of two.

```
DO until done

    /* Communication Module */
    Wait for the connection

    /* Decoder */
    Decode the chromosome and
        filter out unrealistic algorithms

    Construct the scheduling algorithm

    /* Algorithm Evaluation */
    Execute the benchmark program using
        the candidate scheduling algorithm

    /* Communication Module */
    Return the performance result
        to the client

ENDDO
```

Fig. 3.3. Server routine running on the SGI Challenge multiprocessors for evaluating the candidate scheduling algorithms

3.3 Representation of a scheduling algorithm

The parallel loop scheduling algorithms used in existing systems try to ensure that the scheduling overhead is as small as possible by minimizing the number of operations that need to be invoked at each scheduling step. Many loop scheduling algorithms have been proposed that have many similar characteristics. We have taken advantage of these similarities to generalize all of these related scheduling algorithms into a single, unified framework. As a result, it incorporates all of the scheduling operations of the existing algorithms and cannot be as highly optimized as an algorithm implemented for a specific system. In our previous work [12], the performance evaluation was done using simulations which allowed us to adjust the simulated scheduling overhead to compensate for this difference. However, in this study, the performance is evaluated on real machines and each extra operation in the scheduling will degrade the performance of the algorithm. Thus, the generalized scheduling algorithm is at a disadvantage compared to the existing algorithms.

To reduce the extra overhead and make the generalized scheduling algorithm more suitable for real-time performance evaluation, we simplify the algorithm by trimming the number of parameters compared to our earlier generalization [12]. Thus, in this distributed genetic algorithm system, each scheduling algorithm is represented by four parameters:

- chunk_size: the number of iterations that are assigned to a processor at one time. It is represented by bits 4 to 13 of the chromosome so that the maximum size of a chunk is 1023 iterations.
- num_chunk: the number of chunks that are scheduled before calculating a new chunk_size. It is represented by bits 0 to 3 giving a maximum of 15 chunks.
- delta: the rate of decrease in the size of the chunks. It is represented by bits 24 to 31 for a maximum of 255 iterations.
- min_size: the minimum number of iterations within a chunk. It is represented by bits 14 to 23 for a maximum of 1023 iterations per chunk.

These parameters are combined to form the chromosomes that are used in the GA engine, as shown in Figure 3.4.

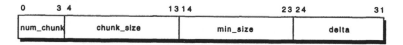

Fig. 3.4. Chromosome representation for the generalized scheduling algorithm

To illustrate this representation, assume that a parallel loop with $N = 400$ iterations is to be executed on a four-processor system. If chunk_size $= 50$, num_chunk $= 4$, delta $= 25$, and min_size $= 20$, then the algorithm schedules the iterations as follows: The first four (num_chunk $= 4$) processors that need work will be assigned 50 iterations each. Then the chunk_size is decreased to 25 iterations (chunk_size $-$ delta $= 50 - 25 = 25$). When the processors complete their assigned iterations, the next four processors (num_chunk $= 4$) will be assigned 25 iterations each. Now the chunk_size should be decreased by 25 iterations, but this would produce a chunk_size of 0 (chunk_size $-$ delta $= 25 - 25 = 0$), which is smaller than min_size $= 20$. Therefore, chunk_size is set to be 20 iterations. The remaining 100 iterations ($400 - 50 \times 4 - 25 \times 4 = 100$) are assigned to the processors in chunks of 20 iterations each.

Although these variables cannot represent all of the existing scheduling algorithms, they do allow characteristics similar to the existing algorithms. Table 3.1 shows how this representation relates to the existing strategies by presenting the appropriate values for the parameters for the existing algorithms based on executing a parallel loop with $N = 400$ iterations on a four-processor system ($P = 4$).

To evaluate the fitness of a chromosome, or, equivalently, the performance of a scheduling algorithm, the values of these parameters are decoded to create the scheduling algorithm. This algorithm is inserted into a benchmark application program that is then run on the multiprocessor system. For instance, the parallel loop in Figure 2.2 will be transformed to the parallel task shown in Figure 3.5.

Table 3.1. Parameter values to approximate the existing scheduling algorithms

Algorithm	CS	SS	GSS	FS	TSS
chunk_size	100	1	100	50	50
delta	0	0	25[†]	25[†]	≈ 4
min_size	100	1	1	1	1
num_chunk	4	400	1	4	1

[†]delta decreases every num_chunk scheduling step(s).

```
LOCK
/* critical section */
IF (num_scheduled_chunks == num_chunk) THEN
    num_scheduled_chunks = 1
    chunk_size = chunk_size - delta
    IF (chunk_size < min_chunk_size) THEN
        chunk_size = min_chunk_size
    ENDIF
ELSE
    num_scheduled_chunks++
ENDIF

i_start = next_i
i_end = i_start + chunk_size
next_i = i_end + 1
/* end of critical section */
UNLOCK

/* execute its share of work */
DO i = i_start, i_end
    a(i) = b(i) + c(i)
ENDDO
```

Fig. 3.5. Code transformation for the parallel loop from Figure 2.2 to include the scheduling code

As was mentioned in the previous subsection, the decoder filters out unrealistic algorithms before it enters the actual evaluation phase. A scheduling algorithm will be rejected unless it satisfies all of the following conditions:

1. The values of chunk_size, min_size, and num_chunk must be greater than or equal to one, for obvious reasons.
2. The value of chunk_size must be greater than or equal to min_size, again for obvious reasons.
3. The value of chunk_size must be less than or equal to three-fourths of the total number of iterations. The reason for having this condition is to

ensure that the algorithm will not allocate all of the iterations to a single processor.

4. The value of **delta** must be less than **chunk_size**/2. This rule ensures that **chunk_size** is decreased at a reasonable rate. This rate is comparable to the rate used in Guided-Self Scheduling and Factoring, as described in Section 2..

In this section, we described the implementation of the distributed genetic algorithm system and how we applied it to the parallel loop scheduling problem. The next section will discuss how well this system performs on solving this problem.

4. Experimental Results

To evaluate the effectiveness of this distributed genetic algorithm system, we employed it to develop scheduling algorithms for three simple benchmark applications that are typical of three different types of parallel loops.

4.1 Generating new scheduling algorithms

The first benchmark application is a simple matrix multiplication routine, shown in Figure 4.1. This routine is parallelized on the outer i-loop. The size of the parallel tasks is large, and the variance in iteration execution times is small. Our benchmark has $N = 400$.

```
DOALL i = 1, N
   DO j = 1, N
      DO k = 1, N
         c(i,j) = c(i,j) + a(i,k) * b(k,j)
      ENDDO
   ENDDO
ENDDO
```

Fig. 4.1. Matrix multiplication benchmark routine

To find a scheduling algorithm for this parallel loop that produces the smallest execution time, ten chromosomes in the distributed genetic algorithm system are allowed to run for twenty generations. We chose ten chromosomes for twenty generations because it seems to be a good mix for finding a solution within a short period of time. We also had the system run three times with different crossover and mutation probabilities. A single run of

Table 4.1. Scheduling parameter values found by the distributed GA system for the matrix multiplication routine

Parameters	Run 1	Run 2	Run 3
chunk_size	105	52	98
delta	7	12	19
min_size	34	43	12
num_chunk	7	14	15
Speedup (P=4)	3.62	3.65	3.61

the distributed genetic algorithm system took from 45 minutes to an hour. Table 4.1 summarizes its findings.

According to these data, all of the values for **num_chunk** are larger than the total number of iterations (N) divided by the **chunk_size**. Therefore, **chunk_size** remains constant throughout the execution of the loop and the parameters **delta** and **min_size** are unused. Thus, the resultant algorithm suggested by the genetic algorithm has the same characteristics as chunk scheduling, and, in this case, a chunk size of about 50 iterations performs slightly better than the other chunk sizes.

The second benchmark is a transitive closure routine, shown in Figure 4.2. It is parallelized at the second loop level (i.e. the j-loop) and the amount of work in each parallel iteration varies significantly, dependent on the input matrix. For this experiment, the input matrix was generated using the random number generator from the SGI system with a 50% chance for the inner-most loop to be executed. We again set $N = 400$ and had ten chromosomes run for twenty generations. The results are shown in Table 4.2.

```
DO i = 1, N
   DOALL j = i, N
      IF ((a(j,i)).eq.1) THEN
         DO k = 1, N
            IF ((a(i,k)).eq. 1) THEN
               a(j,k) = 1
            ENDIF
         ENDDO
      ENDIF
   ENDDO
ENDDO
```

Fig. 4.2. Transitive closure benchmark routine

The resulting algorithms do not perform too well attaining a speedup of only about 1.80 from a four-processor system. The relatively poor performance is due to the randomness of the iteration execution times of the bench-

Table 4.2. Scheduling parameter values found by the distributed GA system for the transitive closure benchmark routine

Parameters	Run 1	Run 2	Run 3
chunk_size	195	106	129
delta	53	36	47
min_size	102	28	109
num_chunk	4	11	2
Speedup (P=4)	1.78	1.80	1.75

mark and the consequent difficulty of finding a suitable scheduling algorithm. The best scheduling algorithm recommended by the genetic algorithm is a chunk scheduling approach with a chunk size of about 100 iterations.

The last benchmark, shown in Figure 4.3, is an adjoint-convolution routine. It is parallelized on the outer-most loop (i.e. the i-loop) and, since the number of iterations in the inner loop depends on the index of the outer loop, the variance in iteration execution times is quite large. For this benchmark,

```
DOALL i = 1, M * M
  DO j = i, M * M
    a(i) = a(i) + x * b(j) * c(i - j)
  ENDDO
ENDDO
```

Fig. 4.3. Adjoint convolution benchmark routine

we set $M = 80$, which means that there are 6400 iterations ($N = 6400$) in the outer loop. Again, we had ten chromosomes run for twenty generations. The results are summarized in Table 4.3.

Table 4.3. Scheduling parameter values found by the distributed GA system for the adjoint convolution benchmark routine

Parameters	Run 1	Run 2	Run 3
chunk_size	344	378	250
delta	66	83	37
min_size	6	71	52
num_chunk	8	8	8
Speedup (P=4)	3.84	3.87	3.7

The GA-generated algorithms for this benchmark all schedule iterations in batches (num_chunk = 8) with the same size (chunk_size) and then decrease the size of a chunk gradually. These GA-generated algorithms are very similar to the factoring strategy except that they have a smaller initial chunk size and a larger number of chunks in a batch. Also, unlike factoring, these algorithms ensure a minimum chunk size.

4.2 Performance comparisons

To determine how well these GA-generated algorithms perform compared to the human-generated algorithms, we plotted the speedup of the best algorithms from the distributed genetic algorithm system (DGAS) along with several of the existing parallel loop scheduling algorithms, including chunk scheduling (CS), self-scheduling (SS), guided self-scheduling (GSS), factoring (FS), and trapezoid self-scheduling (TSS). These comparisons are shown in Figure 4.4.

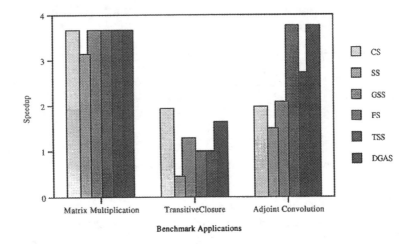

Fig. 4.4. Performance comparison of the GA-generated parallel loop scheduling algorithms with existing scheduling algorithms with $P = 4$ processors

In the case of matrix multiplication, the DGAS recommended a chunk scheduling approach. As seen in Figure 4.4, this GA-generated algorithm performs as well as the other algorithms for this application. The best algorithm for transitive closure is chunk scheduling, and the DGAS suggested a chunk scheduling algorithm with a very large chunk size. This GA-generated algorithm performs worse than the standard chunk scheduling, but it significantly outperforms the other algorithms. For the adjoint convolution benchmark, factoring (FS) performs the best compared to other existing algorithms. Interestingly, the DGAS correctly suggested a factoring-type approach with a decreasing chunk size. To summarize, the distributed genetic algorithm system was able to automatically generate algorithms in a relatively short period of time that perform very well on three representative benchmarks compared to existing human-generated algorithms.

5. Conclusion

We have described the implementation of a distributed genetic algorithm system and how we applied it to the development of parallel loop scheduling algorithms for multiprocessor systems. This distributed genetic algorithm system consists of a workstation and two multiprocessor systems and utilizes current technologies in distributed computing, multithreading, and parallel processing. The effectiveness of this system was evaluated using three small benchmark applications. In all cases, it was able to generate a scheduling algorithm within a short period of time. While these GA-generated algorithms are quite primitive and are not optimized, their performance is comparable to, if not better than, the existing human-designed algorithms. Therefore, we feel that this system could be a valuable tool to assist human-programmers in the design of new parallel loop scheduling algorithms for different classes of parallel loops.

Acknowledgement

This work was supported in part by the National Science Foundation under grant no. MIP-9221900 and equipment grant no. CDA-9414015. This work was performed while K. Yue was with the Department of Computer Science, University of Minnesota.

References

1. C. J. Beckmann and C. D. Polychronopoulos. The effect of scheduling and synchronization overhead on parallel loop performance. In *International Conference on Parallel Processing*, pages II:200–204, 1989.
2. Z. Fang, P.-C. Yew, P. Tang, and C.-Q. Zhu. Dynamic processor self-scheduling for general parallel nested loops. In *International Conference in Parallel Processing*, 1 – 10, August 1987.
3. D. E. Goldberg. *Genetic Algorithms in Search, Optimization, and Machine Learning*. Addison-Wesley, Reading, Massachusetts, 1989.
4. S. F. Hummel, E. Schonberg, and L. E. Flynn. Factoring - a method for scheduling parallel loops. *Communications of the ACM*, 35:90– 101, August 1992.
5. C. Kruskal and A. Weiss. Allocating independent subtasks on parallel processors. *IEEE Transactions on Software Engineering*, pages 1001–1016, Oct. 1985.

6. D. Lilja. Exploiting the parallelism available in loops. *Computer*, 27(2):13 – 26, February 1994.
7. C. Polychronopoulos and D. Kuck. Guided self-scheduling: A practical scheduling scheme for parallel supercomputers. *IEEE Transactions on Computers*, C-36:1425 – 1439, December 1987.
8. Silicon Graphics Inc. *Symmetric Multiprocessing Systems.* Mountain View, California, 1993.
9. W. R. Stevens. *Unix Network Programming.* Prentice-Hall, Englewood Cliffs, New Jersey, 1990.
10. T. H. Tzen and L. M. Ni. Trapezoid self-scheduling: A practical scheduling scheme for parallel compilers. *IEEE Transactions on Parallel and Distributed Systems*, 4:87–97, January 1993.
11. K. Yue and D. Lilja. Parallel loop scheduling for high performance computers. In J. Dongarra, L. Grandinetti, G. Joubert, and J. Kowalik, editors, *High Performance Computing: Issues, Methods, and Applications*, pages 223 – 242. Elsevier Science B.V., Amsterdam, The Netherlands, 1995.
12. K. Yue and D. Lilja. Parameter estimation for a generalized parallel loop scheduling algorithm. In L. Chambers, editor, *Practical Handbook of Genetic Algorithms: New Frontiers*, volume II, pages 155 – 171. CRC Press, 1995.

Prototype Based Supervised Concept Learning Using Genetic Algorithms

Sandip Sen, Leslie Knight, and Kevin Legg

Dept of Mathematical & Computer Sciences, The University of Tulsa, 600 South College Avenue, Tulsa, OK 74104-3189

Summary. Prototypes have been proposed as representation of concepts that are used effectively by humans. Developing computational schemes for generating prototypes from examples, however, has proved to be a difficult problem. We present a novel genetic algorithm based prototype learning system, PLEASE, for constructing appropriate prototypes from classified training instances. After constructing a set of prototypes for each of the possible classes, the class of a new input instance is determined by the nearest prototype to this instance. Attributes are assumed to be ordinal in nature and prototypes are represented as sets of feature-value pairs. A genetic algorithm is used to evolve the number of prototypes per class and their positions on the input space. We present experimental results on a series of artificial problems of varying complexity. PLEASE performs competitively with several nearest neighbor classification algorithms and C4.5 on the problem set. We provide an analysis of the strengths and weaknesses of the current system.

1. Introduction

It has been recognized that a crucial component of intelligent behavior involves the construction and use of concepts from experience. Hence, the computational problem of concept formation from data have received considerable attention in the cognitive sciences and machine learning communities. This body of research have produced a number of different representations of conceptual knowledge, as well as effective mechanisms for drawing inductive inferences using these representations. Researchers in the field of machine learning have used both symbolic and sub-symbolic methods for inductively acquiring concept classification knowledge from a set of pre-classified instances [DeJ90, Mit86, Qui86, RHW86]. Some of the more influential symbolic systems include ID3 [Qui86], version spaces [Mit86], COBWEB [Fis87], AQ [Mic83], and FOIL [Qui90]. The most widely used sub-symbolic systems for concept learning are based either on neural networks [RZ85, RHW86] or on genetic algorithms (GAs) [BGH89, DeJ90, DSG93, Hol86, JM93]. Some of the more frequently used representations in symbolic concept learning systems use logic expressions and programs, decision trees and lists, hierarchical clusters, and rules. Representations used in sub-symbolic systems include networks of simple computational units connected by weighted links and bit strings.

The automatic induction of target concepts from sample data is such a widespread and important problem that it has been studied extensively even outside the discipline of computer science. In addition to the mainstream machine learning approaches mentioned above, there exists a significant body of work on statistical induction schemes [BFO84, CKS+88]. Other researchers have been motivated by the findings of cognitive psychologists [Sch78, SM81] and are using exemplar-based classification schemes [AKA91, Kru92, SW86]. These methods store the set of training instances and classify a new instance by some voting scheme using a given number of nearest stored examples. The exemplar-based approach (also called the nearest neighbor classification algorithm [VO89]) is attractive because of its simple representation, and because it can be used to explain a number of human cognitive phenomena [Sch78, SM81]. Hybrid GA-nearest neighbor algorithms have been developed to learn weights associated with individual attributes [KD91] (used in calculating distance of new instance from the stored instances) and to store only a subset of the set of modified training instances [Che92].

The other form of concept representation that has been commonly posited by cognitive scientists as a theory of human concept learning is that of *prototypes* [Ree72, Smi89]. A prototype is a collection of salient features of a concept. An instance can be classified using prototypes by finding the prototype with which it shares most of its features, and then using the class of that prototype. The exemplar based models of concept classification are more flexible than prototype models, but cannot account for some general aspects of ordinary concepts. Smith [Smi89] concludes that for ordinary concepts like *birds*, we use prototypes rather than storing exemplars; for complex concepts like *students in my class*, however, exemplars can be used to construct prototypes or to directly classify instances. An attractive feature of prototype models is that they produce much more compact concept descriptions compared to exemplar-based models, and hence require much less computational effort to classify new instances. The on-line computation of prototypes, however, has been recognized as a difficult computational problem [Hin86, KM86].

We believe that in addition to possessing psychological plausibility, prototype models can be used to build effective classification techniques for supervised classification problems. We have developed a genetic algorithm (GA) [Hol75] based system, PLEASE, to evolve prototypes from pre-classified instances [SK95]. To evaluate this system, we have defined a set of classification problems involving two classes and two real-valued features. The constructed problems are of varying complexity as measured by the number of prototypes per category required for correct classification and the number of different optimal solutions for the problem. In order to get a measure for the effectiveness of PLEASE in inducing the target concepts, we compared it against two sets of algorithms. The first algorithm chosen for comparison is C4.5 [Qui93], a descendant of ID3, and perhaps the most recognized piece of work in inductive machine learning. We also chose a set of exemplar-based algorithms [AKA91]. While the choice of C4.5 was obviously to compare the performance of PLEASE against the performance of the most recognized system in the field, the choice of exemplar-based algorithms were motivated by the curiosity of finding out how the psychologically plausible algorithms fared compared to each other.

Experimental results show that our prototype-based system performed competitively with both C4.5 and the exemplar-based algorithms on most problems with very low error rates. The generated concept descriptions were also orders of magnitude more compact than the number of training instances stored by the exemplar-based methods. The number of prototypes stored were also less than the number of nodes in the decision-trees produced by C4.5.

We also discuss the shortcomings of our proposed system by using a particular problem in our suite of test problems. An analysis of the relatively poor performance of PLEASE on this domain suggests some further modifications to the system.

The rest of the paper is organized as follows: Section 2. presents a menagerie of possible application areas of supervised concept learning research; Section 3. discusses alternate representations for prototype based classification; Section 4. describes the GA used to evolve these prototypes; Section 5. describes the classification problem set used to evaluate our system; Section 6. compares experimental results of our system with the nearest neighbor algorithms, and Section 7. presents the shortcomings of the current system and future research directions to address these problems.

2. Application Areas

Classification problems are ubiquitous in nature, and are encountered in almost every practical and theoretical discipline. We would like to stress the general applicability of the concept learning paradigm, and hence will not focus on any one classification problem in this work. However, to corroborate our claim that our system can be applied to a wide variety of real-life application domains, we list below a number of interesting classification problems gleaned from the University of California, Irvine repository of machine learning databases [MA92]:

Medical applications: The following is a list of some medical applications of classification:
  - classifying breast-cancer patients
  - classifying echocardiogram traces
  - classifying hepatitis patients
  - classifying patients with Adult Respiratory Distress syndrome (ARDS)
Business applications: The following lists some applications of classification in business:
  - screening credit card applications
  - predicting housing prices in Boston suburbs
  - predicting final settlements in labor negotiations in Canadian industry

– analyzing student loans

Scientific applications: In the following we list some scientific classification problems:

– classifying of radar returns from the ionosphere
– identifying promoter and splice-junction gene sequences
– classify solar flares that occur in a 24 hour period
– classifying low resolution spectrometer data

Industrial and Engineering applications: The following are some of the possible applications of classification techniques in the industrial and engineering disciplines:

– predicting automobile fuel consumption rates
– analyzing design of Pittsburgh bridges
– classifying computer hardware performance in terms of cycle time, memory size, etc.
– identifying types of glass from their oxide contents
– fault diagnosis of electro-mechanical devices
– classifying mushrooms as poisonous or edible in terms of their physical characteristics
– analysis of launch temperature Vs. O-ring stress on USA Space shuttle Challenger
– categorization of Landsat Satellite images

Our work on developing more accurate concept descriptions with relatively small number of training instances will help develop more effective applications for the above-mentioned problem domains.

3. Prototype Based Classification

Usually prototypes are constructed from descriptions of individual instances of a given class by abstracting the more frequent properties of these instances [KM86]. A prototype is often represented in a slot-filler structure containing default attribute values, relationships between attributes, and weights on attributes. Prototypes are often linked in hierarchies. Other possible representations for prototypes include production systems [HHNT86] and connectionist networks. In most cases, only one prototype is constructed per class[1]. The similarity of an input instance to a prototype is calculated from their respective attribute values using a "contrast rule": attributes not having common values are weighted and subtracted from the weighted sum of the attributes having common values. This method of classification explains typicality effects seen in most natural concepts (e.g., some fish are more easily recognized as fish than others). Using a single prototype per class, however,

---

[1] It is widely believed that an ordinary concept definition consist of a prototype constructed from perceptually salient and easy to describe features and a *core* made up of more accurate but less accessible features [SMR84].

is not sufficient to learn linearly non-separable categories. Since most practical classification problems are likely to be linearly non-separable, we allow multiple prototypes per class.

The classification problems used in this paper are defined on real-valued attributes. Accordingly, the calculation of the similarity metric and the representation of prototypes are adapted for real-valued problems. Actually, our representation of prototypes is identical to that of exemplars: a prototype is represented as a list of feature values together with its associated class. Note that whereas Skalak [Ska94] chooses prototypes from the given set of training examples (his methods are more accurately described as exemplar-based methods), prototypes developed by PLEASE are in general different from any of the training instances.

The similarity metric used is the Euclidean distance between the input instance and the prototype in the space defined by the features as follows: Let $P_{ij}, i = \{1, \ldots, n\}$, $j = \{1, \ldots, m\}$ be the value of the j attribute of the ith prototype, and I_{kj} be the value of the jth attribute of the kth instance. Then the nearest prototype to the kth instance is calculated as

$$\arg \min_i \sum_{j=1}^{m} (P_{ij} - I_{kj})^2.$$

Note that our prototype model is similar to the exemplar model in the nearest neighbor calculation. The difference is that the prototypes are, in general, distinct from any examples seen by the system, and the number of prototypes per class is restricted to a small number (10 in this paper). Our model is similar to some prototype models which calculate similarity from a distance measure in some underlying psychological space rather than measuring it by common and distinctive properties [She74]. It should be noted that our use of prototypes to represent target concepts is distinctly different from most of the work on Genetic Based Machine Learning systems, where the focus is on learning rule sets to describe target concepts [DSG93, JM93, GS93].

4. Prototype Representation and Manipulation in PLEASE

In the PLEASE system, each structure in the GA population represents a complete description of the target concept. Hence, PLEASE is a Pitt-style [DeJ90, Smi80] genetic-based machine learning system. We chose the Pitt approach over the Michigan approach [Hol86] because of the availability of all training data before learning is initiated. We believe that the Pitt approach seems to be better suited for batch-mode learning (where all training instances are available before learning is initiated) and for static domains.

The GABL system have successfully used a Pitt approach for learning classification rules on discrete-valued domains [DSG93]. The difference between PLEASE and a typical Pitt system is that a structure consists of a set of prototypes instead of a set of rules.

Each prototype in a structure is represented as a set of feature values. Prototypes belonging to the same class are placed adjacent to each other and prototypes belonging to different classes are separated by a marker, $\|$. Let us denote the jth feature value of the ith prototype belonging to the kth population structure by P_{ij}^k. Then given a problem with two classes and two features, the kth structure in the population with two prototypes belonging to the first class and three prototypes belonging to the second class is represented as:

$$P_{11}^k P_{12}^k \ P_{21}^k P_{22}^k \ \| \ P_{31}^k P_{32}^k \ P_{41}^k P_{42}^k \ P_{51}^k P_{52}^k.$$

Note that the representation can analogously be extended for problems with more than two classes and with many attributes.

Traditional GAs use bit-string representations for population structures [Hol75]. The choice of real valued representation has recently received increasing attention in the GA community [Gol91, JM91] and has been effectively used for supervised concept classification problems [KD91, CS94]. Since the number of prototypes per structure varies, we have a variable-length representation. But we do not allow the length of a structure to grow above a fixed upper bound (predetermined maximum number of prototypes).

Each structure is used to classify all instances in the training set. An instance is correctly classified if its class is the same as the prototype nearest to it in the structure being evaluated. The number of misclassifications by a structure is used as a measure of its fitness. We formulate the prototype learning problem as a function minimization problem, where we are searching for structures that reduce the number of misclassifications on the training set. We have used a rank-based selection method [Whi89] in PLEASE.

4.1 Basic structure manipulation operators

Mutation is used to replace the current value of one feature of a prototype by a randomly generated number in the domain of the feature. Thus mutation allows for a drastic change in the position of a prototype along one of the dimensions in the input space (each dimension corresponds to one attribute of the problem). Mutation, in general, takes place very infrequently. The probability of mutation is set by a parameter p_{mut}. A *creep* operator is used to displace the current position of a prototype on the input space by a small amount [KD91]. The creep operator is applied with a probability p_{creep} and changes the values of all the features of a prototype[2] by a small

---

[2] In our experiments, we found faster convergence when all the feature values were changed by the creep operator rather than changing only one feature at a time.

amount, δ. This translates the prototypes parallel to one of the diagonals of the hypercube representing the input space.

A two point crossover operator is used in PLEASE. The crossover points can fall anywhere within a structure. When two parents are selected for crossover, two crossover points are first randomly selected on one of the parents. Now, two points are chosen on the other parent which match up semantically with the previously chosen points on the first parent. For example, if a crossover point on the first parent is in between prototypes belonging to category X, the corresponding crossover point in the the second parent should also fall between prototypes for category X. Similarly, if the crossover point in the first parent falls between the second and third features of a prototype belonging to category Y, the corresponding crossover point in the the second parent should also fall between the second and third features of a prototype belonging to category Y. After crossover points are selected in both parents, the portions of the chromosomes in between these points are swapped between the parents to produce two offsprings. Any crossover that results in chromosomes with empty categories or with a number of prototypes for a category exceeding a limit, is disallowed. The following example demonstrates a valid two-point crossover as described above (crossover points are marked by the symbol $\updownarrow$):

$$
\begin{array}{rl}
\text{Parent 1:} & P_{11}^k \updownarrow P_{12}^k \;\; P_{21}^k P_{22}^k \parallel P_{31}^k P_{32}^k \;\; P_{41}^k P_{42}^k \updownarrow P_{51}^k P_{52}^k \\
\text{Parent 2:} & P_{11}^l \updownarrow P_{12}^l \parallel P_{21}^l P_{22}^l \updownarrow P_{31}^l P_{32}^l \\
\text{Offspring 1:} & P_{11}^k P_{12}^l \parallel P_{21}^l P_{22}^l \;\; P_{51}^k P_{52}^k \\
\text{Offspring 2:} & P_{11}^l P_{12}^k \;\; P_{21}^k P_{22}^k \parallel P_{31}^k P_{32}^k \;\; P_{41}^k P_{42}^k \;\; P_{31}^l P_{32}^l.
\end{array}
$$

4.2 Advanced structure manipulation operators

We also implemented a number of additional structure manipulation operators to enhance the performance of PLEASE. Most of these operators utilized the special structure of the concept learning problem where there are only two possible classifications. Classification problems with more than two categories can readily be decomposed into as many concept learning problems as there are classes [Mla93].

To implement the advanced operators we keep additional information with each population structure. For each prototype, we maintained a set of indices of the training set instances for which that prototype was the nearest prototype. The set was divided into two subsets: one for instances whose classification matches the classification of the prototype ($\mathcal{C}$), and the other for which the classifications were different ($\mathcal{I}$). The *classification accuracy of a prototype* is the percentage of time it correctly predicts the class of an input instance when it is the nearest prototype to that input, and is calculated as $\frac{\mathcal{C}}{\mathcal{C}+\mathcal{I}} \times 100$.

The advanced operators that we have implemented are the following:

swap: If the classification accuracy of a prototype was below acc_{low} (we have used a value of 25%), then transfer the prototype to the other category in the structure. This makes sense because if a prototype, currently in the -ve category, was matching many more +ve instances than -ve instances, then it will be obviously beneficial to switch this prototype to the +ve category, and vice versa.

split: If the classification accuracy of a prototype, P was between acc_{low} and acc_{high} then we deleted this prototype from the structure and introduced two new prototypes in the two categories. The prototype put into the -ve category was constructed by averaging the attribute values of the -ve classification training set instances to which P was the closest prototype; the prototype to be put into the +ve category was constructed analogously. The motivation behind this operator was that if a prototype is found in a region of the input space such that it is close to both +ve and -ve training set instances, it suggests that that part of the space needs finer resolution for classification. By putting new +ve and -ve prototypes in the midst of instances of the corresponding classification, we would likely increase the classification accuracy of the entire structure.

compress: Because split introduces new prototypes into structures, the population structures tend to increase in size over time. To prevent structures from getting too large, we introduced the compress operators, which will replace two prototypes of the same classification with a new prototype at the midpoint of these prototypes (and having the same classification). This replacement is done only if the two prototypes are within a distance $d_{compress}$ (we have used a value of 0.05) of each other. Since only two closely located prototypes of the same category are replaced with a third prototype of the same category, we expect a decrease in the size of the structure with no significant change in classification accuracy.

Since keeping all the additional information with a structure consumes space and time, we applied this additional operators to only the top 10% of the population in every generation.

A run is terminated if a given number of generations has been completed or if the best solution generated does not change over G generations.

5. Problem Set

To evaluate PLEASE, we designed a set of problems of varying complexity. These problems were defined on two continuous attributes, x and y, each having a range of $[0, 1]$, producing an input space of unit square area. We formulated a number of different problems by labeling different regions of this unit square with one of two possible categories, 1 or 0. The area of the region allocated to each of the categories is equal for all the problems used in this paper. These regions were chosen so that the classification problem can be solved with a specific number of appropriately positioned prototypes for each categories. Different problems require different number of prototypes per category. Given a particular problem, we randomly generated a set of points from the input space and then labeled each point with the category associated with the region containing that point. These sets of points were then divided into training and testing sets.

We present the problems used in the experiments in this paper in Figures 5.1 and 5.2. The problems are labeled as N/M, where N and M stand for the minimum number of prototypes in the two categories required to accurately solve the classification problem. The darkened regions in each of the figures are associated with category 0, and the rest of the square is associated with category 1. The prototypes required to solve the problem are labeled '-' and '+' for categories 0 and 1 respectively.

An important point to note about the problem set presented in Figures 5.1 and 5.2 is that whereas an infinite number of minimal solutions[3] exist for the 1/1, 2/1, 2/2, 3/2, 3/3 and 4/4 problems, unique minimal solutions exist for problems 3/1 and 4/1 problems. The problem set, therefore, consists of problems of varying complexity, both in terms of the minimum number of prototypes required per category to solve the classification problems, as well as in terms of the number of minimal solutions existing for the problems.

6. Experimental Results

In this section, we present results from experiments conducted to solve the previously mentioned classification problems using PLEASE, C4.5, and the exemplar-based algorithms. For each of the problems, 1000 points were randomly generated from the input space and classified according to the region from which they were drawn. We used a five-fold cross validation, with each of the five training/testing set splits containing 800 training instances and 200 testing instances. The constraint imposed on training/test split was that both the categories were equally represented in the training and test sets. Additionally, within a category, all the regions were also equally represented. Results obtained with the C4.5 and exemplar-based algorithms are averaged over these five training/testing set splits for each of the problems. Our PLEASE system uses GAs for learning prototypes, and the performance of the latter depends on the initial random population. So the PLEASE system was run on each training/testing split with 5 different random initializations. The results of PLEASE on each problem, therefore, is averaged over 25 runs.

The exemplar-based algorithms used for comparison are IB1, IB2, and IB3 (Aha, Kibler, & Albert, 1991). IB1 stores all training instances. IB2

---

[3] Minimal solutions refer to solutions that require the minimum number of prototypes for each category; any of the problems can also be solved with a larger number of prototypes in infinitely many ways.

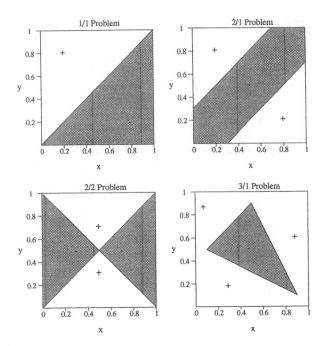

Fig. 5.1. The classification problem set; part 1

Table 6.1. Average percentage of incorrect classifications by PLEASE, C4.5 and the exemplar-based algorithms on test sets for different problems

Problem	PLEASE	IB1	IB2	IB3	C4.5
1/1	0.6	1.5	2.8	7.6	3.7
2/1	1.37	1.7	5.4	14.3	4.3
2/2	1.32	2.4	5.3	23.5	5.1
3/1	13.84	4.6	8	32.7	12.5
3/3	4.54	3.2	10.1	21.34	5.04
4/1	2.42	2.9	7.3	17	5.5
4/4	4.34	5.22	10.75	34.5	7.3

stores a training instance only if it is incorrectly classified by already stored instances. IB3 maintains a classification record of stored instances and discards instances that fails a significance test to detect good classifiers. IB2 and IB3 produces much more compact concept descriptions than IB1, but their classification accuracy is often not as good.

The parameters used for the PLEASE system in these experiments are as follows: population size = 150, number of generations = 200, $p_{mut} = 0.0001$, $p_{creep} = 0.3$, $\delta = 0.1$, selection bias = 1.7, $G = 25$, maximum number of prototypes per category = 10. Table 6.1 contains the average of the percentage

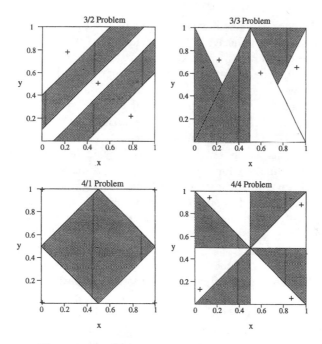

Fig. 5.2. The classification problem set; part 2

error rates of the exemplar-based algorithms, C4.5, and PLEASE on the problem set. From the table we find that for a majority of the problems, PLEASE was able to consistently discover a set of prototypes that achieved very high classification accuracy. A two-sample t procedure shows that the test set performance differences between PLEASE and IB1 are significant at the 95% confidence level except for 2/1, 4/1, and 4/4 problems. Among the other problems, half of the time PLEASE performed better, and for the other half, IB1 performed better. IB1 achieves this higher accuracy by storing about 50 times the number of instances as the prototypes developed by PLEASE. PLEASE consistently outperforms IB2 and IB3, which stores considerably less number of instances.

Though C4.5 performed slightly better on training instances in the some of the problems, these differences were not statistically significant. On the other hand, a two-sample t procedure shows that with at least 99% confidence level it can be stated that PLEASE produced lower error rates on almost all of the test instances than C4.5. The other notable difference between C4.5 and PLEASE was their relative performance on the training and test cases. Training and test set errors were much more consistent in PLEASE than in C4.5. In fact, for the C4.5 system, test errors were approximately 5 to 8 times higher than errors on training instances. Since test set performance is more important in supervised classification problems, PLEASE seems to be able to better represent the underlying target concept.

Solutions produced by PLEASE contain, in general, more prototypes than required by the minimal solutions to the corresponding problems. Some of the solutions, however, were indeed minimal solutions. The average number of prototypes used for the two categories for some of the test problems are as follows: 1.93 and 1.72 for 1/1, 4.03 and 2.23 for 2/1, 2.93 and 3.3 for 2/2, 4.52 and 4.04 for 4/1, 4.32 and 4.52 for 4/4. The number of prototypes in the best solutions can possibly be reduced by including a penalty term in the evaluation function for more prototypes. We have not done that yet because we did not want to distort the fitness landscape. As IB1 stores all training instances, the data compression obtained with the PLEASE system is significantly better. The number of instances stored by IB2 and IB3 varied between 40 and 120. The decision trees produced by C4.5 for these problems contained, on the average, 44.2, 57.4, 69, and 70.2 nodes for the 1/1, 2/1, 3/1, and 4/1 problems respectively. Hence, the data compression obtained with the PLEASE system is significantly better compared to that obtained with C4.5.

We will use a typical solution for the 4/1 problem to analyze both the nature of the solutions generated by PLEASE, and the classification errors made by such a solution. Figure 6.2 presents the 1000 data points for the 4/1 problem and also a typical solution produced by PLEASE together with the training and test set misclassifications of that solution. For the sake of clarity we have also drawn the lines outlining the embedded square for this problem. The placement of the prototypes in the solution generated by PLEASE is found to approximate the prototype locations in the minimal solution. This close approximation to the unique minimal solution produces very low classification errors, 3 on the training set and 3 on the test set. An interesting characteristic of the solutions generated by PLEASE for most problems is the nature of the misclassifications produced. There are no "blatant" errors, i.e., all errors are located close to the boundary of separation between the categories (the 3/2 problem is an exception, and we discuss it in detail below).

Results from the 3/2 problem exposed some shortcomings of our system (see Figure 6.4). We found a wide variation of the types of solutions generated by PLEASE for this problem. The 3/2 problem is a difficult problem as even the well-known decision tree based system C4.5 [Qui93] misclassified 12.5% of the test instances on the average. From Figure 6.4 we find that the solution produced by PLEASE for this problem almost completely misclassified the points in the middle "bar" associated with category 1. The solution developed is similar to a solution for the 2/1 problem! The prototypes in the middle are so located that each of them correctly classifies a number of points in category 0 and misclassified a number of points in category 1. The advanced operators does not help, either because the classification accuracies of the prototypes are such that the operator is not invoked, or because the new prototypes created

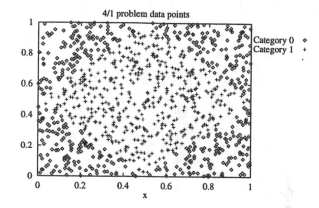

Fig. 6.1. The 4/1 problem

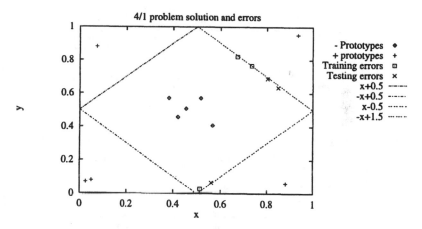

Fig. 6.2. The solutions generated by PLEASE

do not improve performance. We believe that further experimentation with fine tuning the acc_{low} and acc_{high} parameters can help alleviate some of these problems.

We ran an additional set of experiments using PLEASE and with or without the advanced structure manipulation operators. An interesting observation from these set of experiments is that these advanced operators helped improve performance on more difficult problems, but also resulted in decreased classification accuracy for simpler problems.

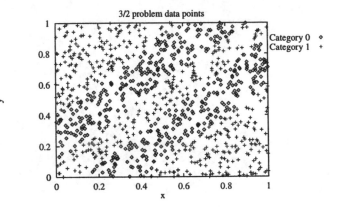

Fig. 6.3. The 3/2 problem

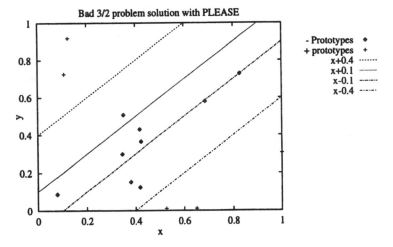

Fig. 6.4. The solutions generated by PLEASE

7. Conclusions

This paper shows that using genetic algorithms to learn prototypes to be used for concept description can be an effective means for addressing supervised classification problems. PLEASE is a novel solution to the difficult problem of on-line computation of prototypes from training data [Hin86, KM86].

This paper contains an experimental evaluation of our proposed classification system, PLEASE, on a set of real-valued classification problems of varying complexity. PLEASE produces highly accurate solutions to most of these problems and performs competitively with the exemplar-based algorithms. In particular, the performance of PLEASE is comparable to or better than C4.5 and exemplar-based learning algorithms on a majority of the test set problems. Our prototype based classification approach is more difficult to train than the exemplar-based algorithms, but because of significantly more compact description of target concepts, is faster in classifying test instances.

We plan to develop a representation for structured and nominal attributes for use in PLEASE. Other planned extensions include using prototypes with a subset of attributes, developing a prototype add (delete) operator that randomly selects a class in the chromosome and then adds (deletes) a prototype to (from) that class, and seeding of the population by chromosomes constructed from possibly useful prototypes. We also plan to evaluate PLEASE for noise immunity and develop a prototype-based Michigan-style classifier system for incremental concept learning.

References

[AKA91] David W. Aha, Dennis Kibler, and Marc K. Albert. Instance-based learning algorithms. *Machine Learning*, 6(1), 1991.

[BFO84] L. Breiman, J.H. Friedman, and R.A. Olshen. *Classification and Regression Trees*. Wadsworth, Belmont, CA, 1984.

[BGH89] L.B. Booker, D.E. Goldberg, and J.H. Holland. Classifier systems and genetic algorithms. *Artificial Intelligence*, 40:235–282, 1989.

[Che92] K.J. Cherkauer. Genetic search for nearest-neighbor exemplars. In *Proceedings of the Fourth Midwest Artificial Intelligence and Cognitive Science Society Conference*, pages 87–91, 1992.

[CKS+88] P. Cheeseman, J. Kelly, M. Self, J. Stutz, W. Taylor, and D. Freeman. Autoclass: A bayesian classification system. In *Proceedings of the Fifth International Conference on Machine Learning*, pages 54–64, 1988.

[CS94] A.L. Corcoran and S. Sen. Using real-valued genetic algorithms to evolve rule sets for classification. In *Proceedings of the IEEE Conference on Evolutionary Computation*, pages 120–124, 1994.

[DeJ90] Kenneth A. DeJong. Genetic-algorithm-based learning. In Y. Kodratoff and R.S. Michalski, editors, *Machine Learning, Volume III*. Morgan Kaufmann, Los Alamos, CA, 1990.

[DSG93] Kenneth A. DeJong, William M. Spears, and Diana F. Gordon. Using genetic algorithms for concept learning. *Machine Learning*, 13:161–197, 1993.

[Fis87] Douglas H. Fisher. Knowledge acquisition via incremental conceptual clustering. *Machine Learning*, 2:139–172, 1987.

[Gol91] David Goldberg. Real-coded genetic algorithms, virtual alphabets, and blocking. *Complex Systems*, 5:139–168, 1991.

[GS93] David Perry Greene and Stephen F. Smith. Competition-based induction of decision models from examples. *Machine Learning*, 13:229–258, 1993.

[HHNT86] John H. Holland, K.J. Holyoak, R.E. Nisbett, and P.R. Thagard. *Induction: Processes of Inferences, Learning, and Discovery.* MIT Press, Cambridge, MA, 1986.

[Hin86] D.L. Hintzman. "schema abstraction" in a multiple trace memory model. *Psychological Review*, 93:411–428, 1986.

[Hol75] John H. Holland. *Adpatation in natural and artificial systems.* University of Michigan Press, Ann Arbor, MI, 1975.

[Hol86] John H. Holland. Escaping brittleness: the possibilities of general-purpose learning algorithms applied to parallel rule-based systems. In R.S. Michalski, J.G. Carbonell, and T. M. Mitchell, editors, *Machine Learning, an artificial intelligence approach: Volume II.* Morgan Kaufmann, Los Alamos, CA, 1986.

[JM91] Cezary Z. Janikow and Zbigniew Michalewicz. An experimental comparison of binary and floating point representations in genetic algorithms. In *Proceedings of the 4th International Conference on Genetic Algorithms*, pages 31–36, San Mateo, CA, 1991. Morgan Kaufman.

[JM93] Cezary Z. Janikow and Zbigniew Michalewicz. A knowledge intensive genetic algorithm for supervised learning. *Machine Learning*, 13:198–228, 1993.

[KD91] James D. Kelly and Lawrence Davis. A hybrid genetic algorithm for classification. In *Proceedings of the International Joint Conference on Artificial Intelligence*, pages 645–650, 1991.

[KM86] D. Kahneman and D.T. Miller. Norm theory: Comparing reality to its alternatives. *Psychological Review*, 93:136–153, 1986.

[Kru92] J.K. Kruschke. ALCOVE: An exemplar-based connectionist model of category learning. *Psychological Review*, 99:22–44, 1992.

[MA92] P.M. Murphy and D.W. Aha. UCI repository of machine learning databases [machine-readable data repository], 1992.

[Mic83] R.S. Michalski. A theory and methodology of inductive learning. In R.S. Michalski, J. Carbonell, and T. Mitchell, editors, *Machine Learning: An Artificial Intelligence Approach.* Morgan Kaufman, Los Alamos, CA, 1983.

[Mit86] T. Mitchell. Generalization as search. *Artificial Intelligence*, 18:203–226, 1986.

[Mla93] Dunja Mladenić. Combinatorial optimization in inductive concept learning. In *Proceedings of the Tenth International Conference on Machine Learning*, pages 205–211, 1993.

[Qui86] Ross J. Quinlan. Induction of decision trees. *Machine Learning*, 1:81–106, 1986.

[Qui90] Ross J. Quinlan. Learning logical definitions from relations. *Machine Learning*, 5(3):239–266, 1990.

[Qui93] Ross J. Quinlan. *C4.5: Programs for Machine Learning.* Morgan Kaufmann, San Mateo, California, 1993.

[Ree72] S.K. Reed. Pattern recognition and categorization. *Cognitive Psychology*, 3:382–407, 1972.

[RHW86] D.E. Rumelhart, G.E. Hinton, and R.J. Williams. Learning internal representations by error propagation. In D.E. Rumelhart and J.L. McClelland, editors, *Parallel Distributed Processing*, volume 1. MIT Press, Cambridge, MA, 1986.

[RZ85] David E. Rumelhart and David Zipser. Feature discovery by competitive learning. *Cognitive Science*, 9(1), 1985.

[Sch78] D. L. Medin & M.M. Schaffer. Context theory of classification learning. *Psychological Review*, 85:207–238, 1978.

[She74] R.N. Shepard. Representation of structure in similarity data: Problems and prospects. *Psychometrika*, 39:373–421, 1974.

[SK95] Sandip Sen and Leslie Knight. A genetic prototype learner. In *Proceedings of the International Joint Conference on Artificial Intelligence*, 1995.

[Ska94] David B. Skalak. Prototype and feature selection by sampling and random mutation hill climbing algorithms. In *Proceedings of the Eleventh International Conference on Machine Learning*, pages 293–301, 1994.

[SM81] E.E. Smith and D.L. Medin. *Categories and concepts*. Harvard University Press, Cambridge, MA, 1981.

[Smi80] Steve F. Smith. *A learning system based on genetic adaptive algorithms*. PhD thesis, University of Pittsburgh, 1980. (Dissertation Abstracts International, 41, 4582B; University Microfilms No. 81-12638).

[Smi89] Edward E. Smith. Concepts and induction. In Michael I. Posner, editor, *Foundations of Cognitive Science*. MIT Press, Cambridge, MA, 1989.

[SMR84] E.E. Smith, D.L. Medin, and L.J. Rips. A psychological approach to concepts: Comments on rey's "concepts and stereotypes". *Cognition*, 17:265–274, 1984.

[SW86] C. Stanfill and D. Waltz. Toward memory-based reasoning. *Communications of the ACM*, 29:1213–1228, 1986.

[VO89] S. Vosniadou and A. Ortony. *Similarity and Analogical Reasoning*. Cambridge University Press, Cambridge, MA, 1989.

[Whi89] D. Whitley. The genitor algorithm and selection pressure: Why rank-based allocation of reproductive trials is best. In *Proceedings of the 3rd International Conference on Genetic Algorithms*, pages 116–121, San Mateo, CA, 1989. Morgan Kaufman.

Prototyping Intelligent Vehicle Modules Using Evolutionary Algorithms

Shumeet Baluja, Rahul Sukthankar, and John Hancock

School of Computer Science, Carnegie Mellon University, Pittsburgh, PA 15213

Summary. Intelligent vehicles must make real-time tactical level decisions to drive in mixed traffic environments. SAPIENT is a reasoning system that combines high-level task goals with low-level sensor constraints to control simulated and (ultimately) real vehicles like the Carnegie Mellon Navlab robot vans.

SAPIENT consists of a number of reasoning modules whose outputs are combined using a voting scheme. The behavior of these modules is directly dependent on a large number of parameters both internal and external to the modules. Without carefully setting these parameters, it is difficult to assess whether the reasoning modules can interact correctly; furthermore, selecting good values for these parameters manually is tedious and error-prone. We use an evolutionary algorithm, termed Population-Based Incremental Learning, to automatically set each module's parameters. This allows us to determine whether the combination of chosen modules is well suited for the desired task, enables the rapid integration of new modules into existing SAPIENT configurations, and provides an automated way to find good parameter settings.

1. Introduction

The task of driving can be characterized as consisting of three levels: strategic, tactical and operational [13]. At the highest (strategic) level, a route is planned and goals are determined; at the intermediate (tactical) level, maneuvers are selected to achieve short-term objectives — such as deciding whether to pass a blocking vehicle; and at the lowest (operational) level, these maneuvers are translated into control operations.

Mobile robot research has successfully addressed the three levels to different degrees. Strategic-level planners [17, 23] have advanced from research projects to commercial products. The operational level has been investigated for many decades, resulting in systems that range from semi-autonomous vehicle control [7, 12] to autonomous driving in a variety of situations [5, 18, 14]. Substantial progress in autonomous navigation in simulated domains has also been reported in recent years [16, 4, 15]. However, the decisions required at the tactical level are difficult and a general solution remains elusive.

Consider the typical tactical decision scenario depicted in Figure 1.1:

Our vehicle (A) is in the right lane of a divided highway, approaching the chosen exit. Unfortunately, a slow car (B) blocks our lane, preventing us from moving at our preferred velocity. Our desire to pass the slow car conflicts with our reluctance to miss the exit. The correct decision in this case depends not only on the distance to the exit, but also on the traffic configuration in the area. Even if the distance to the exit is sufficient for a pass, there may be no

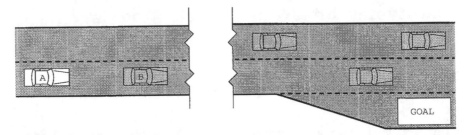

Fig. 1.1. Car A is approaching its desired exit behind a slow vehicle B. Should Car A attempt to pass?

suitable gaps in the right lane ahead before the exit. SAPIENT, described in Section 3., is a collection of intelligent vehicle algorithms designed to drive the Carnegie Mellon Navlab [22, 10] in situations similar to the given scenario. SAPIENT has a distributed architecture which enables researchers to quickly add new reasoning modules to an existing configuration, but it does not address the problem of reconfiguring the parameters in the new system. We present an evolutionary algorithm, Population-Based Incremental Learning (PBIL), that automatically searches this parameter space and learns to drive vehicles in traffic.

2. SHIVA

Simulation is essential in developing intelligent vehicle systems because testing new algorithms in real traffic is expensive, risky and potentially disastrous. SHIVA[1] (Simulated Highways for Intelligent Vehicle Algorithms) [21, 20] is a kinematic micro-simulation of vehicles moving and interacting on a user-defined stretch of roadway that models the elements of the tactical driving domain most useful to intelligent vehicle designers. The vehicles can be equipped with simulated human drivers as well as sensors and algorithms for automated control. These algorithms influence the vehicles' motion through simulated commands to the accelerator, brake, and steering wheel. SHIVA's user interface provides facilities for visualizing and influencing the interactions between vehicles (see Figure 2.1). The internal structure of the simulator is comprehensively covered in [21] and details of the design tools may be found in [20].

SHIVA's architecture is open-ended, enabling researchers to simulate interactions between a variety of vehicle configurations. All vehicles can be functionally represented as consisting of three subsystems: perception, cognition, and control (see Figure 2.2).

---

[1] For further information, see <http://www.cs.cmu.edu/~rahuls/shiva.html>

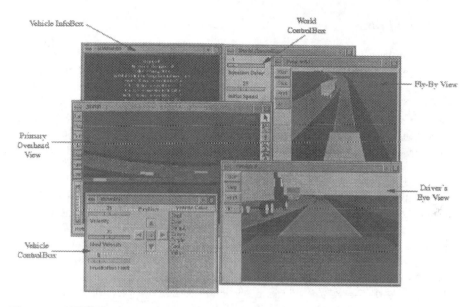

Fig. 2.1. SHIVA: A design and simulation tool for developing intelligent vehicle algorithms.

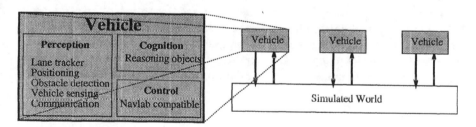

Fig. 2.2. Each vehicle is composed of three subsystems which interact with the simulated world.

2.1 Perception

The perception subsystem consists of a suite of simulated functional sensors (e.g. global positioning systems, range-sensors, lane-trackers), whose outputs are similar to real perception modules implemented on the Navlab vehicles. SHIVA vehicles use these sensors to get information about the road geometry and surrounding traffic. Vehicles may control the sensors directly, activating and panning the sensors as needed, encouraging active perception. Some sensors also model occlusion and noise, forcing cognition routines to be realistic in their input assumptions. Two perception components are particularly relevant to this paper: the lane tracker and the car tracker.

The lane tracker assumes a pure-pursuit [24] model of road-following. This means that the lane tracker suggests a steering arc that will bring the vehicle to the center of the lane after traveling a (velocity dependent) *lookahead distance*. The lane tracker may also be directed to steer the vehicle towards an arbitrary lateral offset on the road. Thus, lane changing is implemented by smoothly varying the lateral position of the pure-pursuit point from the center of one lane to the center of the desired adjacent lane [11]. It is important to note that the actual lateral offset of the vehicle always lags the current position of its pure-pursuit point.

Car tracking is a two-step process. In the first phase, the sensor determines the nearest visible vehicle in its range and field of view. In the second, the sensed vehicle's position is transformed into *road coordinates* (i.e. relative lateral and longitudinal offsets). This allows the tactical reasoning algorithms to remain invariant over changes in road curvature. Car trackers scanning different areas of the road (e.g. front-right, rear-right) are activated as needed during tactical maneuvers to provide relevant information about surrounding traffic.

2.2 Cognition

While a variety of cognition modules have been developed in SHIVA, this paper is only concerned with two types: rule-based reasoning and SAPIENT. The rule-based reasoning system, which was manually designed, is implemented as a monolithic decision tree. An internal state reflects the current mode of operation (lane changing, lane tracking, seeking an exit, etc.) and hand-crafted rules are used to generate actions (steering command and velocity changes) and state transitions. For example, a rule included in car passing maneuvers is:

"Initiate a left lane change if the vehicle ahead is moving slower than $f(v)$ m/s, and is closer than $h(v)$, and if the lane to your left is marked for legal travel, and if there are no vehicles in that lane within $g(v)$ meters, and if the desired right-exit is further than $e(x, y, v)$ meters."

where: $f(v)$ is the desired car following velocity, $h(v)$ is the desired car following distance (headway), $g(v)$ is the required gap size for entering an adjacent lane, and $e(x, y, v)$ is a distance threshold to the exit based on current lane, distance to exit and velocity. As the maneuver is initiated, the vehicle moves from the *lane tracking* to the *lane changing* state. While this system performs well on many scenarios, it suffers from four disadvantages: 1) as the example above illustrates, realistic rules require the designer to account for many factors; 2) modification of the rules is difficult since a small change in desired behavior can require many non-local modifications; 3) hand-coded rules perform poorly in unanticipated situations; 4) implementing new features requires one to consider an exponential number of interactions with existing rules. Similar problems were reported by Cremer *et al.* [4] in their monolithic state-machine implementation for scenario control. To address some of these problems, we have developed a distributed reasoning architecture, SAPIENT, which is discussed in Section 3.

2.3 Control

The control subsystem is compatible with the controller available on the Carnegie Mellon Navlab II robot testbed vehicle, and only allows vehicles to control desired velocity and steering curvature. Denying control over acceleration prevents simulated vehicles from performing operations such as platooning, but ensures that systems developed in simulation can be directly ported onto existing hardware.

3. SAPIENT

SAPIENT (**S**ituational **A**wareness **P**lanner **I**mplementing **E**ffective **N**avigation in **T**raffic) [19] is a reasoning system designed to solve tactical driving problems. To overcome deficiencies with the monolithic reasoning systems described in Section 2.2, SAPIENT partitions the driving task into many independent aspects. Each aspect is represented by an independent module known as a *reasoning object*.

3.1 Reasoning objects

Wherever possible, each *reasoning object* represents a physical entity relevant to the driving task (e.g. car ahead, upcoming exit). Similarly, different aspects of the vehicle's self-state (e.g. how velocity compares to desired velocity) are also represented as individual reasoning objects. Every reasoning object takes inputs from one or more sensors (e.g. the reasoning object for the vehicle ahead monitors forward-facing car tracking sensors).

Each reasoning object tracks relevant attributes of the appropriate entity. Some aspects of the tactical situation can be represented using stateless models of the entity (e.g. speed limits) while others require information about the past (e.g. lane changing). In SAPIENT, each reasoning object is responsible for maintaining the relevant state information. The following reasoning objects were used in the scenarios described in this paper:

Reasoning Object	Monitors
Inertia	None: blindly favors constant velocity, same lane.
Velocity	Self state: current and desired velocity
Lane	Self state: lane keeping, lane changing
Exit	Self state: desire to exit; distance to exit
Lead car	Position, velocity and size of car ahead
Front-right car	Position, velocity and size of front-right car
Front-left car	Position, velocity and size of front-left car
Back-right car	Position, velocity and size of back-right car
Back-left car	Position, velocity and size of back-left car

Reasoning objects do not communicate with each other, and are activated and destroyed in response to sensed events and higher level commands. For example, the Exit Object activates only when the desired exit is nearby, and Sensed Car Objects are destroyed when the vehicle being tracked moves out of the region of interest. Each reasoning object examines the repercussions of each *action* (see Section 3.2) as it would affect the appropriate entity. Thus an Exit Object analyzes a possible right lane change only in terms of its impact on the chance of making the desired exit, and ignores the possible interactions with vehicles in the right lane (this is taken care of by other reasoning objects). Every reasoning object then presents its recommendations about the desirability of each proposed maneuver. For this to work, all reasoning objects must share a common output representation. In SAPIENT, every object votes over a predetermined set of actions.

3.2 Actions

At the tactical level, all actions have a *longitudinal* and a *lateral* component. This choice of coordinate frames allows reasoning objects to be invariant over the underlying road geometry as far as possible. Intuitively, longitudinal commands correspond to speeding up or braking, while lateral commands map to lane changes. More complex maneuvers are created by combining these basic actions. Since the scales in the two dimensions are greatly different, we chose to encode longitudinal motion in terms of changes in velocity, and lateral motion as changes in displacement. Thus the null action represents maintaining speed at the current lane offset.

Tactical maneuvers (such as lane changing) are composed by concatenating several basic actions. Reasoning objects indicate their preference for a basic action by assigning a vote to that action. The magnitude of the vote corresponds to the intensity of the preference and its sign indicates approval or disapproval. Each reasoning object must assign some vote to every action in the action space. This information is expressed in the *action matrix*. For the experiments reported in this paper, we used the following simple 3×3 action matrix:

decelerate-left	nil-left	accelerate-left
decelerate-nil	nil-nil	accelerate-nil
decelerate-right	nil-right	accelerate-right

If smoother control is desired, an action matrix with more rows and/or columns may be used.

Since different reasoning objects can return different recommendations for the same action, conflicts must be resolved and a good action selected. SAPI-ENT uses a voting arbiter to perform this integration. During arbitration, the votes in each reasoning object's action matrix are multiplied by a scalar weight, and the resulting matrices are summed together to create the *cumulative action matrix*. The action with the highest cumulative vote is selected for execution in the next time-step. This action is sent to the controller and converted into actuator commands (steering and velocity control).

3.3 Parameters

As described in Section 3.1, different reasoning objects use different internal algorithms. Each reasoning object's output depends on a variety of *internal parameters* (e.g. thresholds, gains, etc.). The outputs are then scaled by *external parameters* (e.g. weights).

When a new reasoning object is being implemented, it is difficult to determine whether a vehicle's poor performance should be attributed to a bad choice of parameters, a bug within the new module or, more seriously, to a poor representation scheme (inadequate configuration of reasoning objects). To overcome this difficulty, we have implemented a method for automatically configuring the parameter space. A total of twenty parameters, both internal and external, were selected for the tests described here, and each parameter was discretized into eight values (represented as a three-bit string). For internal parameters, whose values are expected to remain within a certain small range, we selected a linear mapping (where the three bit string represented integers from 0 to 7); for the external parameters, we used an exponential representation (with the three bit string mapping to weights of 0 to 128). The latter representation increases the range of possible weights at the cost of sacrificing resolution at the higher magnitudes. A representation with more bits per parameter would allow finer tuning but increase the learning times. Empirically, we found that three bits per parameter allowed good solutions to be rapidly discovered. The encoding is illustrated in Figure 3.1. In the next section, we describe the evolutionary algorithm used for the learning task.

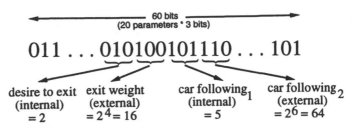

Fig. 3.1. The three-bit encoding scheme used to represent parameters: internal parameters are linearly scaled while external ones are exponentially scaled.

4. Population-Based Incremental Learning

This section provides a brief introduction to Population-Based Incremental Learning (PBIL), and its relation to genetic algorithms (GAs). PBIL is a combination of genetic algorithms and competitive learning [1, 3]. The PBIL algorithm attempts to explicitly maintain statistics about the search space and uses them to direct its exploration. The object of the algorithm is to create a real valued probability vector which, when sampled, reveals high quality solution vectors with high probability. For example, if a good solution can be encoded as a string of alternating 0's and 1's, a suitable final probability vector would be 0.01, 0.99, 0.01, 0.99, etc. The full algorithm is shown in Figure 4.1.

Initially, each element of the probability vector is initialized to 0.5. Sampling from this vector yields random solution vectors because the probability of generating a 0 or 1 is equal. As search progresses, the values in the probability vector gradually shift to represent high evaluation solution vectors through the following process. A number of solution vectors are generated based upon the probabilities specified in the probability vector. The probability vector is pushed towards the generated solution vector with the highest evaluation. After the probability vector is updated, a new set of solution vectors is produced by sampling from the updated probability vector, and the cycle is continued. As the search progresses, entries in the probability vector move away from their initial settings of 0.5 towards either 0.0 or 1.0.

One key feature of the early generations of genetic optimization is the parallelism in the search; many diverse points are represented in the population of points during the early generations. When the population is diverse, crossover is an effective means of search, since it provides a method to explore novel solutions by combining different members of the population. Since a GA uses a full population, while PBIL only uses a single probability vector, PBIL may seem to have less expressive power than a GA. A GA can *represent* a large number of points simultaneously; however, a traditional single population GA will not be able to *maintain* a large number of diverse points. Over

```
****** Initialize Probability Vector ******
for i := 1 to LENGTH do P[i] = 0.5;

while (NOT termination condition)
        ****** Generate Samples ******
        for i := 1 to SAMPLES do
                sample_vectors[i] := generate_sample_vector_according_to_probabilities(P);
                evaluations[i] := Evaluate_Solution( sample_vectors[i]; );
        best_vector := find_vector_with_best_evaluation( sample_vectors, evaluations );

        ****** Update Probability Vector towards best solution ******
        for i := 1 to LENGTH do
                P[i] := P[i] * (1.0 – LR) + best_vector[i] * (LR);

        ****** Mutate Probability Vector ******
        for i := 1 to LENGTH do
                if (random (0,1) < MUT_PROBABILITY) then
                        if (random (0,1) > 0.5) then mutate_direction := 1;
                        else mutate_direction := 0;
                        P[i] := P[i] * (1.0 – MUT_SHIFT) + mutate_direction * (MUT_SHIFT);


USER DEFINED CONSTANTS (Values Used in this Study):
SAMPLES: the number of vectors generated before update of the probability vector (100)
LR: the learning rate, how fast to exploit the search performed (0.1).
LENGTH: the number of bits in a generated vector (3 * 20)
MUT_PROBABILITY: the probability of a mutation occuring in each position (0.02).
MUT_SHIFT: the amount a mutation alters the value in the bit position (0.05).
```

Fig. 4.1. PBIL algorithm, explicit preservation of best solution from one generation to next is not shown.

a number of generations, sampling errors cause the population to converge around a single point. This phenomenon is summarized below:

"... the theorem [Fundamental Theorem of Genetic Algorithms [8]] assumes an infinitely large population size. In a finite size population, even when there is no selective advantage for either of two competing alternatives ... the population will converge to one alternative or the other in finite time (De Jong, 1975; Goldberg & Segrest, ICGA-2). This problem of finite populations is so important that geneticists have given it a special name, genetic drift. Stochastic errors tend to accumulate, ultimately causing the population to converge to one alternative or another" [9].

Diversity in the population is crucial for GAs. By maintaining a population of solutions, the GA is able — in theory at least — to maintain samples in many different regions. Crossover is used to merge these different solutions. However, when the population converges, this deprives crossover of the diversity it needs to be an effective search operator. When this happens, crossover begins to behave like a mutation operator that is sensitive to the convergence of the value of each bit [6]. If all individuals in the population converge at some bit position, crossover leaves those bits unaltered. At bit positions where individuals have not converged, crossover will effectively mutate values in those positions. Therefore, crossover creates new individuals that differ from the individuals it combines only at the bit positions where the mated individuals disagree. This is analogous to PBIL which creates new trial solutions that differ mainly in bit positions where prior good performers have disagreed. More details can be found in [3].

Some of the problems with diversity loss can be addressed by using GAs which employ several independently evolving populations with infrequent interactions. PBIL has also been extended in similar directions with promising results [2]. However, for the tasks explored in this study, these advanced techniques were not needed.

As an example of how the PBIL algorithm works, we can examine the values in the probability vector through multiple generations. Consider the following, simple maximization problem:

$$\text{Eval} = \frac{1.0}{|366503875925.0 - X|}$$

where $0 \leq X < 2^{40}$. Note that 366503875925 is represented in binary as a string of 20 pairs of alternating '01'. The evolution of the probability vector is shown in Figure 4.2. Note that the most significant bits are pinned to either 0 or 1 very quickly, while the least significant bits are pinned last. This is because during the early portions of the search, the most significant bits yield more information about high evaluation regions of the search space than the least significant bits.

The probabilistic generation of solution vectors does not guarantee the creation of a good solution vector in every iteration. This problem is exacer-

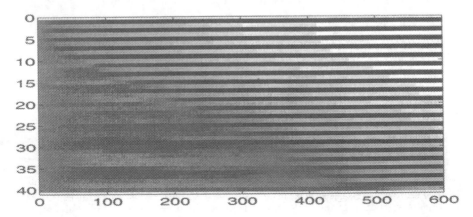

Fig. 4.2. Y axis: bit position in probability vector. X axis: generations. Evolution of the probability vector over successive generations. The value of each element is displayed as a greyscale value. Black and white represent a high probability of generating a 0 and 1 respectively (with grey mapping to intermediate probabilities). Position 0 is the most significant, position 40 is the least. Note that PBIL pins the most significant elements of the probability vector early.

bated by the small population sizes used in these experiments. Therefore, in order to avoid moving towards unproductive areas of the search space, the best vector from the previous population is included in the current population (by replacing the worst member of the current population). This solution vector is only used if the current generation does not produce a better solution vector. In GA literature, this technique of preserving the best solution vector from one generation to the next is termed *elitist selection*, and is used to prevent the loss of good solutions once they are found.

Our application challenges PBIL in a number of ways. First, since a vehicle's decisions depend on the behavior of other vehicles which are not under its control, each simulation can produce a different evaluation for the same bit string. We evaluate each set of vehicle parameters multiple times to compensate for the stochastic nature of the environment. Second, the PBIL algorithm is never exposed to all possible traffic situations (thus making it impossible to estimate the "true" performance of a PBIL string). Third, since each evaluation takes considerable time to simulate, minimizing the total number of evaluations is important.

5. Training Specifics

All of the tests described below were performed on the track shown in Figure 5.1, known as the *Cyclotron*. While this highway configuration is not

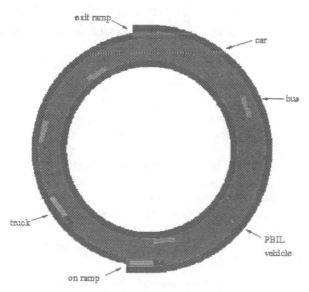

Fig. 5.1. The *cyclotron* test track

encountered in real-life, it has several advantages as a testbed:

1. It is topologically identical to a highway with equally spaced exits.
2. Taking the nth exit is equivalent to traveling n laps of the course.
3. One can create challenging traffic interactions at the entry and exit merges with only a small number of vehicles.
4. The entire track can be displayed on a workstation screen.

For training, each scenario was initialized with one PBIL vehicle, and eight rule-based cars (with hand-crafted decision trees). The PBIL car was directed to take the second exit (1.5 revolutions) while the other cars had goals of zero to five laps. Whenever the total number of vehicles on the track dropped below nine, a new vehicle was injected at the entry ramp (with the restriction that there was always exactly one PBIL vehicle on the course).

Whenever a PBIL vehicle left the scenario (upon taking an exit, or crashing 10 times), its evaluation was computed based on statistics collected during its run. This score was used by the PBIL algorithm to update the probability vector — thus creating better PBIL vehicles in the next generation.

While driving performance is often subjective, all good drivers should display at least the following characteristics: they should drive without colliding with other cars, try to take the correct exit, maintain their desired velocity, and drive without straddling the lane markers. Additionally, they should always recommend some course of action, even in hopeless situations.

We encoded the above heuristics as an evaluation function to be maximized:

$$\text{Eval} = \begin{aligned}&-(10000 \times \text{all-veto}) - (1000 \times \text{num-crashes}) - (500 \times \text{if-wrong-exit})\\&-(0.02 \times \text{accum-speed-deviation}) - (0.02 \times \text{accum-lane-deviation})\\&+(\text{dist-traveled})\end{aligned}$$

where:

- **all-veto** indicates that the PBIL vehicle has extreme objections to all possible actions. With good parameters, this should never happen.
- **num-crashes** is the number of collisions involving the PBIL vehicle.
- **if-wrong-exit** is a flag, which is true if and only if the PBIL vehicle exited prematurely, or otherwise missed its designated exit.
- **accum-speed-deviation** is the difference between desired and actual velocities, integrated over the entire run.
- **accum-lane-deviation** is the deviation from the center of a lane, integrated over the entire run.
- **dist-traveled** is the length of the run, in meters; this incremental reward for partial completion helps learning.

While the evaluation function is a reasonable measure of performance, it is important to note that there can be cases when a "good" driver becomes involved in unavoidable crashes; conversely, favorable circumstances may enable "bad" vehicles to score well on an easy scenario. To minimize the effects of such cases, we tested each PBIL string in the population on a set of four scenarios. In addition to light traffic, these scenarios also included some pathological cases with broken-down vehicles obstructing one or more lanes. Adding a wider variety of scenarios, including those in which the traffic congestion levels are higher, has the potential to reveal more comprehensive evaluations of parameter settings. However, the addition of more training scenarios will lead to longer training times.

6. Results

We performed a series of experiments using a variety of population sizes, evaluation functions and initial conditions. The evaluation of vehicles using the learned parameters in each case were found to be consistent. This indicates that our algorithms are tolerant of small changes in evaluation function and environmental conditions, and that PBIL is reliably able to optimize parameter sets in this domain. Figure 6.1 shows the results of one such evolutionary experiment with a population size of 100. Also shown are the results from a second experiment with a population size of 20, using the same evaluation function (see Figure 6.2). For space considerations, only two experiments are shown; however, these experiments have been replicated from a variety of initial starting conditions and parameter settings.

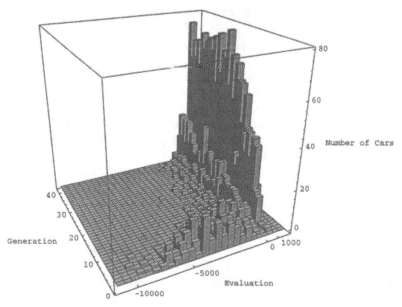

Fig. 6.1. 3-D Histogram showing increase of high-scoring PBIL strings over successive generations. Population size is 100 cars in each generation.

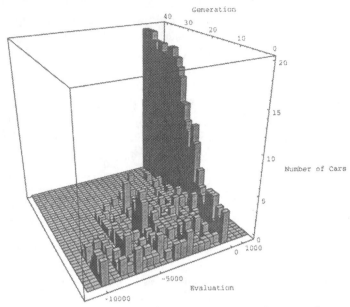

Fig. 6.2. 3-D Histogram showing 20 cars in each generation.

These 3-D histograms display the distribution of vehicles scoring a certain evaluation for each generation. It is clear that as the parameters evolve in successive generations, the average performance of vehicles increases and the variance of evaluations within a generation decreases. In the experiments with population size 100, good performance of some vehicles in the population is achieved early (by generation 5) although consistently good evaluations are not observed until generation 15. The number of vehicles scoring poor evaluations drops rapidly until generation 10, after which there are only occasional low scores. The PBIL strings converge to a stable set of parameters and by the last generation, the majority of the PBIL vehicles are able to circle the track, take the proper exit, and avoid crashes in all four scenarios. In the experiments with population size 20, more generations are required before good parameter settings are found; however, fewer evaluations are performed in each generation. Also note that the run with population size 20 does not find as good solutions as that with population size 100. By the end of the run, the majority of evaluations for the size 20 population ranged between 0–500; for the size 100 population, between 1000–1500.

Finally, it should also be noted that even cars created in the final generation are not guaranteed to drive perfectly. This is because the parameters are generated by sampling the probability vector. Therefore, it is possible, though unlikely in later generations, to create cars with bad sets of parameters. Furthermore, not all accidents are avoidable; they may be caused by dangerous maneuvers made by the other vehicles in response to the difficult traffic situations that often arise in tactical driving domains.

To visually display the learned behavior of the cars, we created obstacle-avoidance test tracks with numerous stopped cars. The path of a car which successfully navigates around these cars is shown in Figure 6.3. While all of the stopped cars are avoided, the trajectory followed shows some undesirable oscillations. This is caused by limitations in the reasoning objects used in this test. Since the obstacle avoidance modules were stateless, they did not consistently follow a single course of action. More sophisticated modules which maintain state will overcome these problems.

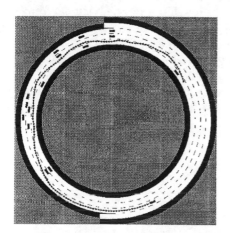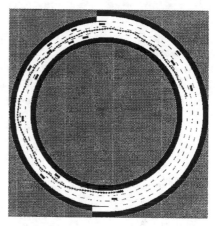

Fig. 6.3. The *Obstacle Avoidance* test tracks. Note that the best vehicles are very aggressive, often avoiding the stopped cars with only narrow margins.

7. Conclusion and Future Directions

Our experiments have demonstrated: (1) The potential for intelligent behavior in the tactical driving domain using a set of distributed reasoning modules. (2) The ability of evolutionary algorithms to automatically configure a *collection* of these modules for addressing their *combined* task.

In this study, we used a very simple evaluation function. By changing the objective function, we can change the learned behavior of the vehicles. By introducing alternative objective functions, we plan to extend this study in at least three directions. First, since the cars created in this study often perform dangerous maneuvers, we will incorporate penalties for overly aggressive behavior. For example, by introducing penalties for coming too close to another car, the cars should learn to maintain a safety cushion. Second, for automated highways, we would like the cars to exhibit altruistic behavior. In a collection of PBIL vehicles, optimizing a *shared* evaluation function (such as highway throughput) may encourage cooperation. Finally, we are developing reasoning objects to address additional complications which will arise when these vehicles are deployed in the real world, such as complex vehicle dynamics and noisy sensors.

Acknowledgements

The authors would like to acknowledge the valuable discussions with Dean Pomerleau and Chuck Thorpe which helped to shape this work. Shumeet Baluja is supported by a graduate student fellowship from the National Aeronautics and Space Administration, administered by the Lyndon B. Johnson Space Center, Houston, Texas. This work is also partially supported by the Automated Highway System project, under agreement DTFH61-94-X-00001. The views and conclusions contained in this document are those of the authors and should not be interpreted as representing the official policies, either expressed or implied, of NASA or the AHS Consortium.

References

1. S. Baluja. Population-based incremental learning: A method for integrating genetic search based function optimization and competitive learning. Technical Report CMU-CS-94-163, Computer Science, Carnegie Mellon, 1994.
2. S. Baluja. Genetic algorithms and explicit search statistics. In *Advances in Neural Information Processing Systems 9* (to appear), 1997.
3. S. Baluja and R. Caruana. Removing the genetics from the standard genetic algorithm. In A. Prieditis and S. Russell, editors, *Proc. International Conference on Machine Learning (ML-95)*, pages 38–46. Morgan Kaufmann Publishers, 1995.
4. J. Cremer, J. Kearney, Y. Papelis, and R. Romano. The software architecture for scenario control in the Iowa driving simulator. In *Proceedings of the 4th Computer Generated Forces and Behavioral Representation*, May 1994.
5. E. Dickmanns and A. Zapp. A curvature-based scheme for improving road vehicle guidance by computer vision. In *Proceedings of the SPIE Conference on Mobile Robots*, 1986.
6. L. Eshelman. The CHC adaptive search algorithm: How to have safe search when engaging in nontraditional genetic recombination. In Rawlins, editor, *Foundations of Genetic Algorithms*, pages 265–283. Morgan Kaufmann Publishers, 1991.
7. K. Gardels. Automatic car controls for electronic highways. Technical Report GMR-276, General Motors Research Labs, June 1960.
8. D. Goldberg. *Genetic Algorithms in Search, Optimization and Machine Learning*. Addison-Wesley, Reading, MA, 1989.
9. D. Goldberg and J. Richardson. Genetic algorithms with sharing for multimodal function optimization. In Grefenstette, editor, *Proceedings of the Second International Conference on Genetic Algorithms*, pages 41–49, San Mateo, CA., 1987. Morgan Kaufmann Publishers.
10. T. Jochem, D. Pomerleau, B. Kumar, and J. Armstrong. PANS: A portable navigation platform. In *Proceedings of IEEE Intelligent Vehicles*, 1995.

11. T. Jochem, D. Pomerleau, and C. Thorpe. Vision guided lane transitions. In *Proceedings of IEEE Intelligent Vehicles*, 1995.
12. I. Masaki, editor. *Vision-Based Vehicle Guidance*. Springer-Verlag, 1992.
13. J. Michon. A critical view of driver behavior models: What do we know, what should we do? In L. Evans and R. Schwing, editors, *Human Behavior and Traffic Safety*. Plenum, 1985.
14. D. Pomerleau. *Neural Network Perception for Mobile Robot Guidance*. PhD thesis, Carnegie Mellon University, February 1992.
15. A. Ram, R. Arkin, G. Boone, and M. Pearce. Using genetic algorithms to learn reactive control parameters for autonomous robotic navigation. *Adaptive Behavior*, 2(3):277–305, 1994.
16. D. Reece. *Selective Perception for Robot Driving*. PhD thesis, Carnegie Mellon University, May 1992.
17. J. Rillings and R. Betsold. Advanced driver information systems. *IEEE Transactions on Vehicular Technology*, 40(1), February 1991.
18. R. Sukthankar. RACCOON: A Real-time Autonomous Car Chaser Operating Optimally at Night. In *Proceedings of IEEE Intelligent Vehicles*, 1993.
19. R. Sukthankar, J. Hancock, S. Baluja, D. Pomerleau, and C. Thorpe. Adaptive intelligent vehicle modules for tactical driving. In *Proceedings of AAAI workshop on Adaptive Intelligent Agents*, 1996.
20. R. Sukthankar, J. Hancock, D. Pomerleau, and C. Thorpe. A simulation and design system for tactical driving algorithms. In *Proceedings of AI, Simulation and Planning in High Autonomy Systems*, 1996.
21. R. Sukthankar, D. Pomerleau, and C. Thorpe. SHIVA: Simulated highways for intelligent vehicle algorithms. In *Proceedings of IEEE Intelligent Vehicles*, 1995.
22. C. Thorpe, M. Hebert, T. Kanade, and S. Shafer. Vision and navigation for the Carnegie Mellon NAVLAB. *IEEE Transactions on PAMI*, 10(3), 1988.
23. R. von Tomkewitsch. Dynamic route guidance and interactive transport management with ALI-Scout. *IEEE Transactions on Vehicular Technology*, 40(1):45–50, February 1991.
24. R. Wallace, A. Stentz, C. Thorpe, W. Moravec, H. Whittaker, and T. Kanade. First results in robot road-following. In *Proceedings of the IJCAI*, 1985.

Gate-Level Evolvable Hardware: Empirical Study and Application

Hitoshi Iba, Masaya Iwata, and Tetsuya Higuchi

1-1-4, Umezono, Tsukuba, Ibaraki 305, Japan, Electrotechnical Laboratory

Summary. Evolvable Hardware (EHW) is a hardware which modifies its own hardware structure according to the environmental changes. EHW is implemented on a programmable logic device (PLD), whose architecture can be altered by downloading a binary bit string, i.e. *architecture bits*. The architecture bits are adaptively acquired by genetic algorithms (GA). This paper describes the fundamental principle of the gate-level EHW and its improvement by the MDL-based fitness evaluation. The effectiveness of our approach is shown by comparative experiments and a successful application. We also discuss the current extension of EHW and related works.

1. Introduction

1.1 Evolvable hardware

Evolvable Hardware (EHW) is a hardware built on a software-reconfigurable logic device, such as PLD (Programmable Logic Device) and FPGAs (Field Programmable Gate Arrays). EHW architecture can be reconfigured through the evolutionary method so as to adapt to the new environment. If hardware errors occur or a new hardware functionality is required, EHW can alter its own hardware structure in order to accommodate such changes.

There is a clear distinction between a conventional hardware (CHW) and EHW. A designer can begin to design a CHW only after its detailed specification is given. In this sense, CHW is a top-down approach. However, EHW is applicable even when no hardware specification is known beforehand. EHW implementation is determined through a genetic learning in a bottom-up way. Thus, EHW will be applied totally differently from CHW. EHW is suitable for problem domains where both on-line adaptation and real-time response are required.

The basic idea of EHW is to regard the architecture bits of a PLD as a chromosome for GA (see Fig. 1.1). The hardware structure is adaptively searched by GA. These architecture bits, i.e. the GA chromosome, are downloaded onto a PLD, on and after the genetic learning. Therefore, EHW can be considered as an on-line adaptive hardware.

The rest of this paper is structured as follows. Section 2 describes the fundamental principle of EHW and its improvement by means of MDL-based evaluation. Section 3 shows the comparative experiments in learning Boolean functions and discusses the results. Thereafter we apply EHW to a pattern recognition. Section 5 discusses the current extension of EHW and related works, followed by some conclusions.

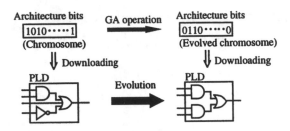

Fig. 1.1. Evolvable Hardware (EHW)

2. Evolving Hardware by GA

2.1 Fundamental principle of gate-level EHW

This section describes the genetic learning method for evolving a hardware. We call our evolutionary method as a *gate-level EHW*, because PLD gates used are fairly primitive ones such as AND gates.

2.1.1 Programmable logic device (PLD).
A PLD consists of *logic cells* and a *fuse array* (see Fig. 2.1). In addition, architecture bits determine the architecture of the PLD. These bits are assumed to be stored in an architecture bit register (ABR). Each link of the fuse array corresponds to a bit in the ABR.

The fuse array determines the interconnection between the device inputs and the logic cell. It also specifies the logic cell's AND-term inputs. If a link on a particular row of the fuse array is switched on, which is indicated by a black dot in Fig. 2.1, then the corresponding input signal is connected to the row. In the architecture bits, these black and white dots are represented by 1 and 0 respectively.

Consider the example PLD shown in Fig. 2.1. The first row indicates that I_0 and $\overline{I_2}$ are connected by an AND-term, which generates $I_0\overline{I_2}$. Similarly, the second row generates I_1. These AND-terms are connected by an OR gate. Thus, the resultant output is $O_0 = I_0\overline{I_2} + I_1$.

As mentioned above, both of the fuse array and the functionality of the logic cell are represented in a binary string. The key idea of EHW is to regard this binary bit string as a chromosome for the sake of GA-based adaptive search.

The hardware structure we actually use is a FPLA device, which is a commercial PLD (Fig. 2.2). This architecture mainly consists of an AND and OR arrays. A vertical line of the OR array corresponds to a logic cell in Fig. 2.1.

2.1.2 Variable length chromosome GA (VGA).
In our earlier works, the architecture bits were regarded as the GA chromosome and the chromosome length was fixed. In spite of this simple representation, the hardware

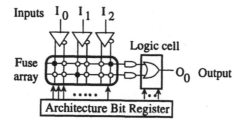

Fig. 2.1. A simplified PLD (Programmable Logic Device) Structure

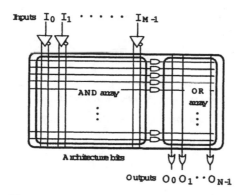

Fig. 2.2. A FPLA Architecture for EHW

evolution was successful for combinatorial logic circuits (e.g. 6-multiplexer) and sequential logic circuits (e.g. 4-state machine, 3-bit counter [Higuchi94]).

However, this straightforward representation had a serious limitation in the hardware evolution. All the fuse array bits should have been included in the genotype, even when effective bits in the fuse array were only a few. This made the chromosome too long to be effectively searched by evolution.

Therefore, we have introduced a variable length chromosome called *VGA*, i.e. Variable length chromosome GA [Kajitani95]. VGA is expected to evolve a large circuit more quickly. The chromosome length of VGA is smaller than the previous GA, especially when evolving a circuit with large inputs. This is because VGA can deal with a part of architecture bits, which effectively determine the hardware structure. Because of this short chromosome, VGA can increase the maximum circuit size and establish an efficient adaptive search.

The coding method of VGA is as follows. An allele in a chromosome consists of a location and a connection type. The location is the position of the allele in the fuse array. For example, the fuse array in Fig. 2.3 (a) has 14 locations as shown in Fig. 2.3 (b). Therefore, the locations of the connected points in Fig. 2.3 (a) are denoted as 0, 4, 8, 9, 13 and 14. The connection type defines the input to be either positive or negative. For example, the connection type at location 0 is 1, i.e. the positive input. Thus, the chromosome for Fig. 2.3 (a) is (0,1) (4,1) (8,2) (9,1) (13,1) (14,1).

We use the roulette wheel selection strategy. Recombination operators are *cut* and *splice*, which are used in the messy GA [Goldberg93]. A mutation operator is applied so as to change the values of the location and the connection type randomly. Splice operator concatenates two chromosomes.

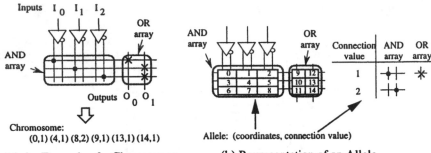

(a) An Example of a Chromosome   (b) Representation of an Allele

Fig. 2.3. Chromosome Representation of Variable Length Chromosome GA

2.2 MDL-based improvement

The fitness of GA is basically evaluated in terms of the output correctness for the training data. In addition, we introduce an MDL (Minimum Description Length) based fitness [Rissanen89] for evolving EHW. As can be seen in later (Section 3.2), the robustness of the generated hardware is achieved as a result of this improvement.

2.2.1 Complexity-based fitness evaluation. Complexity-based fitness is grounded on a *"simplicity criterion"*, which is defined as a limitation on the complexity of the model class that may be instantiated when estimating a particular function. For example, when one is performing a polynomial fit, it seems fairly apparent that the degree of the polynomial must be less than the number of data points. Simplicity criteria have been studied by statisticians for many years.

The complexity of an algorithm can be measured by the length of its minimal description in some language. The old but vague intuition of Occam's razor can be formulated as the *minimum description length criterion* [Rissanen89], i.e., given some data, the most probable model is the model that minimizes the sum:–

$$MDL(model) = desc\_len(data\ given\ model) + desc\_len(model) \longrightarrow min.$$
(2.1)

desc_len(data given model) is the code length of the data when encoded using the model as a predictor for the data. The sum MDL(*model*) represents the tradeoff between residual error (i.e., the first term) and model complexity (i.e., the second term) including a structure estimation term for the final model. The final model (with the minimal MDL) is optimum in the sense of being a consistent estimate of the number of parameters while achieving the minimum error.

The complexity-based fitness evaluation can be introduced in order to control genetic algorithms (GA) search strategies. For instance, when applying GAs to the classification of genetic sequences, [Konagaya93] employed the MDL principle for GA fitness in order to avoid overlearning caused by the statistical fluctuations. They presented a GA-based methodology for learning stochastic motifs from given genetic sequences. A stochastic motif is a probabilistic mapping from a genetic sequence (which has been drawn from a finite alphabet) to a number of categories (cytochrome, globin, trypsin, etc). They employed Rissanen's MDL principle in selecting an optimal hypothesis.

When applying the MDL principle to genetic programming, redundant structures should be pruned as much as possible, but at the same time premature convergence (i.e., premature loss of genotypic diversity) should be avoided [Zhang95]. Zhang proposed a dynamic control to fix the error factor at each generation and change the complexity factor adaptively with respect to the error. In [Iba94,96], MDL-based fitness functions were applied successfully to system identification problems by using the implemented system STROGANOFF. The results showed that MDL-based fitness evaluation works well for tree structures in STROGANOFF, which controls GP-based tree search.

2.2.2 MDL-based fitness for EHW. We have introduced the above MDL criterion into the GA fitness evaluation. The purpose is to establish a robust learning method for EHW. In general, the greater the number of "don't care" [1] inputs, the more robust (i.e. noise-insensitive) the evolved hardware. Thus, we regard the number of "don't care" inputs as an index of MDL.

More formally, the MDL value for our EHW is written as follows:

$$MDL = A_c \log(C+1) + (1 - A_c) \log(E+1), \qquad (2.2)$$

where C denotes the complexity of the EHW. E is the error rate of the EHW's output. Usually, A_c is increased with generations (see Table 3.5).

The C value (i.e. the complexity of the EHW) determines the performance of the MDL. We introduce three types of C definitions as follows:-

---

[1] We call an input "don't care" if it is not included in the output expression. For instance, if $O = I_1 + I_2$ in case of a PLD shown in Fig. 2.1, then I_0 is a "don't care" input.

$$C_1 = \sum_i |AND_{Oi}|, \tag{2.3}$$

$$C_2 = |AND| \times |OR|, \tag{2.4}$$

$$C_3 = \sum_i |AND_{Oi}| \times |OR_{Oi}|. \tag{2.5}$$

Where $|AND_O|$ and $|OR_O|$ are the numbers of ANDs and ORs connected to the output O. $|AND|$ ($|OR|$) is the number of ANDs (ORs) on the AND (OR) array. Consider Fig. 2.3(a) for instance. ANDs and ORs are represented as black dots and $\times$ marks in the figure. The values of C_1, C_2 and C_3 are 3 ($= 1+2$), 9 ($= 3 \times 3$) and 5 ($= 1 \times 1 + 2 \times 2$) respectively, because $|AND_{Oo}|$, $|OR_{Oo}|$, $|AND_{O1}|$, $|OR_{O1}|$, $|AND|$, and $|OR|$ are 1, 1, 2, 2, 3 and 3.

The definition of C_1 is not very precise because it does not include the information of OR gates. On the other hand, C_2 and C_3 are expected to give more exact MDL values. We will show the comparative experiments by using these different MDL definitions in section 3.1

The above MDL value is normalized so that it satisfies $0 \leq \text{MDL} \leq 1$. That is, the GA fitness is defined to be:

$$\text{Fitness} = 1 - \text{MDL}. \tag{2.6}$$

3. Experimental Results

3.1 Comparative studies

In order to evaluate the performance of EHW, this section describes the comparative experiments in Boolean concept learning based on PAC (Probably Approximately Correctly) learnability theory.

Although earlier algorithmic approaches to Boolean concept learning such as decision trees or enumeration [Anthony92] proved to be sound and complete, they suffered from computational complexity. Alternatively several stochastic or evolutionary methods have been proposed, which aim at improving efficiency by using probabilistic search at the expense of completeness. However there have been few comparative studies between their performances from the viewpoint of computational learning theory [Anthony92]. We compare the performance of EHW with those by neural networks (NN) [Rumelhart86], classifier systems (CS) [Wilson87] and adaptive logic networks (ALN) [Armstrong79].

The theoretical background for our experiments is as follows. Let N be the number of attributes and K the number of literals needed to write down the smallest DNF (Disjunctive Normal Form) description of the target concept. Let ϵ be the maximum percentage error that can be tolerated during the testing task. The number of learning examples we used is given by the following formula:-

Table 3.1. Test Functions

Name	description	attributes	terms	#training data
dnf3	random DNF	32	6	1650
	$x_1 x_2 x_6 x_8 x_{25} x_{28} \overline{x}_{29} \vee x_2 x_9 x_{14} x_{16}\, \overline{x}_{22}\, \overline{x}_{25} \vee x_1 \overline{x}_4\, \overline{x}_{19}\, x_{22} x_{27} x_{28}$ $\vee \overline{x}_2 x_{10} x_{14} \overline{x}_{21}\, \overline{x}_{24} \vee x_{11} x_{17} x_{19} x_{21} \overline{x}_{25} \vee \overline{x}_1\, \overline{x}_4 x_{13} \overline{x}_{25}$			
mx6	6-multiplexor	16	4	720
	$x_{13} x_{16} x_1 \vee \overline{x}_{13} x_{16} x_7 \vee x_{13} \overline{x}_{16} x_4 \vee \overline{x}_{13}\, \overline{x}_{16} x_{10}$			
par4	4-parity	16	8	1280
	$x_1 \oplus x_5 \oplus x_9 \oplus x_{13}$	(where $\oplus$ is the XOR operator)		

Table 3.2. Parameters for Classifier Systems

Population Size	400
Crossover Rate	12%
Crossover TYPE	One-Point
Mutation Rate	0.1%
Payoff Quantity (R)	1000
Decay by Error (e)	80%
Bias for # (G)	4.0
	Boole [Wilson87]

$$\frac{K \times log_2 N}{\epsilon}. \tag{3.1}$$

Qualitatively the formula indicates that we require more training examples as the complexity of the concept increases or the error decreases [Pagallo90].

In our experiments we set $\epsilon = 10\%$ and used 2000 examples to test classification performance. Thus 90% ($2000 \times 0.9 = 1800$ examples) correctness of testing data is the expected learning success rate.

We used 3 problems (target concepts) shown in Table 3.1 [Pagallo90]; dnf3 (randomly generated DNF, 32 attributes, 6 terms), mx6 (6-multiplexor, 16 attributes with 10 irrelevant attributes), and par4 (4-parity, 16 attributes with 12 irrelevant attributes).

The number of training data is derived from equation (3.1); i.e. 1280 ($= \frac{32 \times log_2 16}{0.1}$) training data are given for par4[2].

All methods, i.e. CS, ALN, NN and EHW, were run according to standard operational criteria. The parameters shown in Tables 3.2, 3.3, 3.4 and 3.5 were used for each method.

These parameters were chosen to obtain the most effective learning results after several experimental runs. Learning was terminated after convergence is attained. Thus the numbers of iterations needed in training phases differ for

---

[2] The correct number for mx6 is 480 according to the equation (3.1). However, we used the number given in the original table [Pagallo90,p.91].

Table 3.3. Parameters for ALN

Initial Nodes		29999
Node Types	AND, OR, LEFT, RIGHT	
		[Armstrong79]

Table 3.4. Parameters for Neural Networks

Learning Rate	0.01
Momentum	0.5
# of Hidden Layers	1
# of Hidden Nodes	4 (3 for dnf3)
	[Rumelhart86]

Table 3.5. Parameters for EHW

Population Size		100
Maximum Generation		2000
Initial Chromosome Length		100
Cut and Splice Rate		0.01
Mutation Rate		0.01
A_C Coefficnet	Gen:0–1000	0.0
	Gen:1001–1500	0.025
	Gen:1501–2000	0.05
		[Higuchi94]

the 4 methods; i.e. $O(10000)$ for NN, $O(1000)$ for CS, $O(100)$ for ALN and $O(1000)$ for EHW. However, this number did not necessarily reflect the computational complexity, because each iteration included qualitatively different computations. We executed several independent runs for each test function.

We conducted experiments with the learning of both noiseless and noisy Boolean concepts. In noisy environments, learning attribute values are inverted from 1 to 0 or from 0 to 1 (with a probability less than 5%).

Tables 3.6 and 3.7 show the averages and the standard deviations of correctness for training and testing data for ten runs. Note that following equation (3.1)the success rate for this Boolean learning is expected to be above 90%. Although we cannot make any concluding remarks as to which is the method, the following points should be emphasized:

1. **Neural Networks (NN)**
   NN copes with noise relatively successfully; i.e. so called "graceful degradation" was observed. However, in noiseless cases (i.e. 0% noise), NN dose not always succeed in learning the training data. NN shows poor results for mx6 or par4. Although it is widely believed that NN performs boolean

Table 3.6. Learning Performances (CS, ALN, and NN)

Noise	Func.	CS Train		CS Test		ALN Train		ALN Test	
0%	mx6	100.0	0.0	100.0	0.0	100.0	0.0	98.9	0.7
	par4	100.0	0.0	100.0	0.0	100.0	0.0	98.6	1.3
	dnf3	90.0	1.8	87.8	3.1	100.0	0.0	87.6	1.6
2%	mx6	100.0	0.0	100.0	0.0	96.4	0.6	95.5	3.5
	par4	98.2	2.3	97.1	3.3	92.6	0.8	99.9	0.3
	dnf3	71.0	27.7	66.2	31.9	96.4	0.6	86.4	1.2
5%	mx6	98.3	2.5	98.4	2.4	90.9	1.2	99.8	0.4
	par4	44.8	2.2	36.6	1.8	74.4	6.7	71.5	23.7
	dnf3	27.0	29.0	20.3	32.2	90.1	0.8	89.9	2.5

Noise	Func.	NN Train		NN Test	
0%	mx6	99.0	1.2	98.7	1.4
	par4	89.1	12.7	85.9	18.3
	dnf3	96.7	0.8	92.7	3.0
2%	mx6	95.7	1.0	95.6	1.0
	par4	84.8	11.4	81.9	17.2
	dnf3	94.5	1.4	92.4	1.7
5%	mx6	90.7	1.1	90.1	0.5
	par4	76.8	8.8	74.7	13.1
	dnf3	92.0	0.8	90.7	0.8

Table 3.7. Learning Performances (EHW w and w/o MDL)

Noise	Func.	EHW(w/o MDL) Train		EHW(w/o MDL) Test		EHW(w MDL) Train		EHW(w MDL) Test	
0%	mx6	95.90	6.96	94.75	8.75	98.61	2.16	97.95	2.98
	par4	70.26	7.35	62.95	11.05	68.74	4.25	62.11	6.31
	dnf3	91.57	2.44	91.05	2.65	90.28	3.42	89.72	3.20
2%	mx6	94.11	5.16	96.45	3.89	94.68	4.29	96.17	6.29
	par4	68.74	6.48	70.91	11.18	74.09	6.75	71.50	10.64
	dnf3	88.56	2.11	88.19	2.93	87.10	2.67	85.28	3.64
5%	mx6	88.17	5.02	96.03	4.93	86.33	8.91	92.20	5.43
	par4	63.92	1.88	54.05	4.66	64.22	3.47	60.62	6.58
	dnf3	86.75	1.39	84.93	2.29	88.43	1.84	87.90	3.51

concept learning well, no significant superiority of NN was observed. This is because the distributed representations prevent NN from distinguishing between relevant and irrelevant attributes for mx6 and par4. Dnf3 is a hard problem for NN.

2. **Classifier Systems (CS)**

CS is superior to the other methods for mx6 and par4. CS can cope with the irrelevant attributes. Actually CS is successful in acquiring a perfect set of rules for mx6. For instance, the acquired rules are as follows:-

Condition ($x_1\ x_2\ x_3\ \cdots\ x_{16}$)	Action	Strength
# # # 0 # # # # # # # # 1 # # 0	0	5620
# # # # # # 0 # # # # # 0 # # 1	0	5526
# # # # # # # # # 1 # # 0 # # 0	1	5512
# # # # # # 1 # # # # # 0 # # 1	1	5503
1 # # # # # # # # # # # 1 # # 1	1	4222
0 # # # # # # # # # # # 1 # # 1	0	4090
# # # 1 # # # # # # # # 1 # # 0	1	3633
# # # # # # # # # 0 # # 0 # # 0	0	3060

Notice that these rules express the concept of mx6 by using significant bits ($x_1, x_4, x_7, x_{10}, x_{13}, x_{16}$) and ignoring the irrelevant attributes. On the other hand, CS fails to solve dnf3. This is because it is difficult for CS to represent the concept of dnf3 in the form of classifier rules. So many classifier rules are required to express 0-valued actions for dnf3 whereas # (wild-card) works very well for par4 and mx6. Therefore the rule size is an important factor for CS. For mx6 and par4, 400 rules were enough. On the other hand, $O(1000)$ rules were necessary for dnf3.

CS has poor records abruptly when the noise level exceeds 2%. Considering their high deviations, the performance of CS is not stable; that is, results of CS are likely to be influenced by noise.

3. **Adaptive Logic Networks (ALN)**

ALN performs better for all 3 tests in noiseless cases. However, considering that the average performance is below 90% for dnf3, ALN was not successful in generalizing the training data. This results from the fact that ALN simply memorizes part of the training data, and lacks the ability to generalize. For these reasons, ALN, in general, requires a large number of initial nodes (for instance, [Armstrong79] used $O(60000)$). Although the final node size might well be reasonable ($O(100)$), a small number of initial nodes results in failure. It should be noted that ALN's performance is heavily dependent upon the problem size. For example, ALN failed to solve par5 (5-bit parity problem with 27 irrelevant bits) Besides, as can be seen in mx6 (2% and 5% noise) and par4 (0% and 2%), overfitting phenomena were observed for ALN.

4. **EHW without MDL**

EHW gave satisfactory results for mx6, whereas it failed in learning par4 and dnf3. This is because the representation used by EHW is not suitable for expressing a long disjunctive clause. For instance, in one run the best expression acquired at the 1991th generation was as follows:-

$$x_0\overline{x_1}\overline{x_4}x_5x_6\overline{x_8}x_9\overline{x_{10}}\overline{x_{15}} \lor \overline{x_0}\overline{x_4}\overline{x_8}x_{12} \lor x_2\overline{x_3}x_4x_5\overline{x_{10}}x_{12}x_{13}\overline{x_{15}} \lor$$

$$\overline{x_0}\overline{x_4}x_7x_8\overline{x_{12}} \lor x_0x_1x_2\overline{x_3}x_4x_6\overline{x_9}x_{10}x_{11}\overline{x_{12}}x_{13}\overline{x_{15}} \lor$$

$$x_0x_2\overline{x_4}x_7\overline{x_8}\overline{x_{12}} \lor x_0\overline{x_4}x_8\overline{x_{10}}x_{12} \lor x_0x_4\overline{x_8}x_{12} \lor x_0x_4x_8\overline{x_{12}} \lor$$

$$\overline{x_0}x_4x_8x_{12} \lor x_0\overline{x_4}x_6\overline{x_8}x_9x_{11}\overline{x_{12}} \lor \overline{x_2}\overline{x_5}x_6x_8\overline{x_{10}}x_{11}x_{12}. \qquad (3.2)$$

The correctness of this expression was 84.70%. The chromosome length is 200. Considering that the correct par4 has 8 terms in a principle disjunctive canonical form and that each term consists of 4 variables, EHW seems to have fallen in a local extreme in this case.

In noisy environments, EHW performed slightly better than CS. However, the results were not necessarily satisfactory.

5. **EHW with MDL** The performance of MDL-based EHW was better than non-MDL EHW in almost all cases. The generalization effect, i.e. the avoidance of overfitting, can be seen in mx6. For example, the typical expression acquired by non-MDL EHW was as follows:-

$$x_{10}\,\overline{x_{13}}\,\overline{x_{16}} \lor \overline{x_1}\,\overline{x_2}\,\overline{x_3}\,x_4\,\overline{x_8}\,\overline{x_{12}}\,x_{13}\,x_{14}\,\overline{x_{16}} \lor$$

$$x_1\,\overline{x_2}\,x_3\,\overline{x_5}\,x_6\,x_7\,\overline{x_9}\,\overline{x_{13}}\,x_{15} \lor x_1\,x_4\,\overline{x_5}\,\overline{x_{12}}\,x_{13}\,\overline{x_{14}} \lor$$

$$x_3\,x_5\,x_6\,x_7\,\overline{x_9}\,\overline{x_{10}}\,\overline{x_{12}}\,x_{16} \lor \overline{x_2}\,\overline{x_4}\,x_7\,x_8\,\overline{x_{11}}\,x_{14}\,\overline{x_{15}}\,x_{16} \lor$$

$$x_1\,x_{13}\,x_{16} \lor x_7\,x_8\,\overline{x_9}\,\overline{x_{12}}\,\overline{x_{13}}\,\overline{x_{14}} \lor \overline{x_2}\,\overline{x_3}\,x_4\,\overline{x_5}\,\overline{x_6}\,x_7\,x_9\,x_{13}\,x_{14}\,x_{16} \lor$$

$$x_4\,x_{13}\,\overline{x_{14}} \lor x_7\,\overline{x_{13}}\,x_{16} \lor \overline{x_2}\,\overline{x_3}\,\overline{x_4}\,x_5\,x_6\,\overline{x_7}\,\overline{x_{10}}\,\overline{x_{12}}\,\overline{x_{14}}\,\overline{x_{15}}, \qquad (3.3)$$

whereas MDL-based EHW resulted in the following expression,

$$x_{13}\,x_{16}\,x_1 \lor \overline{x_{13}}\,x_{16}\,x_7 \lor x_{13}\,\overline{x_{16}}\,x_4 \lor \overline{x_{13}}\,\overline{x_{16}}\,x_{10}. \qquad (3.4)$$

Although both expressions gave 100% correct outputs for the training data, yet non-MDL EHW did not succeed in learning the target expression, i.e. the correctness of the equation (3.3) was 99.4% for the testing data. Note that the equation (3.4) is equivalent to the definition of mx6 (see Table 3.1). This shows the more compact expression has been acquired by means of MDL-based fitness, which clearly shows the success of complexity-based evaluation of EHW.

In summary, we cannot conclude that EHW is superior to the other learning methods. The methods we compared can be classified roughly into analog approaches (NN) and into digital approaches (ALN, CS). As we have observed, both approaches have their own merits and demerits. In order to improve the EHW learning ability, we are currently working on the extension of EHW, which integrates analog and digital approaches.

3.2 Application to pattern recognition

3.2.1 Motivation. EHW has been applied to high-speed pattern recognition in order to establish a robust system in noisy environments [Iwata96]. This ability, i.e. robustness, seems to be the main feature of ANN. ANN is mostly run in a software-based way, i.e. executed by a workstation. Thus, current ANN may have difficulty with real-time processing because of the speed limit of the software-based execution.

Another desirable feature of EHW is its readability. The learned result by EHW is expressed as a Boolean function, whereas ANN represents it as thresholds and weights. Thus, the acquired result of EHW is more easily understood than that of ANN. We believe that this understandable feature leads to wider usage of EHW in industrial applications.

For the sake of achieving flexible recognition capability, it is necessary to cope with a pattern which is classifiable not by a linear function, but by a non-linear function. We have conducted an experiment in learning the exclusive-OR problem in order to check the above capability. From the simulation result, we confirmed that EHW can learn non-linear functions successfully [Higuchi95]. In other words, EHW is supposed to fulfill the minimum requirement towards the robust pattern recognition.

3.2.2 Procedure of pattern recognition. The pattern recognition procedure consists of two phases as shown in Fig. 3.1. The first is the learning phase of training patterns. The training patterns are genetically learned by EHW. We use VGA and MDL-based fitness described in section 2. The second phase is the recognition of test patterns. Our aim is the noise-insensitive pattern recognition.

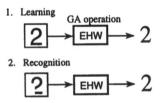

Fig. 3.1. Procedure of Pattern Recognition using EHW

3.2.3 The pattern recognition system. We have developed the pattern recognition system (Fig. 3.2). The organization of the system is shown in Fig. 3.3. It consists of an EHW board including 4 FPGA chips (Xilinx 4025), a DOS/V machine, and an input tablet for drawing patterns. The DOS/V machine handles GA operations, the control of EHW board and the display of patterns. The PLD on FPGA is reconfigurable, which means that the system can be used as a universal EHW system.

In the EHW board, there are four FPGA (hatched area in the figure), board control registers, and SRAM which stores the configuration data of FPGA. In the EHW, a circuit chromosome is realized by an ABR (architecture bit register) and a PLD. The ABR stores architecture bits. The PLD has the architecture of FPLA device (Fig. 2.2). In this figure, there are K populations, i.e. K pairs of an ABR and a PLD in a FPGA chip. In the first version of this system, we designed a genetically reconfigurable hardware device with four FPGAs.

Fig. 3.2. Pattern Recognition System using EHW

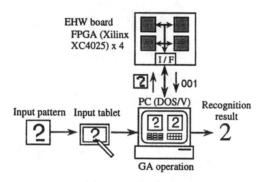

Fig. 3.3. Block Diagram of Pattern Recognition System

3.2.4 Experiment. We have conducted the experiment in recognizing binary patterns of 8×8 pixels. They are 30 input patterns of 64 bits in the training set as shown in Fig. 3.4. Three patterns exactly represent numerical characters (i.e. 0, 1, and 2). The other 27 patterns represent the same numerical characters with noises (i.e. 5 bits are randomly flipped). The outputs of EHW consists of 3 bits; each bit corresponds to one of three characters. The initial length of a chromosome is 100. The probability of the cut and splice operators is 0.1. The mutation probability is 0.01. The line number of AND array in the PLD is 24. The test data set consists of 30 patterns, which are generated with random noises (i.e. less than 5 bits are flipped randomly).

For different learning methods were examined, i.e. MDL-based EHW with three types of MDL definitions (MDL1, MDL2 and MDL3 which correspond to equations (2.4), (2.5), and (2.5), respectively) and non-MDL EHW. The recognition result of the test set is plotted in Fig. 3.5. From the figure, it is clear that MDL-based EHWs give better performance for noisy patterns than EHW without MDL.

An important feature of EHW is that the resultant expression can be represented by a simple Boolean function. For example, in one run, learned results in case of MDL3 are $O_0 = I_{34}I_{38}, O_1 = \overline{I_{22}I_{38}} + \overline{I_{13}}$, and $O_2 = I_{37}$, where $I_i (0 \leq i \leq 63)$ indicates the location of the pixel in the pattern and O_i is the recognition output for the pattern of letter i. Clearly, the results obtained by EHW are easier to understand, compared with ANN.

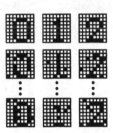

Fig. 3.4. Training Patterns

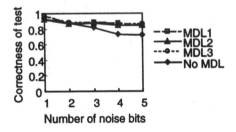

Fig. 3.5. Recognition Result of Test Set

4. Discussions and Conclusion

4.1 Related works

Research on EHW was initiated independently in Japan and in Switzerland around 1992 (for recent overviews see [Higuchi94] and [Marchal94], respectively). Since then, the interest is growing rapidly. EVOLVE95, the first international workshop on the evolvable hardware, was held in Lausanne October 1995.

The research can be roughly classified into two groups: engineering-oriented approach and embryology-oriented approach. The engineering-oriented approach aims at developing a machine which can change its own hardware structure. It also tries to develop a new methodology for hardware design, i.e. the hardware design without human designers. The embryology-oriented approach aims at developing a machine which can self-reproduce or self-repair itself. Most of the researches in this direction are based on two-dimensional cellular automata.

The next subsections briefly overlook the case studies in both directions.

4.1.1 Engineering-oriented approach. Since 1992, ETL has conducted a research on the gate-level evolution and developed two application systems. One is the prototypical welding robot, in which the control part can be taken over by EHW when hardware error occurs. EHW learns the target circuit by GA without any knowledge on the circuit. Another application is the flexible pattern recognition which realizes the robustness (i.e. noise immunity). NNs learn a noise-insensitive function by adjusting their weights and thresholds of neuron units. On the other hand, EHW implements such a function directly in the hardware as a result of the genetic learning.

Thompson at University of Sussex attempts a gate-level evolution for the robot controller. For example, he evolved a 4kHz oscillator and a finite-state machine for the sake of achieving the wall-avoidance behavior of a robot. The oscillator consisted of about 100 gates. The functionalities of the gates and their interconnections were determined by the GA [Thompson95].

Hemmi at ATR evolves the hardware structure specified in the hardware description, i.e. the hardware description language (HDL), by using genetic programming. The HDL used is SFL (Structured Function description Language), which is a part of the LSI CAD system, PARTHENON. This means that once such a description is obtained, the real hardware can be manufactured by PARTHENON [Hemmi94]. Hardware evolution proceeds as follows. The grammar rules of SFL are given at first. Applying grammar rules results in a hardware description. This application process is represented by a binary tree, i.e. the tree designates the application ordering. Genetic programming is used to evolve a desirable tree. Circuits such as adders were successfully obtained through this adaptive search.

4.1.2 Embryology-oriented approach. Ongoing research at the Swiss Federal Institute of Technology aims at developing an FPGA, which can self-reproduce or self-repair (see [Marchal94]). A most interesting point is that the hardware description is represented by a Binary Decision Diagram (BDD). These BDDs are treated as genomes for the GA. Each logical block of the FPGA reads as a part of the genome, which describes the functionality and is reconfigurable. If some block is damaged, the genome can be used to do a sort of self-repair. In other words, one of the spare logical blocks will be reconfigured according to the information of the damaged block.

Another embryology-oriented approach is de Garis' work at ATR. His goal is to evolve neural networks using a two-dimensional cellular automata machine (MIT CAM8 machine) for the sake of building an artificial brain. The neural network is formed as the trail on the two-dimensional cellular automata by evolving the state-transition rules [deGaris94].

4.2 Current extension

The EHW described so far is based on the hardware evolution at *gate-level*, in the sense that each gene of a chromosome corresponds to a primitive gate such as an AND gate or an OR gates. However, because of the limitation of the GA execution time, the size of a circuit allowable for this gate-level evolution may not be large enough for practical applications. If hardware is genetically synthesized from higher-level functions (e.g. adder, subtracter, sine generator, etc.), more practical facilities can be provided by EHWs.

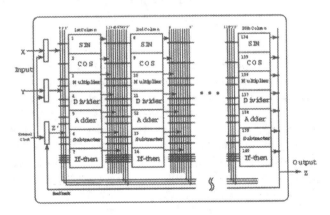

Fig. 4.1. The FPGA Model for Function Level Evolutions

Recently ETL initiated a function-level evolution approach, in which each gene corresponds to a real function such as a floating multiplication and a sine function. The function-level evolution using the FPGA model is described in Figure 4.1. The FPGA model consists of 20 columns, each containing seven hardware functions; an adder, a subtracter, an if-then, a sine generator, a cosine generator, a multiplier, and a divider. Columns are interconnected by crossbar switches. The crossbars designate inputs to columns. In addition to these columns, a state register maintains a past output for dealing with temporal data. There are two inputs and one output. Data handled with this FPGA are floating point numbers.

Function-level evolution is expected to attain the satisfactory performance comparable to the neural networks. We have confirmed the validity of this model by experimenting with various problems, such as the classification of iris data and distinguishing two interwined spirals (for details of these experiments, refer to [Murakawa96]).

References

[Anthony92] Anthony, M. and Biggs, N., Computational Learning Theory, Cambridge Tracts in Theoretical Computer Science 30, Cambridge, 1992

[Armstrong79] Armstrong, W.W. and Gecsei, J., Adaptation Algorithms for Binary Tree Networks, *IEEE TR. SMC*, SMC-9, No.5, 1979

[deGaris94] de Garis, H., An Artificial Brain – ATR's CAM-Brain Project Aims to Build/Evolve an Artificial Brain with a Million Neural Net Modules, Inside a Trillion Cell Cellular Atutomata Machine *New Generation Computing* (OHMSHA.LTD and Springer-Verlag) Vol. 12, pp 215–221, 1994

[Goldberg93] Goldberg, D. et al., Rapid Accurate Optimization of Difficult Problems using Fast Messy Genetic Algorithms, Proc. 5th Int. Joint Conf. on Genetic Algorithms (ICGA93), 1993

[Hemmi94] Hemmi, H., Mizoguchi, J. and Shimohara, K., Development and evolution of hardware behaviors, *Proc. of the fourth international workshop on the synthesis and simulation of living systems* eds. R A Brooks and P Maes , MIT Press, pp 371–376, 1994

[Higuchi94] Higuchi, T. et al., Evolvable Hardware with Genetic Learning, in Massively Parallel Artificial Intelligence (eds. H. Kitano), MIT Press, pp 398–421, 1994.

[Iba94] Iba, H., deGaris, H. and Sato, T., Genetic Programming using a Minimum Description Length Principle, in *Advances in Genetic Programming*, (ed. Kenneth E. Kinnear, Jr.), MIT Press, pp.265–284, 1994

[Iba96] Iba, H., deGaris, H. and Sato, T., Numerical Approach to Genetic Programming for System Identification, in *Evolutionary Computation*, vol.3, no.4, 1996

[Iwata96] Iwata, M., Kajitani, I., Yamada, H., Iba, H. and Higuchi, T., A Pattern Recognition System Using Evolvable Hardware, in *Proc. of Parallel Problem Solving from Nature IV* (Berlin, Germany: Springer-Verlag), 1996

[Kajitani95] Kajitani, I. et al., Variable Length Chromosome GA for Evolvable Hardware in *Proc. of 3rd Int. Conf. on Evolutionary Computation (ICEC96)*, 1996 .

[Konagaya93] Konagaya, A. and Kondo, H., Stochastic Motif Extraction using a Genetic Algorithm with the MDL Principle, in *Hawaii Int. Conf. of Computer Systems*, 1993

[Marchal94] Marchal, P., Piguet, C., Mange, D., Stauffer, A. and Durand, S., Embryological development on silicon, *Proc. of the fourth international workshop on the synthesis and simulation of living systems* eds. R A Brooks and P Maes (Cambridge, Mass.:MIT Press) pp 365–370, 1994

[Murakawa96] Murakawa, M., Yoshizawa, S., Kajitani, I., Furuya, T., Iwata, M. and Higuchi, T., Hardware Evolution at Function Levels, *Proc. of Parallel Problem Solving from Nature IV* (Berlin, Germany: Springer-Verlag), 1996

[Pagallo90] Pagallo, G. and Hausslear, D. Boolean Feature Discovery in Empirical Learning, *Machine Learning*, vol.5, 1990

[Rissanen 89] Rissanen, J., Stochastic Complexity in Statistical Inquiry, World Scientific, 1989

[Rumelhart86] Rumelhart, D.E. and McClelland, J.L. Parallel Distributed Processing, MIT Press, 1986

[Thompson95] Thompson, A., Evolving electronic robot controllers that exploit hardware resources" *Proc. of the third European Conference on Artificial Life* (Berlin, Germany: Springer-Verlag) pp 640–656

[Wilson87] Wilson, S.W., Classifier Systems and the Animat Problem, *Machine Learning*,vol.2, no.3, 1987

[Zhang95] Zhang, B.-T. and Mühlenbein, H., Balancing Accuracy and Parsimony in Genetic Programming, in *Evolutionary Computation*, vol.3, no.1, pp.17–38, 1995

Physical Design of VLSI Circuits and the Application of Genetic Algorithms

Jens Lienig

Tanner Research, 180 North Vinedo Avenue, Pasadena, CA 91107

Summary. The task of VLSI physical design is to produce the layout of an integrated circuit. New performance requirements are becoming increasingly dominant in today's sub-micron regimes requiring new physical design algorithms. Genetic algorithms have been increasingly successful when applied in VLSI physical design in the last 10 years. Genetic algorithms for VLSI physical design are reviewed in general. In addition, a specific parallel genetic algorithm is presented for the routing problem in VLSI circuits.

1. Introduction

Electronic design automation is concerned with the design and production of VLSI systems. One of the important steps in creating a VLSI circuit is its physical design. The input to the physical design step is a logical representation of the system under design. The output of this step is the layout of a physical package that optimally or near-optimally realizes the logical representation. Physical design problems are generally combinatorial in nature.

What makes electronic design automation problems particularly difficult compared to traditional combinatorial optimization problems is that the number of elements that must be handled can be quite large – a circuit can be easily composed of over one million gates. For this reason, design automation practitioners have a strong tradition of quickly considering and adapting new and alternative solution techniques. For example, simulated annealing is an optimization technique that emulates the annealing of crystals. This combinatorial optimization method was first proposed in the literature in 1983 and by the following year, the major design automation conferences had multiple sessions on simulated annealing for design automation [35]. Early adoption was again repeated on neural networks [36] which simulate the organizing principles of nervous systems.

Although design automation did not immediately add genetic algorithms to its basic tool chest, genetic algorithms have been consistently used in the field for the last ten years.

The purpose of this chapter is to provide the reader with an up-to-date overview of genetic algorithm applications for the VLSI physical design process. In Section 2, below, we first briefly describe the VLSI physical design process. We present a systematic review of genetic algorithms that have been successfully applied to the physical design process in Section 3. In Section 4 we present a parallel genetic algorithm for the channel and switchbox routing problem in VLSI circuits. We conclude with a summary in Section 5.

2. Overview of VLSI Physical Design

The major steps in a typical VLSI design process are shown in Figure 2.1. In particular, the substeps of physical design are given.

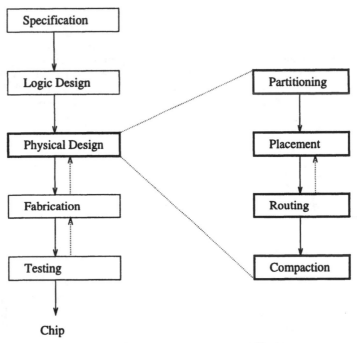

Fig. 2.1. The design process of VLSI chips

The physical design phase is an important part of this process. Its input is generally a logical description of the circuit, often in the form of a netlist. The task of the physical design step is to produce a layout, an assignment of geometric coordinates to the circuit components, either in the plane or in a specified number of planar layers. The layout must satisfy the requirements of the fabrication technology (sufficient spacing between components of the circuit, and so on) and should minimize certain cost criteria (the lengths of the interconnections, etc.).

Due to its complexity, the physical layout problem is generally divided into subproblems which can be solved sequentially. These subproblems are still NP-hard, but they reduce the practical complexity to a manageable level. The physical design problem is usually decomposed into the following subproblems: *partitioning, placement, routing* and *compaction*.

Partitioning is the task of dividing a circuit into smaller parts in order to reduce the problem size. The circuit is often divided into portions that are implemented separately. The goal is to partition the circuit such that the sizes of the parts are within prescribed ranges and the complexity of the interconnections between these parts is minimized.

Placement assigns the cells of the circuit to their geometrical locations on the chip. (A cell may be a single transistor, an adder or subcircuit, etc.) The objective of placement depends on the design style. In *standard cell design* (where all cells have the same width and are placed in rows) and in *macro cell design* (where cells have different sizes and are placed irregularly), the goal is to minimize the total layout area of the chip. In *gate-matrix design* (where all cells are placed in a matrix pattern) the objective of placement is to ensure routability and to minimize congestion of the interconnections between the cells.

Routing follows the placement phase. It determines the paths of the interconnections between the cells laid out during the placement procedure. The goal is to connect all pins that belong to the same net, subject to certain quality constraints (such as minimizing the lengths of interconnections, and so on) and routing constraints (interconnections must not short-circuit or cross one another, etc.).

Compaction is often the final step in the physical layout design. It transforms the symbolic layout (produced by the preceding steps) into a mask layout, the geometric mask features on the silicon. The objective of compaction is to minimize the size of the resulting circuit layout.

3. Application of Genetic Algorithms to VLSI Physical Design

3.1 Partitioning

To our knowledge, the first evolution-based algorithm for solving the partitioning problem was published in [37]. In contrast to previous heuristic algorithms that usually optimize on only one constraint, this approach is capable of handling both a number of constraints and a number of objectives. The algorithm, which does not include a crossover operator, yields multi-way partitions with fairly balanced sizes and a small number of pins for each part. The presented strategy has a reasonable execution speed that is similar to other published heuristic approaches.

In [21] and [22], different coding schemes for the problem of circuit partitioning are investigated to find the most suitable coding. The proposed genetic algorithm is specifically tailored for the partitioning of circuits with complex bit-slice components using a special two-step coding of partitions. The algorithm consists of a crossover and a mutation operator and a deterministic improvement strategy.

The genetic algorithm in [9] is based on a population structure that involves subpopulations which have their isolated evolution occasionally interrupted by inter-population communication. Although the investigated problems are from actual VLSI design efforts, comparisons with other approaches and runtimes are not presented.

A hybrid genetic algorithm for the ratio-cut partitioning problem is presented in [4]. Here the problem is formulated in terms of a hypergraph. The genetic encoding is a binary string, where each subset of the circuit's components has a corresponding location on the string. Before the genetic algorithm is executed, the ordering of these genes is determined by a depth-first search to improve the performance of the genetic algorithm. Traditional crossover and mutation operators are combined with a fast partitioning heuristic applied to each offspring as an improvement operator. The performance of the algorithm is compared with two other partitioning approaches, using benchmark data sets. Averaged over all benchmarks, the presented algorithm achieves better results than the other approaches, while having a similar amount of runtime for smaller graphs and less runtime for the largest graphs.

Another hybrid approach is published in [42]. It combines a simulated annealing method with a genetic algorithm. The main motivation for this approach is the parallelization of the simulated annealing strategy by replacing its single solution search process with a population-based approach using a genetic algorithm. The two benchmark results that are presented are not compared with other state-of-the-art approaches.

3.2 Placement

As mentioned in Section 2, the placement procedure is responsible for the assignment of the circuit's cells to their locations on the chip. According to the variation in size and location of these cells, placement algorithms can be divided into algorithms for standard cell design, macro cell design and gate-matrix design.

After the pioneering work of [6], further applications of genetic algorithms [39], [40] and evolution strategies [24], [25] for standard cell placement have been presented. These approaches produce high-quality placements of real-world VLSI circuits that can compete with sophisticated simulated annealing-based placement strategies. However, the published runtimes are not as competitive (up to 6 hours [24] and up to 12 hours [40]).

In [33], the runtime has been reduced significantly by using a parallel implementation of a genetic algorithm that runs on a distributed network of workstations. The total population is split over different processors and a migration mechanism is used to exchange genetic material between them. While the placement results are similar to a serial genetic algorithm, an almost linear speedup can be achieved with this method.

We next discuss three investigations that use genetic algorithms for macro cell placement [5], [11], [12]. The approach in [5] is based on a two-dimensional bitmap representation of the macro cell placement problem. Another representation scheme, a binary tree, is applied in [11]. In [12], a combination of a genetic algorithm with a simulated annealing-based optimization strategy is presented. The experimental results suggest that a mixed strategy performs better than a pure genetic algorithm for the macro cell placement problem. The results are better or comparable to previously published results of placement benchmarks. However, the runtime is not as competitive.

An application of a genetic algorithm for the placement of gate-matrix design has been published in [41]. The approach uses the Genesis package [19] as the basic genetic algorithm. This package is modified with a special algorithm for constructing permutations that considers only a small subset of the solution space. The results are compared with only one previously published algorithm. The runtime is in the order of minutes (up to 1 hour).

3.3 Routing

Routing is the process of connecting pins subject to a set of routing constraints. VLSI routing is usually divided into *global routing* (to assign nets into certain routing regions) and *detailed routing* (to assign nets to exact positions inside a routing region).

To our knowledge, only one genetic algorithm for global routing has been reported [13]. The algorithm is based on a two-phase router. In the first phase, a genetic algorithm for the Steiner problem in a graph is used to generate a number of distinct, alternative routes for each net. Then, in a second phase, another genetic algorithm selects a specific route for each net (among the alternatives given from phase one), such that the overall layout area is minimized. The router is superior to TimberWolfMC [38], a state-of-the-art global router, with respect to solution quality, while being inferior with respect to runtime.

According to the position of the pins, detailed routing can be separated into *channel routing* (pins are only located on two parallel sides of the routing area) and *switchbox routing* (pins are placed on all four sides of the routing area).

Several papers have been published in which genetic algorithm-derived strategies are applied to the unrestrictive[1] channel routing problem [15], [17], [30], [32].

In [32], a rip-up-and-rerouter is presented which is based on a probabilistic rerouting of nets of one routing structure. However, the routing is done by a deterministic Lee algorithm [27] and main components of genetic algorithms, such as the crossover of different individuals, are not applied. Results are presented only for one channel routing benchmark. No runtime for this example is given.

The router in [15] combines the steepest descent method with features of genetic algorithms. The crossover operator is restricted to the exchange of

---

[1] Approaches for the restrictive channel routing problem (where all vertical net segments are located on one layer and all horizontal segments are placed on a second layer) cannot be applied to real-world VLSI channel routing problems and thus, won't be considered here.

entire nets and the mutation procedure performs only the creation of new individuals. The presented results are limited to simple VLSI problems, and no runtime remarks are made.

The genetic algorithm for channel routing published in [30] is based on a problem-specific representation scheme, i.e. individuals are coded in three-dimensional chromosomes with integer representation. The genetic operators are also specifically developed for the channel routing problem. The results are either qualitatively similar to or better than the best published results for channel routing benchmarks. The runtime of the algorithm (in the range of 1...50 minutes) is not as competitive.

The algorithms in [15], [32] are also applied to switchbox routing. While the router in [15] is not usable for large switchbox routing benchmarks, the algorithm in [32] can compete with other switchbox routing algorithms. However, the runtime is not given.

A different genetic algorithm for switchbox routing is presented in [31]. Similar to [30], the genotype is essentially a lattice corresponding the coordinate points of the layout. Crossover and mutation are performed in terms of interconnection segments. The algorithm assumes that the switchbox is extendable in both directions. Subsequently, these extensions are reduced with the goal to reach the fixed size of the switchbox. While more costly in runtime, on numerous benchmark examples the genetic algorithm produces solutions with equal or better routing characteristics (netlength, number of vias) than the previously best published results.

3.4 Compaction

As mentioned in Section 2, compaction transforms the symbolic layout to a mask layout with the goal of minimizing the size of the resulting circuit layout.

To the best of our knowledge, the only application of a genetic algorithm for compaction has been advanced by Fourman [14]. He describes two prototypes of genetic algorithms that perform compaction of a symbolic circuit layout. Although his results are limited to very simple layout structures, he does propose a new problem-specific representation for layout design that includes constraints of the compaction process.

4. A Parallel Genetic Algorithm for the VLSI Routing Problem

In the following, we present a parallel genetic algorithm to solve the VLSI channel and switchbox routing problems with the objective of satisfying crosstalk constraints for the nets. This approach is an extension of [30] in which a sequential genetic algorithm was applied to channel routing problems. In [28] we first introduced the parallel genetic approach and extensively investigated its main parameters.

4.1 Introductory remarks

As mentioned earlier, interconnection routing is one of the major tasks in the physical design of VLSI circuits. Pins that belong to the same net are connected subject to a set of routing constraints. Channel and switchbox routing are the two most common routing problems in VLSI circuits. Simple examples of a channel routing problem and a switchbox routing problem are shown in Figure 4.1.

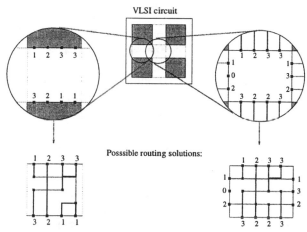

Fig. 4.1. The VLSI channel (left) and switchbox (right) routing problem and possible routing solutions

Our motives for developing a parallel genetic algorithm for the detailed routing problem have been threefold. First, almost all previously published detailed routing strategies only consider physical constraints, such as the netlength. However, with further minimization in VLSI design, new electrical constraints, such as crosstalk, are becoming dominant and need to be addressed. Second, today's typical computer-aided design environment consists of a number of workstations connected together by a high-speed local network. Although many VLSI routing systems make use of the network to share files or design databases, none of the known routing programs (evolution-based or deterministic algorithms) use this distributed computer resource to parallelize and speed up their work. Third, all published genetic algorithms that address the routing problem are sequential approaches, i.e., *one* population evolves by means of genetic operators. However, recent publications (e.g. [2],[26]) clearly indicate that parallel genetic algorithms with isolated evolving subpopulations (that exchange individuals from time to time) perform better than sequential approaches.

We present a parallel genetic algorithm for detailed routing, called GAP (Genetic Algorithm with Punctuated equilibria), that runs on a distributed network of workstations. Our approach considers routing quality characteristics such as the the netlength, the number of connections between layers, and the importance of crosstalk between neighboring interconnections. Due to variable weight factors, these routing objectives can be easily adjusted to the requirements of a given VLSI technology. Furthermore, on many benchmark examples, the router produces better results than the best of those previously published.

4.2 Problem description

The VLSI routing problem is defined as follows. Consider a rectangular routing region with *pins* located on two parallel boundaries (*channel*) or four boundaries (*switchbox*) (see Figure 4.1). The pins that belong to the same net need to be connected subject to certain constraints and quality factors. The interconnections need to be made inside the boundaries of the routing region on a symbolic routing area consisting of horizontal *rows* and vertical *columns*.

We define a *segment* to be an uninterrupted horizontal or vertical part of a net. (Thus, any connection between two pins will consist of one or more net segments.) A connection between two net segments from different layers is called a *via*. The overall length of all segments of one net to connect its pins is defined as its *netlength*.

In sub-micron regimes, crosstalk results mainly from coupled capacitance between adjacent (parallel routed) interconnections. The shorter the length of these parallel routed segments, the better the performance of the circuit.

Thus, the following three factors (which are to be minimized) are used in this work to assess the quality of the routing:

— netlength,
— number of vias, and
— crosstalk.

4.3 Overview of the parallel genetic algorithm

Different ways exist to parallelize a genetic algorithm [2]. However, most of these methods result only in a speed-up of the algorithm without qualitative improvements to the problem solutions. To gain better problem solutions, we use the theory of *punctuated equilibria* to design a parallel genetic algorithm [7],[10]. A genetic algorithm with punctuated equilibria is a parallel genetic algorithm in which independent *subpopulations* of individuals with their own *fitness functions* evolve in isolation except for an exchange of individuals (*migration*) when a state of equilibrium throughout all the subpopulations has been reached (see Figure 4.2). Previous research has shown genetic algorithms with such punctuated equilibria to have superior performance when compared to sequential genetic approaches [7],[9].

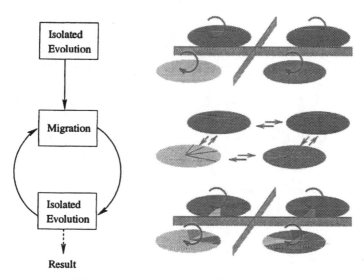

Fig. 4.2. Punctuated equilibria model

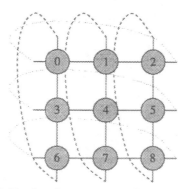

Fig. 4.3. Neighborhood structure with nine subpopulations

The parallel structure of our algorithm for the case of nine processors is shown in Figure 4.3. We assign a set of n individuals (problem solutions) to each of the N processors, for a *total population* size of $n \times N$. The set assigned to each processor, c, is its subpopulation, $\mathcal{P}_c$. The processors are connected

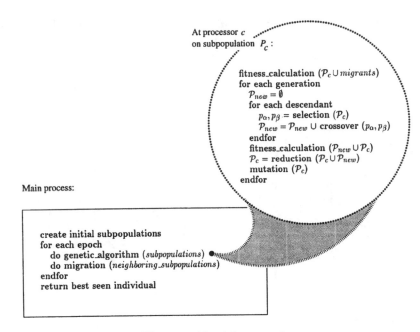

At processor c
on subpopulation P_c:

fitness_calculation ($P_c \cup migrants$)
for each generation
    $P_{new} = \emptyset$
    for each descendant
        $p_\alpha, p_\beta :=$ selection (P_c)
        $P_{new} = P_{new} \cup$ crossover (p_α, p_β)
    endfor
    fitness_calculation ($P_{new} \cup P_c$)
    $P_c =$ reduction ($P_c \cup P_{new}$)
    mutation (P_c)
endfor

Main process:

create initial subpopulations
for each epoch
    do genetic_algorithm (*subpopulations*)
    do migration (*neighboring_subpopulations*)
endfor
return best seen individual

Fig. 4.4. Algorithm overview

by an interconnection network with a torus topology. Thus, each processor (subpopulation) has exactly four *neighbors*.

The genetic algorithm used by each processor and the main process that steers the parallel execution are presented in Figure 4.4. First, the main process creates an initial subpopulation at each processor. This initial subpopulation consists of randomly constructed (i.e., not optimized) routing solutions. They are designed by a random routing strategy which connects net points in an arbitrary order with randomly placed interconnections. (See [30] for a detailed description of our random routing strategy.) The main process consists of a given number of *epochs*. During an epoch, each processor, disjointly and in parallel, executes the sequential genetic algorithm on its subpopulation for a certain number of generations (*epoch length*). Afterwards, each subpopulations exchanges a specific number of individuals (*migrants*) with its four neighbors. The process continues with the separate evolution of each subpopulation during the next epoch. At the end of the process, the best individual that exists constitutes our final routing solution.

The following section briefly describes the genetic operators used by each processor to evolve its subpopulation.

4.4 Genetic operators

Fitness Calculation. The fitness $F(p)$ of each individual $p \in \mathcal{P}_c$ is calculated to assess the quality of its interconnections relative to the rest of the sub-population $\mathcal{P}_c$. The following factors are taken into account (with different weights) when determining $F(p)$:

- overall netlength of p,
- number of vias of p, and
- the length of adjacent, parallel interconnections (crosstalk).

After the evaluation of $F(p)$ for all individuals of the subpopulation $\mathcal{P}_c$ these values are scaled linearly [18], in order to control the relative range of fitness in the subpopulation.

Selection. Our selection strategy, which is responsible for choosing the mates for the crossover procedure, is stochastic sampling with replacement [18]. That means any individual $p_i \in \mathcal{P}_c$ is selected with a probability proportional to its fitness value.

Crossover. During a crossover, two individuals are combined to create a descendant. Our crossover operator is a 1-point crossover operator [18] that gives high-quality routing parts of the mates an increased probability of being transferred intact to their descendant.

Crossover is performed in terms of wire segments. A randomly positioned line (crossline) perpendicular to the edges of the routing area divides this area into two sections, playing the role of the crosspoint. This line can be either horizontally or vertically placed. For example, interconnection segments *exclusively* on the upper side of the crossline are inherited from the first parent, and segments *exclusively* on the lower side of the crossline are inherited from the second parent. Segments intersecting the crossline are newly created within the descendant by means of our random routing strategy [30].

(A more detailed description of our crossover operator is given in [30].)

Reduction. Our reduction strategy simply chooses the $|\mathcal{P}_c|$ fittest individuals of $(\mathcal{P}_c \cup \mathcal{P}_{new})$ to survive as $\mathcal{P}_c$ into the next generation.

Mutation. The mutation operator performs random modifications on an individual (to overcome local optima) by applying the random routing strategy [30] on randomly selected interconnections.

4.5 Experimental results

Our algorithm, called GAP, has been implemented on a network of SPARC workstations (SunOS and Solaris systems). The parallel computation environment is provided by the Mentat system, an object-oriented parallel processing system [20].

4.5.1 Parameter settings. The main parameters (see Section 4.3) were
set to:

Individuals per subpopulation :	50
Descendants per subpopulation :	20
Number of subpopulations :	9
Number of epochs :	10
Epoch length (generations) :	50

Two randomly selected migrants were sent to each of the four neighbors
in each epoch.

(See [28] for a detailed discussion of these parameters as well as a comparison with a sequential genetic approach.)

4.5.2 Comparison of GAP to other routing algorithms. Any application of a genetic algorithm should focus on a comparison to solution techniques that have been acknowledged as effective by that application's community. Here we compare the results of GAP with the best known results of
other algorithms for channel and switchbox routing benchmarks (see Table
4.1). The other routing algorithms do not consider crosstalk, and thus can
only be compared with our routing results regarding netlength and number
of vias.

We ran our algorithm 50 times per benchmark with different initialization of the random number generator. Table 4.1 presents the best-ever-seen
results for all algorithms. We note that for GAP the best-ever-seen quality
was achieved in at least 50 percent of the program executions.

It can be seen that our results are qualitatively similar to or better than
the best known results from popular channel and switchbox routers published
for these benchmarks. The layout of Burstein's difficult switchbox achieved
with our algorithm is depicted in Figure 4.5.

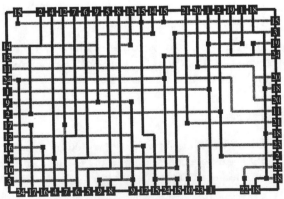

Fig. 4.5. Our routing solution of Burstein's difficult switchbox

Table 4.1. Comparison of GAP with the best-known results for some benchmark channels (upper half) and switchboxes (lower half)

Bench-mark	Algorithm	Columns	Rows	Net-length	Vias	Time (sec)
Joo6_13	WEAVER[23]	18	7	167	29	312
	PACKER[16]	18	6	167	25	710
	SAR[1]	18	6	166	25	70
	GAP	18	6	164	22	172
Joo6_16	WEAVER[23]	11	7	121	21	220
	Monreale[15]	11	7	120	19	?
	GAP	11	6	115	15	207
Burstein's Difficult Switchbox	WEAVER[23]	23	15	531	41	1508
	BEAVER[a][8]	23	15	547	44	1
	PACKER[16]	23	15	546	45	56
	GAP	23	15	538	36	1831
Dense Switchbox	Silk[32]	16	17	516	29	?
	SAR[1]	16	17	519	31	150
	GAP	16	17	516	29	2380
Augmented Dense Switchbox	BEAVER[a][8]	16	18	529	31	1
	PACKER[16]	16	18	529	32	31
	SAR[1]	16	18	529	31	205
	GAP	16	18	529	29	2281

[a] BEAVER's number of vias has been adjusted.

4.5.3 Reduction of crosstalk. By adjusting the value of the weight for crosstalk (see Section 4.4), our algorithm can also optimize the interconnections regarding crosstalk. The results presented in [29] show that an increase of the weight for crosstalk leads to significantly fewer parallel routed net segments. Hence, our router can construct solutions which contain minimal coupling capacitances between interconnections – an increasingly significant consideration in sub-micron VLSI design.

However, as discussed in [29], the minimization of crosstalk leads in general to an increase in both the netlength and the number of vias. It has to be decided by the user to which optimization goals he/she gives priority.

5. Summary

We presented a systematic review of genetic algorithm investigations for the VLSI physical design process. These contributions are generally different than standard genetic algorithm investigations. One difference is that the genetic operators for physical design algorithms are typically very problem-specific. This specificity occurs because of the extreme importance of determining very high quality solutions – therefore expert information on the likely form of solutions is included as much as possible. Combinations of genetic algorithms with other optimization strategies are no longer an exception. Once the "high quality regions" are identified by a genetic algorithm, the application of local search routines are often the only way to ensure effective runtimes. Another difference is the concern for robustness. There exists a rich collection of design automation benchmarks (e.g., [34]) and for a solution method to be accepted, it must be demonstrated to work consistently well on those benchmarks.

We also presented a parallel genetic algorithm for the channel and switch-box routing problem. Our results are qualitatively similar to or better than the best known results from popular channel and switchbox routers. In addition, our algorithm is able to significantly reduce the occurance of crosstalk.

Genetic algorithms have a very large potential within physical design of VLSI circuits. The problems encountered in this field are extremely complex which is exactly the situation in which the performance of a genetic algorithm compares best to that of other methods. However, genetic-algorithm-based approaches are of practical interest to the VLSI community only if they are competitive with the acknowledged existing approaches with respect to performance and runtime. This chapter gave a review of the current situation in this field with the purpose of stimulating and guiding further applications of genetic algorithms.

References

1. A. Acan and Z. Ünver, "Switchbox Routing by Simulated Annealing: SAR," in *Proc. IEEE International Symposium on Circuits and Systems*, vol. 4, pp. 1985-1988, 1992.
2. P. Adamidis, *Review of Parallel Genetic Algorithms*, Technical Report, Aristotle University of Thessaloniki, 199
3. H. B. Bakoglu, *Circuits, Interconnections, and Packaging for VLSI*, Reading, MA: Addison-Wesley, 1990.
4. T. N. Bui and B. R. Moon, "A Fast and Stable Hybrid Genetic Algorithm for the Ratio-Cut Partitioning Problem on Hypergraphs", *Proc. of the ACM-IEEE Design Automation Conference*, pp. 664-669, 1994.
5. H. Chan, P. Mazumder and K. Shahookar, "Macro-Cell and Module Placement by Genetic Adaptive Search with Bitmap-Represented Chromosome," *Integration, The VLSI Journal*, vol. 12, no. 1, pp. 49-77, Nov. 1991.
6. J. P. Cohoon and W. D. Paris, "Genetic Placement," *IEEE Trans. on Computer-Aided Design*, vol. 6, no. 6, pp. 956-964, Nov. 1987.
7. J. P. Cohoon, S. U. Hedge, W. N. Martin, and D. S. Richards, "Punctuated Equilibria: A Parallel Genetic Algorithm," *Proc. Second International Conference on Genetic Algorithms*, pp. 148-154, 1987.
8. J. P. Cohoon and P. L. Heck, "BEAVER: A Computational-Geometry-Based Tool for Switchbox Routing," *IEEE Trans. on Computer-Aided Design*, vol. 7, no. 6, pp. 684-697, 1988.

9. J. P. Cohoon, W. N. Martin, and D. S. Richards, "Genetic Algorithms and Punctuated Equilibria in VLSI," *Parallel Problem Solving from Nature*, H. P. Schwefel and R. Männer, eds., Lecture Notes in Computer Science, vol. 496, Berlin: Springer Verlag, pp. 134-144, 1991.

10. N. Eldredge and S. J. Gould, "Punctuated Equilibria: An Alternative to Phyletic Gradualism," *Models of Paleobiology*, T. J. M. Schopf, ed., San Francisco, CA: Freeman, Cooper and Co., pp. 82-115, 1972.

11. H. Esbensen, "A Genetic Algorithm for Macro Cell Placement," *Proc. of the European Design Automation Conference*, pp. 52-57, Sept. 1992.

12. H. Esbensen and P. Mazumder, "SAGA: A Unification of the Genetic Algorithm with Simulated Annealing and its Application to Macro-Cell Placement," *Proc. of the 7th International Conference on VLSI Design*, pp. 211-214, Jan. 1994.

13. H. Esbensen, "A Macro-Cell Global Router Based on Two Genetic Algorithms" *Proc. of the European Design Automation Conference*, pp. 428-433, Sept. 1994.

14. M. P. Fourman, "Compaction of Symbolic Layout using Genetic Algorithms," *Proc. of the First International Conference on Genetic Algorithms*, pp. 141-153, 1985.

15. M. Geraci, P. Orlando, F. Sorbello and G. Vasallo, "A Genetic Algorithm for the Routing of VLSI Circuits," *Euro Asic '91*, Parigi 27-31 Maggio, Los Alamitos, CA: IEEE Computer Society Press, pp. 218-223, 1991.

16. S. H. Gerez and O. E. Herrmann, "Switchbox Routing by Stepwise Reshaping," *IEEE Trans. on Computer-Aided Design*, vol. 8, no. 12, pp. 1350-1361, 1989.

17. N. Göckel, G. Pudelko, R. Drechsler, B. Becker, "A Hybrid Genetic Algorithm for the Channel Routing Problem," *Proceedings of the 1996 IEEE International Symposium on Circuits and Systems, ISCAS-96*, pp. 675-678, 1996.

18. D. E. Goldberg, *Genetic Algorithms in Search, Optimization, and Machine Learning*, Reading, MA: Addison-Wesley, 1989.

19. J. J. Grefenstette and N. N. Schraudolph, *A User's Guide to GENESIS 1.2 UCSC*, CSE Dept., University of California, San Diego, 1987.

20. Homepage: "http://www.cs.virginia.edu/~mentat/".

21. M. Hulin, "Analysis of Schema Distributions," *Proc. of the Fourth International Conference on Genetic Algorithms*, pp. 204-209, 1991.

22. M. Hulin, "Circuit Partitioning with Genetic Algorithms Using a Coding Scheme to Preserve the Structure of a Circuit," *Parallel Problem Solving from Nature*, H. P. Schwefel and R. Männer, eds., Lecture Notes in Computer Science, vol. 496, Berlin: Springer Verlag, pp. 75-79, 1991.

23. R. Joobbani, *An Artificial Intelligence Approach to VLSI Routing*, Boston, MA: Kluwer Academic Publishers, 1986.

24. R. M. Kling and P. Banerjee, "ESP: Placement by Simulated Evolution," *IEEE Trans. on Computer-Aided Design*, vol. 8, no. 3, pp. 245-256, March 1989.

25. R. M. Kling and P. Banerjee, "Optimization by Simulated Evolution with Applications to Standard Cell Placement," *Proc. of the 27th ACM-IEEE Design Automation Conference*, pp. 20-25, 1990.

26. B. Kröger, *Parallel Genetic Algorithms for Solving the Two-Dimensional Bin Packing Problem* (in German), Ph.D. Thesis, University of Osnabrück, 1993.

27. C. Y. Lee, "An Algorithm for Path Connections and its Applications," *IRE-Trans. on Electronic Computers*, pp. 346-365, 1961.

28. J. Lienig, "A Parallel Genetic Algorithm for Two Detailed Routing Problems", *Proceedings of the 1996 IEEE International Symposium on Circuits and Systems, ISCAS-96*, pp. 508-511, 1996.

29. J. Lienig, "Channel and Switchbox Routing with Minimized Crosstalk – A Parallel Genetic Approach", *Proceedings of the 10th International Conference on VLSI Design*, pp.27–31, Jan. 1997.
30. J. Lienig and K. Thulasiraman, "A Genetic Algorithm for Channel Routing in VLSI Circuits," *Evolutionary Computation*, vol. 1, no. 4, pp. 293-311, 1994.
31. J. Lienig and K. Thulasiraman, "GASBOR: A Genetic Algorithm for Switchbox Routing in Integrated Circuits," *Progress in Evolutionary Computation*, X. Yao, ed., Lecture Notes in Artificial Intelligence, vol. 956, Berlin: Springer Verlag, pp. 187-200, 1995.
32. Y.-L. Lin, Y.-C. Hsu and F.-S. Tsai, "SILK: A Simulated Evolution Router," *IEEE Trans. on Computer-Aided Design*, vol. 8, no. 10, pp. 1108-1114, Oct. 1989.
33. S. Mohan and P. Mazumder, "Wolverines: Standard Cell Placement on a Network of Workstations," *IEEE Trans. on Computer-Aided Design*, vol. 12, no. 9, pp. 1312-1326, Sept. 1993.
34. B. T. Preas, "Benchmarks for Cell-based Layout Systems," *Proc. of the ACM-IEEE Design Automation Conference*, pp. 319-320, 1987.
35. *Proc. of the ACM-IEEE Design Automation Conference*, 1984.
36. *Proc. of the ACM-IEEE Design Automation Conference*, 1987.
37. Y. Saab and V. Rao, "An Evolution-Based Approach to Partitioning ASIC Systems," *Proc. of the ACM-IEEE Design Automation Conference*, pp. 767-770, 1989.
38. C. Sechen, *VLSI Placement and Global Routing Using Simulated Annealing*, Boston, MA: Kluwer Academic Publishers, 1988.
39. K. Shahookar and P. Mazumder, "GASP - A Genetic Algorithm for Standard Cell Placement," *Proc. of the European Design Automation Conference*, pp. 660-664, 1990.
40. K. Shahookar and P. Mazumder, "A Genetic Approach to Standard Cell Placement using Meta-Genetic Parameter Optimization", *IEEE Trans. on Computer-Aided Design*, vol. 9, no. 5, pp. 500-511, May 1990.
41. K. Shahookar, W. Khamisani, P. Mazumder and S. M. Reddy, "Genetic Beam Search for Gate Matrix Layout," *Proc. of the 6th International Conference on VLSI Design*, pp. 208-213, Jan. 1993.
42. J. M. Varanelli and J. P. Cohoon, "Population-Oriented Simulated Annealing: A Genetic/Thermodynamic Hybrid Approach to Optimization," *Proc. of the Sixth International Conference on Genetic Algorithms*, pp. 174-181, 1995.

Statistical Generalization of Performance-Related Heuristics for Knowledge-Lean Applications

Arthur Ieumwananonthachai and Benjamin W. Wah

Department of Electrical and Computer Engineering and the Coordinated Science Laboratory, University of Illinois at Urbana-Champaign, 1308 West Main Street, Urbana, IL 61801

Summary. In this chapter, we present new results on the automated generalization of performance-related heuristics learned for knowledge-lean applications. By first applying genetics-based learning to learn new heuristics for some small subsets of test cases in a problem space, we study methods to generalize these heuristics to unlearned subdomains of test cases. Our method uses a new statistical metric called probability of win. By assessing the performance of heuristics in a range-independent and distribution-independent manner, we can compare heuristics across problem subdomains in a consistent manner. To illustrate our approach, we show experimental results on generalizing heuristics learned for sequential circuit testing, VLSI cell placement and routing, branch-and-bound search, and blind equalization. We show that generalization can lead to new and robust heuristics that perform better than the original heuristics across test cases of different characteristics.

1. Introduction

Heuristics or heuristic methods (HMs), in general terms, are "Strategies using readily accessible though loosely applicable information to control problem-solving processes in human being and machines" [12]. They exist as problem solving procedures in problem solvers to find (usually) suboptimal solutions for many engineering applications. Since their design depends on user experience and is rather ad hoc, it is desirable to acquire them automatically by machine learning.

We make the following assumptions in this chapter. First, we assume that the applications are *knowledge-lean*, implying that domain knowledge for credit assignment is missing. In this class of applications, we are interested to learn and generalize *performance-related* HMs whose goal is to find solutions with the best numerical performance. Examples of targeted HMs and applications include symbolic formulae for decision making in a branch-and-bound search and a set of numerical parameters used in a simulated annealing package for placement and routing of VLSI circuits. (See Section 4.).

Second, we assume that performance of a HM is characterized by one or more statistical metrics and is obtained by evaluating multiple test cases (noisy evaluations). We further assume that a HM may have different performance distributions across different subsets of test cases in the problem space, thereby disallowing the use of performance metrics such as the average.

For example, given two heuristic methods HM_1 and HM_2 and two subsets of test cases TC_1 and TC_2, assume that cost is the performance measure of HMs. Suppose after testing HM_1 on the two subsets of test cases, we found its average costs be 10 and 100 units, respectively. Similarly, we got 150 and 5 units for HM_2 on the two subsets of test cases. It will be difficult to say whether HM_1 is better than HM_2 in terms of cost, and which HM should be used as a general HM for all test cases in the problem domain.

Third, we assume that heuristics used in generalization are learned by a genetics-based learning method [18, 8]. This is a form of learning by induction that involves applying genetic algorithms [3] to machine learning problems. There are two steps involved in this learning method:

- *Generation* and *selection* of HMs that can better solve test cases used in learning, as compared to the best existing (*baseline*) HMs;
- *Generalization* of the selected HMs to test cases not seen in learning with the same high level of performance as compared to that of the baseline HMs.

As illustrated in Figure 1.1, these two steps are generally separated in genetics-based learning.

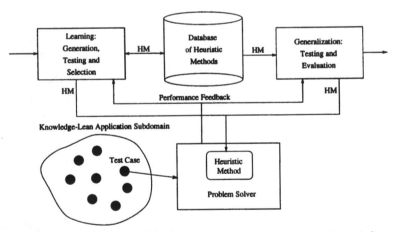

Fig. 1.1. Learning and generalization in knowledge-lean applications is based on evaluating a heuristic method on a test case and on observing its performance feedback

In this chapter, we study *statistical generalization of HMs across test cases of an application with different performance distributions*. The problem is illustrated in Figure 1.2 in which we show three heuristic methods and three subsets (or subdomains) of test cases in an application domain. Let $p_{i,j}$ be the performance of HM_i on Subdomain j, in which we assume that the

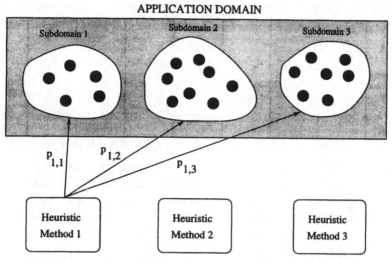

Fig. 1.2. Performance of a heuristic method may vary significantly across different subsets of test cases in an application domain, making it difficult to combine these numbers into a single performance number

performance of an HM in a subdomain can be aggregated into a single value. When an HM behaves differently across different subdomains of test cases, it will not be possible to aggregate its performance values across subdomains into a single number. Further, when one HM performs better than another HM in one subdomain but worse in another, we need to develop a method to differentiate HMs with high performance from those with low performance across all test cases in the application.

Generalization is important because learning time is often limited, and only a small set of test cases can be evaluated during learning. Generalization in many existing genetics-based learning systems [8, 3, 2] is a post-learning verification phase that simply verifies the generalizability of the learned HMs by evaluating them on a new set of test cases. This approach is suitable when test cases used in learning are representatives of *all* the test cases targeted by the HM. When test cases used in generalization have different characteristics, the HMs learned cannot be generalized.

To compare HMs bearing different performance distributions across different subsets of test cases in an application, we need to develop a performance metric that is independent of the actual distributions. We propose in this chapter a new metric called *probability of win* that measures the probability that a particular HM is better than another randomly chosen HM from a set of learned HMs for a given subset of test cases. Since probabilities are between 0 and 1, we eliminate the dependence of HMs on actual performance distributions. Using this metric, we can verify whether a HM is generalizable across test cases of different performance distributions. Our approach can be summarized as follows:

- Partition the domain of test cases into *subdomains* in such a way that performance values in a subdomain are *independent and identically distributed* (i.i.d.).
- Develop conditions under which a HM can be considered to perform well across multiple subdomains. In contrast to studies in artificial intelligence [6], we do not modify a HM in order to generalize it across subdomains. Rather, we test certain conditions to see if a HM is generalizable.

This chapter is divided into five sections. Section 2. defines problem space and its partitioning into subdomains. We propose in Section 3. a new metric called probability of win and a new generalization strategy. Section 4. reports our experimental results on four real-world applications — circuit testing, VLSI cell placement and routing, branch-and-bound search and blind equalization. Conclusions are drawn in Section 5.

2. Problem Domains and Subdomains

Given an application problem consisting of a collection of test cases, the first task in learning and generalization is to classify the test cases into *domains* such that a unique HM can be designed for each [13]. This classification step is domain specific and is generally carried out by experts in the area.

For instance, consider the problem of generating test patterns to test VLSI circuits. Previous experience shows that sequential circuits require tests that are different from those of combinatorial circuits. Consequently, we can consider combinatorial circuits and sequential circuits as two different problem domains.

In comparing the performance of HMs in a problem domain, it is necessary to aggregate their performance values into a small number of performance metrics (such as average or maximum). Computing these aggregate metrics is not meaningful when performance values are of different ranges and distributions across different subsets of test cases in the domain. In this case, we need to decompose the domain into smaller partitions so that quantitative comparison of performance of HMs in a partition is possible. We define a *problem subdomain* as a partitioning of the domain of test cases such that performance values of a HM in a subdomain are i.i.d. Under this condition, it is meaningful to compute the average performance of test cases in a subdomain. It is important to point out that performance values may need to be normalized with respect to those of the baseline HM before aggregated.

We need to know the attributes of an application in order to classify its test cases, and a set of decision rules to identify the subdomain to which a test case belongs. For example, in learning new decomposition HMs in a branch-and-bound search for solving a traveling-salesman problem (Section 4.), we can treat graph connectivity as an attribute to classify graphs into subdomains.

In some applications, it may be difficult to determine the subdomain to which a test case belongs. This is true because the available attributes may not be well defined or may be too large to be useful. For instance, in test-pattern generation for sequential circuits, there are many attributes that can be used to characterize circuits (such as length of the longest path and maximum number of fan-in's and fan-out's). However, none of these attributes is a clear winner.

When we do not know the attributes to classify test cases into subdomains, we can treat each test case as a subdomain by itself. This works well when the HM to be learned has a random component: by using different random seeds in the HM, we can obtain statistically valid performance values of the HM on a test case. We have used this approach in the two circuit-related applications discussed in Section 4. and have chosen each circuit as an independent subdomain for learning.

After applying learning to find good HMs for each subdomain, we need to compare their performance across subdomains. This may be difficult because test cases in different subdomains of a domain may have different performance distributions, even though they can be evaluated by a common HM. As a result, the performance of test cases cannot be compared statistically. For instance, we cannot use the average metric when performance values are dependent or have multiple distributions.

As an example, Table 2.1 shows the average and maximum fault coverages of two HMs used in a test-pattern generator to test sequential circuits. The data indicate that we cannot average their fault coverages across the two circuits as the performance distribution of HM_{101} across the two circuits is not the same as that of HM_{535}.

Table 2.1. Maximum and average fault coverages of two HMs used in a test-pattern generator with different random seeds

Circuit	HM	Maximum FC	Average FC
S444	101	60.3	28.5
	535	86.3	84.8
S1196	101	94.9	94.2
	535	93.6	93.1

It should now be clear that there can be many subdomains in an application, and learning can only be performed on a small number of them. Consequently, it is important to generalize HMs learned for a small number of subdomains to unlearned subdomains. In some situations, multiple HMs may have to be identified and applied together at a higher cost to find a high quality solution.

3. Generalization of Heuristic Methods Learned

Since learning can only cover a small subset of a problem space, it is necessary to generalize HMs developed to test cases not studied in learning. When test cases used in learning have the same performance distribution as those used in generalization, generalization simply involves verifying the performance results. However, as illustrated in the last section, test cases used in generalization may have different performance distribution for two reasons: (a) A learned HM has different performance distributions across subdomains. (b) The baseline HM used in normalization has different performance distributions across subdomains. In either case, performance values after normalization will have different distributions across subdomains. This leads us to develop a generalization strategy that can compare HMs across different subdomains with different performance distributions.

The goal of generalization is somewhat vague: we like to find one or more HMs that perform well most of the time across multiple subdomains as compared to the baseline HM (if it exists). To achieve this goal, two issues are apparent here.

- How to compare the performance of HMs within a subdomain in a range-independent and distribution-independent fashion? Here, we need to evaluate and generalize the performance of a HM in a single subdomain in a range-independent and distribution-independent way.
- How to define the notion that one HM performs well across multiple subdomains?

Our method to address these two issues involves a new metric called probability of win. Informally, *probability of win* is a range-independent metric that evaluates the probability that the *true mean performance* of a HM in one subdomain is better than the true mean performance of another randomly selected HM in the same subdomain. It is important to point out that the HMs used in computing the probability of win are found by learning; hence, they already perform well within a subdomain. Further, probabilities of win are in the range zero to one, independent of the number of HMs evaluated and the distribution of performance values.

3.1 Performance evaluation within a subdomain

There are many ways to address the first issue raised above, and solutions to the second issue depend on the solution to the first. For instance, scaling and normalization of performance values is a possible way to compare performance in a distribution-independent manner; however, this may lead to new inconsistencies [18]. Another way is to rank HMs by their performance values and use the average ranks of HMs for comparison. This does not work well because it does not account for actual differences in performance values, and two HMs with very close or very different performance may differ only by one in their ranks. Further, the maximum rank of HMs depends on the number of HMs evaluated, thereby biasing the average ranks of individual HMs. In this section, we propose a metric called probability of win to select good HMs within a subdomain.

$P_{win}(h_i, d_m)$, the *probability-of-win* of HM h_i in subdomain d_m, is defined as the probability that the true mean of h_i (on one performance measure[1]) is better than the true mean of HM h_j randomly selected from the pool. When h_i is applied on test cases in d_m, we have

$$P_{win}(h_i, d_m) = \frac{\sum_{j \neq i} P\left[\mu_i^m > \mu_j^m | \hat{\mu}_i^m, \hat{\sigma}_i^m, n_i^m, \hat{\mu}_j^m, \hat{\sigma}_j^m, n_j^m\right]}{|s| - 1}, \qquad (3.1)$$

where $|s|$ is the number of HMs under consideration, and n_i^m, $\hat{\sigma}_i^m$, $\hat{\mu}_i^m$, and μ_i^m are, respectively, the number of tests, sample standard deviation, sample mean, and true mean of h_i in d_m.

Since we are using the average performance metric, it is a good approximation to use the normal distribution as a distribution of the sample average. The probability that h_i is better than h_j in d_m can now be computed as follows.

$$P\left[\mu_i^m > \mu_j^m \,|\, \hat{\mu}_i^m, \hat{\sigma}_i^m, n_i^m, \hat{\mu}_j^m, \hat{\sigma}_j^m, n_j^m\right] \approx \Phi\left[\frac{\hat{\mu}_i^m - \hat{\mu}_j^m}{\sqrt{\hat{\sigma}_i^{m2}/n_i^m + \hat{\sigma}_j^{m2}/n_j^m}}\right]$$

where $\Phi(x)$ is the cumulative distribution function for the $N(0,1)$ distribution.

To illustrate the concept, we show in Table 3.1 the probabilities of win of four HMs tested to various degrees. Note that P_{win} is not only related to the sample mean but also depends on the sample variance and number of tests performed. Further, the probability that h_i is better than h_j and the probability that h_j is better than h_i are both counted in the evaluation. Hence, the average of P_{win} over all HMs in a subdomain ($= \sum_i P_{win}(h_i, d_m)/|s|$) will be 0.5.

P_{win} defined in (3.1) is range-independent and distribution-independent because all performance values are transformed into probabilities between 0 and 1 independent of the number of HMs evaluated and the distribution of performance values. It assumes that all HMs are i.i.d. and takes into account uncertainty in their sample averages (by using their variances); hence, it is better than simple scaling that only compresses performance averages into a

---

[1] Due to space limitation, we do not consider issues dealing with multiple performance measures in this chapter.

Table 3.1. Probabilities of win of four HMs in d_m

h_i	$\hat{\mu}_i$	$\hat{\sigma}_i$	n_i	$P_{win}(h_i, d_m)$
1	43.2	13.5	10	0.4787
2	46.2	6.4	12	0.7976
3	44.9	2.5	10	0.6006
4	33.6	25.9	8	0.1231

range between 0 and 1. It is also important to point out that the HMs used in computing P_{win} are found by learning; hence, they already perform well within a subdomain.

3.2 Performance evaluation across subdomains

One of the major difficulties in handling multiple subdomains is that it may be difficult to aggregate performance values statistically from different subdomains, and to define the notion that one HM performs better than another across multiple subdomains. For instance, it is not meaningful to find an average of random numbers from two different distributions. We address this problem using P_{win} defined in the last subsection.

First, we assume that when HM h is applied over multiple subdomains in partition Π_p of subdomains, all subdomain are equally likely. Here, we compute P_{win} of h over subdomains in Π_p as the average P_{win} of h over all subdomains in Π_p.

$$P_{win}(h, \Pi_p) = \frac{\sum_{d \in \Pi_p} P_{win}(h, d)}{|\Pi_p|}, \qquad (3.2)$$

where Π_p is the p'th partition of subdomains in the problem domain. The HM picked is the one that maximizes (3.2). When subdomains are not equally likely but with known relative weights, we can compute P_{win} as a weighted average instead of (3.2). HMs picked using 3.2 generally wins with a high probability across most of the subdomains in Π_p but occasionally may not perform well in a few subdomains.

Second, we consider the problem of finding a good HM across multiple subdomains in Π_p as a multi-objective optimization problem. In this case, evaluating HMs based on a combined objective function (such as the average P_{win} in (3.2) may lead to inconsistent conclusions. To alleviate such inconsistencies, we should treat each subdomain independently and find a common HM across all subdomains in Π_p satisfying some common constraints. For example, let δ be the allowable deviation of P_{win} of any chosen HM from q_{win}^m, the maximum P_{win} in subdomain m. Generalization, therefore, amounts to finding h that satisfies the following constraints for every subdomain $m \in \Pi_p$.

$$P_{win}(h, m) \geq (q_{win}^m - \delta) \qquad \forall m \in \Pi_p \qquad (3.3)$$

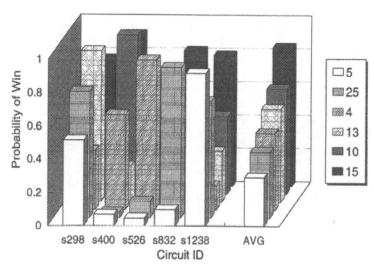

Fig. 3.1. P_{win} of six HMs across five subdomains in the test-pattern generation problem

Here, δ may need to refined if there are too many or too few HMs satisfying the constraints.

To illustrate the generalization procedure, consider the test-pattern generation problem discussed in Section 2. Assume that learning had been performed on five circuits (subdomains), and that the six best HMs from each subdomain were reported. After full evaluation of the 30 HMs (initialized by ten random seeds) across all five subdomains, we computed P_{win} of each HM in every subdomain. Figure 3.1 shows the probabilities of win of six of these HMs. If we generalize HMs based on (3.2), then HM_{15} will be picked since it has the highest average P_{win}. Likewise, if we generalize using (3.3), we will also select HM_{15}. Note that in this example, no one HM is the best across all subdomains.

4. Experimental Results

To illustrate the generalization procedure described in Section 3., we present in this section results on generalization for two applications in VLSI design and branch-and-bound search. These results were obtained using TEACHER [18], a genetics-based learning system that implements our proposed generalization strategy. The parameters used during learning are shown in Table 4.1.

Table 4.1. Genetic-algorithm parameters used in our learning system. (*# HMs Verified at Termination* is the number of HMs selected for verification at the end of the last generation)

Application	CRIS	Timber-Wolf	Branch-and-Bound
Number of Generations	10	10	10
Duration of a Generation	100	100	160
# Active HMs in each Gen.	30	30	40
New HMs Generated in each Gen.	20	20	30
Crossover Rate	0.45	0.45	0.5
Mutation Rate	0.35	0.35	0.17
Random Generation Rate	0.20	0.20	0.33
# HMs Verified at Termination	20	20	20

4.1 HM for sequential circuit testing

The first application is based on CRIS [15], a genetic-algorithm software package for generating patterns to test sequential VLSI circuits. CRIS mutates an input test sequence continuously and analyzes the mutated vectors in selecting a test set. Since many copies of a circuit may be manufactured, it is desirable to obtain as high a fault coverage as possible, and computational cost is of secondary importance.

In our experiments, we used sequential circuits from the ISCAS89 benchmarks [1] plus several other larger circuits. We treat each circuit as an individual subdomain. Since we want one common HM for all circuits, we assume that all circuits are from one domain.

CRIS in our experiments is treated as a black-box problem solver, as we have minimal knowledge in its design. A HM targeted for improvement is a set of eight parameters used in CRIS (Table 4.2). Note that parameter P_8 is a random seed, implying that CRIS can be run multiple times using different random seeds in order to obtain better fault coverages. (In our experiments, we used a fixed sequence of ten random seeds.) Our goal is to develop one common HM that can be applied across all the benchmark circuits and that has similar or better fault coverages as compared to those of the original CRIS. Note that in the original CRIS, the HM used for each circuit is unique and was tuned manually. The advantage of having one HM is that it can be applied to new circuits without further manual tuning.

In our experiments on CRIS, we chose five circuits as our learning subdomains. In each of these subdomains, we used TEACHER [18] to test CRIS 1000 times (divided into 10 generations) with different HMs. A HM in learning is represented as a tuple of the first seven parameters in Table 4.2. The majority of time was spent in testing the HMs generated, since the time to generate a HM is very small (involving the crossover or mutation of sets of seven parameters). At the end of learning, we picked the top twenty HMs

Table 4.2. Parameters of CRIS treated as a HM in learning and in generalization. (The type, range, and step of each parameter were given to us by the designer of CRIS. The default parameters were not given to us as they are circuit-dependent)

Parameter	Range	Step	Definition	New Value
P_1	1-10	1	related to the number of stages in a flip flop	1
P_2	1-40	1	sensitivity of state change of a flip flop	12
P_3	1-40	1	survival rate of a test sequence in next generation	38
P_4	0.1-10.0	0.1	number of test vec. concat. to form a new vec.	7.06
P_5	50-800	10	number of useless trials before quitting	623
P_6	1-20	1	number of generations	1
P_7	0.1-1.0	0.1	how genes are spliced in GA	0.1
P_8	Integer	1	seed for random number generator	-

in each subdomain and evaluated them fully by initializing CRIS using ten different random seeds (P_8 in Table 4.2). We then selected the top five HMs from each subdomain, resulting in a total of 25 HMs supplied to the generalization phase. We evaluated the 25 HMs fully (each with 10 random seeds) on the five subdomains used in learning and five new subdomains. We then selected one generalized HM to be used across all the ten circuits (based on (3.2)). The HM found is shown in the last column in Table 4.2.

Table 4.3 shows the costs and qualities in applying our generalized HM learned for CRIS (see Table 4.2) and compares them to the results of CRIS [15] and HITEC [10], the latter is a deterministic search algorithm that is often used as a benchmark algorithm. We do not have the cost figures of CRIS because they were not published. The designer of CRIS hand tuned the parameters for each circuit; hence, the time (or cost) for obtaining these parameters are very large. Note that the maximum fault coverages reported were based on ten runs of the underlying problem solver, implying that the computational cost is ten times of the average cost. Recall that we like to obtain the maximum coverage of a circuit, and that computational cost is a secondary issue in circuit testing. Table 4.4 summarizes the results shown in Table 4.3.

Our results show that our generalization procedure can discover new HMs that are better than the original HMs in 16 out of 22 circuits in terms of the maximum fault coverage, and in 11 out of 22 circuits in terms of the average fault coverage. Our results are significant in the following aspects:

– new faults detected by our generalized HMs were not discovered by previous methods;

Table 4.3. Performance of HMs in terms of computational cost and fault coverage for CRIS. (Learned subdomains for CRIS are marked by "*" and generalized subdomains by "+"). Performance of HITEC is from the literature [16, 11]. Costs of our experiments are running times in seconds on a Sun SparcStation 10/51; costs of HITEC are running times in seconds on a Sun SparcStation SLC [14] (around 4-6 times slower than a Sun SparcStation 10/51)

Circuit ID	Total Faults	Fault Coverage		Cost	CRIS Generalized HM		
		HITEC	CRIS	HITEC	Avg. FC	Max. FC	Avg. Cost
*s298	308	86.0	82.1	15984.0	84.7	86.4	10.9
s344	342	95.9	93.7	4788.0	96.1	96.2	21.8
s349	350	95.7	–	3132.0	95.6	95.7	21.9
+s382	399	90.9	68.6	43200.0	72.4	87.0	7.2
s386	384	81.7	76.0	61.8	77.5	78.9	3.5
*s400	426	89.9	84.7	43560.0	71.2	85.7	8.4
s444	474	87.3	83.7	57960.0	79.8	85.4	9.3
*s526	555	65.7	77.1	168480.0	70.0	77.1	10.0
s641	467	86.5	85.2	1080.0	85.0	86.1	19.5
+s713	581	81.9	81.7	91.2	81.3	81.9	23.0
s820	850	95.6	53.1	5796.0	44.7	46.7	51.3
*s832	870	93.9	42.5	6336.0	44.1	45.6	44.6
s1196	1242	99.7	95.0	91.8	92.0	94.1	20.0
*s1238	1355	94.6	90.7	132.0	88.2	89.2	23.0
s1488	1486	97.0	91.2	12960.0	94.1	95.2	85.6
+s1494	1506	96.4	90.1	6876.0	93.2	94.1	85.5
s1423	1515	40.0	77.0	–	82.0	88.3	210.4
+s5378	4603	70.3	65.8	–	65.3	69.9	501.8
s35932	39094	89.3	88.2	13680.0	77.9	78.4	4265.7
am2910	2573	85.0	83.0	–	83.7	85.2	307.6
+div16	2147	72.0	75.0	–	79.1	81.0	149.9
tc100	1979	80.6	70.8	–	72.6	75.9	163.8

Table 4.4. Summary of wins and losses in applying our generalized HM for CRIS on 22 circuits when compared to the performance of HITEC, CRIS, and the best of CRIS and HITEC. (Not all circuits were tested by HITEC and CRIS)

Our HM wins/ties with respect to the following	CRIS Generalized HM					
	Max. Fault Coverage			Avg. Fault Coverage		
	Wins	Ties	Losses	Wins	Ties	Losses
HITEC	6	2	14	4	0	18
CRIS	16	1	5	11	0	10
Best of HITEC and CRIS	5	3	14	3	0	9

– only one HM (rather than many circuit-dependent HMs in the original CRIS) was found for all circuits.

Table 4.4 also indicates that HITEC is still better than our new generalized HM for CRIS in most of the circuits (in 14 out of 22 in term of the maximal fault coverage, and in 18 out of 22 in term of the average fault coverage). This happens because our generalized HM is bounded by the limitations in CRIS and our HM generator for CRIS. Such limitations cannot be overcome without generating more powerful HMs in our HM generator or using better test-pattern generators like HITEC as our baseline problem solver.

4.2 HM for VLSI placement and routing

In our second application, we use TimberWolf[17] as our problem solver. This is a software package based on simulated annealing (SA) [7] to place and route various circuit components on a piece of silicon. Its goal is to minimize the chip area needed while satisfying constraints such as the number of layers of poly-silicon for routing and the maximum signal delay through any path. Its operations can be divided into three steps: placement, global routing, and detailed routing.

The placement and routing problem is NP-hard; hence, heuristics are generally used. SA used in TimberWolf is an efficient method to randomly search the space of possible placements.

Although in theory SA converges asymptotically to the global optimum with probability one, the results generated in finite time are usually suboptimal. Consequently, there is a trade-off between the quality of a result and the cost (or computational time) of obtaining it. In TimberWolf version 6.0, the version we have studied, there are two parameters to control the running time (which indirectly control the quality of the result): *fast-n* and *slow-n*. The larger the *fast-n* is, the shorter time SA will run. In contrast, the larger the *slow-n* is, the longer time SA will run. Of course, only one of these parameters can be used at any time.

TimberWolf has six major components: *cost function, generate function, initial temperature, temperature decrement, equilibrium condition,* and *stopping criterion.* Many parameters in these components have been tuned manually. However, their settings are generally heuristic because we lack domain knowledge to set them optimally. In Table 4.5, we list the parameters we have focused in this study. Our goal is to illustrate the power of our learning and generalization procedures and to show improved quality and reduced cost for the placement and routing of large circuits, despite the fact that only small circuits were used in learning.

In our experiments, we used seven benchmark circuits [9] (*s298, s420, fract, primary1, struct, primary2, industrial1*) that were mostly from ftp.mcnc.org in /pub/benchmark. We studied only the standard-cell placement problem, noting that other kinds of placement can be studied in a similar fashion. We used *fast-n* values of 1, 5, and 10, respectively.

We first applied TEACHER to learn good HMs for circuits *s298* with *fast-n* of 1, *s420* with *fast-n* of 5, and *primary1* with *fast-n* of 10, each of which was taken as a learning subdomain. We used a fixed sequence of ten

Table 4.5. Parameters of TimberWolf (Version 6) used in the original HM and after learning and generalization

Para-meter	Range	Step	Meaning	Original	New
P_1	0.1 - 2.5	0.1	vertical path weight for estimating the cost function	1.0	0.958
P_2	0.1 - 2.5	0.1	vertical wire weight for estimating the cost function	1.0	0.232
P_3	3 - 10	1	orientation ratio	6	10
P_4	0.33 - 2.0	0.1	range limiter window change ratio	1.0	1.30
P_5	10.0 - 35.0	1.0	high temperature finishing point	23.0	10.04
P_6	50.0 - 99.0	1.0	intermediate temperature	81.0	63.70
P_7	100.0 - 150.0	1.0	low temperature finishing point	125.0	125.55
P_8	130.0 - 180.0	1.0	final iteration temperature	155.0	147.99
P_9	0.29 - 0.59	0.01	critical ratio that determines acceptance probability	0.44	0.333
P_{10}	0.01 - 0.12	0.01	temperature for controller turn off	0.06	0.112
P_{11}	integer	1	seed for the random number generator	-	-

random seeds (P_{11} in Table 4.5) in each subdomain to find the statistical performance of a HM. Each learning experiment involved 1000 applications of TimberWolf divided into ten generations. Based on the best 30 HMs (10 from each subdomain), we applied our generalization procedure to obtain one generalized HM. This generalized HM as well as the default HM are shown in Table 4.5.

Figure 4.1 plots the quality (higher quality in the y-axis means reduced chip area averaged over 10 runs using the defined random seeds) and cost (average execution time of TimberWolf) between the generalized HM and the default HM on all seven circuits with *fast-n* of 1, 5, and 10, respectively. Note that all performance values in Figure 4.1 are normalized with respect to those of *fast-n* of 10, and that the positive (resp., negative) portion of the x-axes shows the fractional improvement (resp., degradation) in computational cost with respect to the baseline HM using *fast-n* of 10 for the same circuit. Each arrow in this figure points from the average performance of the default HM to the average performance of the generalized HM.

The equation for computing the normalized symmetric cost is as follows. Let C_{new}, C_{base} and C_{sym}^{norm} be, respectively, the costs of the new HM, the cost of the baseline HM, and the normalized symmetric cost.

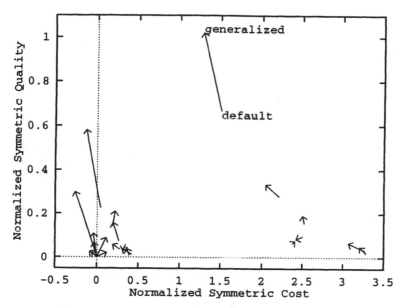

Fig. 4.1. Comparison of normalized average performance between the default and the generalized HMs. The plots are normalized with respect to the performance of applying the baseline HM on each circuit using *fast-n* = 10. (See (4.1))

$$
C_{sym}^{norm} = \begin{cases} \frac{C_{new}}{C_{base}} - 1 & \text{if } C_{new} \geq C_{base}, \\ 1 - \frac{C_{base}}{C_{new}} & \text{if } C_{new} < C_{base} \end{cases} \tag{4.1}
$$

The reason for using the above equation is to avoid uneven compression of the ratio C_{new}/C_{base}. This ratio is between 0 and 1 when $C_{new} < C_{base}$, but is between 1 and ∞ when $C_{new} > C_{base}$. (4.1) allows increases in cost to be normalized in the range between 0 and ∞, and decreases to be normalized in the range between 0 and $-\infty$. The normalized symmetric quality in the y-axis is computed in a similar way.

Among the 22 test cases, the generalized HM has worse quality than that of the default in only two instances, and has worse cost in 4 out of 22 cases. We see in Figure 4.1 that most of the arrows point in a left-upward direction, implying improved quality and reduced cost. Note that these experiments are meant to illustrate the power of our generalization procedure. We expect to see more improvement as we learn other functions and parameters in TimberWolf. Further, improvements in TimberWolf are important as the system is actually used in industry.

4.3 Branch-and-bound search

A branch-and-bound search algorithm is a systematic method for traversing a
search tree or search graph in order to find a solution that optimizes a given
objective while satisfying the given constraints. It decomposes a problem
into smaller subproblems and repeatedly decomposes them until a solution
is found or infeasibility is proved. Each subproblem is represented by a node
in the search tree/graph.

The algorithm has four sets of HMs: (a) *Selection HM* for selecting a
search node for expansion based on a sequence of selection keys for ordering
search nodes; (b) *Decomposition HM* (or branching mechanism) for expanding a search node into descendants using operators to expand (or transform)
a search node into child nodes; (c) *Pruning HM* for pruning inferior nodes
in order to trim potentially poor subtrees; and (d) *Termination HM* for determining when to stop. In this subsection, we apply learning to find new
decomposition HMs for expanding a search node into descendants.

We illustrate our method on three applications: traveling salesman problem (TSP) on incompletely connected graphs mapped on a two-dimensional
plane, vertex-cover problem (VC), and knapsack problem (KS). The second
problem can be solved by a polynomial-time approximation algorithm with
guaranteed performance deviations from optimal solutions, and the last can
be solved by a pseudo polynomial-time approximation algorithm. Hence, we
expect that improvements due to learning are likely for the first two problems
and not likely for the last. Table 4.6 shows the parameters used in generating
a *test case* in each application. We assume that each problem constitutes one
domain.

The problem solver here is a branch-and-bound algorithm, and a test
case is considered solved when its optimal solution is found. Note that the

Table 4.6. Generation of test cases for learning and generalization of decomposition
HMs in a branch-and-bound search (each has 12 subdomains)

Application	Subdomain Attributes
VC	• Connectivity of vertices is $(0.05 - 0.6)$ with step size 0.05 • Number of vertices is between 16 and 45
TSP	• Distributions of 8-18 cities ($U(0,100)$ on both X and Y axes, $N(50,12.5)$ on both axes, or $U(0,100)$ and $N(50,12.5)$ on different axes) • Graph connectivity of cities is $(0.1, 0.2, 0.3,$ or $1.0)$
KS	• Range of both profits and weights is $\{(100\text{-}1000), (100\text{-}200), (100\text{-}105)\}$ • σ^2 of profit/weight ratio is $(1.05, 1.5, 10, 100)$ • 13-60 objects in the knapsack

Table 4.7. Original and generalized decomposition HMs used in a branch-and-bound search (l: number of uncovered edges or live degree of a vertex; n: average live degree of all neighbors; Δl: difference between l of parent node and l of current node; c: length of current partial tour; m: minimum length to complete current tour; p: profit of object; w: weight of object)

Application	Original HM	Generalized HM
VC	l	$1000\,l + n - \Delta l$
TSP	c	$m\,c$
KS	p/w	p/w

decomposition HM studied is only a component of the branch-and-bound algorithm.

We have used well-known decomposition HMs developed for these applications as our baseline HMs (see Table 4.7). The normalized cost of a candidate decomposition HM is defined in terms of its *average symmetric speedup* (see Eq. (4.1) in Section 4.2), which is related to the number of nodes expanded by a branch-and-bound search using the baseline HM and that using the new HM. Note that we do not need to measure quality as both the new and existing HMs when applied in a branch-and-bound search look for the optimal solution.

In our experiments, we selected six subdomains in each application for learning. We performed learning in each subdomain using 1,600 tests, selected

Table 4.8. Results of generalization for VC, TSP, and KS. (In the results on generalization, numbers with "*" are the ones learned; only one common HM is generalized to all 12 subdomains)

Subdomain	Subdomain Performance (Sym-SU)					
	Learning			Generalization		
	VC	TSP	KS	VC	TSP	KS
1	0.218	0.072*	0.000*	0.070	0.417	0.000
2	0.283*	0.004	0.000*	0.638	0.036	0.000
3	0.031	0.082*	0.000	0.241	0.144	0.000
4	0.068*	0.225	0.000	0.078	0.155	0.000
5	0.054	0.005*	0.000	0.073	0.131	0.000
6	0.060*	0.061*	0.000*	0.020	0.364	0.000
7	0.017	0.139	0.000*	−0.013	1.161	0.000
8	0.049*	0.155	0.000	−0.004	0.101	0.000
9	0.016	−0.010	0.000*	−0.018	0.108	0.000
10	−0.000*	0.054	0.000	−0.000	0.008	0.000
11	−0.011	0.090*	0.000	−0.019	0.022	0.000
12	0.028*	0.083*	0.000*	−0.010	0.131	0.000
Average	0.068	0.080	0.000	0.088	0.231	0.00

the top five HMs in each subdomain, fully verified them on all the learned subdomains, and selected one final HM to be used across all the subdomains. (See (3.2).) Table 4.8 summarizes the generalization and validation results.

We show in our results the average symmetric speedup of the top HM learned in each subdomain and the normalized cost of learning, where the latter was computed as the ratio of the total CPU time for learning and the harmonic mean of the CPU times required by the baseline HM on test cases used in learning. The results show that a new HM learned specifically for a subdomain has around 1-35% improvement in its average symmetric speedups and 3,000-16,000 times in learning costs.

Table 4.8 also shows the average symmetric speedups of the generalized HMs. We picked six subdomains randomly for learning. After learning and fully verifying the five top HMs in each subdomain, we applied (3.2) to identify one top HM to be used across all the twelve subdomains. Our results show that we have between 0-8% improvement in average symmetric speedups using the generalized HMs. Note that these results are worse than those obtained by learning, Moreover, the baseline HM is the best HM for solving the knapsack problem.

The second part of Table 4.8 shows the average symmetric speedups when we validate the generalized HMs on larger test cases. These test cases generally require 10-50 times more nodes expanded than those used earlier. Surprisingly, our results show better improvement (9-23%). It is interesting to point out that six of the twelve subdomains with high degree of connectivity in the vertex-cover problem have slowdowns. This is a clear indication that these subdomains should be grouped in a different domain and learned separately.

Table 4.7 shows the new decomposition HMs learned for the three applications that lists the variables used in the HMs. Note that we have included constants in our HMs in learning; an example of which is shown in the HM learned for the vertex-cover problem. This formula can be interpreted as using l as the primary key for deciding which node to include in the covered set. If the l's of two alternatives are different, then the remaining terms in the formula $(n - \Delta l)$ are insignificant. On the other hand, when the l's are the same, then we use $(n - \Delta l)$ as a tie breaker.

In short, our results show that reasonable improvements can be obtained by generalization of learned HMs. We anticipate further improvements by

- learning and generalizing new pruning HMs in a depth-first search,
- partitioning the problem space into a number of domains and learning a new HM for each, and
- identifying attributes that help explain why one HM performs well in one subdomain but not in others.

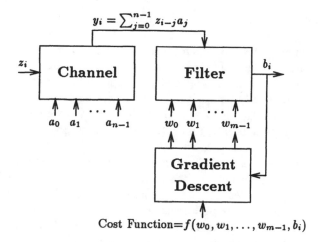

$$y_i = \sum_{j=0}^{n-1} z_{i-j}a_j$$

Cost Function$=f(w_0, w_1, \ldots, w_{m-1}, b_i)$

Fig. 4.2. Blind equalization process for recovering input data stream for n-th order channel and m-th order filter

Table 4.9. Summary of average symmetric improvements in terms of number of accumulated errors for the learned cost function over ten subdomains. (b_i in Figure 4.2 is the instantaneous value of b)

Average Symmetric Improvement				Original HM	New HM
Average	Std.Dev.	Maximum	Minimum		
0.153	0.395	0.694	-0.465	$b^3 - b$	$4b^3 - 2b^2 Sign(b) - b$

4.4 Blind equalization

Our last application is on applying genetic algorithms to learn a cost function in blind equalization. Our goal is to minimize the number of accumulated errors for a sequence of input data corrupted in transmission (Figure 4.2). The process is equivalent to adjusting the weights of an FIR filter using gradient descent in order to minimize the value of a cost function, which is defined in term of the weights of the filter and its current output.

In this application, we define a test case as multiple random sequences of data of fixed length passing through a fixed channel and a blind equalizer with given random initial weights. We further define a subdomain to be all test cases with the same channel specification. In our experiments, we attempt to cover all possible third-order channels: from relatively easy ones ($|a_i| > \sum_{i \neq j} |a_j|$ where a_i is the i-th weight of the channel) to the hardest one ($a_i = a_j$ for all i and j).

Table 4.9 shows the average symmetric improvements in terms of number of accumulated errors for HM_{base} (CMA 2-2) [4] and the new HM found after learning and generalization.

5. Conclusions

In this chapter, we have presented a method for generalizing performance-related heuristics learned by genetics-based learning for knowledge-lean applications. We have focused on a class of heuristic methods (HMs) whose performance is evaluated statistically by applying them on multiple test cases. Due to a lack of domain-knowledge for improving such heuristics, we have used a genetics-based learning paradigm (a generate-and-test method) to learn new HMs.

One of the major problems in performance evaluation of heuristics is that a HM may have different performance distributions across different sets of test cases in an application. This renders it impossible to use statistical metrics, such as average, to compare their performance.

We have proposed in this chapter a new metric called probability of win to characterize the performance of heuristics. This metric evaluates the probability that the mean performance of a HM is better than the mean performance of another randomly chosen HM in a set of learned HMs on a common set of test cases. The only requirement on the choice of test cases in evaluating probabilities of win is that each HM, when evaluated on the test cases, produces a set of independent and identically distributed performance results. We define such a set of test cases as a subdomain. Since probabilities of win are between 0 and 1, we can compare them across subdomains in generalizing HMs.

We have developed TEACHER [5], an integrated system that incorporates the learning and generalization method presented in this chapter. The system is relatively easy to use: the design of an interface between an application program and TEACHER usually takes less than two weeks to complete.

We have applied TEACHER [18], a genetics-based learning system that incorporates our generalization method, on four engineering applications and found very good improvements over existing HMs. These applications are hard to improve because they have been studied and tuned extensively by many others before. In each case, we have found very good improvements over existing HMs for these applications. These demonstrate that learning and generalization is important in refining heuristics used in many application problem solvers.

Acknowledgments

The authors would like to thank Mr. Yong-Cheng Li for interfacing TEACHER to TimberWolf and for collecting some preliminary results in Section 4.. This research was supported partially by National Science Foundation Grants MIP 92-18715 and MIP 96-32316 and by National Aeronautics and Space Administration Contract NAG 1-613.

References

1. F. Brglez, D. Bryan, and K. Kozminski, "Combinatorial profiles of sequential benchmark circuits," in *Int'l Symposium on Circuits and Systems*, pp. 1929–1934, May 1989.
2. C. M. Fonseca and P. J. Fleming, "Genetic algorithms for multiobjective optimization: Formulation, discussion, and generalization," in *Proc. of the Fifth Int'l Conf. on Genetic Algorithms*, (Morgan Kaufman), pp. 416–423, Int'l Soc. for Genetic Algorithms, June 1993.
3. D. E. Goldberg and J. H. Holland, "Genetic algorithms and machine learning," *Machine Learning*, vol. 3, pp. 95–100, Oct. 1988.
4. S. Haykin, *Blind Deconvolution*. Englewood Cliffs, NJ: Prentice Hall, 1994.
5. A. Ieumwananonthachai and B. W. Wah, "TEACHER – an automated system for learning knowledge-lean heuristics," Tech. Rep. CRHC-95-08, Center for Reliable and High Performance Computing, Coordinated Science Laboratory, Univ. of Illinois, Urbana, IL, March 1995.
6. C. Z. Janikow, "A knowledge-intensive genetic algorithm for supervised learning," *Machine Learning*, vol. 13, no. 2-3, pp. 189–228, 1993.
7. S. Kirkpatrick, J. C. D. Gelatt, and M. P. Vecchi, "Optimization by simulated annealing," *Science*, vol. 220, pp. 671–680, May 1983.
8. J. R. Koza, *Genetic Programming*. Cambridge, MA: The MIT Press, 1992.
9. LayoutSynth92, *International Workshop on Layout Synthesis*.   ftp site: mcnc.mcnc.org in directory /pub/benchmark, 1992.
10. T. M. Niermann and J. H. Patel, "HITEC: A test generation package for sequential circuits," in *European Design Automation Conference*, pp. 214–218, 1991.
11. T. M. Niermann and J. H. Patel, "HITEC: A test generation package for sequential circuits," in *European Design Automation Conference*, pp. 214–218, 1991.
12. J. Pearl, *Heuristics–Intelligent Search Strategies for Computer Problem Solving*. Reading, MA: Addison-Wesley, 1984.
13. C. L. Ramsey and J. J. Grefenstette, "Case-based initialization of genetic algorithms," in *Proc. of the Fifth Int'l Conf. on Genetic Algorithms*, (Morgan Kaufman), pp. 84–91, Int'l Soc. for Genetic Algorithms, June 1993.
14. E. M. Rudnick, J. H. Patel, G. S. Greenstein, and T. M. Niermann, "Sequential circuit test generation in a genetic algorithm framework," in *Proc. Design Automation Conf.*, ACM/IEEE, June 1994.
15. D. G. Saab, Y. G. Saab, and J. A. Abraham, "CRIS: A test cultivation program for sequential VLSI circuits," in *Proc. of Int'l Conf. on Computer Aided Design*, (Santa Clara, CA), pp. 216–219, IEEE, Nov. 1992.
16. D. G. Saab, Y. G. Saab, and J. A. Abraham, "CRIS: A test cultivation program for sequential VLSI circuits," in *Proc. of Int'l Conf. on Computer Aided Design*, (Santa Clara, CA), pp. 216–219, IEEE, Nov. 8-12 1992.
17. C. Sechen, *VLSI Placement and Global Routing Using Simulated Annealing*. Boston, MA: Kluwer Academic Publishers, 1988.
18. B. W. Wah, A. Ieumwananonthachai, L. C. Chu, and A. Aizawa, "Genetics-based learning of new heuristics: Rational scheduling of experiments and generalization," *IEEE Trans. on Knowledge and Data Engineering*, vol. 7, pp. 763–785, Oct. 1995.

Part IV

Electrical, Control, and Signal Processing

Optimal Scheduling of Thermal Power Generation Using Evolutionary Algorithms

Dipankar Dasgupta

Mathematical Science Department, Room No. 378, Dunn Hall, The University of Memphis, Memphis, TN 38152

Summary. The optimal scheduling of power generation is a major problem in Electricity generating industry. This scheduling problem is complex because of many constraints which can not be violated while finding optimal or near-optimal scheduling. A real-time scheduling of generating units involves the selection of units for for operation to meet the consumer requirement. These decisions are to be taken so as to minimize the sum of startup, banking and expected running costs subject to the demand and spinning_reserve, and satisfying the minimum down-time and up-time constraints of generating units. So the objective of the scheduling is to generate power to meet the load with minimum cost of generation. We tackled the problem with heuristic knowledge (load forecast) and genetic search technique. The paper examines the feasibility of using genetic algorithms, and reports results in determining a near-optimal commitment order of thermal units in a studied power system.

1. Introduction

In power industries, fuel expenses constitute a significant part of the overall generation costs. In general, there exist different types of thermal power units based on fuel used (e.g coal, natural gas, oil), with different production costs, generating capacities and characteristics. Figure 1.1 shows the block diagram of a simple power system coupling the generating units and different end users. The system usually operates under continuous variation of consumer load demand. This demand for electricity exhibits such large variations between weekdays and weekends, and between peak and off-peak hours that it is not economical to keep all the generating units continuously on-line. So the demand and the reserve requirement impose global constraints in coupling all active generating units, while the different operating characteristics of each unit constitute local constraints. Thus determining which units should be kept on-line and which ones should not, constitutes a difficult problem for operators seeking to minimize the system operational cost.

Thus unit commitment (UC) problem belongs to the class of complex combinatorial optimization problem. Several mathematical programming techniques have been proposed [13, 4] to solve these time-dependent unit commitment problems. They typically include Complete Priority Ordering (CPO) and Heuristic methods, Dynamic Programming (DP) [9, 12], Method of Local Variations, Mixed Integer Programming, Lagrangian Relaxation [2, 14, 22], Branch and Bound method [4], Bender Decomposition [1], etc.

Among all these methods, dynamic programming methods based on a priority list have been used most extensively throughout the power industry. However, different strategies for selecting a set of units from priority list have been adopted with dynamic programming to limit the search space and execution time. They include DP-SC (Dynamic Programming-Sequential Combination) [19], DP-TC (Dynamic Programming-Truncated Combination) [18], DP-STC [19] which is a combination of DP-SC and DP-TC approaches, DP-VW (Variable Window-Truncated Dynamic Programming) [16], and a neural-based method DP-ANN [17].

Recently, some researchers have suggested Artificial Intelligence based techniques to supplement the limitation of mathematical programming methods. These include simulated annealing [21], expert systems [15], heuristic rule-based systems [20] and neural networks [17]; these hybrid approaches have demonstrated some improvement in solving unit commitment problems. However, heuristic and expert system based mathematical approaches require a lot of operator interaction which is troublesome and time-consuming for even a medium-size utility [8].

This paper presents an evolutionary algorithm to solve the unit commitment problem. The main purpose of using the evolutionary approach is to replace classical solution methods with a population-based global search procedure which has some distinct advantages. EAs are different from the above-mentioned classical methods in three ways [6]: they work with a coding of the parameter set rather than with actual parameters and work equally with discrete and continuous functions; and they use probabilistic transition rules. These differences ensure that the success of a genetic algorithm is usually not related to the semantics of any particular problem.

The paper will first give a brief description of the unit commitment problem and then th use of a GA to the problem of optimizing commitment order of thermal units in (single area) power systems.

2. Problem Description

Operating under the present competitive environment, unit commitment has become increasingly important in the power industry, because significant savings can be accrued from sound commitment decisions. A scheduler to do this

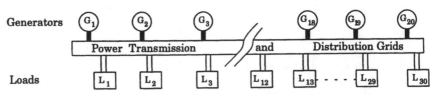

Fig. 1.1. A simple block diagram showing a power system

must indicate which of the generating units are to be committed (on or off) in every time interval during the scheduling time horizon. This decision must take into account load forecast information and the economic implications of the startup or shutdown of various units. The transition between their commitment states must satisfy the operating (minimum up and down time) constraints.

So there are two types of constraints associated with unit commitment problems. The first type is the global one resulting from load requirement that couple all the generating units during each time period. The second type of constraint is local, representing the different operating restrictions of the individual unit.

In order to maintain a certain degree of reliability, some stand-by capacity is necessary that can immediately take over when a running unit breaks down or unexpected load occurs. Hence the amount of spinning reserve is an important factor in ensuring an uninterrupted power supply. There are several spinning reserve policies that have usually been adopted: a fixed percentage of the forecast peak demand at every time period, a variable reserve, a reserve slightly greater than the output of the most heavily loaded unit, or a probabilistic reserve constraint known as *unit commitment risk* may be used to ensure better system reliability. But there should be no significant excess of reserve capacity in an economic commitment. However, spinning reserve constraints only provide the lower bound since the total capacities of the committed units may not exactly match the load and spinning reserve.

2.1 Objective function

The objective (or cost) function of the unit commitment problem is to determine the state of each unit u_i^t (0 or 1) at each time period t, where unit number $i = 1 \ldots U_{max}$, and time periods $t = 1 \ldots T_{max}$, so that the overall operation cost is a minimum within the scheduling time horizon.

$$min \sum_{t=1}^{T_{max}} \sum_{i=1}^{U_{max}} [u_i^t (AFLC)_i + u_i^t(1 - u_i^{t-1}) S_i(x_i^t)$$

$$+u_i^{t-1}(1 - u_i^t) D_i]\ldots..(1)$$

For each committed unit, the cost involved is the *start-up* cost (S_i) and the *Average Full Load Cost (AFLC)* per MWh, according to the unit's maximum capacity

$$such \ that \ \sum_{i=1}^{U_{max}} P_i^{max} \geq R^t + L^t, \quad \ldots\ldots\ldots\ldots\ldots\ldots(2)$$

where P_i^{max} is the maximum output capacity of unit i, L^t is the demand and R^t is the spinning reserve in time period t. The above objective function should satisfy minimum *up-time* and *down-time* constraints of generating units.

The start-up cost is expressed as a function of the number of hours (x_i^t) the unit has been down and the shut-down cost is considered as a fixed amount (D_i) for each unit per shut-down, and these state transition costs are applied in the period when the unit is committed or taken off-line respectively [9].

However, unit commitment decisions based solely on unit-$AFLC$ usually do not provide sufficient information about the impact of system load conditions on how efficiently (eg. fully) the committed units being utilized while determining the optimal or a near-optimal commitment [11]. An index is used to measure the utility of each commitment decision, called $Utility\ Factor = \frac{Load-Reserve\ Requirements}{Total\ committed\ output}$, while satisfying the global constraint (equ. (2)). This factor helps to compensate the deficiency associated with over committed decisions based soly on the classical $AFLC$ of units. So during the performance evaluation, commitment decisions having a low utility factor are penalized accordingly.

3. Implementation Details

The implementation proposed here uses a simple genetic search technique for determining the optimal (a least total cost solution) or a near-optimal commitment schedule for a given study period.

The short term commitment is considered with a 24-hour time horizon, which may be repeated using the load profile of each day. Since the system load varies substantially over a 24-hour period and the cost of operation over this time span depends on the timing and frequency of unit's start-ups and shut-downs, this commitment problem is generally viewed as a multi-period problem where the commitment horizon is divided into a number of periods of shorter length (usually a 1-hour commitment interval).

In order to use a genetic algorithm for this problem, the first step is to encode the commitment space. If the whole planning horizon (24-hour) is encoded in a chromosome by the concatenation of commitment spaces of all time periods, it appears to be possible to determine a complete commitment order for the whole span at a time by performing a global search using the genetic algorithm. But this approach makes the problem hard for the genetic algorithms to solve for the following reasons:

Firstly, with the increased number of generating units, the length of the encoded chromosome increases in a higher ratio; for example, a system with 20 generating units, the chromosome length becomes 480 bits when the scheduling span is 24 hours. But with a longer encoded string, genetic algorithms find difficulty in reaching to a near-optimal solution; since a genetic search exploits schemata representing hyperplanes and an increase the size of encoding increases the amount of space the algorithm has to explore in order to find good schemata [10]. Secondly, since the optimal population size is a function of the string length for better schema processing [7], so with the increased string length a bigger size of population is needed. But this will result in both memory and computational overhead. Thirdly, in addition to the above drawbacks such encoding makes the problem space highly epistasis[1]. When the epistasis is extremely high, gene values are so dependent on each other that unless a complete set of unique values is found simultaneously, no substantial fitness improvements can be noticed [5]. In this case, since the commitment decisions of previous hours have strong effect on the decision in successive hours, so encoding these dependent decisions in a single string makes the representation highly epistasis. This increases the complexity of search space and makes it difficult for a genetic search to find a near-optimal solution in a reasonable time.

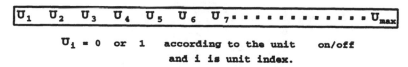

$U_i = 0$ or 1 according to the unit   on/off
and i is unit index.

Fig. 3.1. Chromosomal representation of unit commitment decisions

In this work, the problem is considered as a multi-period process as in practice and a simple genetic algorithm is used for commitment scheduling. Each chromosome is encoded in the form of a position-dependent genes (bit string) representing the number of thermal units available in the system, and the allele value at loci give the state (on/off) of the units as a commitment decision at each time period, shown in figure 3.1.

Unit commitment decisions satisfying load-reserve requirement and the operating constraints of units are regarded as feasible solutions, and any violation of the constraints is penalized through a penalty function. So the raw fitness function is formulated here using a weighted sum of the objective function and values of the penalty function based on the number of constraints violated and the extent of these violations. By choosing suitable weights for the penalty function, it is possible to find a near-optimal solution to the problem. In our case,

Fitness function = Objective function + Penalty function ($L - R$,
Utility Factor, min_up, min_down),

where $L - R$ is the load-reserve requirements, *min_up* and *min_down* are minimum-up and minimum-down time constraints of the units.

---

[1] A biological term that states the amount of interdependency among genes encoding the chromosome.

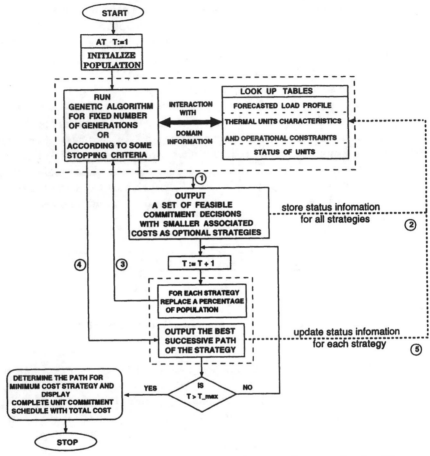

Fig. 3.2. A flow diagram for unit commitment using genetic algorithms

In our implementation here, the following scaling technique is used to normalize fitness and to produce a non-negative figure of merit.

$$Scaled\ Fitness\,(I) = (M - I)/(M - N),$$

where I is the individual raw fitness, M and N are the relative maximum and minimum fitnesses respectively among all individuals in the current generation. The goal is to maximize the scaled fitness in order to minimize the cost (or objective) function. The scaled fitness is used to determine the probability of selecting of members in the population for breeding.

The flow chart of implementing the genetic algorithm for unit commitment is shown in figure 3.2. For better understanding of the diagram, some of the flow lines connected to the repeatedly used program module (GA routines) and domain information (look-up tables) are nuberbered accordance with the sequence of execution. The genetic-based unit commitment program starts with a random initial population (at T=1) and computes the fitness of each individual (commitment decision) using the forecasted load demand at each period, and the operating constraints of the units (using lookup tables). Each time the genetic optimizer is called, it runs for a fixed number of generations or until the best individual remains unchanged for a long time (here 100 successive generations).

Since the unit commitment problem is time-dependent, these piecewise approaches of working forward in time and retaining the best decision, can not be guaranteed to find the optimal commitment schedule. The reason for this is that a decision with significantly higher costs during the early hours of scheduling could lead to significant savings later and may produce a lower overall cost commitment schedule.

In order to make the genetic-based unit commitment program robust in finding near-optimal solutions, a number of feasible commitment decisions (less than or equal to a predefined value S) with smaller associated costs are saved at each time period. These strategies[2] determine how many possible alternative paths are available at each period for finding the overall operation cost. Figure 3.3 illustrates the different paths available, where one path converges to the other in midway and another stopped because it could not find path within the allocated resources. The selection of S is affective in economical scheduling (in finding an optimal solution), memory requirement and computation time. In order to save computation time the same strategies are carried forward to the next period if the load remains unaltered or varies slightly in the current period such that load-reserve requirements are satisfied by all strategies. If a strategy cannot meet the demand of present period, the genetic optimization process is performed for the period to find a feasible successor paths with smaller cost. This approach increases the likelihood of finding the path of minimum cumulative cost.

These temporary commitment strategies used to update the status information of the units (up-time/down-time counter) to keep track of the units in service or shutdown for a number of successive hours. In the next time period, half of the population is replaced by randomly generated individuals to introduce diversity in the population so that the search for new commitment strategy can proceed according to the load demand. The purpose of keeping half of the previous population is that in most situations the load varies slightly in some successive time intervals and the previous better individuals (commitment strategies) are likely to perform well in the current period. However, if there is a drastic change in load demand, newly generated individuals can explore the commitment space for finding the best solution. The iterative process continues for each period in the scheduling horizon,

---

[2] A strategy is a sequence of commitment decisions from the starting period to the current period with its accumulated cost.

and the accumulated cost associated with each commitment strategy gives the overall cost for the commitment path.

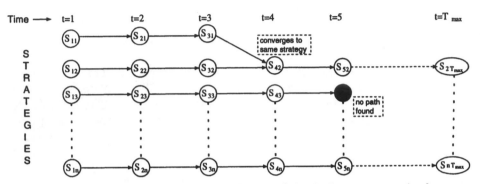

Fig. 3.3. A number of alternative path (strategy) for finding a near optimal commitment order

In a time period, if a unit is to be decommitted due to a decrease in load demand, and because of that

-if its minimum-up time constraint is violated, then the unit is considered to remain committed (in banking state) until its minimum-up period is completed.

-also to tackle sharp rises in load demand in the next period, a look-ahead mechanism is incorporated which decides that the unit will remain committed (in banking state) even though it represents an uneconomical decision for the current period.

During these multi-period optimization processes, if in a particular period no feasible solution (strategy) is found, the process is repeated so that at least one feasible solution is found before shifting to the next period. However, in our test example such repetitions are required only in a few occasions in later periods.

4. Experimental Results

The genetic algorithm-based unit commitment program is applied to an example problem which consists of 10 thermal units. The capacities, costs and operating constraints vary greatly among the various generating units in these test systems. Different type of load profiles are tested which represent typical operating circumstances in the studied power system. We have considered a short term scheduling where the time horizon is 24-hour and scheduling for an entire day is done in advance which may be repeated using load profile of each day for a long-term scheduling.

In these experiments, the spinning reserve requirement is assumed to be 10% of the expected hourly peak load. A program implementing the algorithm has been run on a SUN (sparc) workstation under UNIX 4.1.1 operating system. The experiment is conducted with a population size of 250 using different crossover and mutation rates. For the result reported here (shown in a tabular form), a crossover probability of 78% and mutation rate of 15% were used along with a stochastic remainder selection scheme [3] for reproduction. We also used an elitist scheme which passes the best individual unaltered to the succeeding generation. Each run was allowed to continue up to 500 generations and the strategy path with minimum cumulative cost gives a near-optimal commitment for the whole scheduling period.

Table 4.1. Characteristics and initial state of the thermal units

Unit No.	Maximum Capacity (MW)	Min. Up Time (hr)	Min. Down Time (hr)	Initial Status (hr)	St_Up cost b_1	b_2	b_3	Sh_Down Cost	APLC
1	60	3	1	-1	85	20.588	0.2	15	15.3
2	80	3	1	-1	101	20.594	0.2	25	16
3	100	4	2	1	114	22.57	0.2	40	20.2
4	120	4	2	5	94	10.65	0.18	32	20.2
5	150	5	3	-7	113	18.639	0.18	29	25.6
6	280	5	2	3	176	27.568	0.15	42	30.5
7	520	8	4	-5	267	34.749	0.09	75	32.5
8	150	4	2	3	282	45.749	0.09	49	26.0
9	320	5	2	-6	187	38.617	0.130	70	25.8
10	200	5	2	-3	227	26.641	0.11	62	27.0

(-) indicates unit is down for hours and positive otherwise.

*We used start-up cost $= b_{1,i}(1 - e^{-b_{3,i} \cdot (x_i^t)}) + b_{2,i}$.

For this example, table-4.1 gives the characteristics of and the initial states of the generating units. Table-4.2 gives commitment schedule for two cases, which were run independently. In first case, one best solution is saved at each time period and in second case, multiple least cost strategies are saved for determining the minimum cost path. A comparison shows that substantial reduction in overall cost can be achieved when the best commitment schedule is determined from multiple least cost strategies. In this table, the second column gives the hourly load demand, the third column shows total requirement after adding spinning reserve, the rest columns give the total output capacity (in MW) of the committed units and the state of units in each case, where '1' is used to indicate a unit is committed, '0' to indicate that a unit is decommitted.

The genetic-based unit-commitment system have been tested under different operating conditions in order to evaluate the algorithm's performance.

It is observed that the scheduling which produces optimal power output does not always give the overall minimum cost scheduling, and also the minimum cost scheduling is very sensitive to the system parameters and the operating constraints of the generating units.

Table 4.2. Unit commitment schedules determined by the genetic algorithm

Time in (hr)	Load Demand (MW)	Load + Reserve (MW)	CASE - 1 Committed Output (MW)	CASE - 1 State of units	CASE - 2 Committed Output (MW)	CASE - 2 State of Units
1	1459.00	1677.85	1700.00	1011111110	1710.00	1110011111
2	1372.00	1577.80	1710.00	1110111011	1860.00	1110111111
3	1299.00	1493.85	1710.00	1110111011	1550.00	1111101011
4	1280.00	1472.00	1490.00	0111101011	1490.00	0111101011
5	1271.00	1461.65	1470.00	1011101011	1470.00	1011101011
6	1314.00	1511.10	1550.00	1111101011	1550.00	1111101011
7	1372.00	1577.80	1600.00	1101101111	1580.00	1110101111
8	1314.00	1511.10	1600.00	1101101111	1520.00	0110101111
9	1271.00	1461.65	1500.00	1111101110	1500.00	1111101110
10	1242.00	1428.30	1630.00	1111011110	1500.00	1111101110
11	1197.00	1376.55	1420.00	0111011010	1380.00	1011110111
12	1182.00	1359.30	1360.00	1111011001	1360.00	1101110111
13	1154.00	1327.10	1330.00	1001111001	1360.00	1101110111
14	1138.00	1308.70	1410.00	1101111001	1310.00	1111110011
15	1124.00	1292.60	1310.00	1111110011	1310.00	1111110011
16	1095.00	1259.25	1280.00	0110110111	1260.00	1111110110
17	1066.00	1225.90	1260.00	1010110111	1260.00	1111110110
18	1037.00	1192.55	1260.00	1010110111	1200.00	0111110110
19	993.00	1141.95	1180.00	1111100111	1180.00	1111100111
20	978.00	1124.70	1200.00	1111001010	1180.00	1111100111
21	963.00	1107.45	1320.00	0101011010	1180.00	1111100111
22	1022.00	1175.30	1210.00	1101011100	1180.00	1111100111
23	1081.00	1243.15	1340.00	1110111100	1250.00	0111110011
24	1459.00	1677.85	1900.00	1011111111	1680.00	1111011011
			Cumulative scheduling Cost=940101.65		Cumulative scheduling Cost = 877854.32	
			The difference in cost is approximately 7% in these two cases.			

Notes:
CASE – 1 When only best strategy is saved at each hour.
CASE – 2 Best of five least cost strategies are aved.

5. Conclusions

The power unit scheduling is a highly-constrained decision-making problem, and traditional methods make several assumptions to solve the problem. Most of these traditional methods require well-defined performance indices and explicitly use unit selection list or priority order list for determining the commitment decisions. The major advantages of using GAs are that they can eliminate some limitations of mathematical programming methods. Particularly, the GA-based unit commitment scheduler evaluates the priority of the units dynamically considering the system parameters, operating constraints and the load profile at each time period while evolving near-optimal schedules. Though global optimality is desirable, but in most practical purposes near-optimal (or good feasible) solutions are generally sufficient. This evolutionary approach attempts to find the best schedule from a set of good feasible commitment decisions. Also the method presented in this section can include more of the constraints that are encountered in real-world applications of this type.

This study suggests that the GA-based method for short-term unit commitment is a feasible alternative approach and is easy to implement. One disadvantage of this approach is the computational time needed to evaluate the population in each generation, but since genetic algorithms can efficiently be implemented in a highly parallel fashion, this drawback becomes less significant with its implementation in a parallel machine environment. Further research should address a number of issues to solve the commitment problem: Experiments should be carried out with large power systems having hundreds of units in multiple areas. One possible approach may be to use indirect encoding or use of some grammar rule (as used in other GA applications) for representing a cluster of units in a chromosome. Also in a large power plant, the unit-commitment task may be formulated as a multi-objective constrined optimization problem; where it is necessary to take into account not only the operational cost but also the emission of pollutant and other environmental factors as mutually conflicting objectives.

References

1. L. F. B. Baptistella and J. C. Geromel. A decomposition approach to problem of unit commitment schedule for hydrothermal systems. *IEE Proceedings- Part-C*, 127(6):250–258, November 1980.
2. Jonathan F. Bard. Short-Term Scheduling of Thermal-Electric Generators using Lagrangian Relaxation. *Operations Research*, 36(5):756–766, Sept/Oct. 1988.
3. L. B. Booker. *Intelligent behavior as an adaptation to the task environment.* PhD thesis, Computer Science, University of Michigan, Ann Arbor, U. S. A, 1982.
4. Arthur I. Cohen and Miki Yoshimura. A Branch-and-Bound Algorithm for Unit Commitment. *IEEE Transactions on Power Apparatus and Systems.*, PAS-102(2):444–449, February 1983.
5. Yuval Davidor. Epistasis Variance: Suitability of a Representation to Genetic Algorithms. *Complex Systems*, 4:369–383, 1990.
6. David E. Goldberg. *Genetic Algorithms in Search, Optimization and Machine Learning.* Addison-Wesley., first edition, 1989.
7. David E. Goldberg. Sizing populations for serial and parallel genetic algorithms. In *International Conference on Genetic Algorithms (ICGA-89)*, pages 70–79, 1989.
8. A. R. Hamdam and K. Mohamed-Nor. Integrating an expert system into a thermal unit-commitment algorithm. *IEE Proceedings-C*, 138(6):553–559, November 1991.
9. W. J. Hobbs, G. Hermon, S. Warner, and G. B. Sheble. An Advanced Dynamic Programming approach for Unit Commitment. *IEEE Transactions on Power Systems*, 3(3):1201–1205, August 1988.
10. John H. Holland. *Adaptation in Natural and Artificial Systems.* University of Michigan press, Ann Arbor, 1975.
11. Fred N. Lee. The application of commitment utilization factor (CUF) to thermal unit commitment. *IEEE Transactions on Power Systems.*, 6(2):691–698, May 1991.

12. P. G. Lowery. Generating Unit Commitment by Dynamic Programming. *IEEE Transactions on Power Apparatus and Systems*, PAS-85(5):422–426, May 1966.
13. A. Merlin and P. Sandrin. A New Method for Unit Commitment at Electricité de France. *IEEE Transactions on Power Apparatus and Systems*, PAS-102(5):1218–1225, May 1983.
14. John A. Muckstadt and Sherri A. Koenig. An Application of Lagrangian Relaxation to Scheduling in Power Generation Systems. *Operations Research*, 25(1):387–403, Jan/Feb 1977.
15. Sasan Mukhtari, Jagjit Singh, and Bruce Wollenberg. A Unit Commitment Expert System. *IEEE Transactions on Power Systems*, 3(1):272–277, February 1988.
16. Z. Ouyang and S. M. Shahidehpour. An Intelligent Dynamic Programming for Unit Commitment Application. *IEEE Transactions on Power Systems*, 6(3):1203–1209, August 1991.
17. Z. Ouyang and S. M. Shahidehpour. A Hybrid Artificial Neural Network-Dynamic Programming approach to Unit Commitment. *IEEE Transactions on Power Systems*, 7(1):236–242, February 1992.
18. C. K. Pang and H. C. Chen. Optimal Short-term Thermal Unit Commitment. *IEEE Transaction on Power Apparatus and Systems*, PAS-95(4):1336–1346, July/August 1976.
19. C. K. Pang, G. B. Sheble', and F. Albuyeh. Evaluation of dynamic programming based methods and multiple area representation for thermal unit commitments. *IEEE Transactions on Power Apparatus and Systems*, PAS-100(3):1212–1218, March 1981.
20. S. K. Tong, S. M. Shahidehpour, and Z. Ouyang. A Heuristic Short-term Unit Commitment. *IEEE Transactions on Power Systems*, 6(3):1210–1216, August 1991.
21. F. Zhuang and F. D. Galiana. Unit Commitment by Simulated Annealing. *IEEE Transactions on Power Systems*, 5(1):311–317, February 1990.
22. Fulin Zhuang and F. D. Galiana. Towards a more rigorous and practical Unit Commitment by Lagrangian Relaxation. *IEEE Transactions on Power Systems*, 3(2):763–773, May 1988.

Genetic Algorithms and Genetic Programming for Control

Dimitris C. Dracopoulos

Brunel University, Department of Computer Science, London

Summary. The use of genetic algorithms and genetic programming in control engineering has started to expand in the last few years. This is mainly for two reasons: the physical cost of implementing a known control algorithm and the difficulty to find such an algorithm for complex plants [56]. Broadly, evolutionary algorithms for control can be classified as either "pure" evolutionary or hybrid architectures. This chapter reviews some of the most successful applications of genetic algorithms and genetic programming in control engineering and outlines some general principles behind such applications. Demonstration of the power of these techniques is given, by describing how a hybrid genetic controller can be applied to the control of complex (even chaotic) dynamic systems. In particular, the detumbling and attitude control problems of a satellite are considered.

1. Evolutionary Techniques for Intelligent Control

The numerous published work on the use of evolutionary algorithms for control applications can be categorised as:

1. evolutionary techniques as pure controllers
   a) genetic algorithms
   b) evolutionary programming
   c) genetic programming
2. hybrid methods (mostly used with neural networks and fuzzy logic)
   a) independent blocks in a "mixed" control architecture
   b) for enhancing or optimising the performance of other controller types

Designed controllers based entirely on evolutionary methods use all three evolutionary paradigms: genetic algorithms (GAs) as introduced by Holland in [27], evolutionary programming introduced by Fogel [21, 22] and revisited among others in [20], and the relatively recent field of genetic programming [33, 34] (see section 3.).

In contrast with this, the second general methodology is the use of these algorithms in hybrid control architectures. This can be in combination with neural, fuzzy or even classical control theory designs. Such "mixed" controllers utilise Darwinian-selection based algorithms either as independent building blocks in a controller, or as optimisers (or simply enhancements) for the performance of a known controller. The success of the latter approach lies especially in the fact that it can use previously known stability results of the controller, while achieving improved overall performance. Methods in 2(a) can or cannot use previous stability results depending on the type of the evolutionary block. Controllers in category 1, on the other hand, have to prove from scratch their stability, something which is not always easy.

All of the aforementioned control techniques belong to the family of intelligent controllers. Intelligent control [57] aims in the control of processes (plants) over large ranges of uncertainties due to large variations in parameter values, environmental conditions and inputs. Although traditional adaptive control has similar aims, an intelligent control system must be able to operate well, even when the range of uncertainty is substantially greater than that which can be tolerated by algorithms for adaptive systems. The main objective with intelligent control is to design a system with acceptable performance characteristics over a very wide range of uncertainty [2, 4, 46].

Generally, complex systems are characterized by poor models, high dimensionality of the decision space, distributed sensors and decision makers, high noise levels, multiple performance criteria, complex information patterns, overwhelming amounts of data and stringent performance requirements. We can broadly classify the difficulties that arise in these systems into three categories for which established methods are often insufficient [47]. The first is computational complexity, the second is the presence of nonlinear systems with many degrees of freedom, and the third is uncertainty. The third category includes model uncertainties, parameter uncertainties, disturbances and noise. The greater the ability to deal successfully with the above difficulties, the more intelligent is the control system. Qualitatively, a system which includes the ability to sense its environment, process the information to reduce the uncertainty, plan, generate and execute control actions, constitutes an intelligent control system.

The next two sections concentrate in the description of successful genetic algorithms and genetic programming control applications. Evolutionary programming control applications can be found in [19, 20]. The last part of the chapter focus on how a hybrid method can be applied to the control of a complex dynamic system. Specifically the satellite attitude control problem is considered.

2. Recent GA-based Control Applications

Since their very first development by Holland, genetic algorithms have been applied to a variety of control problems. The interested reader is referred to the standard textbook [24] (or other textbooks e.g. [41]) for more details. Here however, some more recent advanced control applications using GAs are briefly described[1]. Once more, these belong to categories 1 and 2 as summarised in the previous section.

---

[1] Due to the space limitations only a part of the most recent interesting work that has come to the attention of the author is presented.

In [43], Mondada and Floreano show how neural networks and genetic algorithms can be combined for the development of control structures in autonomous agents. Holter et al. [28] use a neuro-genetic approach for manufacturing controllers, while in [5], NN and GAs enhance the performance of ATM networks. The identification and control of a distillation plant is addressed in [23] and Whitley et al. [58] discusses genetic control in a reinforcement learning fashion mode.

Fuzzy logic and genetic algorithms are "mixed" to design a flight controller in [35]. Wong and Feng [59] discuss the parameter tuning of an inverted pendulum fuzzy controller by a GA, while the evolution of a rule based fuzzy boat rudder controller is given in [7]. Another genetic-fuzzy approach for both online and offline control is given in [39, 40]. In [31], the fuzzy GA-based computer-aided system design methodology for rapid prototyping of control systems is demonstrated. More genetic-fuzzy control combinations can be found in [26] (tuning of a fuzzy controller) and in a similar fashion [1] (chemical reaction control) and [53] (autonomous mobile robot navigation).

The optimisation of parameters in a $NO(x)$ emmision control using genetic algorithms is shown in [42]. Derivation of new optimising algorithms for control by combining traditional optimisation theory and genetic algorithms can be found in [9, 49].

In [60] the genetic design of multivariable adaptive digital tracking PID controllers for space control is presented. Similarly, the optimisation of classical PID controllers for nonlinear plants is shown in [37] (adaptive hydro-generator governor) and [55] (pH neutralisation process). The use of genetic algorithms for identification (the location of poles and zeros of a physical system), and their use for the design of a discrete time pole placement adaptive controller is described in [36].

The application of a genetic algorithm to the production planning and control manufacturing, after decomposition of the initial problem to a series of scheduling problems, is discussed in [32]. In [45] a genetic algorithm instrumentation control approach is introduced. Porter and Mohamed [51] consider a two-level hierarchy of genetic algorithms for minimum time control problems, while [18] addresses the evolution of prototype control rules with application to the cart-pole balancing problem. Some other interesting genetic control applications (hybrid or not) can be found in [6, 8, 17, 29, 30, 52, 54] but space limitations cannot allow their discussion.

A comparison between simulated annealing and genetic algorithm techniques for choosing source locations in control systems is attempted in [3]. A more general discussion and brief assessment of the application of AI-technologies in standard monitoring and control systems (SMCS) shipboard operational requirements can be found in [16].

3. Genetic programming

The relatively recent development of Genetic programming (GP) [33, 34] offers an ideal candidate for controller designs, due to the direct matching of its individual structures and control rules. One particular problem to the application of GP to control problems is the consideration of stability issues. One can either try to prove the stability of a derived controller after the end of a GP run, or somehow try to incorporate the stability behaviour of a controller in the fitness measure [11].

Another major problem and topic of research nowadays, is the time required by GP to derive its solutions. Although research in the algorithmic convergence speedup of the GP process is substantial, another recent attempt to face this problem is parallel GP implementations [15].

So far, genetic programming has been applied to a small number of control problems but without major consideration of stability issues. Some of the GP successful control applications include the solution to classical control theory problems, the broom balancing problem and the truck backer upper problem [33] while some newer include the articulated figure motion control [25]. However, the current "explosion" of GP applications to control problems makes it difficult to give a more extensive list of such approaches.

4. The Adaptive Neuro-genetic Control Architecture

In the rest of this chapter, the power of genetic hybrid control methods is demonstrated by examining a specific neuro-genetic adaptive control design. The hybrid controller is used for the control of a complex chaotic dynamic system. In particular the satellite attitude control problem is considered.

The adaptive neuro-genetic architecture introduced in [14] is a hybrid method combining neural networks and genetic algorithms. Its aim is to achieve adaptive control for nonlinear dynamic systems. The method was developed to face ill-posed inverse kinematics problems where direct inverse control cannot be applied. The neuro-genetic architecture is a two part architecture, of which the first part is a locally predictive network (LPN), a neuromodel that learns to predict a future system state a short time ahead given the few last states and control inputs 4.1.

Earlier results on neuromodelling [12, 13], showed that this is a possible. The time scale for such a prediction might be as high as 0.5–1 sec. With this predictive capability a large number of hypothetical control inputs can be evaluated. The adaptive control problem can be reduced now to the optimization task of finding the control law which lead to improved system performance in the near future. This optimization is achieved by using a zero-order optimization algorithm, a *genetic algorithm*, to explore the space of hypothetical control inputs at any given moment and the LPN predictor

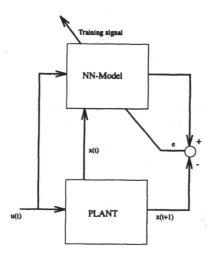

Fig. 4.1. Building the neuromodel for the plant

to evaluate any member of the current population (of hypothetical control inputs).

Normally the evaluation phase (computing the fitness of a member of the population) is the most time consuming aspect of the genetic algorithm, but a hardware neural network implementation is possible for this phase, thus allowing the evaluation of many thousands of sets of control inputs over the relevant prediction interval which may be a significant fraction of a second.

Of itself the genetic algorithm is very simple. The control signals are represented by binary strings, with simple bit string manipulations for crossover and mutation. A variety of implementations could be used for the genetic algorithm ranging from execution by a serial processor, to gate arrays with additional memory and processors (to provide a stack, for example). Given that the evaluation time per member of the population is very fast, the rest of the genetic algorithm is quite simple. The particular implementation chosen would depend heavily on the system to be controlled and the speed of events in the real world. The genetic algorithm used throughout this paper is shown in Figures 4.2, 4.3. In all the examples presented here a population size of 50 was used having 100 generations as the maximum number of generations. For more details in the actual genetic algorithm implementation see [14].

The overall operation of the neuro-genetic architecture, briefly described by the sequence of events (with $t = k\Delta t$ and $\Delta t = 0.01$ in the simulations of this paper), is (Figure 4.4):

At step $k - 1$: Controller applies $u^*(k - 1)$—LPN predicts $x^q(k)$.

At times between $k - 1$ and k: Based on this prediction the genetic algorithm tries to choose $u^*(k)$ so as to optimize $x(k + 1)$. The genetic algorithm

```
population_size := 50;
generation := 1;
Initialize population with random binary strings;
while generation ≤ 100 do
    Find the two best individuals of current population;
    Copy the two best individuals to new population;
    for i = 1 to population_size − 2 step 2 do
        Select two individuals based on fitness;
        Probabilistically mutate the two individuals;
        Probabilistically perform crossover;
        if crossover_performed then
            Copy the two offspring into new population;
        else
            Copy the two (mutated) individuals into new
            population;
        endif
    endfor
    generation := generation + 1;
endwhile
```

Fig. 4.2. Pseudocode of the Genetic Algorithm for the Attitude Control Problem.

```
Select()
    j := 1; sector := 0;
    number := random();
    while sector ≤ number and j ≤ population_size do
        sector := sector + fitness_of_individual_j;
        j := j + 1;
    endwhile
    return individual_{j−1};
end
```

Fig. 4.3. Select procedure of the Genetic Algorithm for the Attitude Control Problem.

uses the LPN to predict the result $x^q(k+1)$ of hypothetical control inputs $u^h(k)$, and hence evaluate the fitness of $u^h(k)$.

At step k: Controller applies $u^*(k)$. LPN predicts $x^q(k+1)$.

. . .

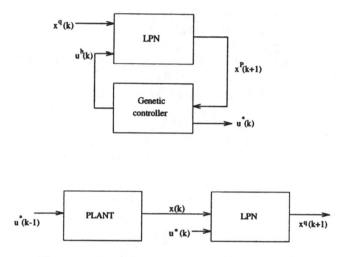

Fig. 4.4. The adaptive neuro-genetic architecture

From the above description of the neuro-genetic architecture it is seen that the method does not require knowledge of the plant dynamics, in analytic form, but is adaptive to unknown changes of the plant, by observing its input-output behaviour. The resulting controller is general in nature, and provides self adaptive nonlinear control without linearisation of any system component.

Compared with the OGY original method[2] the control architecture proposed in [14] can effectively control a chaotic system starting from any system state. In contrast the OGY method in order to start control has to wait until the system trajectory passes through an unstable periodic orbit [10, 48]. In addition, the OGY method requires some knowledge of the plant dynamics (i.e. unstable periodic orbits) while the method described here requires no minimum knowledge of the plant dynamics.

Here the neuro-genetic architecture is tested on the attitude control problem, in the case where external forces are acting upon the dynamic system (satellite). The external forces are chosen so as if left to themselves they lead the system in a chaotic motion. Such a chaotic motion is extremelly difficult to be controlled.

---

[2] The OGY method named after Ott, Grebogi and Yorke who devised it in 1990, was the first method for "controlling chaos".

4.1 The attitude control problem

The orientation control of a rigid body has important applications from pointing and slewing of aircraft, helicopter, spacecraft and satellites, to the orientation control of a rigid object held by a single or multiple robot arms.

Attitude control is the process of achieving and maintaining an orientation in space by performing attitude maneuvers (an attitude maneuver is the process of reorienting a satellite or spacecraft from one attitude to another). The attitude of the body is relative to some external reference frame. This reference frame may be either inertially fixed, or slowly rotating, as in the case of Earth-oriented satellites. The attitude control problem lies into reorienting the satellite towards a desired orientation and at the same time detumble it (lead all or some of its angular velocities to zero).

The attitude control problem [14] is described by two systems of first order differential equations. The first set describes the rotational motion of the rigid body:

$$
\begin{aligned}
L &= I_x \dot{\omega}_1 - (I_y - I_z)\omega_2\omega_3 \\
M &= I_y \dot{\omega}_2 - (I_z - I_x)\omega_3\omega_1 \\
N &= I_z \dot{\omega}_3 - (I_x - I_y)\omega_1\omega_2
\end{aligned}
\tag{4.1}
$$

where I_x, I_y, I_z are principal moments of inertia, $\omega_1, \omega_2, \omega_3$ are the angular velocities of the rigid body about the x, y, z body axes and L, M, N are the torques about the x, y, z axes respectively.

The second set of differential equations describing the attitude control problem defines the orientation of the body relative to an inertial frame:

$$
\begin{aligned}
\dot{\Phi} &= \omega_1 + \omega_2 \sin\Phi \tan\Theta + \omega_3 \cos\Phi \tan\Theta \\
\dot{\Theta} &= \omega_2 \cos\Phi - \omega_3 \sin\Phi \\
\dot{\Psi} &= (\omega_2 \sin\Phi + \omega_3 \cos\Phi)\sec\Theta
\end{aligned}
\tag{4.2}
$$

where Φ, Θ, Ψ are the angles given by three consecutive rotations about the x, y, z body axes respectively.

There are several cases where freely rotating rigid bodies can exhibit chaotic behaviour [38, 44]. The first case is when one of the moments (L, M or N), varies periodically in time. The second case is where one has parametric excitation through time-periodic changes in the principal inertias, for example, $I_y = I_0 + B \cos \Omega t$.

The third case comes from the schemes based on linear or quadratic feedback, which have been proposed to stabilize rigid body spacecraft attitude, described by the Euler equations (4.1). A simple such scheme employs jets to impart torque according to suitable linear combinations of the sensed angular velocities ω about the body-fixed principal axes.

If the torque feedback matrix for the nonlinear system is denoted by $\mathbf{A}$, so that $\mathbf{G} = (\mathbf{L}, \mathbf{M}, \mathbf{N}) = \mathbf{A}\omega$, the equations (4.1) become

$$\begin{aligned}
(\mathbf{A}\omega)_1 &= I_x\dot\omega_1 - (I_y - I_z)\omega_2\omega_3 \\
(\mathbf{A}\omega)_2 &= I_y\dot\omega_2 - (I_z - I_x)\omega_3\omega_1 \\
(\mathbf{A}\omega)_3 &= I_z\dot\omega_3 - (I_x - I_y)\omega_1\omega_2
\end{aligned} \tag{4.3}$$

It has been noticed [38, 50] that for certain choices of I_x, I_y, I_z and $\mathbf{A}$ equations (4.3) exhibit both strange attractors and limit cycles. Since these linear feedback rigid body motion equations are slightly more complicated than Lorenz's equations [38], this conclusion is not surprising. The interpretation of strange attractor motion, in this case, is that the body executes a wobbly spin first about one, then about the other of a conjugate pair of directions fixed relative to the body axes (and symmetric about an axis). The limiting spin magnitudes and directions define rest points of the system, called eye-attractors because of their appearance. Figure 4.5 shows the Poicaré map of the double strange attractor of equations (4.3) in the $x-z$ plane for $I_x = 3$, $I_y = 2$, $I_z = 1$ and

$$\mathbf{A} = \begin{pmatrix} -1.2 & 0 & \sqrt6/2 \\ 0 & 0.35 & 0 \\ -\sqrt6 & 0 & -0.4 \end{pmatrix} \tag{4.4}$$

The attractor of an orbit is determined by the location of the initial point of that orbit.

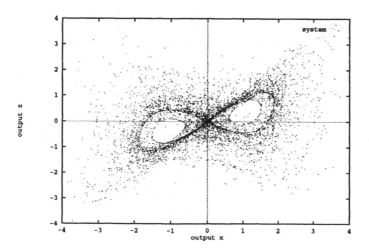

Fig. 4.5. Euler system. Poincaré section through the $x-z$ plane

In the following two sections results of the neuro-genetic architecture applied to the attitude control problem are presented. External forces make the dynamics of the system chaotic and similar to the chaotic system described above. In all the simulation results, every time that the system dynamics change in an unknown way, it will be assumed that a neuromodel describing the new dynamics of the plant has been trained according to the methodology described in [13]

4.2 Spin stabilization of a satellite from a random initial position

Assume, that for some unknown reason (damage), a satellite with specified dynamics, changes its characteristics. Its moments of inertia, become $I_x = 1160, I_y = 23300$ and $I_z = 24000$.

During the period, where the system changes dynamics, unknown forces lead it to the $(\omega_1, \omega_2, \omega_3) = (3, 2, 1)$ and $(\Phi, \Theta, \Psi) = (2, 1, 3)$ state. The goal is to detumble the satellite about the x, y body axes, spin it about the z body axis, and reorient it, so that $\Theta = 0, \Phi = 0$.

The application of the Genetic adaptive controller architecture, described by Figures 4.1a, 4.1b leads to the situation described by Figures 4.6a, 4.6b. Figure 4.6a shows the evolution in time, of the angular velocities ω_1, ω_2 about the x, y body axes respectively. It is easily shown that the Genetic controller, soon leads both of these angular velocities to the prespecified value of zero. The third angular velocity is arbitrary, since it was not in the control objectives. Consequently the control torque N is not subject to evolutionary pressure.

Figure 4.6b show the reorientation of the satellite for the angles Φ, Θ after the application of the controller. While these are becoming zero, the satellite is rotating about its z axis. It should be noted, that the controller not only leads the system to a desired state, but it maintains this state afterwards.

4.3 Satellite attitude control subject to external forces leading to chaos

Assume that a satellite changes its characteristic moments of inertia to $I_x = 1160, I_y = 23300$ and $I_z = 24000$. External forces are asked upon it for $t > 0$. The system is described by dynamic equations (4.2) and

$$
\begin{aligned}
I_x\dot{\omega}_1 - (I_y - I_z)\omega_2\omega_3 &= N_1 + L \\
I_y\dot{\omega}_2 - (I_z - I_x)\omega_3\omega_1 &= N_2 + M \\
I_z\dot{\omega}_3 - (I_x - I_y)\omega_1\omega_2 &= N_3 + N
\end{aligned}
\tag{4.5}
$$

$$\tag{4.6}$$

where the torques N_1, N_2, N_3 produced by the external forces are given by the equations

$$
\begin{aligned}
N_1 &= -1200\omega_1 + 1000 \cdot \frac{\sqrt{6}}{2}\omega_3 \\
N_2 &= 350\omega_2 \\
N_3 &= -1000 \cdot \sqrt{6}\omega_1 - 400\omega_3
\end{aligned}
\tag{4.7}
$$

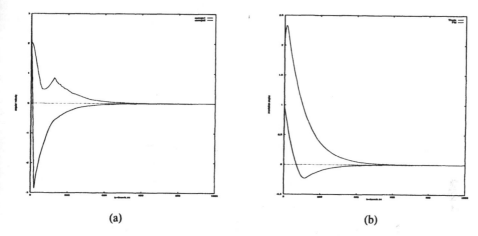

(a)   (b)

Fig. 4.6. (a) Angular velocities ω_1, ω_2 for the satellite after the application of the Genetic Controller. (b) Orientation angles Θ, Φ for the satellite after the application of the Genetic Controller. Initial conditions $(\omega_1, \omega_2, \omega_3) = (3, 2, 1)$ and $(\Phi, \Theta, \Psi) = (2, 1, 3)$

The above is a system in which the externally imposed torques N_1, N_2, N_3 would left to themselves result in a chaotic motion, while the thrust vector $G = (L, M, N)$ is trying to control this chaotic motion and lead the system into a prespecified state (control of chaos). In this case, it is assumed that the initial state is $(\omega_1, \omega_2, \omega_3) = (1.3, 3.0, 2.8)$ and $(\Phi, \Theta, \Psi) = (2, 1, 3)$ while the desired target state is $(\omega_1, \omega_2, \omega_3) = (0, 0, 0)$ and $(\Phi, \Theta, \Psi) = (0, 0, 0)$. Since the goal from the previous simulation is different, a new objective function is defined as follows:

$$U(t) = |\dot{\Phi} + \Phi| + |\Phi| + |\dot{\Theta} + \Theta| + |\Theta| + |\dot{\Psi} + \Psi| + |\Psi| \qquad (4.8)$$

Figure 4.7a show the evolution in time, of the angular velocities $\omega_1, \omega_2, \omega_3$ about the x, y, z body axes respectively. The Genetic controller, soon leads these angular velocities to the prespecified value of zero despite the fact that during the control process the external perturbing torques were as large as 44% of the maximum available control torques. As soon as this is achieved, the controller maintains these angular velocities.

In Figure 4.7b the reorientation of the satellite for the angles Φ, Θ, Ψ after the application of the controller is shown. Again it can be noticed, that the controller not only reorients the satellite, but it maintains this reorientation.

5. Summary

In this chapter an attempt to cover some recent work in the application of evolutionary algorithms in control engineering was presented. As shown, some

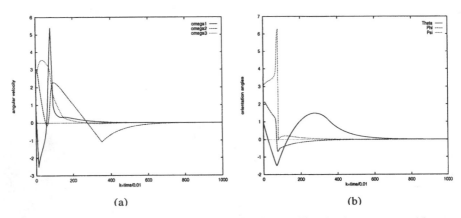

Fig. 4.7. (a) Angular velocities $\omega_1, \omega_2, \omega_3$ for the satellite, in the presence of forces trying to lead it in a chaotic motion, after the application of the Genetic Controller. (b) Orientation angles Θ, Φ, Ψ for the satellite, in the presence of forces trying to lead it in a chaotic motion, after the application of the Genetic Controller. Initial conditions $(\omega_1, \omega_2, \omega_3) = (1.3, 3.0, 2.8)$ and $(\Phi, \Theta, \Psi) = (2, 1, 3)$

evolutionary hybrid methods are very powerful and successful even for the control of complex nonlinear dynamic systems.

Although many general control applications have been successfully addressed by evolutionary techniques, more work is required in order for such methods to be accepted by traditional control theorists. Especially theoretical results of stability for these architectures are essential.

Acknowledgements

This work was supported by BRIEF award no 302.

References

1. C. L. Karr abd S. K. Sharma, W. J. Hatcher, and T. R. Harper. Fuzzy control of an exothermic chemical reaction using genetic algorithms. *Engineering Applications of Artificial Intelligence*, 6(6):575–582, 1993.
2. Karl Johan Astrom and Bjorn Wittenmark. *Adaptive Control*. Addison-Wesley, first edition, 1989.
3. K. H. Baek and S. J. Elliott. Natural algorithms for choosing source locations in active control systems. *Journal of sound and vibration*, 186(2):245–267, 1995.
4. Behnam Bavarian. Introduction to neural networks for intelligent control. *IEEE Control Systems Magazine*, pages 3–7, April 1988.
5. L. D. Chou and J. L. C. Wu. Parameter adjustment using neural network based genetic algorithms for guaranteed qos in atm networks. *IEICE Transactions on Communications*, E78B(4):572–579, 1995.

6. M. Colombetti and M. Dorigo. Learning to control an autonomous robot by distributed genetic algorithms. In J. A. Meyer, H. L. Roitblat, and S. W. Wilson, editors, *From Animals to Animats 2: Proceedings of the Second International Conference on Simulation of Adaptive Behavior*. MIT Press, 1993.

7. M. G. Cooper. Evolving a rule based fuzzy controller. *Simulation*, 65(1):67–72, 1995.

8. Yuval Davidor. A genetic algorithm applied to robot trajectory generation. In Lawrence Davis, editor, *Handbook of genetic algorithms*, chapter 12, pages 144–165. Van Nostrand Reinhold, 1991.

9. A. K. Dhingra and B. H. Lee. Multiobjective design of actively controlled structures using a hybrid optimization method. *International journal for numerical methods in engineering*, 38(20):3383–3401, 1995.

10. William L. Ditto and Louis M. Pecora. Mastering chaos. *Scientific American*, August 1993.

11. Dimitris C. Dracopoulos. Evolutionary control of a space shuttle. paper in preparation while this chapter is written, 1996.

12. Dimitris C. Dracopoulos and Antonia J. Jones. Modeling dynamic systems. In *1st World Congress on Neural Networks Proceedings*, pages 289–292. INNS/Erlbaum Press, 1993.

13. Dimitris C. Dracopoulos and Antonia J. Jones. Neuromodels of analytic dynamic systems. *Neural Computing & Applications*, 1(4):268–279, 1993.

14. Dimitris C. Dracopoulos and Antonia J. Jones. Neuro-genetic adaptive attitude control. *Neural Computing & Applications*, 2(4):183–204, 1994.

15. Dimitris C. Dracopoulos and Simon Kent. Speeding up genetic programming: A parallel implementation using BSP. In John R. Koza, David E. Goldberg, David B. Fogel, and Rick L. Riolo, editors, *Genetic Programming 1996: Proceedings of the First Annual Conference*. Stanford University, MIT Press, 28–31 July 1996.

16. D. L. Fairhead and C. C. Hall. Intelligent machinery control integration. *Naval Engineers Journal*, 107(5):51–57, 1995.

17. T. C. Fogarty and L. Bull. Optimizing individual control rules and multiple communicating rule based control systems with parallel distributed genetic algorithms. *IEE Proceedings Control Theory and Applications*, 142(3):211–215, 1995.

18. T. C. Fogarty and R. Huang. Evolving prototype control rules for a dynamic system. *Knowledge Based Systems*, 7(2):142–145, 1994.

19. D. B. Fogel. Applying evolutionary programming to selected control problems. *Computers & Mathematics with Applications*, 27(11):89–104, 1994.

20. D. B. Fogel. Evolutionary programming-an introduction and some current directions. *Statistics and Computing*, 4(2):113–129, 1994.

21. L. J. Fogel. *On the Organisation of Intellect*. PhD thesis, UCLA, 1962.

22. L. J. Fogel, A. J. Owens, and M. J. Walsh. *Artificial Intelligence through Simulated Evolution*. John Wiley, 1966.

23. D. Gariglio, J. Heidepriem, and A. Helget. Identification and control of a simulated distillation plant using connectionist and evolutionary techniques. *Simulation*, 63(6):393–404, 1994.

24. David E. Goldberg. *Genetic Algorithms in Search, Optimization and Machine Learning*. Addison Wesley, 1989.

25. L. Gritz and J. K. Hahn. Genetic programming for articulated figure motion. *Journal of Visualization and Computer Animation*, 6(3):129–142, 1995.

26. F. Herrera, M. Lozano, and J. L. Verdegay. Tuning fuzzy logic controllers by genetic algorithms. *International Journal of Approximate Reasoning*, 12(3–4):229–315, 1995.

27. John H. Holland. *Adaptation in Natural and Artificial Systems*. The University of Michigan Press, 1975.
28. T. Holter, X. Q. Yao, L. C. Rabelo, A. Jones, and Y. W. Yih. Integration of neural networks and genetic algorithms for an intelligent manufacturing controller. *Computers & Industrial Engineering*, 29:211–215, 1995.
29. C. L. Karr and S. K. Sharma. An adaptive process control system based on fuzzy logic and genetic algorithms. In *Proceedings of the 1994 American Control Conference*, pages 2470–2474, 1994.
30. S. Kawaji and K. Ogasawara. Swing up control of a pendulum using genetic algorithms. In M. Peshkin, editor, *Proceedings of the 33rd IEEE Conference on Decision and Control*, pages 3530–3532. IEEE Service Center, 1994.
31. J. W. Kim, Y. K. Moon, and B. P. Zeigler. Designing fuzzy net controllers using genetic algorithms. *IEEE Control Systems Magazine*, 15(3):66–72, 1995.
32. H. Kopfer, I. Rixen, and C. Bierwirth. Integrating genetic algorithms in production planning and control. *Wirtschaftsinformatik*, 37(6):571–580, 1995.
33. John R. Koza. *Genetic Programming*. MIT Press, 1982.
34. John R. Koza. *Genetic Programming II*. MIT Press, 1994.
35. K. Krishnakumar, P. Gonsalves, A. Satyadas, and G. Zacharias. Hybrid fuzzy logic flight controller synthesis via pilot modeling. *Journal of Guidance Control and Dynamics*, 18(5):1098–1105, 1995.
36. K. Kristinsson and G. A. Dumont. Systems indentification and control using genetic algorithms. *IEEE Transactions on Systems Man and Cybernetics*, 22(5):1033–1046, 1992.
37. J. E. Lansberry and L. Wozniak. Adaptive hydrogenerator governor tuning with a genetic algorithm. *IEEE Transactions on Energy Conversion*, 9(1):179–185, 1994.
38. R. B. Leipnik and T. A. Newton. Double strange attractors in rigid body motion with linear feedback control. *Physics Letters*, 86A:63–67, 1981.
39. D. A. Linkens and H. O. Nyongesa. Genetic algorithms for fuzzy control. 1. offline system development and application. *IEE Proceedings Control Theory and Applications*, 142(3):161–176, 1995.
40. D. A. Linkens and H. O. Genetic algorithms for fuzzy control. 2. online system development and application. *IEE Proceedings Control Theory and Applications*, 142(3):177–185, 1995.
41. Z. Michalewicz. *Genetic Algorithms + Data Structures = Evolution Programs*. Springer Verlag, 1996.
42. Y. Miyamoto, T. Miyatake, S. Kurosaka, and Y. Mori. A parameter tuning for dynamic simulation of power plants using genetic algorithms. *Electrical Engineering in Japan*, 115(1):104–115, 1995.
43. F. Mondada and D. Floreano. Evolution of neural control-structures - some experiments on mobile robots. *Robotics and Autonomous Systems*, 16(2-4):183–195, 1995.
44. Francis C. Moon. *Chaotic and Fractal Dynamics*. John Wiley and Sons, 1992.
45. J. H. Moore. Artificial intelligence programming with labview-genetic algorithms for instrumentation control and optimization. *Computer methods and programs in Biomedicine*, 47(1):73–79, 1995.
46. Kumpati S. Narendra and Anuradha M. Annaswamy. *Stable Adaptive Systems*. Prentice Hall, 1989.
47. Kumpati S. Narendra and Snehasis Mukhopadhyay. Intelligent control using neural networks. *IEEE Control Systems Magazine*, pages 11–18, April 1992.
48. E. Ott, C. Grebogi, and James Yorke. Controlling chaos. *Physical Review Letters*, 64(11), 1990.

49. R. J. Patton and G. P. Liu. Robust control design via eigenstructure assignment genetic algorithms and gradient based optimization. *IEE Proceedings Control Theory and Applications*, 141(3):202–208, 1994.
50. George E. Piper and Harry G. Kwatny. Complicated dynamics in spacecraft attitude control systems. *Journal of Guidance, Control and Dynamics*, 15(4):825–831, July-August 1992.
51. B. Porter and S. S. Mohamed. Genetic design of minimum time controllers. *Electronic Letters*, 29(21):1897–1898, 1993.
52. M. J. Schutten and D. A. Torrey. Genetic algorithms for control of power converters. In *IEEE Power Electronics Specialists Conference Records*, pages 1321–1326. 26th Annual IEEE Power Electronics Specialists Conference, Atlanta, GA, IEEE Service Center, 18–22 June 1995.
53. D. D. Sobrino, J. G. Casao, and C. G. Sanchez. Genetic processing of the sensory information. *Sensors and Actuators A-Physical*, 37–38:255–259, 1993.
54. A. Varsek, T. Urbancic, and B. Filipic. Genetic algorithms in controller design and tuning. *IEEE Transactions on Systems Man and Cybernetics*, 23(5):1330–1339, 1993.
55. P. Wang and D. P. Kwok. Optimal design of pid process controllers based on genetic algorithms. *Control Engineering Practice*, 2(4):641–648, 1994.
56. Paul Werbos. Neurocontrol and related techniques. In A. Maren, editor, *Handbook of Neural Computing Applications*, pages 345–381. Academic Press, 1990.
57. David A. White and Donald A. Sofge, editors. *Handbook of Intelligent Control*. Van Nostrand Reinhold, 1992.
58. D. Whitley, S. Dominic, R. Das, and C. W. Anderson. Genetic reinforcement learning for neurocontrol problems. *Machine Learning*, 13(2–3):259–284, 1993.
59. C. C. Wong and S. M. Feng. Switching type fuzzy controller design by genetic algorithms. *Fuzzy sets and systems*, 74(2):175–185, 1995.
60. W. Zuo. Multivariable adaptive control for a space station using genetic algorithms. *IEE Proceedings Control Theory and Applications*, 142(2):81–87, 1995.

Global Structure Evolution
and Local Parameter Learning
for Control System Model Reductions

Yun Li, Kay Chen Tan, and Mingrui Gong

Centre for Systems and Control, and Department of Electronics and Electrical Engineering, University of Glasgow, Glasgow G12 8LT, United Kingdom

Summary. This chapter develops a Boltzmann learning refined evolution method to perform model reduction for systems and control engineering applications. The evolutionary technique offers the global search power from the 'generational' Darwinism combined with 'biological' Lamarckism. The evolution is further enhanced by interactive fine-learning realised by Boltzmann selection in a simulated annealing manner. This hybrid evolution program overcomes the well-known problems of chromosome stagnation and weak local exploration of a pure evolutionary algorithm. The use of one-integer-one-parameter coding scheme reduces chromosome length and improves efficiency dramatically. Enabled by a control gene as a structural switch, this indirectly guided reduction method is capable of simultaneously recommending both an optimal order number and corresponding parameters. Such order flexibility and the insight it provides into the system behaviour cannot be offered by existing conventional model reduction techniques. The evolutionary approach is uniformly applicable to both continuous and discrete time systems in both the time and the frequency domains. Three examples involving process and aircraft model reductions verify that the approach not only offers higher quality and tractability than conventional methods, but also requires no a priori starting points.

List of Symbols

$\|\cdot\|$	Euclidean metric (l_2 norm by default in this chapter)
a_i, b_i	Coefficients of the denominator and numerator of an original model, respectively
β	Annealing coefficient (inverse Boltzmann 'learning rate')
C_k, C	A current chromosome and a chromosome mutated from C_k, respectively
c_i, d_i	Coefficients of the denominator and numerator of a reduced model, respectively
$\Delta(s)$	Denominator of the transfer function elements of a fighter aircraft model
$\delta(s)$	Stick force of a fighter aircraft model
f	The discrete frequency index
$G(s), G(z)$	Transfer function of an original plant model in continuous and discrete times, respectively
$G_\theta(s), G_\theta(z)$	Transfer function of a reduced-order model in continuous and discrete times, respectively
g, g_{max}	The evolving number and the maximum number of generations, respectively
$h(t)$	Output discrepancy between an original and a reduced model
$J(\theta), J_{pq}(\theta)$	Time and frequency domain cost in the total weighted quadratic error

$J_{RMS}(\theta)$	Time and frequency domain cost in the root mean-square error
k	Index number of a chromosome randomly selected for learning
l, k	Order of a reduced model; order of its numerator
m, n	Order of an original model; order of its numerator
N	The window size for finite time/frequency truncation
$n_{zcr}(s)$	Normal acceleration of a fighter aircraft model
p, q	The numbers of outputs and inputs of a multivariable system, respectively
$q(s)$	Pitch rate of a fighter aircraft model
θ	A parameter set of a reduced-order model
s	Argument of the Laplace transform
t	The time, both continuous and discrete
T, T_{ini}, T_{final}	Annealing temperature, its initial value and its final value, respectively
$u(t)$	Unit step function
ω	The angular frequency
$W(j\omega), w(t)$	Frequency and time domain weighting functions, respectively
z	The Z-domain argument and time domain unit forward-shift operator

1. Introduction

For many modern control schemes, such as optimal or H_∞ based control, it is usually required to perform plant model or controller order reduction prior to or during the process of design. The order reduction problem is the problem of approximating, as closely as possible, the dynamics of a high-order system by a reduced-order linear time-invariant (LTI) model, while retaining the important structural and dynamic properties of the original system. It is usually a multimodal optimisation problem in a multidimensional space.

So far, model reduction methods have been mainly based on conventional optimisation techniques either in the time domain or in the frequency domain alone [17]. For example, in the time domain, Anderson [2] used a 'geometric approach' to obtain reduced-order models of minimised quadratic errors. Other time domain methods include those based on Powell's directional optimisation [20] and those utilising an 'error polynomial' that matches the 'Markov parameters' for discrete-time systems [19]. In the frequency domain, Levy [9] established a complex curve-fitting technique to minimise quadratic errors of a single-input and single-output (SISO) transfer function. This was later formalised by Elliott and Wolovich [4] and extended to multiple-input and multiple-output (MIMO) systems [5].

These model reduction methods are mainly based on Bayesian type parameter estimation, its extended maximum likelihood version and their modified recursive versions. The optimal estimation is essentially guided by a priori directions on the error surface of the cost function and is thus noise-prone. For this, the cost must be differentiable or well-behaved and the order of the reduced model be known a priori. To use such methods often requires a 'good' excitation [17]. These conditions may not often be satisfied if the reduced model is required to meet some practical constraints or the original model was obtained from a noisy environment. Although such directed optimisation techniques still serve as the major optimisation tool in control systems engineering, following an a priori direction may not lead to globally optimised model parameters [16], as the unknown multidimensional error surface is usually multimodal. Further, it is difficult to identify an optimal order or structure if two or more separate parameters (such as repeated poles) have a similar effect on the system output. To achieve good reduction tractability and quality, optimisation methods that rarely rely on a 'wise' selection of initial conditions or a priori parameters of the reduced model are needed.

Based on a posteriori information obtained from trial-and-errors, evolutionary algorithms (EAs) require no direct guidance and thus no stringent conditions on the cost function [14]. Therefore, the index function can be constructed in a way that satisfies the need of engineering systems most and not the need of analytical or numerical tools. For example, in design automation facilitated by EAs, a design specification index can include all practical constraints and logic terms of decision making [10, 13]. The evolutionary methods have proven to be very powerful in solving difficult non- polynomial problems usually found in engineering practice. They improve tractability and robustness in global optimisations by slightly trading off precision in a nondeterministic polynomial (NP) manner. Thus, compared with the non-NP exhaustive search or dynamic programming techniques, EAs offer an exponentially reduced search time. Compared with Powell's technique, Renders and Bersini [16] have shown that EAs offer double accuracy and reliability in multidimensional optimisations. The evolutionary methods, in their various forms, have been successfully applied to systems and control engineering problems [6, 7, 11, 12, 15, 18].

In this chapter, a local learning enhanced evolutionary technique for model reduction has been developed. It provides a uniform approach to continuous and discrete-time system reductions in both the time and frequency domains. The formulation of model reduction problems is shown in Section 2. The evolutionary algorithm is presented in Section 3. Section 4 illustrates and validates the proposed reduction methods by three examples, in both the time and frequency domains. Conclusions and further work are highlighted in Section 5.

2. The Problem of Optimal Model Reduction for Control Systems

2.1 Reduction based on a time domain cost

Consider an m-th order discrete-time system described in Z-transform by

$$G(z) = \frac{b_n z^n + b_{n-1} z^{n-1} + \dots + b_0}{z^m + a_{m-1} z^{m-1} + \dots + a_0} \tag{2.1}$$

where $n < m$ for a causal discrete-time system. Suppose a reduced-order model approximating this system is represented by

$$G_\theta(z) = \frac{d_k z^k + d_{k-1} z^{k-1} + \ldots + d_0}{c_l z^l + c_{l-1} z^{l-1} + \ldots + c_0} \qquad (2.2)$$

where

$$0 < k < l \leq m - 1 \qquad (2.3)$$

Alternatively, the numerators and denominators of (2.1) and (2.2) may be expressed in factorised forms, which are omitted here for simplicity. Note that the coefficients of the most significant order terms, $c_l z^l$ and $d_k z^k$, are not pre-normalised to 1, so that these coefficients are allowed to be zero in an optimal reduction. Note also that, however, coefficients of high-order terms may not automatically become zero in conventional optimisations, since relatively higher-order models tend to offer relatively lower errors and conventional techniques are poor in approximating zero parameters.

To formulate the reduction problem, first define the parametric vector corresponding to G_θ as

$$\theta = [c_0, \ldots, c_l, d_0, \ldots, d_k]^T \qquad (2.4)$$

Since the time history data of system responses to a discrete Dirac impulse signal may be insufficiently short, step inputs are often used for model reductions in the time domain. Thus suppose both the original and the reduced models are excited by the identical unit step signal $u(t)$. The discrepancy between their responses is given by

$$h(t) = [G(z) - G_\theta(z)]u(t) \qquad (2.5)$$

This discrepancy is usually used to assess the closeness of the two models. Here the Z-transform argument z is also used to represent the unit forward-shift operator in the time domain.

The reduced model may thus be obtained by minimising the error norm $\|h(t)\|$ or a cost in the form of a weighted effective error energy

$$J(\theta) = \|w(t)h(t)\|^2 = \sum_{t=1}^{N} [w(t)h(t)]^2 = N J_{RMS}^2(\theta) \qquad (2.6)$$

where $w(t)$ is the weighting function. The standard deviation is represented by the root mean-square (RMS) error, $J_{RMS}(\theta)$. Here the Euclidean distance in the l_2 space is measured from $t = 1$, for a causal system, to $t = N$, the window size stretching to the steady-state. Equation (2.6) also means that, if the variance is regarded as the cost to minimise, the use of $J(\theta)$ will penalise the window size N in the same way as $J^2_{RMS}(\theta)$.

If lower order models are preferred and the cost of the order needs to be treated in the same degree as the error norm, for example, (2.6) may take the form of

$$J(\theta) = l\|w(t)h(t)\| \qquad (2.7)$$

where the range of k and l are constrained by (2.3). If the order number needs to be treated distinctively, it can form a second objective in addition to (2.6). The cost formulations can also be extended to represent continuous-time and multivariable systems. They are omitted here for simplicity. It is worth noting that linear metric equivalence implies control application oriented norms, such as the 'worst-case' ∞-norm, may be used instead, although quadratic errors or the l_2 norms are often used as a measure of fitting quality in model reduction applications. Note also that from (2.5)

$$w(t)h(t) = w(t)[G(z) - G_\theta(z)]u(t) = [G(z) - G_\theta(z)]w(t) \qquad (2.8)$$

This means that the weighting function $w(t)$ can, in effect, be used as the excitation function and thus the norm can be obtained directly from the response of the 'difference model' $[G(z) - G_\theta(z)]$ to this input.

2.2 Reduction based on a frequency domain cost

For some control engineering applications, the model reduction accuracy needs to be measured by a frequency gain discrepancy. Consider an original model as given by

$$G(s) = \frac{b_n s^n + b_{n-1}s^{n-1} + \dots + b_0}{s^m + a_{m-1}s^{m-1} + \dots + a_0} \qquad (2.9)$$

The transfer function representing a reduced model is

$$G_\theta(s) = \frac{d_k s^k + d_{k-1}s^{k-1} + \dots + d_0}{c_l s^l + c_{l-1}s^{l-1} + \dots + c_0} \qquad (2.10)$$

where $n < m$ for a causal system and k and l satisfy (2.3). In the frequency domain, an error transfer function may be defined by

$$H(s) = G(s) - G_\theta(s) \qquad (2.11)$$

For a SISO system, an order reduction problem is to find an optimal $G_\theta(s)$ such that it minimises a cost function, such as

$$J(\theta) = l\|W(j\omega_f)H(j\omega_f)\|^2 = l\sum_{f=1}^{N}[W(j\omega_f)H(j\omega_f)]^2 = lNJ^2_{RMS}(\theta) \quad (2.12)$$

Here $W(j\omega)$ is the frequency weighting function that allows the fitting errors in the chosen parts of the frequency range to be emphasised if needed. It can be seen from (2.11) and (2.12) that this function can also be used as a frequency excitation function directly.

Similarly, for a MIMO system, a weighted cost summing up all squared Euclidean error norms

$$J(\theta) = l \sum_p \sum_q J_{pq}^2(\theta) \tag{2.13}$$

can be used, where

$$J_{pq}(\theta) = \|W_{pq}(j\omega_f)H_{pq}(j\omega_f)\| \tag{2.14}$$

with $W_{pq}(j\omega)$ being a weighting element. Here $H_{pq}(j\omega)$ is an element of the error transfer function matrix, and $G_{\theta pq}(j\omega)$ terms contained in (2.11) are usually pre-set to share an identical denominator. The metric window size needs to cover the interested frequency range for the required reduction. Note that, instead of using the summation of Euclidean norms, the Frobenius or the H_∞ norm may also be used for MIMO systems.

3. Global Evolution and Inheritance with Local Learning

As discussed in the introduction, existing model reduction methods are based on conventional, directed optimisation techniques, which may not lead to globally optimised models. In this section, an evolutionary model reduction technique is developed to overcome tractability difficulties encountered in conventional methods. The most widely applied evolutionary algorithm is the genetic algorithm (GA), a coding version of EAs where the 'genetic codes' enable possible representation and adjustment of a system structure and parameters of the structure in the same evolution process. The coded 'chromosome' length and search structure can be freed up if a 'messy-GA' is used. If cellular encoding is used, the messy-GA will become 'genetic programming'. Note that, for engineering applications, coding using a finite string-length will limit resolutions or ranges of the real-valued system parameters, but the use of an additional 'control gene' in coding can provide a dimension for switching between possible system structures [10, 15]. If no coding is used, the mutation operation may be implemented by Monte Carlo perturbations [10, 18]. In such circumstances, a GA will become an 'evolution strategy', which additionally employs an adaptive mutation scheme [15]. This technique further shrinks to 'evolutionary programming', if no parameter recombination or internal adaptive model for crossover and mutation are used.

All the above evolutionary techniques are based on a multi-point a posteriori trial-and- error process, guided by the Darwinian-Wallace principle of 'survival-of-the-fittest'. The search power of an EA mainly lies in its crossover and mutation operators, which provide the diversity of candidate 'species' (i.e., candidates of reduced models) and provide varying search domains. However, it is well-known that pure EAs are weak in local exploration and are thus poor in finding exact optima [8, 10, 14, 16, 18]. The underlying reason of this is that, in an EA, there is a lack of 'biological diversity' resulting from interactions with, and thus direct learning from, the evolution environment, termed the 'Lamarckism inheritance' [1].

For improving the EA performance, it is intended to combine the 'generational' optimisation power of crossover of an evolving population with the Lamarckism 'inheritance' of learning individuals in each generation. 'Positive mutations' can result from learning also by trial-and-error, requiring no derivative information. Here Boltzmann type of learning is realised by simulated annealing (SA) [18] as shown in Figure 3.1, which asserts a probability of retaining possibly correct search directions [14].

An existing chromosome C_k (a potential parameter set) may be replaced by its mutant chromosome C with a probability given by the Boltzmann selection criterion

$$P(C \leftarrow C_k | C_k) = \min \left\{ \exp \left[-\frac{J(C) - J(C_k)}{k_B T} \right], 1 \right\} \qquad (3.1)$$

where C may be slightly inferior. Here k_B may be the Boltzmann's constant, but can be set to an artificial value in the annealing process. It is set to 5×10^{-6} in this chapter for a fine annealing decision, coupled with the 'annealing temperature' T, which decreases from T_{ini} exponentially at the rate of β^{j-1}. Here $\beta < 1$ is the annealing factor, inversely similar to the learning rate in gradient-guided recursive learning algorithms [11] and the integer $j \in [1, j_{max}]$ is the annealing cycle index. The decreasing temperature in Figure 3.1 implies that the learning mechanism will move close to hillclimbing when the generational evolution progresses. The final temperature T_{final} is determined by how tight the fine-tuning should be bounded at the end of the learning process. Here, a fast annealing scheme ($\beta = 30\%$) is used, because the major task of optimisation is undertaken by the global evolution. For a final learning tolerance given by $T_{final} = 1$ and an initial tolerance by $T_{ini} = 10^5$, the total number of annealing trials will be:

$$j_{max} = \lceil \log_\beta \frac{T_{final}}{T_{ini}} \rceil = 10 \qquad (3.2)$$

Note that the binary coding scheme in a normal GA uses one integer, and thus one memory block, to represent one binary code but not one parameter. This coding inefficiency can be dramatically improved by 'one-integer-one-parameter' coding [10]. Here, for example, a single-precision integer gene represents a parameter valued between -32768 and $+32767$. For a floating-point parameter, a decimal point is inserted in decoding, to give the required number of significant digits after the decimal point. If finer resolution coding is needed, long-integers can be used or a range coding scheme [15] may be

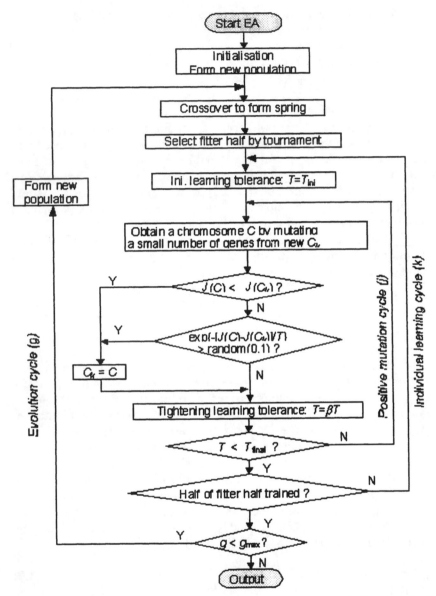

Fig. 3.1. Flow chart of an EA using 'positive mutations' with a Boltzmann learning schedule

added. The single or double precision integer coding scheme saves the chromosome length (and memory requirement) 16 or 8 times and thus also reduced program overhead substantially. It is also more efficient in overcoming the Hamming cliff effect than the Gray code.

For the majority of problems studied in this chapter, there are about 10 parameters to be optimised and thus the population size of the EA is chosen as 100. The SA realised mutation is applied at a probability rate of 10%, through Monte Carlo perturbations. Due to increased mutation activity, the crossover (or recombination) rate can be reduced in a similar way to the evolution strategy (or evolutionary programming) techniques [10, 14]. Here a randomly positioned single-point crossover is applied at a rate of 50% to the 'parent population'. In Figure 3.1, the tournament selection scheme is employed for a rapid reproduction, where two random chromosomes compete once for survival. Half of the winning 50% chromosomes resulting from the tournament will be further trained for Lamarckian heredity. The mature spring (25 in total) will be mixed with the winning individuals (50 in total) and 25% of the parents (25 in total) to form a new population. This population formation mechanism is similar to a 'steady-state GA', which reserves some possibly useful genes and creates a generation gap of 0.25. It helps to save the function evaluation time and tends to maintain a diversity of the species for global optima. Coupled with fast direct Lamarckian inheritance, off-line reduced models this way can be quickly refined on-line for adaptive control.

4. Model Reduction by Evolution and Learning

4.1 A time domain example

Consider the fourth-order discrete-time system studied by Warwick [19] and later by Xue and Atherton [20] as follows:

$$G(z) = \frac{0.3124z^3 - 0.5743z^2 + 0.3879z - 0.0889}{z^4 - 3.233z^3 + 3.9869z^2 - 2.2209z + 0.4723} \tag{4.1}$$

In a control engineering application, it was required to used a second-order causal model to approximate this discrete-time system. The learning evolutionary algorithm charted in Figure 3.1 has been applied here to solve this model reduction problem. The parameter sets in the initial population were generated randomly. The parameter evolution process for the reduced model is shown in Figure 4.1. The algorithm was coded in Turbo Pascal 7.0 and it took, in average, 2.5 minutes on a 50 MHz 80486 DX2 processor to complete 100 global evolutionary cycles coupled with local learning.

The RMS error and the parameters of the reduced system obtained by the learning EA are shown in Table 4.1. For comparison, the parameters of the reduced models and the resulting RMS errors obtained by Warwick [19] and Xue and Atherton [20] using conventional optimisation tools are also shown in Table 4.1. Note that the latter required a good initial starting point via transforming the discrete-time system to a continuous-time system before Powell's optimisation technique was employed [20]. It can be seen

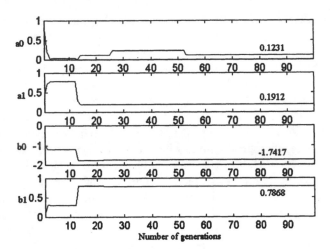

Fig. 4.1. Convergence traces of the parameter set with the lowest fitting error

Table 4.1. The parameter sets and RMS errors obtained by various methods for (4.1)

Parameters	Warwick (1984)	Xue and Atherton (1993)	Boltzmann Learning EA
c_0	0.3124	0.1299	0.1231
c_1	-0.0298	0.1820	0.1912
d_0	-1.7369	-1.7431	-1.7417
d_1	0.7773	0.7877	0.7868
$c_2 = 1, d_2 = 0$ (reducing to a causal second-order model is required here)			
RMS error	0.1884	0.0835	0.0822

that the proposed method performs much better than the 'error polynomial' approach of Warwick [19]. Compared with the RMS error obtained by Xue and Atherton [20], the learning EA offers a slightly (1.6%) improved fitting quality without the need of a priori starting points, although the use of such knowledge usually leads to a faster convergence [13]. The step responses of the original system and the evolved reduced model are plotted in Figure 4.2, which serves to validate the reduced model.

4.2 Frequency domain reduction of an aircraft model

The 8th-order transfer functions of a 1-input and 2-output continuous-time model of a fighter aircraft studied by Bacon and Schmidt [3] and later by Gong and Murray-Smith [5] are given below:

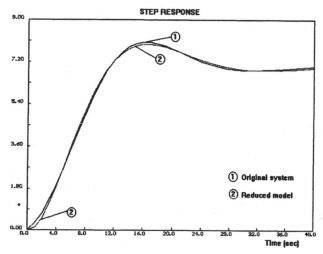

Fig. 4.2. Step responses of the original and the evolved lower-order models

$$\left[\begin{array}{c} \frac{q(s)}{\delta(s)} \frac{n_{zcr}(s)}{\delta(s)} \end{array} \right] = \Delta^{-1}(s)$$

$$\left[\begin{array}{c} 5.26s(s+0.0103)(s+0.5)(s+1.887)(s+13.986) \\ 1.34s(s+0.00066)(s+49.99)(s+0.5)(s+1.887)(s+13.986) \end{array} \right] \quad (4.2)$$

where

$$\Delta(s) = (s+0.418)(s+1.34)(s^2+0.00264s+0.006724)$$
$$(s^2+3.3916s+7.7284)(s^2+33.0576s+290.3611) \quad (4.3)$$

This example represents the longitudinal responses in terms of pitch rate $q(s)$ and the normal acceleration $n_{zcr}(s)$ to elevator stick force $\delta(s)$. It is noted that the dynamics in phugoid and lateral spiral modes are covered by the frequency range $[0.5, 15]$ rad s^{-1} and thus this range has been used in model reduction studies [5].

A reduced model here is required to be of the third-order in the application of a controller design. The hybrid EA has been applied to the original model to evolve reduced models for 10 minutes on a 100 MHz Pentium PC. The results are shown in the middle row of Table 4.2. To compare, the transfer functions and RMS errors of a reduced model obtained by Gong and Murray-Smith [5] using the complex curve-fitting technique are shown in the first row of Table 4.2. It can be derived that the global evolution and local learning provide an improved reduction quality by 7.4% with respect to (2.13).

The frequency responses of the original system, the reduced model by Gong and Murray-Smith [5] and the reduced model by evolution and learning are compared in Figure 4.3. Clearly the EA performs better than the

Table 4.2. Reduced-order transfer function and standard deviation matrices for the nonlinear system of (4.2)–(4.3)

Orders	Transfer function matrices	$[J_{RMS}(\theta)]$	$\sum\limits_{pq} J^2_{RMS}(\theta)$	RMS of $H(j\omega)$
3 (Gong and Murray-Smith 1993)	$\begin{bmatrix} \dfrac{0.18s^2 + 0.54s + 0.39}{s^3 + 2.96s^2 + 10.78s + 6.8} \\[2ex] \dfrac{0.2s^2 + 2.03s + 5.53}{s^3 + 2.96s^2 + 10.78s + 6.8} \end{bmatrix}$	$\begin{bmatrix} 0.0155 \\ 0.2427 \end{bmatrix}$	0.05914	$\begin{bmatrix} 32.5549 \\ 53.9169 \end{bmatrix}$
3 (fixed)	$\begin{bmatrix} \dfrac{0.15s^2 + 1.23s + 0.84}{s^3 + 5.7s^2 + 19.76s + 18.92} \\[2ex] \dfrac{2.49s + 13.61}{s^3 + 5.7s^2 + 19.76s + 18.92} \end{bmatrix}$	$\begin{bmatrix} 0.0144 \\ 0.2336 \end{bmatrix}$	0.05478	$\begin{bmatrix} 30.4882 \\ 10.0567 \end{bmatrix}$
1 (variable)	$\begin{bmatrix} \dfrac{5.77}{96.7s + 103} \\[2ex] \dfrac{79.4}{96.7s + 103} \end{bmatrix}$	$\begin{bmatrix} 0.0285 \\ 0.2355 \end{bmatrix}$	0.05627	$\begin{bmatrix} 39.5149 \\ 69.6141 \end{bmatrix}$

conventional optimiser. Note that, although the cost minimised was based on the magnitudes (and not phases) of error transfers, the phase discrepancy of the evolved model is also smaller than that of the conventionally reduced model in the interested frequency range. Note also that, if minimising phase discrepancy is required in an application, it can be easily incorporated in the cost function.

4.3 An open-order reduction example

The previous examples are based on fixed-order reductions. The coding version of an EA can, however, allow the order number to be encoded and thus to be optimised in the evolution process. This is done by the 'control gene' that acts as a 'structure switch'. It varies the number of the highest order of a candidate model and thus switches on and off the coefficients of the terms that have a higher order. Here the range of the order coded should satisfy (2.3). In a rapid simulation exercise as part of an integrated aircraft control system design, the order number is penalised in the same way as the total magnitude variance. Thus the order number is coded, as a control gene, in the parameter set for optimisation and is also included in the cost as shown in (2.13).

For the aircraft system of (4.2) and (4.3), the control gene is coded by an integer valued between 0 and 7 for reduction from 8. Suppose at one evolution stage its 'allele format' (value of the gene) is 2. Then this control gene will switch off the coefficients of the denominator terms of an order higher than or equal to 3. It will also switch off the coefficients of the numerator of order higher than or equal to 2. Running the learning EA to optimise the

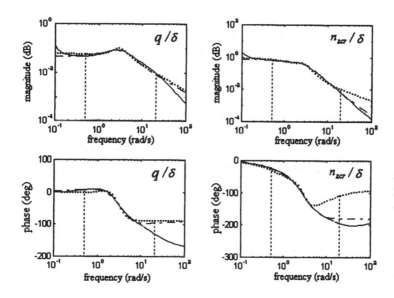

Fig. 4.3. Magnitude and phase plots. Original $G(s)$ { - }; Reduced order $G_\theta(s)$ {....} by Gong and Murray-Smith [5]; Reduced order $G_\theta(s)$ {-.-.} by the learning EA

system order and coefficients in the same process took 15 minutes (about 50% more than the previous fixed-order reduction), although the number of parameters being optimised were doubled. This indicates the advantage of an NP algorithm. For this variable-order reduction task, the evolution revealed that a first-order reduced model should be used and the reduced model is shown in the last row of Table 4.2. It can be seen that it offers a reduction quality almost as high as the fixed third-order model and better than that obtained by Gong and Murray-Smith [5]. It is not surprising to see that the EA did not recommend a second or higher-order model for the cost governed by (2.13) because of the high accuracy obtained by the first-order model.

4.4 Reduction of a process model from an 'infinite-order'

Evolutionary methods have also been applied to fitting physical nonlinear plants with differential equations structured by nonlinear dominant dynamics [18]. In the design of many modern control schemes, however, a simple linear model is often required. Similarly, EAs can be employed to reduce the equivalent 'order' of a nonlinear system, which is in effect an infinite-order linear system. The difference between a reduction from an infinite-order and that from a finite-order lies in the need of a set operating point of the nonlinear system. Around the operating point, data of $G(j\omega)$ in (2.11) may be obtained by 'small' perturbations. Note that information on the interested frequency range for reduction can be used to determine the 'clock' that generates pseudo random binary sequences (PRBS) such that the PRBS weighting input may act as additive excitations mainly covering the interested frequency points.

Now consider a liquid-level regulation system that simulates mass balance and heat balance dynamics widely found in chemical or dairy plants. The coupled nonlinear system is described by the state-space equations

$$
\begin{bmatrix} \dot{h}_1 \\ \dot{h}_2 \end{bmatrix} = \begin{bmatrix} -sgn(h_1-h_2)\frac{C_1 a_1}{A}\sqrt{2g|h_1-h2|} \\ sgn(h_1-h_2)\frac{C_1 a_1}{A}\sqrt{2g|h_1-h2|} - \frac{C_2 a_2}{A}\sqrt{2g|h_2-H_3|} \end{bmatrix}
$$
$$
+ \begin{bmatrix} \frac{1}{A} & 0 \\ 0 & \frac{1}{A} \end{bmatrix} \begin{bmatrix} q_1 \\ q_2 \end{bmatrix} \tag{4.4}
$$

where h_1 and h_2 are the liquid levels of Tanks 1 and 2, respectively; q_1 and q_2 input flow rates mapped from the pump voltages; $C_1 = 0.53$ and $C_2 = 0.63$ discharge constants; $a_1 = 0.396$ cm^2 and $a_2 = 0.385$ cm^2 orifice and connecting pipe cross-sectional areas; $A = 100$ cm^2 is the cross-sectional area of both tanks; and $g = 981$ cm/sec^2 the gravitational constant. There is a practical constraint imposed on this system by its physical structure, being $h_{1,2} \geq H_3 = 3$ cm, the minimum liquid level bounded by the height of the orifices.

For an application, the operating point of Tank 1 needs to be set at 10 cm and that of Tank 2 at 8 cm. This can be achieved by applying step input voltages to set the steady-state levels. Then additive PRBS signals are applied. In this work, the magnitude of the PRBS signals are at $\pm$ 5% of the set voltages. The evolutionary algorithm used to generate a reduced model with an optimally reduced order number. Giving the order trade-off in the form of (2.13) and (2.14), the EA tends to suggest a first-order reduced model as the overall choice. At the end of the evolution, the best model and its corresponding RMS errors are shown in the first row of Table 4.3. For a fixed-order linearisation problem, the results of a second-order linearised model are shown in the last row of Table 4.3 [18]. Note that the first-order model obtained by the joint order optimisation approach here offers an even smaller overall RMS error than the second-order model. This implies that the nonlinear system behaves more like a first-order system at the operating point than a second-order one. The benefit of such insight into the system behaviour and the flexibility of the reduced order offered by the EA based optimisation method cannot be matched by existing conventional model order reduction techniques.

5. Conclusion and Further Work

This chapter has developed a local tuning enhanced evolutionary approach to model order reduction. The hybrid search provides both global conver-

Table 4.3. Reduced-order transfer function and standard deviation matrices for the nonlinear system of (4.4)

Orders	Transfer function matrices	$[J_{RMS}(\theta)]$
1 (variable)	$\dfrac{1}{56.5s + 0.743}\begin{bmatrix} 0.955 & 0.612 \\ 0.633 & 0.668 \end{bmatrix}$	$\begin{bmatrix} 0.0833 & 0.0649 \\ 0.0638 & 0.0562 \end{bmatrix}$
2 (Tan et al, 1996)	$\Delta^{-1}(s)\begin{bmatrix} 4.1041s + 0.7688 & 0.4805 \\ 0.4945 & 5.3153s + 0.5305 \end{bmatrix}$ $\Delta(s) = s^2 + 49.05s + 0.5385$	$\begin{bmatrix} 0.1126 & 0.0641 \\ 0.0643 & 0.0475 \end{bmatrix}$

gence by the EA and individual learning by the Boltzmann selection. The Lamarckism provides fast generational inheritance and learning offers fine-tuning. Both techniques are important for on-line applications, such as in model-reference adaptive control and in self-tuning control, where model refinements and controller redesigns need to be carried out in real-time.

The EA based reduction method is generic and applicable to both discrete and continuous- time systems by minimising square errors between the original and reduced models. It is applicable to both SISO and MIMO systems in both time and frequency domains. The model reduction examples show that this indirectly guided optimising method provides a superior performance to that of existing methods and offers optimised orders in the same process of parameter fitting.

Currently, the enhanced evolutionary method is further developed at Glasgow, to incorporate an adaptive feature in the training mechanism for more efficient learning and for on-line adaptation. Such methods are also being applied to structural identification in physical parametric modelling of nonlinear engineering systems. Further results will be reported in due course.

Acknowledgements

The EPSRC grant awarded to the first author ('Evolutionary programming for nonlinear control', GR/K24987) is gratefully acknowledged. The second and third authors would like to thank University of Glasgow, the CVCP and the EPSRC for financial support.

References

1. Aboitiz, F. 1992. 'Mechanisms of adaptive evolution - Darwinism and Lamarckism restated', *Medical Hypotheses*, 38 (3), 194-202.
2. Anderson, J. H., 1967. 'Geometric approach to the reduction of dynamical systems', *IEE Proc. Control Theory Appl.*, Pt. D, 114(7), 1014-1018.

3. Bacon, B. J., and Schmidt, D. K., 1988. 'Fundamental approach to equivalent system analysis', J. Guidance, Control and Dynamics, 11 (6), 527-534.
4. Elliott, H., and Wolovich, W. A., 1980. 'A frequency domain model reduction procedure', Automatica, 16, 167-177.
5. Gong, M. R., and Murray-Smith, D. J., 1993. 'Model reduction by an extended complex curve-fitting approach', Trans. Inst. Measurement and Control, 15 (4), 188-198.
6. Gray, G. J., Li, Y. Li, Murray-Smith, D. J., and Sharman, K. C., 1996. 'Structural system identification using genetic programming and a block diagram oriented simulation tool', Electron. Lett., 32(15), 1422-1424.
7. Kristinsson, K., and Dumont G.A., 1992. 'System identification and control using genetic algorithms', IEEE Trans. Syst., Man and Cyber., 22 (5), 1033-1046.
8. Kwong, S., Ng, A. C. L., and Man K. F., 1995. 'Improving local search in genetic algorithms for numerical global optimization using modified GRID-point search technique', Proc 1st IEE/IEEE Int. Conf. on Genetic Algorithms in Eng. Syst.: Innovations and Appl., Sheffield, UK, 419-423.
9. Levy, E. C., 1959. 'Complex-curve fitting', IRE Trans. on Automatic Control, 4, 37-44.
10. Li, Y., 1996. Neural and Evolutionary Computing IV, Fourth Year Course Notes (1JYX), Department of Electronics and Electrical Engineering, University of Glasgow.
11. Li, Y., and Häußler, A., 1996. 'Artificial evolution of neural networks and its application to feedback control', Artificial Intelligence in Engineering, 10(2), 143-152.
12. Li, Y., and Ng, K. C., 1996. 'A uniform approach to model-based fuzzy control system design and structural optimisation', in Genetic Algorithms and Soft Computing, F. Herrera and J. L. Verdegay (Eds.), Physica-Verlag Series on Studies in Fuzziness, Vol. 8, 129-151.
13. Li, Y., Ng, K. C., Murray-Smith, D. J., Gray, G. J., and Sharman K. C., 1996. 'Genetic algorithm automated approach to design of sliding mode control systems', Int. J. Control, 63(4), 721-739.
14. Michalewicz, Z., 1994. Genetic Algorithms + Data Structures = Evolution Programs, Springer-Verlag, Berlin, 2nd Ed.
15. Ng, K. C., 1995. Switching control systems and their design automation via genetic algorithms, Ph.D. Thesis, Department of Electronics and Electrical Engineering, University of Glasgow.
16. Renders, J. M., and Bersini, H., 1994. 'Hybridizing genetic algorithms with hill-climbing methods for global optimisation: two possible ways', Proc 1st IEEE Int. Conf. Evolutionary Computation, First IEEE World Cong. Computational Intelligence, Orlando, 1, 312-317.
17. Söderström, T., and P. Stoica (1989). System Identification. Prentice-Hall International, London.
18. Tan, K. C., Gong, M., and Li, Y., 1996. 'Evolutionary linearisation in the frequency domain', Electron. Lett., 32(1), 74-76.
19. Warwick, K., 1984. 'A new approach to reduced-order modeling', Proc. IEE Control Theory and Applications, Pt D, 131 (2), 74-78.
20. Xue, D., and Atherton D. P., 1993. 'An optimal reduction algorithm for discrete-time systems', Proc. IFAC 12th Triennial World Congress, Sydney, Australia, 821-824.

Adaptive Recursive Filtering Using Evolutionary Algorithms

Michael S. White and Stuart J. Flockton

Department of Physics, Royal Holloway, University of London, Egham, Surrey
TW20 0EX, United Kingdom

Summary. Adaptive digital filters have been used for several decades to model
systems whose properties are *a priori* unknown. Pole-zero modeling using an output
error criterion involves finding an optimum point on a (potentially) multimodal
error surface, a problem for which there is no entirely satisfactory solution. In this
chapter we discuss previous work on the application of genetic-type algorithms to
this task and describe our own work developing an evolutionary algorithm suited
to the particular problem.

1. Signal Processing and Adaptive Filters

Digital Signal Processing (DSP) is used to transform and analyze data and
signals that are either inherently discrete or have been sampled from analogue
sources. With the availability of cheap but powerful general-purpose computers and custom-designed DSP chips, digital signal processing has come to have
a great impact on many different disciplines from electronic and mechanical
engineering to economics and meteorology. In the field of biomedical engineering, for example, digital filters are used to remove unwanted 'noise' from
electrocardiograms (ECG) while in the area of consumer electronics DSP
techniques have revolutionized the recording and playback of audio material
with the introduction of compact disk and digital audio tape technology. The
design of a conventional digital signal processor, or *filter*, requires *a priori*
knowledge of the statistics of the data to be processed. When this information is inadequate or when the statistical characteristics of the input data are
known to change with time, *adaptive filters* [1, 22] are employed.

Adaptive filters are employed in a great many areas of telecommunica-
tions for such purposes as adaptive equalization, echo cancellation, speech
and image encoding, and noise and interference reduction. Adaptive filters
have the property of self-optimization. They consist, primarily, of a time-
varying filter, characterized by a set of adjustable coefficients and a recursive
algorithm which updates these coefficients as further information concerning
the statistics of the relevant signals is acquired.

A desired response $d(n)$, related in some way to the input signal, is made
available to the adaptive filter. The characteristics of the adaptive filter are
then modified so that its output $\hat{y}(n)$, resembles $d(n)$ as closely as possible.
The difference between the desired and adaptive filter responses is termed
the *error* and is defined as:

$$e(n) = \hat{y}(n) - d(n) \qquad (1.1)$$

Ideally, the adaptive process becomes one of driving the error, $e(n)$ towards zero. In practice, however, this may not always be possible and so an optimization criterion, such as the mean square error or some other measure of fitness, is employed.

Adaptive filters may be divided into *recursive* and *non-recursive* categories depending on their inclusion of a feedback path. The response of non-recursive, or finite impulse-response (FIR) filters is dependent upon only a finite number of previous values of the input signal. Recursive, or infinite impulse-response (IIR) filters, however, have a response which depends upon *all* previous input values, the output being calculated using not only a finite number of previous input values directly, but also one or more previous output values. Many real-world transfer functions require much more verbose descriptions in FIR than in recursive form. The potentially greater computational efficiency of recursive filters over their non-recursive counterparts is, however, tempered by several shortcomings, the most important of which are that the filter is potentially unstable and that there are no wholly satisfactory adaptation algorithms.

There are two main types of adaptive IIR filtering algorithms [17], which differ in the formulation of the *prediction error* used to assess the appropriateness of the current coefficient set during adaptation.

In the *equation-error* approach the error is a linear function of the coefficients. Consequently the mean square error is a quadratic function of the coefficients and has a single global minimum and no local minima. This means that simple gradient-based algorithms can be used for adaptation. However in the presence of noise (which is present in all real problems) equation-error-based algorithms converge to biased estimates of the filter coefficients [17].

The second approach, the *output-error* formulation, adjusts the coefficients of the time-varying digital filter directly in recursive form. The response of an output-error IIR filter is characterized by the *recursive* difference equation:

$$\hat{y}(n) = \sum_{m=1}^{N-1} a_m(n)\hat{y}(n-m) + \sum_{m=0}^{N-1} b_m(n)x(n-m) \qquad (1.2)$$

which depends not only upon delayed samples of the input $x(n)$, but also upon past output samples, $\hat{y}(n-m), m = 1, \ldots, N-1$. The output, $\hat{y}(n)$ is a non-linear function of the coefficients, since the delayed output signals themselves depend upon previous coefficient values. Consequently, the mean square error is *not* a quadratic function of the feedback coefficients (the a's), though it is of the b's, and may have multiple local minima. Adaptive algorithms based on gradient-search methods, such as the widely used LMS may converge to sub-optimal estimates of the filter coefficients if initialized within the basin of attraction of one of these local minima.

Evolutionary algorithms offer a search method which can be resistant to being trapped in local minima, so they provide a possible avenue for the successful adaptation of output-error based recursive digital filters. The rest of this article describes the results of exploring some avenues in this area.

2. System Identification

Many problems in the areas of adaptive control and signal processing can be reduced to system identification. Modeling of dispersive communication channels, the analysis of geophysical data and digital filter synthesis are just a few representative examples of the application of system identification techniques. In this task, an adaptive filter is used to model the behavior of a physical dynamic system. Generally, the nature of the system (or *plant*, using control systems terminology) is unknown and thus it may be regarded as a "black box" having one or more inputs and one or more outputs. Modeling of a single-input, single-output system is illustrated in figure 2.1. In this con-

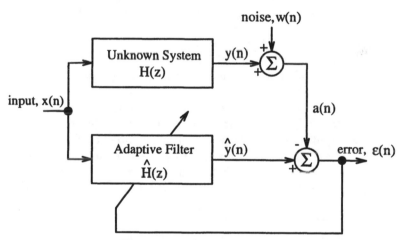

Fig. 2.1. System identification

figuration, the adaptive filter is allowed to adjust itself in order to match its output $\hat{y}(n)$ to that of the unknown system. Many physical systems, however, contain noise. Internal plant noise is generally uncorrelated with the input to the system and is commonly represented at the plant output as additive noise. If the adaptive filter has sufficient degrees of freedom, the input-output response characteristics of the adaptive model will converge to match closely those of the unknown system.

3. Previous and Related Work

The first application of evolutionary optimization techniques to the task of recursive filter adaptation is [3] in which Etter, Hicks and Cho describe an adaptive algorithm based on a simple, bit-string genetic algorithm. Two alternative approaches to the task of system identification based on 'learning algorithms' are discussed in [14] in which Nambiar, Tang and Mars investigated the performance of both Stochastic Learning Automata and GA-based adaptive algorithms. The GA-approach to system identification described used a simple bit-string genetic algorithm to optimize the coefficients of a parallel form filter, each coefficient being quantized to four bits.

In [10], Kristinsson and Dumont describe the implementation of a binary genetic algorithm as an estimator for both discrete and continuous time systems. Their GA employed stochastic selection with a replacement reproductive algorithm and linear ranking was introduced whenever the the percentage involvement of the population (the number of individuals receiving offspring) declined significantly. Crossover was single-point and the mutation operator randomly flipped the state of a bit from 0 to 1 and vice versa. The genetic encoding differed from the straightforward quantization of the direct-form coefficients employed by Etter et al. in that it made use of the fact that the roots of a rational system function are either real or occur in complex conjugate pairs. Stability was maintained by ensuring that poles were only updated within the unit circle.

Patiakin et al. [16] have developed a further technique for IIR adaptive filtering which they term *Darwinian Design*. Based upon the principles of evolutionary design put forward by Dawkins [2], their guided random cumulative-selection procedure has been primarily applied to an adaptive notch filtering scheme for interference cancellation. Starting from a randomly generated initial population, the "filtomorphs" were evaluated and ranked according to cost (proportional to the error). Starting from the best performing pair of candidate filters, offspring were produced by randomly choosing new parameter values (coefficients, gain, delay etc.) within the range delimited by the parameters of the two parent structures. The next two parents in the ordered cost list of filters were allowed to produce a reduced number of offspring. This reproductive process continued until a whole new generation of filters had been produced. The Darwinian adaptation scheme outlined was shown, through simulation experiments, to be broadly competitive with recursive least squares (RLS) when adapting low-order filters, provided the parameter ranges were strictly limited. However the large variance in the results produced by this stochastic adaptive algorithm, sometimes resulting in very poor parameter estimates, was highlighted as an area for future refinement.

Flockton and White [5] reported initial experiments showing the ability of a binary string based GA implementation to converge to the global optimum on a bimodal error surface. They extended this to lattice implementations of higher order filters in [6] and [7]. Comparisons of the performance of a carefully tuned binary encoded GA approach with the floating point GA of Michalewicz and Attia [12] and the Adaptive Simulated Annealing (ASA) algorithm developed by Ingber and Rosen [8] were reported in [19] and generally showed the binary encoded GA to be inferior. Extensive investigations of the use of evolutionary algorithms for the adaptation of recursive lattice filters for a variety of uses are described in [18], which also contains a substantial review of associated work.

4. Floating-Point Representation for the GA

The observations described in [19] as well as empirical evidence such as [9, 11] led to the decision to concentrate on floating point GAs for the problem. This choice raises a number of issues. A great many genetic operators have been developed for use with non-binary representations and the choice of operators to implement is far from straightforward; mutation and crossover schemes for optimization in the continuous domain are particularly numerous.

Possible choices for mutation operators include uniformly-distributed and normally-distributed mutation [13], dynamic mutation (which starts off searching the parameter space uniformly and gradually increases the probability of generating new parameters close to the old value) [9], fixed and variable step-size mutations, etc. The operator adopted in the work described here is the Breeder Genetic Algorithm (BGA) mutation scheme [13] in which one of the 32 points $\pm(2^{-15}A, 2^{-14}A, \ldots, 2^0A)$ (where A defines the mutation range and is, in these simulations, set to $0.1 \times$ coefficient range) is chosen randomly each time. The normalized expected progress of the BGA mutation scheme is of the same order as that of the uniform and normally distributed mutation operators with optimally adapted mutation ranges. Since the information required to optimally adjust the mutation range, the distance from the global optimum, is generally unknown, the mutation range has to be adapted empirically. The expected progress (improvement of the individual due to successive mutations) of non-optimized uniform and normally-distributed mutation operators goes to zero as the individual approaches the optimum. BGA mutation, however, does not require adaptation of the mutation range because it generates points more often in the neighborhood of the given point in parameter space.

Many floating-point crossover/recombination strategies are also possible. They range from discrete recombination operators which interchange whole parameters in a manner analogous to bit-string multi-point and uniform crossover, to operators which take the arithmetic mean of two or more sets of coefficient values. In the absence of any clear guidance in the literature and in the light of results of small-scale comparisons undertaken by the authors which highlighted no clear winner, a simple two-point crossover operator was implemented. For each application of the crossover operator, two parent filter structures were selected at random and identical copies generated. Two cut-points were randomly selected and coefficients lying between these limits were swapped between the offspring. The newly generated lattice filters were then inserted into the population replacing the two parent structures.

The choice of problem representation and operator construction also has a bearing on the issue of operator application rates. Since the mutation operator is generally not very disruptive, quite high mutation levels are possible without significantly degrading the performance of the genetic algorithm; a mutation rate which, on average, resulted in a single mutation event in every filter structure was observed to provide reasonable performance. For the floating-point crossover operator, it was discovered that the GA was largely insensitive to changes in the application rate so long as some crossover was occurring.

A further question that arose was whether the ordering of the filter coefficients in the genetic representation of the lattice structure was significant. Ideally, lattice coefficients which together engender low mean squared error values should be located close together in the genetic representation of the filter so that they are less likely to be disrupted by crossover. Determining which coefficients to associate, though, is by no means a trivial matter since it requires the identification of those coefficients which contribute most beneficially to the response of the lattice filter. In a separate experiment to try to estimate the importance of coefficient ordering, we compared the fraction of successful crossovers generated by different coefficient orderings on several test problems and found that there was no discernible difference between them.

5. Performance Comparison

The final portion of this chapter compares and contrasts the performance of the GA-based adaptive algorithm with conventional recursive-filter optimization techniques. The floating-point genetic adaptive algorithm and the adaptive lattice algorithm of Parikh et al. [15] were investigated using a test suite designed to highlight particular strengths and weaknesses. Since the mode of operation of the GA-based adaptive algorithm and the gradient lattice is vastly dissimilar, a direct comparison between the two approaches based on a single parameter, such as the number of time samples processed, is not very informative. The central motivation for this comparative study was to gain an understanding of the classes of system identification problems to which the individual algorithms are best suited. In particular, it was hoped that the genetic adaptive algorithms would be able to locate good models of the 'unknown' system when the error surface was multimodal as this is an obvious weakness of gradient-based output-error adaptive algorithms. This is not to say that the adaptive lattice and other related algorithms will always fail on such error surfaces, but that they have the potential to do so.

5.1 The test suite

Six system identification problems made up the test suite used for this comparative study. Several of the problems were drawn from the system identification literature, but others were specially created for this work. The examples were selected so as to include problems with the following characteristics:

- unimodal/multi-modal
- no noise/noisy
- low-dimensionality/high-dimensionality

The problems are detailed in the following section.

5.1.1 Test problem 1: Low-dimensional (3), unimodal, no noise. The
first system identification problem was taken from [4]. This example represents the simplest situation considered by Fan and Jenkins, a problem where the adaptive filter is of sufficient order with white noise excitation. The transfer functions of the plant, $H_p(z)$, and adaptive filter, $H_a(z)$, are given by:

$$H_p(z) = \frac{1}{1 - 1.2z^{-1} + 0.6z^{-2}} \tag{5.1}$$

$$H_a(z) = \frac{b_0}{1 + a_1 z^{-1} + a_2 z^{-2}} \tag{5.2}$$

The error surface of this example has a single, global minimum at $(b_0, a_1, a_2) = (1.0, -1.2, 0.6)$, corresponding to zero normalized mean squared error. The input $x(n)$ was a unit variance white Gaussian pseudo-noise sequence.

5.1.2 Test problem 2: Low-dimensional (3), bimodal, no noise. This
problem is the most complex and practical of the four categories investigated by Fan and Jenkins [4]. In this example a reduced-order adaptive filter was used to identify the third-order plant given by:

$$H_p(z) = \frac{1}{(1 - 0.6z^{-3})^3} \tag{5.3}$$

$$H_a(z) = \frac{b_0}{1 + a_1 z^{-1} + a_2 z^{-2}} \tag{5.4}$$

Colored noise inputs were obtained by filtering a white Gaussian pseudo-noise sequence with a finite impulse response filter having the transfer function:

$$H_c(z) = (1 - 0.6z^{-1})^2(1 + 0.6z^{-1})^2 \tag{5.5}$$

The error surface is bimodal, having a global minimum with normalized mean squared error better than -12 dB.

5.1.3 Test problem 3: Low-dimensional (3), bimodal, SNR = 20 dB.
The third test problem was also taken from [4]. This combination of plant
and adaptive filters is an example of the second class of system identification
problems considered by Fan and Jenkins: sufficient order with colored noise
excitation. The transfer functions $H_p(z)$ and $H_a(z)$ are given by:

$$H_p(z) = \frac{1}{1 - 1.4z^{-1} + 0.49z^{-2}} \qquad (5.6)$$

$$H_a(z) = \frac{b_0}{1 + a_1 z^{-1} + a_2 z^{-2}} \qquad (5.7)$$

Colored noise inputs were obtained by filtering a white Gaussian pseudo-noise
sequence with an FIR filter which has a non-zero frequency response over the
entire usable band. The transfer function of the coloring filter, $H_c(z)$ was
given by:

$$H_c(z) = (1 - 0.7z^{-1})^2 (1 + 0.7z^{-1})^2 \qquad (5.8)$$

The error surface of this system identification example is bimodal. The local
minimum at $(b_0, a_1, a_2) = (1.0, -1.4, -0.49)$ has a normalized mean squared
error of 0.9475, and the global minimum at $(1.0, -1.4, 0.49)$ has normalized
mean squared error value zero . Zero mean Gaussian pseudo-noise was added
to the channel output to produce a signal-to-noise ratio of 20 dB.

5.1.4 Test problem 4: High-dimensional (11), unimodal, no noise.
The plant in this test problem was a fifth-order, low-pass Butterworth filter,
with transfer function:

$$H_p(z) = \frac{0.1084 + 0.5419z^{-1} + 1.0837z^{-2} + 1.0837z^{-3} + 0.5419z^{-4} + 0.1084z^{-4}}{1 - 0.9853z^{-1} + 0.9738z^{-2} + 0.3864z^{-3} + 0.1112z^{-4} + 0.0113z^{-5}}$$

$$(5.9)$$

The adaptive filter was of sufficient order producing an error surface with a
single optimum giving zero normalized mean square error.

5.1.5 Test problem 5: High-dimensional (11), bimodal, no noise.
The plant was a sixth-order pole-zero filter with transfer function:

$$H_p(z) = \frac{1.0 - 1.8z^{-2} + 1.04z^{-4} + 0.05z^{-2} + 0.192z^{-6}}{1 - 0.8z^{-2} - 0.17z^{-4} - 0.56z^{-6}} \qquad (5.10)$$

and the adaptive filter was fifth-order. The error surface is bimodal with a
global minimum below the -18 dB NMSE level.

**5.1.6 Test problem 6: High-dimensional (11), bimodal, SNR =
20 dB.** The plant was also a sixth-order filter and had the transfer function:

$$H_p(z) = \frac{1 - 0.4z^{-2} - 0.65z^{-4} + 0.26z^{-6}}{1 - 0.77z^{-2} - 0.8498z^{-4} + 0.6486z^{-6}} \qquad (5.11)$$

The adaptive filter was fifth order. The error surface is bimodal with a non-
zero global minimum. Zero mean Gaussian pseudo-noise was added to the
channel output to give a signal-to-noise ratio of 20 dB.

5.2 The algorithms

5.2.1 Adaptive lattice algorithm. In [15], Parikh et al. describe an adaptive algorithm for recursive filters which are implemented as lattice structures. Just as with the genetic algorithm approach, this adaptive algorithm ensures that stability is maintained throughout the adaptation process. The update equations for the adaptive filter coefficients are computed by using the method of steepest descent to minimize the instantaneous mean squared error with respect to the lattice filter coefficients, κ_i and ν_i.

5.2.2 Genetic adaptive algorithm. The GA-based adaptive algorithm used in these experiments has been described in some detail in the preceding sections. The lattice filter coefficients of each model were represented as floating-point numbers and were manipulated using the two-point crossover and floating-point mutation operator described above. Simple fitness proportional selection coupled with a logarithmic objective function was used and the best model from each generation was always passed on to the next (elitism). The three low-dimensional system identification problems in the test suite were tackled using a population of size 50 whilst the fifth-order problems required a population size of 800. Crossover and mutation rates were set at 0.6 and 0.01, respectively. The input to both the plant and adaptive filters was a Gaussian pseudo-noise sequence, one hundred samples in duration. The same input sequence was used for each adaptive filter evaluation.

5.3 Results obtained with the gradient lattice algorithm

Figure 5.1 presents the results obtained with the gradient lattice algorithm on the system identification test suite. Each graph shows the normalized mean squared error, obtained by averaging over 50 independent runs, plotted against the number of time samples. The low-pass filtering coefficient, ρ was set to 0.4 for all of the experiments and the step-size parameter μ was adjusted for each problem so that convergence (in the sense of cessation of significant updates) had been achieved within less than 10^5 samples. In all simulations, the algorithm was started from a point on the error surface corresponding to an adaptive filter with all coefficients set to zero (the origin of the coefficient space).

The mean squared error surface of the first system identification problem (figure 5.1a) is unimodal and the gradient lattice algorithm therefore experienced very little trouble locating the global minimum. With a step size of 0.01 the algorithm converged to the optimal coefficient values ($a_1 = -0.75$, $a_2 = 0.6$, $b_0 = 1.0$) in approximately 3000 samples. Figure 5.1b shows the results obtained with the gradient lattice algorithm on the second system identification problem. The error surface of this example is bimodal and the gradient-based algorithm clearly failed to converge to anything like the optimal coefficient values. The effect of additive noise on the performance of

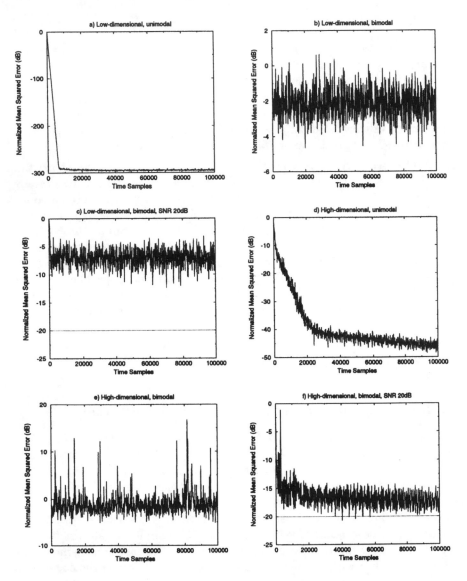

Fig. 5.1. System identification using the gradient lattice algorithm

the gradient lattice algorithm can be observed for a particular test case in figure 5.1c. In this graph, and in figure 5.1f, the noise floor is represented by a dotted line at the -20 dB level. The error surface of the third problem is bimodal and again the algorithm failed to locate the global minimum when started from the origin in coefficient space. Instead the normalized mean squared error fluctuated around a value of approximately -7 dB which is far above the noise floor, indicating that the algorithm had converged to the local minimum.

The results of the higher-order system identification runs are presented in figure 5.1d through to figure 5.1f. The gradient lattice algorithm had no problem identifying the coefficients of the fifth-order Butterworth filter (figure 5.1d). However as the coefficient space is higher-dimensional (11, as opposed to 3), the rate of convergence was much slower than in the first experiment. Figure 5.1e and figure 5.1f present the results of optimization runs on the two higher-order multi-modal error surface examples (test problems 5 and 6). In the first case, the algorithm converged to a normalized mean squared error value of only about -2 dB, indicating that the gradient lattice algorithm had totally failed to locate the global optimum. The final test case, figure 5.1f shows the algorithm converging to a mean NMSE just above the noise floor. Since the problem possesses a bimodal error surface, this indicates that, in this particular test problem, the origin of coefficient space is within the basin of attraction of the global minimum, unlike test problem 5 where the origin was in the basin of attraction of a local minimum. It is the fact that in real problems there is rarely sufficient *a priori* knowledge to indicate: (a) whether an error surface is significantly multimodal; and (b) whether the origin or any other arbitrary starting point in coefficient space is in the basin of attraction of the global minimum; that is the biggest weakness of this class of algorithm.

5.4 Results obtained with the genetic adaptive algorithm

The results of the system identification simulations using the floating-point genetic adaptive algorithm are presented in figure 5.2. Each graph shows the normalized mean squared error of the best model in the population (averaged over 50 independent runs of the algorithm) plotted against generation number. The dotted lines in figure 5.2c and figure 5.2f portray the noise floor at the -20 dB level. The first graph shows the performance of the genetic adaptive algorithm on a low-order, unimodal error surface. The normalized mean squared error of the best model was observed to decrease steadily throughout the course of the run. Since the mutation operator facilitated a degree of fine tuning of the coefficient values, progress near to the global minimum did not slow down down significantly. After 1000 generations had elapsed, the best model in the population returned a NMSE value of approximately -81 dB.

Figure 5.2b and figure 5.2c show the results of the GA-based adaptive algorithm on low-dimensional bimodal error surfaces. In the first case (fig-

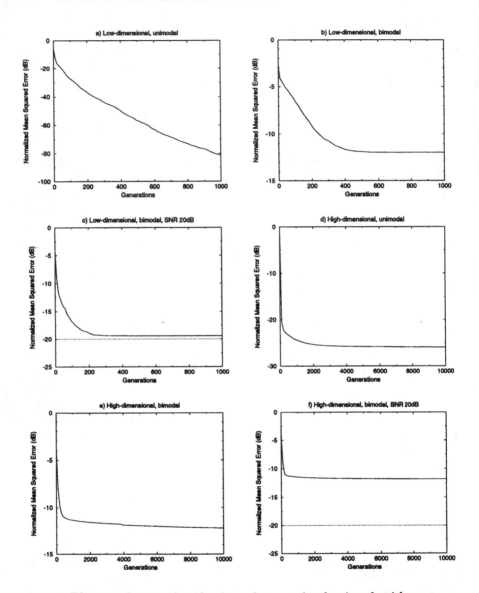

Fig. 5.2. System identification using genetic adaptive algorithm

ure 5.2b) where the adaptive filters were of insufficient order, producing an error surface with a minimum attainable normalized mean squared error of approximately -12 dB, the genetic algorithm required, on average, some four hundred generations to arrive near this optimal value. In all 50 experimental runs on this performance surface, the GA managed to locate the globally optimal solution. Figure 5.2c illustrates the performance of the adaptive algorithm in the presence of measurement noise. The best model produced by the genetic adaptive algorithm quickly approached the noise floor and achieved a normalized mean squared error of -19.4 dB within 300 generations. The slight bias from the -20 dB level can be attributed to slight fluctuations in the noise floor from run to run.

Some higher-order examples appeared to pose greater problems to the genetic algorithm. Despite possessing a unimodal error surface, problem 4 (figure 5.2d) consistently proved difficult for the genetic adaptive algorithm, resulting in a final mean normalized mean square error of only -26 dB. The variance in the best solutions generated for this particular problem was quite large, the worst model produced at the end of a run managed a rather poor -13.3 dB whereas the best a figure of -40.6 dB. In each case, although perfect identification was theoretically possible, the genetic algorithm had a tendency to cease significant improvement after the first few thousand generations had been processed.

Figure 5.2e illustrates the much improved performance of the GA over the adaptive lattice algorithm on problem 5. The genetic adaptive algorithm quickly located a model able to produce a NMSE below -11 dB and continued to make slight improvements over the course of the run. Since this system (and that of the next problem) is undermodeled, the error surface has a non-zero global minimum. The best run produced a normalized mean squared error of -18.6 dB and the worst corresponded to just -5.9 dB. The results for the final test case are shown in figure 5.2e. The average NMSE generated by the genetic adaptive algorithm was just -11.8 dB with best and worst values of -19.8 dB and -3.1 dB. respectively. On only two of the fifty runs did the normalized mean squared error fall to the level of the noise floor so on this particular example the GA was inferior to the gradient algorithm.

6. Conclusions

On the unimodal examples the gradient lattice algorithm won hands-down, quickly locating the optimal filter coefficients which corresponded to the best models of the 'unknown' system. The genetic algorithm was also able to reduce the NMSE very significantly on the low-order, unimodal problem but experienced more difficulty on the fifth-order example despite the larger population and longer simulation duration. Although the shape of the curve in figure 5.2d closely mirrors the results obtained by the adaptive lattice algorithm on the bimodal error surfaces, the root cause is different in each. The

lack of progress by the gradient-based adaptive algorithm can only be due to
the search becoming trapped in a local minimum (or at the very least on a
very flat region of the error surface), but the genetic adaptive algorithm may
become stalled at any point in the search space and not just at local minima. The primary reason for this 'premature convergence' in an algorithm of
the type described here is a severe loss of diversity in the population. When
the population becomes largely homogeneous, the effectiveness of crossover
is severely curtailed and mutation becomes the sole method of exploration.
The problem can be addressed by including within the algorithm schemes
to ensure that population diversity is maintained, but one difficulty of doing this is that the extent of population diversity that is necessary is itself
problem-dependent, meaning that further arbitrary choices have to be made
in the input parameters of the algorithm.

The major weakness of the gradient-based approach is clearly revealed by
the results of simulation experiments on the multimodal error surfaces. In all
but one case the gradient lattice algorithm failed to converge to the optimal
filter coefficients. Starting the adaptive filter from points within the basin
of attraction of the global optimum will of course ensure convergence to the
desired solution, but this is in general not practical. Even if the approximate
location of the global optimum were known in advance, it would not in general
be easy to be sure whether any particular point was in the basin of attraction
of the global minimum, and since the normal reason for using an adaptive
filter is that the desired solution is not known in advance the problem becomes
even more intractable.

In contrast to the behavior of the gradient-based approach, on the low-
order bimodal problems the GA, with its random initialization step, was
able to produce models of the 'unknown' system which were close to optimal
within a few hundred generations (this corresponds to the processing of between 1 and 2 million time samples). Of all the problem classes considered in
this simulation study, the low-dimensional, bimodal examples illustrated the
clearest advantage for the genetic adaptive algorithm over the gradient-based
search technique. As the search space was increased in size, however, the GA
required a vastly increased number of computations and produced a poorer
overall solution.

The performance of an evolutionary algorithm on a particular problem
depends upon a number of inter-related factors. Aside from stochastic differences which affect the solution quality from run to run, the defining parameters have a major role to play in shaping the search dynamics. The quality
of the initial sampling of the performance surface and, hence, the behavior
of the GA during the early stages of the search, for example, is partly controlled by the relative sizes of the population and search space. The trade off
between exploration and exploitation, on the other hand, depends upon the
type and application rates of the genetic operators. Unfortunately, no single
set of GA parameters provides excellent performance over a broad range of

problems. Whilst tuning of the controlling parameters invariably produces improved performance on the problem under investigation, these parameters may not provide the best results when the problem size is scaled up or down. However, for the floating-point GA, which was able to outperform even highly tuned bit-string variants, we were able to find parameter sets which provided reasonably consistent performance over a range of filter sizes.

In summary, GA based approaches offer a worthwhile alternative adaptation scheme for recursive adaptive filters when it is likely that the error surface to be explored is multimodal. However as the order of the recursive filter increases the performance of the GA method falls rapidly; in other words the GA approach does not seem to offer the hoped-for scalability into a very high-dimensional search space. This is almost certainly due to the very complicated shape of the error surface which would demand an excessively large population to characterize it adequately. Hence GAs do not guarantee good solutions but do provide a non-traditional type of adaptive algorithm capable of providing better performance than traditional approaches on some realistic problems.

References

1. Colin F. N. Cowan and Peter M. Grant, editors. *Adaptive Filters*. Englewood Cliffs, NJ: Prentice-Hall Inc., 1985.
2. Richard Dawkins. *The Blind Watchmaker*. Penguin Books, 1991.
3. D. M. Etter, M. J. Hicks, and K. H. Cho. Recursive filter design using an adaptive genetic algorithm. In *Proceedings of the IEEE International Conference on Acoustics, Speech and Signal Processing (ICASSP 82)*, volume 2, pages 635–638. IEEE, 1982.
4. Hong Fan and W. Kenneth Jenkins. A new adaptive IIR filter. *IEEE Transactions on Circuits and Systems*, CAS-33(10):939–947, October 1986.
5. S.J. Flockton and M.S. White. The application of genetic algorithms to infinite impulse response adaptive filters. In *IEE Colloquium Digest 1993/039*, pages 9/1–9/4. IEE, 1993.
6. S.J. Flockton and M.S. White. Application of genetic algorithms to infinite impulse response filters. In D W Ruck, editor, *Science of Artificial Neural Networks II, Orlando, Florida*, volume SPIE–1966, pages 414–419. SPIE, 1993.
7. S.J. Flockton and M.S. White. Pole-zero system identification using genetic algorithms. In S. Forrest, editor, *Proceedings of the Fifth International Conference on Genetic Algorithms, University of Illinois at Urbana-Champaign*, pages 531–535. Morgan Kaufmann, 1993.
8. Lester Ingber and Bruce Rosen. Genetic algorithms and very fast simulated reannealing: A comparison. *Mathematical and Computer Modelling*, 16(11):87–100, 1992.
9. Cezary Z. Janikow and Zbigniew Michalewicz. An experimental comparison of binary and floating-point representations in genetic algorithms. In Richard K. Belew and Lashon B. Booker, editors, *Proceedings of the Fourth International Conference on Genetic Algorithms, University of California, San Diego*, pages 31–36. San Mateo, CA: Morgan Kaufmann, July 13–16 1991.

10. Kristinn Kristinsonn and Guy A. Dumont. System identification and control using genetic algorithms. *IEEE Transactions on Systems, Man, and Cybernetics*, 22(5):1033–1046, September/October 1992.
11. Zbigniew Michalewicz. *Genetic Algorithms + Data Structures = Evolution Programs*. New York: Springer-Verlag, 1992.
12. Zbigniew Michalewicz and F. Attia, Naguib. Genetic algorithm + simulated annealing = GENOCOP II: A tool for nonlinear programming. Technical report, University of North Carolina, Department of Computer Science, 1993.
13. Heinz Mühlenbein and Dirk Schlierkamp-Voosen. Predictive models for the breeder genetic algorithm I. Continuous parameter optimization. *Evolutionary Computation*, 1(1):25–49, Spring 1993.
14. R. Nambiar, C. K. K. Tang, and P. Mars. Genetic and learning automata algorithms for adaptive digital filters. In *Proceedings of the IEEE International Conference on Acoustics, Speech and Signal Processing (ICASSP 92)*. IEEE, 1992.
15. D. Parikh, N. Ahmed, and S. D. Stearns. An adaptive lattice algorithm for recursive filters. *IEEE Transactions on Acoustics, Speech, and Signal Processing*, ASSP-28(1):110–111, February 1980.
16. O. V. Patiakin, B. S. Zhang, G. D. Cain, and J. R. Leigh. An adaptive filter using Darwinian algorithm for system identification and control. In *Proceedings of the IEEE International Conference on Systems, Man and Cybernetics, France*, volume 4, pages 627–631. IEEE, October 1993.
17. John J. Shynk. Adaptive IIR filtering. *IEEE ASSP Magazine*, pages 4–21, April 1989.
18. M.S. White. *Evolutionary Optimization of Recursive Lattice Adaptive Filters*. PhD thesis, University of London, 1996.
19. M.S. White and S.J. Flockton. A comparative study of natural algorithms for adaptive iir filtering. In *Proceedings of the IEE/IEEE Workshop on Natural Algorithms in Signal Processing, Danbury Park, Essex*, pages 22/1–22/8. IEE, 1993.
20. M.S. White and S.J. Flockton. A genetic adaptive algorithm for data equalization. In *Proceedings First IEEE International Conference on Evolutionary Computation, Orlando, Florida*, volume II, pages 665–669. IEEE, 1994.
21. M.S. White and S.J. Flockton. Genetic algorithms for digital signal processing. *Lecture Notes in Computer Science*, 865:291–303, 1994.
22. Bernard Widrow and Samuel D. Stearns. *Adaptive Signal Processing*. Englewood Cliffs, NJ: Prentice-Hall Inc., 1985.

Numerical Techniques for Efficient Sonar Bearing and Range Searching in the Near Field Using Genetic Algorithms

D.J. Edwards and A.J. Keane

[1] Department of Engineering Science, University of Oxford, Parks Road, Oxford, OX1 3PJ, U.K.
[2] Department of Mechanical Engineering, University of Southampton, Highfield, Southampton, SO17 1BJ, U.K.

Summary. This article describes a numerical method that may be used to efficiently locate and track underwater sonar targets in the near-field, with both bearing and range estimation, for the case of very large passive arrays. The approach used has no requirement for *a priori* knowledge about the source and uses only limited information about the receiver array shape. The role of sensor position uncertainty and the consequence of targets always being in the near-field are analysed and the problems associated with the manipulation of large matrices inherent in conventional eigenvalue type algorithms noted. A simpler numerical approach is then presented which reduces the problem to that of search optimization. When using this method the location of a target corresponds to finding the position of the maximum weighted sum of the output from all sensors. Since this search procedure can be dealt with using modern stochastic optimization methods, such as the genetic algorithm, the operational requirement that an acceptable accuracy be achieved in real time can usually be met.

The array studied here consists of 225 elements positioned along a flexible cable towed behind a ship with 3.4m between sensors, giving an effective aperture of 761.6m. For such a long array, the far field assumption used in most beam-forming algorithms is no longer appropriate. The waves emitted by the targets then have to be considered as curved rather than plane. It is shown that, for simulated data, if no significant noise occurs in the same frequency band as the target signal, then bearing and range can be estimated with negligible error. When background noise is present (at -14dB), the target can normally still be located to within 1% in bearing and to around 5% error in range. Array shape uncertainty worsens this performance but it is shown that it is still possible to accurately determine the target bearing and to provide approximate estimates of range in such circumstances. Finally, the ability of the approach to track the sources over a period of time and with evolving array shape is demonstrated.

1. Introduction

In most published beam-forming and target location algorithms, there is an assumption that the sources are far from the array so that a planar wave-front approximation is possible. Moreover, it is also normal to assume that the array transducer positions are known with complete accuracy. In these circumstances eigen-decomposition high resolution methods [11] have proven to be effective means of obtaining bearing estimates of far-field narrow-band sources from noisy measurements. The performance of these algorithms is, in general, severely degraded when the target is in the near-field or there is uncertainty in the transducer positions. There is therefore a need to look again at the particular problems which arise in the case of flexible arrays of the order of a kilometre in length where the targets are effectively still in the near-field of the array for distances greater than 100 km and where sensor positions are never known precisely.

Techniques for estimating the directions-of-arrival (DOA) of signals impinging on antenna arrays have been investigated intensively over the years [10]. Beam-forming [24] is the most widely used processing technique for these arrays and can be used for direction finding by 'steering the beam' until the array output power is maximized (or minimized). The direction at which the power is maximized provides an estimate of the source bearing. In this approach, antenna arrays need to be calibrated to produce reliable DOA estimates. During the calibration process, the array response vector (also called the array manifold) is measured and stored, for a finite number of directions [18]. This response description depends on the amplitude and phase weighting of the signals from each element or sensor of the array and the positions of the elements within the array structure. It has been suggested that the array manifold be stored in the form of a set of vectors, for a grid of DOAs and frequencies spanning the desired field of view and the frequency band of interest [9]. However, very little work seems to have been done for the near-field case. Here, for point sources, the wave-front presented to the array is spherical and offers the additional degree of freedom of range variation. Also, any sensor position uncertainty presents an effective noise source and an investigation into its role in the beam-forming process needs to be undertaken to identify both the nature and magnitude of its effect. It is therefore desirable that a technique be developed that locates targets without prior knowledge of the array manifold or the number of sources, in a way that minimizes the computational effort and number of parameters to be estimated. It is also required that the technique be robust enough to perform satisfactorily when the signals are corrupted by background noise or uncertainty in sensor positions.

2. Signal Conditioning—Filtering and Averaging

When performing sonar target identification and location, in the absence of noise, it is possible to determine the target position from a single set of sensor responses at any instant in time, given that the sensor locations are known. When there is noise present, due to background ocean noise, hydrodynamic flow noise, etc., or perhaps because the sensor locations fluctuate in an ill-determined manner with time, some form of averaging or matched filtering is required to improve the effectiveness of the process. Such averaging can occur over the set of sensors, over a number of different time-scales or both. Initially, a series of sensor readings can be taken at the given sampling rate and a Fast Fourier Transform (FFT) performed on each: this is a form of short term averaging and is normally carried out because the target is thought to generate signals with a specific, unchanging frequency spectrum over the averaging period under consideration. This condition being satisfied, the FFT will highlight the target spectrum and thus improve the signal to noise (S/N) ratio, improving the identification process in terms of both target detection and frequency determination. The FFTs can then be averaged across the array of sensors, although if the precise sensor and target positions are unknown, this may only be carried out on the sensor powers, i.e., phase information is lost in such cases.

If further refinement is required, a series of FFTs on overlapped or subsequent sets of sensor readings can be taken. Of course, such a process is only really useful if coherent averages are taken. This is straightforward if the target remains stationary relative to the array and the mean array shape is unchanging with respect to the time-scales involved, so that the relevant frequency dependent phase corrections can be made: overlap averaging of this kind is quite commonly used for small arrays and can further improve the S/N ratio in the process of target detection and frequency determination, without additional knowledge of either the target position or the array shape. In both these time frames, target location can be performed with varying degrees of success according to the relative magnitude of the noise inputs to the system. Clearly, by preserving the correct phase information in the FFTs during overlap averaging the S/N ratio can be continuously improved with increasing numbers of overlaps provided that the target and mean array positions continue to remain in fixed relationships to each other. If they do not, some progress can nonetheless still be made if their relative motions can be estimated in some way.

This leads on to long term averaging where the target position and mean array shape are both assumed to alter, albeit slowly, with time and some form of weighted moving average of the results of target location are then used. In this time frame, target tracking and model evolution can be performed so that position estimate errors can be further diminished by using generalised knowledge about the behaviour of targets and their ability to manoeuvre within the given time frame [6] alongside hydrodynamic predictions of likely sensor motions. In this article all but the last of these methods of averaging will be investigated, i.e., up to and including overlapped FFTs with simple target modelling and weighted moving averages but without hydrodynamic predictions of sensor motions.

3. Target Identification

Before the locations of targets may be estimated it is, of course, necessary to identify any targets that may be present. For most purposes in sonar processing this means the identification of fixed frequency tones that lie above the noise floor of the signals being analysed. As has just been noted, this leads directly to the use of FFTs to produce power spectral density estimates for each sensor in an array. Then, if there are obvious spikes in the amplitudes of these spectra, a target is considered present. Target identification is thus intimately dependent on the noise level present in the spectra. In general, and as has already been mentioned, this can only be effectively controlled by averaging such spectra, preferably in a coherent fashion. If insufficient sensors or time slices are taken or the data are averaged incoherently (i.e., the phase information is ignored, probably because the sensor and target positions are not known with sufficient accuracy) quiet targets may well not be identified from the background noise. Of course, in the real world any target tone may also disappear or slowly change frequency or new tones may suddenly appear. Here, however, fixed tones that are present throughout the data sets will be used, although one of these will be below the noise floor that may be reached by incoherently averaging the sensor data.

By way of example, consider an array that is attempting to locate an unknown number of possibly low noise targets. The data collected by the sensors can then be written as

$$X(t) = [x_1(t), x_2(t), ...] \qquad (3.1)$$

In order to filter out uncorrelated noise so as to distinguish targets, an FFT on a suitably windowed version of the data is performed first. Since the sampling rate for the data to be used here is set at 512 samples per second, the same number of data is chosen to perform the FFT, i.e., 512 data points are taken from each sensor simultaneously. In the frequency domain the data can be written as

$$Y(f) = [y_1(f), y_2(f), ...] \qquad (3.2)$$

where $y_1(f)$ is the FFT of $x_1(t)$ from sensor one, and so on.

Now each $y(f)$ is an array of 512 elements, with real and imaginary parts for each individual frequency component, ranging from zero frequency upwards, with the first 256 values being unique. For a signal with poor signal to noise ratio, the FFT from a single sensor may not give a clear indication of the frequency spectrum of the target signal. If the range and bearing of the target are known alongside the positions of the sensors, then the signals at each sensor can be coherently added so as to enhance the spectral component (if this condition is not satisfied, the best that can be hoped for is a power addition of the spectral components). When the noise is uncorrelated, integrating the FFT power output from all sensors will enhance the signal while in effect smoothing the random noise component, thus

$$Y_s(f) \; = \; \left| \sum_{m=1}^{M} y_m(f) \right| \tag{3.3}$$

where M is the number of sensors. Further suppression of the noise may be sought by averaging the data over a number of time periods often employing some kind of overlap (commonly 50% overlaps are adopted when using Hanning (or von Hann) windows). Such overlap averaging can be carried out before or after averages are taken across the array. If they are taken before the spatial average and each individual sensor position is unchanging in its mean distance from the target during the overlaps (or the changes are known or estimated), the average can be carried out coherently and the S/N ratio further improved whether or not the spatial averaging is also carried out coherently. Figure 3.1 shows a typical FFT derived from an individual sensor using simulated data, while Figure 3.2 represents the incoherently averaged power signal taken across an array of 225 such sensors.

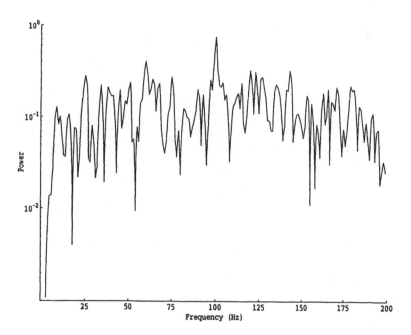

Figure 4.1 shows a further spectrum, where 30 time slices of data (i.e., spanning a 30 second length of simulated record) have been coherently averaged together with 50% overlaps, followed by an incoherent spatial power

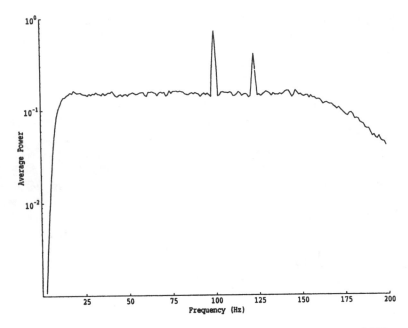

Fig. 3.2. Typical incoherently averaged FFT taken across an array of 225 sensors

average across the array, clearly demonstrating the improved S/N performance that may be obtained (of course, this S/N ratio can be improved still further by averaging coherently across the array elements but this requires knowledge of the relative positions of the target and the elements - the calculations behind Figure 4.1 only require the target to remain stationary with respect to the elements). In Figure 3.1 only one target can be clearly seen, in Figure 3.2 a second is apparent and in Figure 4.1 a third. These three signals are at 100 Hz, 123 Hz and 147 Hz, respectively and, relative to the signal at 100 Hz have levels of 0dB, -5dB and -25dB. The noise floor in the single, incoherently spatially averaged time slice is at -14dB, i.e., above the level of the third signal. Note that these plots are limited to 200 Hz so that they stay well below the Nyquist frequency, which is here 256 Hz.

4. Target Location

Having identified the presence of one or more targets, most analytical techniques in target location then involve the solution of a set of simultaneous equations which generally require the manipulation of matrices. For small

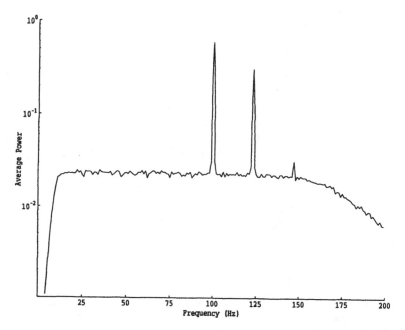

Fig. 4.1. Typical FFT where 30 time slices of data have been coherently averaged together with 50% overlaps, followed by an incoherent spatial power average across the array

array dimensions this represents a reasonable overhead in terms of computational effort and memory requirement. However, as the array size increases this feature becomes less acceptable as the costs in both computation and memory increase with the square of the number of elements. Thus for arrays of hundreds of elements or more, the prime consideration in such work is the tractability of the techniques involved. Additionally, a generalised approach is required that does not limit the solution to one dimension (i.e., bearing) but that can also deal with two- or three-dimensional problems, as is the case for a target in the near-field of a possibly non-linear passive sonar array. Numerical techniques have been used widely in various areas to solve problems which would be difficult or impossible by analytical methods and so attention is now turned to how well a passive sonar system consisting of a linear array of hydrophones, together with appropriate signal processing and numerical techniques can determine the range and bearing of individual targets.

When a spherical wave-front is used for beam forming, the so called array manifold varies along the array. Assume that a spherical wave signal $s(t)$ is

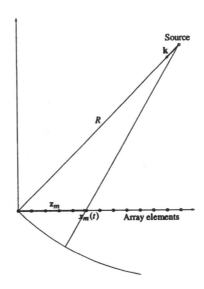

Fig. 4.2. A spherical wave signal $s(t)$ propagating in a direction $-\mathbf{k}$ from the first sensor of an array of M sensors positioned at $\mathbf{z_m}$ and generating signals $x_m(t)$ in these sensors

propagating in a medium at speed v from a source positioned in a direction $-\mathbf{k}$ (a unit vector) from the first sensor in the array, see Figure 4.2.

An array of M sensors is present in the medium: each one is assumed to record the acoustic field at its spatial position with perfect fidelity. The wave form measured at the mth sensor relative to that at the first is denoted by $x_m(t)$ and is given by

$$x_m(t) \;=\; s\left(t + \frac{R - |R\mathbf{k} + \mathbf{z_m}|}{v}\right) + q_m(t) \tag{4.1}$$

where $\mathbf{z_m}$ is the position vector of the mth sensor measured from the first, and $q_m(t)$ is additive noise. This noise may be due to disturbances propagating in the medium or generated internally in the sensor or due to array element location uncertainty, etc.

One of the classical methods used in array processing for finding the bearing of an acoustic source is time delay beam-forming [19]: the outputs of the sensors are summed with weights and delays to form a beam, $B(t)$ where

$$B(t) \;=\; \sum_{m=1}^{M} a_m x_m(t - t_m) \quad . \tag{4.2}$$

The idea behind beam-forming is to align the propagation delays of a signal presumed to be propagating in a direction $\mathbf{k}$ so as to reinforce it. Signals propagating from other directions and also the noise are not reinforced. If the delay t_m is ideally compensated by the signal delay $(R - |R\mathbf{k} + \mathbf{z_m}|)/\mathrm{v}$, the signal would be completely reinforced. The delayed output of each sensor would then be

$$x_m(t - t_m) = s(t) + q_m(t - t_m) \quad . \tag{4.3}$$

The resulting beam output becomes

$$B(t) = Ms(t) + Q(t) \quad . \tag{4.4}$$

Thus the signal amplitude in $B(t)$ is M times that measured at each sensor while the noise amplitude is increased by only $\sqrt{M}$ (assuming the noise outputs from the sensors are uncorrelated). Generally, the energy in the beam $B(t)$ is computed for many directions-of-look, $-\mathbf{k}$, by manipulating the delays t_m. Maxima of this energy, as a function of $\mathbf{k}$ are assumed to correspond to acoustic sources, and the source bearings correspond to the locations of these maxima. For a far field target, the interval between successive delays is constant for a linear array with uniform spacing between elements. For an array of length D, in the case of the example to be considered here 761.6m, the estimation of the boundary of the near-field zone of the array is at $2D^2/\lambda$, or here approximately 77km at 100Hz. For a target at 100km with bearing 90.12° from the first element, the phase difference between the first and last elements is 107.9° at 100Hz (note that all bearings given in this article are absolute and based on a right handed axis system with true north at 0° and depth positive downwards). For a plane wave this phase difference would be 38.3°. This simple example demonstrates the inadequacies of the plane wave assumption in such cases.

If the target is in the near-field and the array sensors are no longer co-linear these ideas can still be used relatively easily. Assuming, for the moment, that the target bearing and range are known, the advancing curved phase front can be compensated for by modifying the phases seen at each sensor, depending on its location. Beam forming can then be carried out on the modified data. Here it is convenient to first average coherently across a set of N overlapped time slices and then coherently across the M elements. The amplitude in the beam at frequency f is then

$$B_m(f) = \left| \sum_{m=1}^{M} \sum_{n=1}^{N} y_{mn}(f) e^{i(2\pi f n \delta t + \phi_m)} \right| \tag{4.5}$$

where $y_{mn}(f)$ is the Fourier transform of the mth sensor signal taken over time slice n, δt is the interval between time slices and ϕ_m is a phase angle correction which allows for the positions of individual sensors relative to the first when compared to that of the target. This is given by

$$\phi_m = \frac{2\pi}{\lambda}\sqrt{(R\cos(\theta) - \Delta x_m)^2 + (R\sin(\theta) - \Delta y_m)^2}\qquad(4.6)$$

where R and θ are the assumed range and bearing of the target from the first sensor in the array and Δx_m and Δy_m are the coordinates of the mth sensor relative to the first one. Target location then requires the maximization of this signal by manipulation of R and θ, i.e., it becomes a two-dimensional search process.

If the target or array moves during the time overlaps used to improve the S/N ratios then estimates must be included to allow for the further phase shifts caused by these motions, i.e., R, θ, Δx_m and Δy_m, and hence ϕ_m, all change with time slice n [7]. The simplest model that can be proposed for such motions is to assume that $dR/dt = \dot{R}$ and $d\theta/dt = \dot{\theta}$ are constant and that Δx_m and Δy_m are known for each n, being taken from array instrumentation (notice also, that suitable choices of $\dot{R}$ and $\dot{\theta}$ will compensate for errors in the instrumentation data to some extent, since these errors can be thought of as being mapped into additional target motions which are then estimated by $\dot{R}$ and $\dot{\theta}$). It is, of course, also possible to model the target's course and speed along with those of the tow vessel and use these instead of $\dot{R}$ and $\dot{\theta}$ if it is thought that these will give more reliable estimates. If the values of $\dot{R}$ and $\dot{\theta}$ are not known *a priori*, as is usually the case, suitable values can be searched for, increasing the number of parameters to be considered from two to four. If it is further desired to acquire targets during such a process, the frequencies being dealt with can also be scanned, adding a fifth variable to be considered. As will be seen later, it is not practicable to investigate even a two-dimensional problem of this type using exhaustive search methods and thus the problem naturally becomes one of array power *optimization*.

5. Array Shape Uncertainty

As has just been noted, the ability to carry out coherent averages is dependent on knowledge of the array shape and how it changes with time. The effects of array element position uncertainty can thus be considered as a further noise input, e.g., treating the distorted array as a linear equally spaced array, the phase front will reflect both the signal delay and the displacement of the array elements.

A considerable amount of work has been undertaken in the field of antenna behaviour on the effect of random and systematic errors in the radiation pattern of the antennas. This work can be directly applied to the case of sonar arrays of uncertain shape. Simplistically, the problem can be addressed in terms of the order of the mean shape of the array. This shape can be described in general terms by a polynomial expansion consisting of linear, quadratic, and higher terms. It is relatively easily seen that a systematic linear phase error down the length of the array is essentially a heading error and simply maps into a bearing error in any target location algorithm. On the other hand, a quadratic component is in effect superimposed on any spherical wave-front that may impinge on the array. This error therefore contributes to an increase (or decrease) in the apparent curvature of the wave-front and hence effectively yields an error in range estimation. Higher order errors generally have major effect in bearing for odd powers and range for even powers. If the array shape can be estimated, however, suitable steps can be taken during the beam forming process to allow for these effects. Moreover, when the array is curved, it is often possible to resolve target bearings without the spatial ambiguity inherent in processing the data from straight arrays.

6. Array Shape Modelling

Because of the problems inherent in processing data when the sensor positions are unknown, it is normal when building long flexible arrays to incorporate shape estimation instrumentation within the array [16]. Typically, compasses are positioned along the array and these may then be used to estimate its shape at any instant. Of course, such compass data is never precisely accurate and noise in this data must also be considered. Here, it will be assumed that a number of compasses are available and that they may be used to help in processing data from the array. To do this a curve fit is applied to the compass data such that the fitted curve has the correct slopes (i.e., bearings) at the compass positions and the correct arc lengths between them. This is most readily achieved by simply fitting independent arcs of circles between each compass that meet tangentially at the compass positions (and which are then independent of the axis system in use). The hydrophones positions can then be readily calculated from the assumed shape. Figure 7.1 shows the apparent compass locations and array shapes for two consecutive positions of an array in a simulated tow sequence separated by a one second interval.

In both cases the array is, in fact, straight and compass errors account for the deviations from this. It can be seen that the compass data changes sufficiently from time slice to time slice for a detectable change in the fitted shape to occur, but that the change is small compared to the array's overall length (note the difference in axis scales).

7. Example Data

Having outlined the background to this problem and suggested how target detection and positioning might be carried out, these ideas are next applied to a set of simulated data. This set consists of 225 individual sensor outputs,

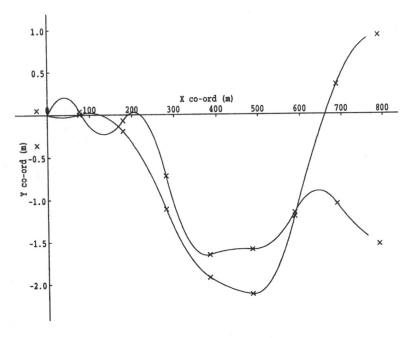

Fig. 7.1. Apparent compass locations (×) and array shapes for two consecutive positions of an array in a simulated tow sequence separated by a one second interval

each sampled at 512 readings per second for a period of 1,400 seconds, plus nine compass outputs read every second (the simulation therefore contains over 160 million individual items of data). During the run there are two targets present. Target one emits two tones: the first at 100 Hz and 0dB and the second at 123 Hz and -5dB. The second target emits a single tone of 147Hz at -25dB. The background noise level is at -14dB, i.e., as per the previously discussed spectra. At all times during the run the targets remain at fixed (absolute) bearings and distances from the ship towing the array: target one is at 90° and 100km (i.e., due east) and target two at 180° and 100km (due south). To begin with all three vessels are travelling south (i.e., on a course of 180°) at 10 knots, then after 600 seconds all ships execute a turn on to a course of 270° (west). Thus, since the array streams out behind the tow ship, the array begins broadside-on to target one and end-on to target two and then, as the turn is carried out, this reverses: the array becomes end-on to target one and broadside to target two, see Figure 7.2.

Note that since the first hydrophone in the array lies 204.43m behind the datum used for positioning the ships, its target bearings and ranges in the subsequent analysis are slightly different from those for the datum and, moreover, change slightly as the ships change course despite the ships maintaining fixed stations to each other.

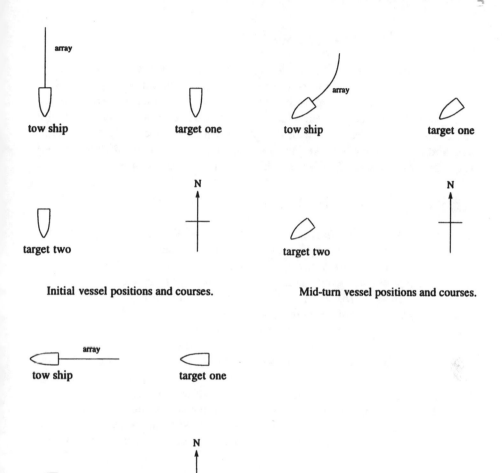

Initial vessel positions and courses.

Mid-turn vessel positions and courses.

Final vessel positions and courses.

Fig. 7.2. Tow ship, array and target manoeuvres during the example data set (a - starting positions and courses, b - mid-turn positions and courses, c - final positions and courses)

8. Simple Target Location Searches

To begin with, a number of simple target location searches will be discussed which illustrate the basic performance of the algorithm being suggested. Subsequent sections then examine the use of a modern stochastic search method when applied to this problem. Throughout these searches it is assumed that the three target tones present in the example data have already been identified and so frequency searches are not considered. Of course, if this were not the case this additional variable would also have to be scanned.

8.1 Noise-free data – linear array

The most basic kind of search that can be proposed simply involves dividing the target space into a large number of search areas and then coherently summing the array powers from a single time slice across all 225 hydrophones at the corners of each search area. For convenience, the space can be divided up in cylindrical coordinates, with the angle varying from 0° to 180° (given the left/right ambiguity inherent in a linear array) and R from 0 to the boundary of the search space to be covered. Table 8.1 contains the results of a number of such searches when applied to the first tone of target one (100Hz) with a set of noise free target data while the array is straight and broadside-on to the target. In this condition the actual range and bearing of the target from the first hydrophone (the datum used here) are 90.1172° and 100.0 km. It is apparent that the step size in θ plays a more important role than that in R. This is due to the relatively slow variation of array gain with angle compared with that in range. Typically, to achieve 10m accuracy for a

Table 8.1. Bearing and range results for various simple search step sizes, linear array, noise-free data

Bearing step size (degs)	Range step size (m)	Bearing (θ) (degs)	Range (R) (m)
0.0001	10	90.1172	100,010
0.001	10	90.117	99,510
0.01	10	90.12	101,910
0.0001	100	90.1172	100,100
0.001	100	90.117	99,450
0.01	100	90.12	102,000
0.001	1,000	90.117	99,000
0.1	10,000	90.1	70,000

target at 100km, a step size of 0.0001° in bearing is needed. If a space of 0° to 180° and range from 0 to 150km is being investigated with such detailed step sizes, several hours are needed to complete an exhaustive search. Therefore, to minimize the computational effort involved, large step sizes can be chosen to perform a coarse search, followed by a reduction in step sizes with searches concentrated in the limited areas identified by the first search.

Thus to begin with, if a step size of 0.1° is chosen for the bearing and 10 km for range, maximum array power is found at 90.1° and 70km. A second iteration can then be limited to the area surrounding this point, but noting that a significant variation in range is still likely. If 0.001° and 100m are then chosen as steps the maximum power is found at 90.117° and 99.45 km. A third iteration with steps of 0.0001° and 10m then yields a target at 90.1172° and 100.010 km. Clearly, in the absence of noise, a simple power maximization technique can give a very accurate bearing and range estimation provided the search is carried out with sufficient resolution. Figure 8.1 shows a contour plot of the signal power corresponding to this target, illustrating how rapid the variation in power is with bearing.

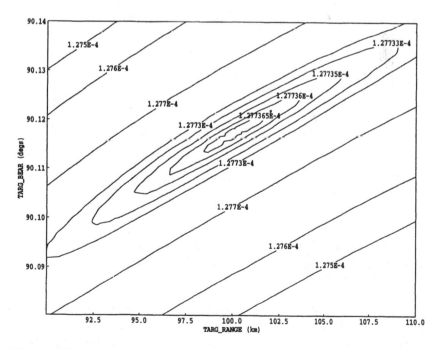

Fig. 8.1. Contour plot of signal strength for the 100 Hz tone of the first target in the vicinity of the target

8.2 Noise-contaminated data – linear array – fixed target positions

When there is noise appearing at the same frequency as the signal, the wavefront is no longer obviously spherical. Figure 8.2 shows typical phase fronts for the two tones from the first target using a single set of 225 FFT readings which are distorted by background noise.

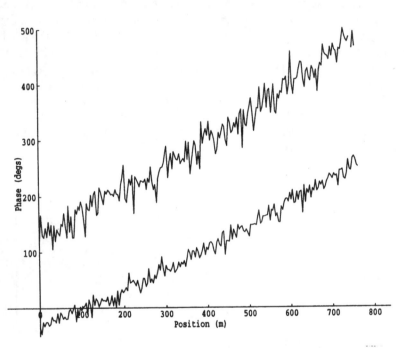

Fig. 8.2. Typical phase fronts for the two tones from the first target when distorted by background noise, using a single set of 225 FFT readings the 100 Hz tone of the first target in the vicinity of the target

Figure 8.3 shows how these phase fronts can be improved by using 30 sets of 50% coherently overlapped FFTs to smooth the data.

With such distorted phase fronts it would be difficult to achieve great accuracy using traditional high performance analytical algorithms, because the assumptions behind these algorithms are no longer valid. Conversely, using the simple three stage search method with the power maximization technique adopted for the noise free data, the bearing errors on the broadside target can be controlled to around 0.5° with a range error of a few kilometres while the quiet target, end-on to the array, can be located to within 2° in bearing but, of course, only poorly in range (which theoretically cannot be resolved at all in the end-on case), see table 8.2. Notice that, since the targets

Table 8.2. Bearing and range results for three target tones, linear array, noise-contaminated data, fixed target positions, exhaustive search

Target freq. (Hz)	True Bearing (θ) (degs)	True Range (R) (m)	Est. bearing (degs)	Est. range (m)
100	90.117	100,000.0	90.68	96,817
123	90.117	100,000.0	90.70	106,411
147	180.0	100,204.4	178.48	4,459

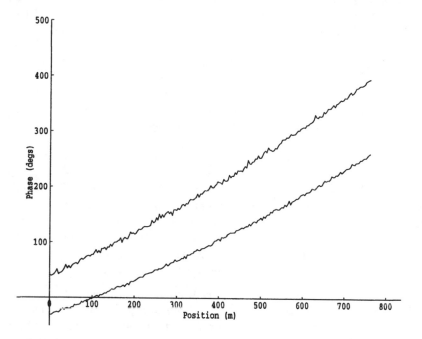

Fig. 8.3. Typical phase fronts for the two tones from the first target improved by using 30 sets of 50% coherently overlapped FFTs to smooth the data

are at the edge of the near-field zone for the array used here, the errors seen in the broadside target ranges are, in reality, as small as can be reasonably expected.

9. Improved Search Algorithms

Having established the basic performance of the proposed power maximization method, attention is next turned to using a modern stochastic search method to speed up the target location process. Up until this point a three stage search has been used that requires 20,000 locations to be examined at each stage to ensure complete accuracy in finding the targets. Such fine meshes are required because of the great sensitivity of the array to bearing changes. The problem posed here is, however, at most only a four dimensional search and thus amenable to optimization methods.

Genetic Algorithm (GA) methods can be used to carry out such searches and much recent research has shown that they are very robust when dealing with noisy or multi-modal problems [13], [14]. They work by maintaining a pool or population of competing estimates which are combined to find improved solutions. In their basic form, each member of the population is represented by a binary string that encodes the variables characterizing the solution. The search progresses by manipulating the strings in the pool to provide new generations of estimates, hopefully with better properties on average than their predecessors. The processes that are used to seek these improved designs are set-up to mimic those of natural selection: hence the method's name. The most commonly used operations are currently: (i) selection according to fitness, i.e., the most promising solutions are given a bigger share of the next generation; (ii) crossover, where portions of two good solutions, chosen at random, are used to form a new estimate, i.e., two parents 'breed' an 'offspring'; (iii) inversion, whereby the genetic encoding of a solution is modified so that subsequent crossover operations affect different aspects of the estimate and (iv) mutation, where small but random changes are arbitrarily introduced into an estimate. In addition, the number of generations and their sizes must be chosen, as must a method for dealing with constraints (usually by application of a penalty function).

The algorithm used here works with up to sixteen bit binary encoding (although parameters that are selected from a number of fixed possibilities use only the minimum necessary number of bits and the number of bits can be reduced by the user for continuous variables). It uses an elitist survival strategy which ensures that the best of each generation always enters the next generation and has optional niche forming to prevent dominance by a few moderately successful designs preventing wide ranging searches. Two penalty functions are available. The main parameters used to control the method may be summarized as:

1. N_{gen} the number of generations allowed (default 10);
2. N_{pop} the population size or number of trials used per generation which is therefore inversely related to the number of generations given a fixed number of trials in total (default 100);
3. P[best] the proportion of the population that survive to the next generation (default 0.8);
4. P[cross] the proportion of the surviving population that are allowed to breed (default 0.8);

5. P[invert] the proportion of this population that have their genetic material re-ordered (default 0.5);

6. P[mutation] the proportion of the new generation's genetic material that is randomly changed (default 0.005);

7. a Proportionality Flag which selects whether the new generation is biased in favour of the most successful members of the previous generation or alternatively if all P[best] survivors are propagated equally (default TRUE) and

8. the Penalty Function choice.

Using a method that incorporates these ideas the following GA based approach can be proposed for the problem of interest here: (i) maximize the array power by searching one half of the bearing space and the range values of interest using a GA search of 10 generations each of 250 evaluations with 12 bit resolution in both variables (i.e., use 2,500 points); (ii) reduce the bearing scan to $\pm 5°$ around the best target found while leaving the range scan as before and search with 10 generations each of 100 evaluations (i.e., another 1,000 steps); (iii) reduce the bearing scan further to $\pm 0.25°$ and the range scan to ± 25km around the best target and carry out a final GA search of 10 generations each of 100 members and with increased resolution of 16 bits (i.e., 1,000 points); (iv) finally, if left-right ambiguity must be resolved repeat this process starting in the other half of the bearing search space.

This search uses only 4,500 points and is able to robustly locate the same results as the previous exhaustive search process, taking a small fraction of the time (the three stage exhaustive search examines 20,000 points). It is worth noting, however, that an exhaustive search is the only way of guaranteeing that all targets have been found to within a given resolution. Therefore, if time permits, it is still to be preferred over other methods. Lastly, it should be noted that search tasks like this can often be handled by parallel processing devices, something the GA is particularly well suited to.

9.1 Noise-contaminated data – array position inaccuracy – fixed target positions

When towing an array of the length considered here, even if the tow ship maintains a steady course, the array will fluctuate in shape. This much has been noted already and a method proposed whereby compass data can be used to provide an estimate of the array shape. Therefore, consideration is next turned to the effects of these uncertainties on the ability of the array to locate targets using the method proposed. Table 9.1 repeats the results presented in table 8.2 but now using sensor positions based on the compass data (rather than the true positions used for table 8.2) along with estimated values of $\dot{R}$ and $\dot{\theta}$ (since although the targets maintain fixed positions, suitable values of these parameters help correct for errors in the array shape estimates).

Table 9.1. Bearing and range results for three target tones, *noisy compass readings*, noise-contaminated data, fixed target positions, GA search

Target freq. (Hz)	True Bearing (θ) (degs)	True Range (R) (m)	Est. bearing (degs)	Est. range (m)
100	90.117	100,000.0	90.35	38,198
123	90.117	100,000.0	90.58	70,485
147	180.0	100,204.4	177.07	3,963

Here, the compass readouts have been set to have random fluctuations of up to ±2° and these give rise to the estimated array shapes such as those already shown in Figure 7.1. As can be seen from the table, the bearings are still accurate to within 0.5° for the two more powerful tones and to within 3.0° for the low noise one. However, the shape errors have significantly degraded the accuracy of the broadside target ranges, which are now predicted only to within a factor of about 2.5. Nonetheless, it should be stressed that since the targets are at the edge of the near-field zone of the array and now the array suffers from position errors, which are known to significantly affect range accuracy, these values are still quite respectable. As will be seen later, by taking a moving average of the target range with time, these range estimates can be further improved upon. It is also worth noting that, when the array is broadside-on to the quiet target, it can also be resolved to within ±0.5° in bearing and a factor of about 3.0 in range.

9.2 Noise-contaminated data – array position inaccuracy – curved array – moving targets

Lastly within this section, attention is turned to the location of targets when the tow and target vessels turn. This leads to significant changes in the array shape which can only be easily deduced from the compass data, i.e., equation (9) can only be used if some estimate of the array curvature is made (of course, given suitable hydrodynamic modelling these estimates might be improved). Moreover, during the period used for averaging, the target ranges and bearings alter slightly in this case. Although these changes are only small, those in bearing are sufficient to affect the location process. Table 9.2 gives the equivalent results to those presented in tables 8.2 and 9.1 but now with the sensor positions based on the compass data *during the turn* and again using estimates of $\dot{R}$ and $\dot{\theta}$ alongside those of R and θ. Here the compass data give rise to estimated array shapes such as that shown in Figure 9.1 and substantial curvature is clearly present.

As can be seen from the table the target bearings are predicted with nearly as great an accuracy as those for the straight array. Moreover, the range estimates for all three target tones are now of similar quality, all remaining within a factor of about 2.5 of the true values. This is as expected, since, during the turn, the array is not end-on or broadside-on to either of the

Table 9.2. Bearing and range results for three target tones, curved array, noisy compass readings, noise-contaminated data, *moving targets*, GA search

Target freq. (Hz)	True Bearing (θ) (degs)	True Range (R) (m)	Est. bearing (degs)	Est. range (m)
100	90.04	99,808.0	90.31	69,965
123	90.04	99,808.0	87.55	241,406
147	180.11	100,070.1	183.62	252,589

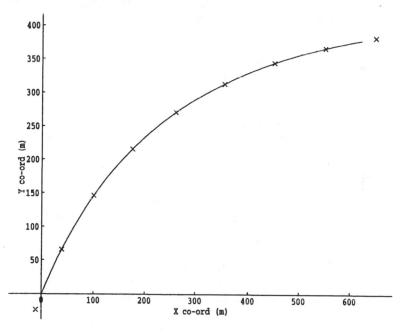

Fig. 9.1. Apparent compass locations ($\times$) and array shape during the turn

targets. Again, as has been noted in the previous sub-section, moving averages of target positions further improve these results. Finally, it should be noted that the four variable search/power optimization problem being solved here is by no means trivial. As will be seen in the next section, attempts to locate the targets using classical, slope ascent methods [23] consistently fail to find adequate solutions. No doubt this is due to the extreme sensitivity of the array power to the bearing selection and also the noisy nature of the data (which means that the function being optimized is not uni-modal or smooth). This difficulty is further compounded by trying to find suitable choices of R and θ.

10. Other Search Methods

Before going on to look at the ability of the proposed GA search method to track targets over extended periods of time it is worth considering whether or not other optimization methods might also be useful in this context. Optimization techniques have been studied for very many years and over that time a considerable number of methods have been proposed. A number of authors have collated these methods into libraries allowing comparison to be made between them [23], [22], [12]; the last of these libraries, called OPTIONS, is used here. This library contains, in addition to the GA, the following methods:

- adaptive random search (Adrans) [23];
- Davidon, Fletcher, Powell strategy (David) [23];
- Fletcher's 1972 method (Fletch) [23];
- Jacobson and Oksman method (Jo) [23];
- Powell's direct search method (PDS) [23];
- Hooke and Jeeves direct search as implemented by Siddall (Seek) [23];
- the simplex strategy of Nelder and Meade as implemented by Siddall (Simplx) [17], [23];
- the method of successive linear approximation (Approx) [23];
- random exploration with shrinkage (Random) [23];
- NAg routine E04UCF, a sequential quadratic programming method (NAg) [5];
- a Bit Climbing algorithm (BClimb) [4];
- a Dynamic Hill Climbing algorithm (DHClimb) [25];
- a Population Based Incremental Learning algorithm (PBIL) [2];
- the routine POWELL taken from the Numerical Recipes cookbook (NumRcp) [20];
- repeated application of a one-dimensional Fibonacci search (FIBO) [22];
- repeated application of a one-dimensional Golden section search (Golden) [22];
- repeated application of a one-dimensional Lagrangian interpolation search (LAGR) [22];
- Hooke and Jeeves direct search as implemented by Schwefel (HandJ) [22];
- Rosenbrock's rotating co-ordinate search (ROSE) [21], [22];
- the strategy of Davies, Swann and Campey with Gram-Schmidt orthogonalization (DSCG) [22];
- the strategy of Davies, Swann and Campey with Palmer orthogonalization (DSCP) [22];
- Powell's strategy of conjugate directions (Powell) [22];
- the Davidon, Fletcher, Powell strategy (DFPS) [22];
- the simplex strategy of Nelder and Meade as implemented by Schwefel (Simplex) [17], [22];
- the complex strategy of Box (Complex) [3], [22];

- Schwefel's two-membered evolution strategy (2MES) [22];
- Schwefel's multi-membered evolution strategy (MMES) [22];
- Simulated Annealing (SA) [15];
- Evolutionary Programming (EP) [8] and
- an evolution Strategy based on the earlier work of Back, Hoffmeister and Schwefel (ES) [1].

Notice that several of the classical methods provided here are common although details of their exact implementations differ. They do not, therefore, always deliver the same results.

To assess the performance of the methods on the basic target location task, each technique has been used to scan the half-space to one side of the array looking from 5 to 200 km and at the three frequencies of the known signals for the model used. These searches have been carried out at 10 different time slices of the data for the uncorrelated background noise model (five when the tow ship is not turning and five when it is) and five when the noise is spherically correlated (and when the ship is not turning). This range of tests gives an understanding of the robustness of the methods (those for the modern stochastic methods have also been averaged over five different random number sequences). The corresponding results are presented in table 10.1. In all cases the optimizers were allocated a single pass of 2,500 trials for these searches (in many cases the search methods use this figure for guidance only and may take considerably more or fewer steps, depending on the algorithm in use).

Table 10.1. Comparative performance of methods across all three signals, coarse search, straight and curved array, coherent and incoherent noise; stochastic methods averaged over five runs

Method	Overall Average			Method	Overall Average		
	% sgth	% st dev	steps		% sgth	% st dev	steps
Adrans	85	27	9153	David	13	12	305
Fletch	14	12	180	Jo	13	14	49
PDS	25	31	891	Seek	36	38	3042
Simplx	22	27	229	Approx	15	14	360
Random	38	31	7844	NAg	10	10	24
BClimb	75	28	2500	DHClimb	87	13	2578
PBIL	78	15	2500	NumRcp	22	29	478
FIBO	71	33	1565	Golden	71	33	1619
LAGR	72	29	544	HandJ	81	29	2134
ROSE	14	12	2225	DSCG	46	34	235
DSCP	45	33	272	Powell	15	14	1151
DFPS	8	8	20	Simplex	59	32	2494
Complex	15	11	78	2MES	65	28	235
MMES	29	33	2478	GA	86	11	2500
SA	81	13	2500	EP	86	10	2500
ES	78	12	2500				

It is apparent from the table that the most reliable methods for locating targets in a broad search area and with varying signal strengths, array shapes and noise models are stochastic in nature, with Dynamic Hill Climbing (DHC), the Genetic Algorithm (GA) and Evolutionary Programming (EP) performing best. Of these the GA and EP (a close relative) have the lowest standard deviations in their results and so are preferred over DHC (which just uses a series of random start points followed by a simple hill climber, and which explains its slightly more varied performance).

Of the classical methods tried the performance of the Schwefel implementations of the Hooke and Jeeves search and the repeated one-dimensional searches (Golden Section, Fibonacci and Lagrange Interpolation) clearly performed best. A more detailed examination of the results than space allows here reveals that the Hooke and Jeeves method deteriorates as the signal being searched for becomes weaker. It also shows that the one-dimensional methods are less affected by this problem but they are clearly are not as robust as the stochastic methods. Such analysis also makes clear that it is much easier for most of the search mechanisms to locate targets when the array is straight.

11. Target Tracking

Finally, the multi-pass GA optimization approach proposed here has been applied to the complete data set and the results used to track the three target signals over time, allowing moving averages of the ranges to be taken. Of course, when tracking targets that have already been located there is no need to search the entire range and bearing space at each time step. It is usually sufficient to use a slightly broader, one stage GA search in an area around the previously located positions, occasionally supplemented by the full three stage search. Here, the one stage search examines ±0.5° in bearing and ±100 km in range with 2,000 evaluations spread over 10 generations of the GA. These searches are carried out for every five seconds of data using five second time slices so that each new optimization uses fresh data. These are then updated by a full three stage search after every 60 seconds of data has been read. Here, both left and right hand signals have been tracked since, when the array becomes curved, the true targets should remain stationary while the false ones veer off and their power levels drop. The results of these searches are shown in Figures 11.1–11.3 and 11.4–11.6, where not only have the located maxima been plotted, but also exponentially time weighted geometric averages of the ranges (using a time constant of 30 seconds). Also shown on these plots are the compass readings from the compasses at either end of the array, which indicate the extent of the array curvature and the duration of the turn.

Figure 11.7 then shows the power levels of the six tracked signals (and repeats the compass data for convenience): notice that when broadside-on

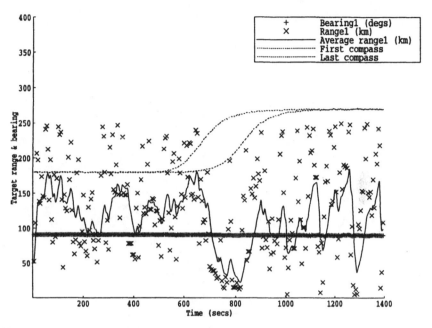

Fig. 11.1. Bearing (+) and range (×) tracks with moving range average (solid line), true targets; also showing first and last compass readings (dotted lines); tone one of target one

to the array, the power traces are less smooth than when end-on, as might be expected, since the phase changes along the array in such cases are then significantly smaller, making the location process less precise.

It is clear from these figures that the false targets can be readily distinguished from the true ones when the array becomes curved: the false target bearings veer as the array changes shape; moreover, the power levels for the false targets are then lower than for the true ones (the steps in the traces occur when the full three stage GA searches relocate the rapidly veering false targets). It can also be seen that, when dealing with data corrupted by noise and when the array position is not known precisely, the true target ranges are rather oscillatory, while their bearings and power levels remain virtually stationary: hence the need for weighted moving averages to smooth the range data.

Overall, it is clearly possible to locate and track all three target tones during a course change when the array has considerable curvature. Moreover, it is also possible to resolve the left/right ambiguity inherent in the signals received by a linear array, by noting the bearing changes and lower powers of the false targets when the array becomes curved. In addition, the target bearing accuracies are of a high standard throughout while those for range are acceptable, especially given the fact that the targets lie at the edge of

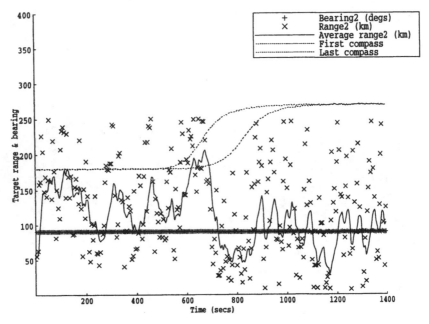

Fig. 11.2. Bearing (+) and range (×) tracks with moving range average (solid line), true targets; also showing first and last compass readings (dotted lines); tone two of target one

the near-field zone of the array and that the array is also subject to shape uncertainties.

12. Conclusions

A numerical technique for bearing and range searching in the near-field for large aperture passive sonar arrays has been presented. The technique gives precise location of targets with noise free data, and still performs well when the data is corrupted by noise. It is based on a simple power maximization approach but one that is corrected for target ranges as well as bearings and also allows for array shape changes. The array shapes are estimated using array compass data and further corrections are made by allowing for range and bearing shifts that occur while taking coherently overlapped FFT averages in the time domain. It is emphasized that, although the target identification work presented here has been carried out using simpler methods (since these are faster when it is known that the targets are stationary with respect to the array), the power maximization technique developed is fully capable of identifying targets even when they are moving and the array is curved in shape.

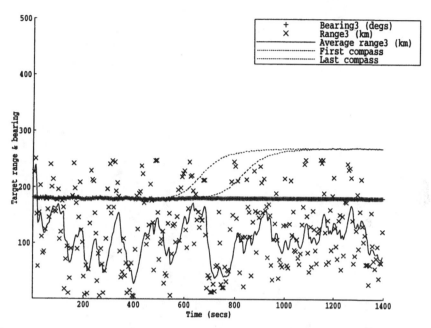

Fig. 11.3. Bearing (+) and range (×) tracks with moving range average (solid line), true targets; also showing first and last compass readings (dotted lines); tone one of target two

This work has established that reasonable target and bearing estimates can be obtained for very long arrays working with targets which are of the order of hundreds of kilometres from the array and lie at the edge of its near-field. The approach used lends itself to generalised optimization and search methods without incurring significant computational overheads and the use of a Genetic Algorithm has been shown to allow rapid range and bearing solutions to be obtained. The approach also allows for target track modelling so that moving averages can be obtained which allow improved range estimates to be obtained. Finally, it is worth noting that further refinements on the ideas presented here would seem to require more sophisticated models of the target and array shape motions over time. For instance, targets could be limited to having known speeds and turning rates or hydrodynamic calculations of the array behaviour could be used to improve on the modelling of its shape.

Acknowledgements

This work was supported by the Defence Research Agency under contract DRA-CB/CEN/9/4/2057/155/A.

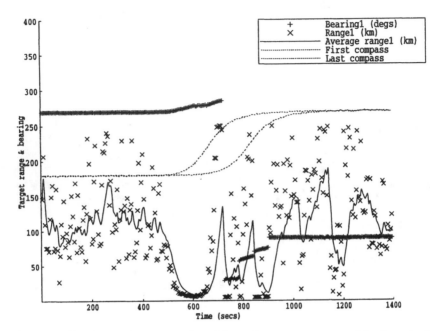

Fig. 11.4. Bearing and range tracks with moving range average, false targets; tone one of target one

References

1. T. Back, F. Hoffmeister, and H.-P. Schwefel. A survey of evolution strategies. In R. K. Belew and L. B. Booker, editors, *Proceedings of the 4th International Conference on Genetic Algorithms (ICGA IV)*, pages 2–9, San Diego, 1991. Morgan Kaufman Publishers, Inc.
2. S. Baluja. *Population-Based Incremental Learning: A Method for Integrating Genetic Search Based Function Optimization and Competitive Learning.* Carnegie Mellon University, 1994.
3. M.J. Box. A new method of constrained optimization and a comparison with other methods. *Comp. J.*, 8:42–52, 1965.
4. L. Davis. Bit-climbing, representational bias, and test suite design. In R. K. Belew and L. B. Booker, editors, *Proceedings of the 4th International Conference on Genetic Algorithms (ICGA IV)*, pages 18–23, San Diego, 1991. Morgan Kaufman Publishers, Inc.
5. NAg E04UCF. *NAg Mark 14 Reference Manual.* Numerical Algorithms Group Ltd., Oxford, 1990.
6. D.J. Edwards and P.M. Terrell. The smoothed step track antenna controller. *Int. J. Satellite Comms.*, 1(1):133–139, 1981.
7. D.D. Feldman and L.J. Griffiths. A projection approach for robust adaptive beamforming. *IEEE Trans. on Signal Processing*, 42(4):867–876, 1994.
8. D.B. Fogel. Applying evolutionary programming to selected traveling salesman problems. *Cybernetics and Systems*, 24(1):27–36, 1993.

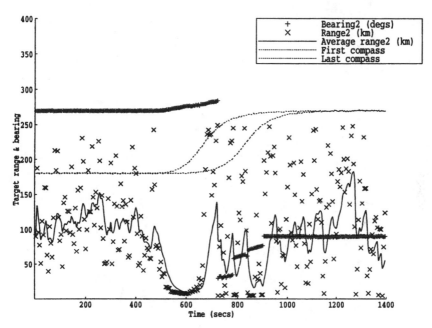

Fig. 11.5. Bearing and range tracks with moving range average, false targets; tone two of target one

9. W.R. Hahn. Optimum signal processing for passive sonar range and bearing estimation. *J. Acoust. Soc. Am.*, 58(1):201–207, 1975.
10. S. Haykin, J.H. Justice, N.L. Ouisley, J.L. Yen, and A.C. Kak. *Array Signal Processing*. Prentice Hall, 1985.
11. D.H. Johnson. The application of spectral estimation methods to bearing estimation problems. *Proc. I.E.E.*, 70(9):1018–1028, 1982.
12. A.J. Keane. *The OPTIONS concept design system user manual.* Dynamics Modelling Ltd., 1993.
13. A.J. Keane. Experiences with optimizers in structural design. In I.C. Parmee, editor, *Proceedings of the Conference on Adaptive Computing in Engineering Design and Control 94*, pages 14–27, Plymouth, U.K., 1994. University of Plymouth, P.E.D.C.
14. A.J. Keane. Passive vibration control via unusual geometries: the application of genetic algorithm optimization to structural design. *J. Sound Vib.*, 185(3):441–453, 1995.
15. S. Kirkpatrick, C.D. Gelatt, Jr., and M.P. Vecchi. Optimization by simulated annealing. *Science*, 220(4598):671–680, 1983.
16. S.J. Marcos. Calibration of a distorted towed array using a propagation vector. *J. Acoust. Soc. Am.*, 93(4)(1):1987–1994, 1993.
17. J.A. Nelder and R. Meade. A simplex method for function minimization. *Computer J.*, 7:308–313, 1965.
18. B.P. Ng, M.H. Er, and C. Kot. Array gain/phase calibration techniques for adaptive beamforming and direction finding. *IEE Proc Radar, Sonar Navigation*, 141(1):25–29, 1994.

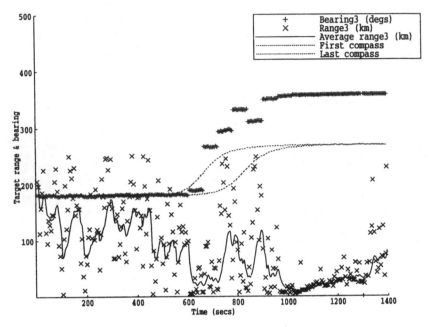

Fig. 11.6. Bearing and range tracks with moving range average, false targets; tone one of target two

19. L.A. Pflug, G.E. Loup, J.W. Loup, R.L. Field, and J.H. Leclere. Time delay estimation for deterministic transients using second and higher order correlations. *J. Acoust. Soc. Am.*, 94(3)(1):1385–1399, 1993.

20. W.H. Press, B.P. Flannery, S.A. Teukolsky, and W.T. Vetterling. *Numerical Recipes: the Art of Scientific Computing.* Cambridge University Press, 1986.

21. H.H. Rosenbrock. An automatic method for finding the greatest or least value of a function. *Comp. J.*, 3:175–184, 1960.

22. H.-P. Schwefel. *Evolution and Optimum Seeking.* John Wiley and Sons, 1995.

23. J.N. Siddall. *Optimal Engineering Design: Principles and Applications.* Marcel Dekker, Inc., 1982.

24. C.J. Tsai and J.F. Yang. Autofocussing technique for adative coherent signal-subspace transformatio beamformers. *IEE Proc Radar, Sonar Navigation*, 141(1):30–36, 1994.

25. D. Yuret and M. de la Maza. Dynamic hill climbing: Overcoming the limitations of optimization techniques. In *Proceedings of the 2nd Turkish Symposium on AI and ANN*, pages 254–260, 1993.

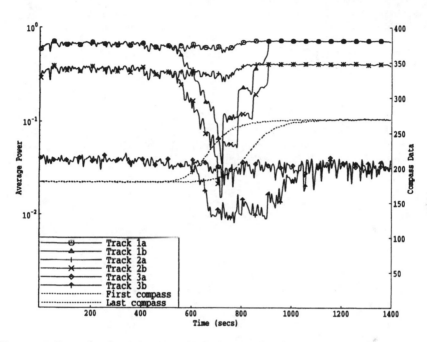

Fig. 11.7. Power levels of the six tracked signals; also showing first and last compass readings

Signal Design for Radar Imaging in Radar Astronomy: Genetic Optimization

Benjamin C. Flores[1], Vladik Kreinovich[2], and Roberto Vasquez[1]

[1] Department of Electrical and Computer Engineering, University of Texas at El Paso, El Paso, TX 79968, USA

[2] Department of Computer Science, University of Texas at El Paso, El Paso, TX 79968, USA

Summary. Radar imaging is an advanced remote sensing technique that maps the reflectivity of distant objects by transmitting modulated signals at radio frequencies and processing the detected echoes. By proper waveform selection, it is currently possible to image the surface of planets or asteroids from Earth with a relatively high degree of resolution, despite the astronomical distances to these objects. Waveforms that are used for radar astronomy are characterized by a large spectral bandwidth and long time duration, which can be obtained by phase modulating a long pulse train. An example of phase modulation is binary phase coding in which each pulse of the train is randomly assigned a phase of 0 or π. The corresponding echo pulses are correlated with the sequence of transmitted pulses to resolve the main features of an object. The correlation of long binary sequences can yield high resolution with significant clutter suppression. However, this process requires the selection of an optimum binary phase code among a significant number of possibilities. For a given code of length N, where N is the number of code elements, the population size is 2^N. Certain members of this population will exhibit good qualities for imaging, while others may behave poorly. In this chapter, we discuss the principles of radar imaging and the implementation of a genetic algorithm to find the "good" member codes from a population of codes of length N. This algorithm was successfully tested for binary phase codes of lengths $N = 13$ and 24 pulses. For the short length $N = 13$ case, in which an optimum code is known to exist, the rate of convergence to this optimum was poor relative to that of an exhaustive search. However, for the longer codes tested ($N = 24$), the genetic algorithm converged much faster than an exhaustive search for the large solution space.

1. Radar Imaging Principles

1.1 Radar concepts

The most basic radar experiment entails the transmission of a signal in the direction of some target, which reflects the signal back toward the radar, and the subsequent detection of the echo signal. In modern systems, the echo signal may be used to describe a target's location (i.e. range), analyze target motion, and to map the scattering features of the target. For a detailed description of the radar signal processing required for these purposes, see e.g., [1, 2, 8, 21, 23, 25, 26, 30, 32]. Target position and motion estimation are traditional priorities for military systems, aircraft sensors, weather monitoring, and traffic monitoring. Target imaging has found many uses in diverse scientific fields such as radio astronomy, automatic object recognition, and remote sensing among others.

Planetary radar astronomy is radar imaging applied to spatial objects within our solar system. A major concern that arises for such an application stems from the large distances to the targets and a system's capability to detect weak echoes from them.[1] Consequently, even with the best antennas and low-noise transceiver systems,[2] the task of successfully detecting an echo is overwhelming.

To conceptually understand the radar imaging process, assume the simple case of a short, monotone (single frequency) pulse of duration T seconds. The signal reflected from a target is received by the radar after a delay time τ_0, which corresponds to the range to the target. However, since the pulse has a duration of T seconds, the reflected signal contains information not only on the parts of the target located at a distance $\frac{c}{2}\tau_0$, but also about its parts located at distances up to $\frac{c}{2}(\tau_0 + T)$. Consequently, we try to make the duration of the signal as small as possible in an effort to simulate an impulse in time, thus increasing the accuracy in range measurements. However, this is done at the expense of signal power. For target detection, the energy E of the transmitted signal must exceed some threshold level E_0 of noise. Realizing that the energy of the transmitted signal is equal to the product $P \cdot T$ of the transmitter power P and the signal duration T, it is evident that a short pulse has very limited energy. Therefore, even with the most powerful transmitter with power P_0, from the condition $P_0 T \geq E_0$, it follows that the duration of the transmitted signal cannot be shorter than E_0/P_0. In applications where the object being imaged is far from the radar, a short pulse does not have nearly as much power that is necessary to properly detect the echo.

Thus, the short monotone pulse is not commonly used for imaging. An alternative is the use of wideband signals, which have large signal bandwidths. Phase or frequency modulating signals over increased bandwidths allows high resolution imaging without the constraint that the duration of the pulse be short. Therefore, only wideband signals are considered for radar astronomy.

1.2 Signal processing

The type of receiver of choice in radar astronomy applications is the correlator receiver. This type of receiver is equivalent to a filter whose impulse response matches the transmitted waveform [25, 30]. Hence, the output of the filter is the autocorrelation of the signal. The general mathematical equation for the autocorrelation of a signal s(t) is defined in Equation 1:

---

[1] The power of a signal that has travelled a distance d toward an object is proportional to $1/d^2$. The power of an echo returning to the radar is proportional $1/d^4$. Thus, for large distances, signal power at any point is very low relative to the transmitted power.

[2] NASA uses the 300 m Arecibo radiotelescope and the Goldstone radio astronomy facility.

$$R(\tau) = \int s(t) \cdot s^*(t - \tau)\, dt \tag{1}$$

where t is time and τ is the time delay of the echo. Since the echo is the response of the signal to the target, a more realistic equation simulating the correlation process is

$$R(\tau) = \int s(t) \cdot r^*(t - \tau)\, dt \tag{2}$$

where $r^*(t - \tau)$ is the complex conjugate of the received echo. Some examples of autocorrelations of binary phase codes of length N=24 are given in *section 4*.

The autocorrelation is characteristically a maximum at $\tau = 0$, which corresponds to zero delay, and is also conjugate symmetric about this point. The maximum of the main lobe is proportional to the signal energy. The autocorrelation of an optimum signal would have an extremely narrow main lobe and sidelobes[3] as low and uniform as possible. This condition corresponds to high *resolution* capability with insignificant *clutter*. This is explained in more detail below.

Resolution is a measure of the minimum distance between two scatterers at which they can still be detected individually. The generally accepted definition is that it is equal to the width of the main lobe of the autocorrelation. High resolution (corresponding to a smaller numerical value for resolution in time) leads to images with fine detail. In contrast, clutter refers to sidelobes that lead to ambiguities in the positions of scatterers. The sidelobes of the autocorrelation are in a sense *self noise* since they are noise components inherent to the signal being used. Any signal will generate some amount of self noise. For this reason, much effort has been placed into minimizing sidelobes. The higher these sidelobes are, the more clutter an image will have. Using waveforms with low, as well as uniform sidelobe structures, eliminates spurious false peaks.

1.3 Range-Doppler radar imaging

Suppose that we transmit a signal $s(t)$ with some form of phase modulation. Let $r(t - \tau_0)$ represent the reflected signal received at time τ_0. The power of this signal is proportional to the reflectivity (radar cross section) of a scattering feature (center) located at a distance $d = c\tau_0/2$ from the radar. If the target has many scattering centers, we would first receive the signal reflected from the scattering center of the target that is the closest to the radar. Subsequently, we would get the signal reflected from the more distant parts of the target. The result of this is a superposition of the returns from all the scatterers of the target. The correlation of this echo with a replica

---

[3] A sidelobe is one of the lobes surrounding the main response of the target.

of the transmitted signal will give us a profile of the target in which each scattering center is resolved. This profile is a one-dimensional image of the target's reflectivity versus range. Rotational motion between the radar and target allows us to obtain a two-dimensional image of a target by further separating the scatterers in a Doppler dimension that is perpendicular to the line of sight. The position of a scatterer in this dimension is related to its speed (relative to the radar).

The echo signal $r(t - \tau_0)$ characterizes the total reflective properties of all the points of a target that are located at a distance $c\tau_0/2$ from the radar. On a spherical planet, such points form a circle equidistant to the Earth on the planet's surface as shown in Figure 1.1. To obtain an image of the planet's surface, we must take into consideration the planet's rotation. Different points on the circle mentioned above have different speeds. Thus, their responses can be separated by their Doppler effect[4] changes in frequency. This imaging process takes into consideration not only the time delay of the radar signal (that corresponds to *range*), but also the change in frequency (that corresponds to cross-range). This mapping approach is called *range-Doppler radar imaging*. For a detailed description of radar imaging, see [5, 19, 24].

1.4 Binary phase codes

Random waveforms are used for applications that demand high resolution because of the desirable autocorrelation properties they exhibit. For instance, the autocorrelation of white noise, which is a waveform with random amplitude, frequency, and phase, is the Dirac delta function [12]. A delta function has a zero width and lacks sidelobe structure. In this sense, white noise is an ideal signal for imaging. Unfortunately, white noise has an infinite bandwidth, which makes the implementation of a matched filter a formidable task. However, it is possible, to transmit a signal that has a random parameter, such as frequency or phase, with some restrictions. Random frequency-hopped waveforms, which are randomly modulated in frequency and have a finite bandwidth, are currently under development, but require relatively expensive systems. Random phase modulation, and in particular, binary phase coding, is a more appealing option.

In order to remedy the power and duration constraints, we consider the processing of long binary sequences with the correlator receiver and an averaging procedure to enable the precise imaging of targets with close scatterers. These sequences are characterized by a large time-bandwidth product[5]

---

[4] The Doppler effect refers to a shift in the center frequency of the echo signal with respect to the reference due to the target's motion relative to the observer.

[5] Time-bandwidth product refers to the product of the duration of the signal and the signal bandwidth. Large bandwidths allow fine resolution, and large durations allow the transmission of more power. Therefore, waveforms with a large time-bandwidth product enable high resolution imaging of targets at far distances.

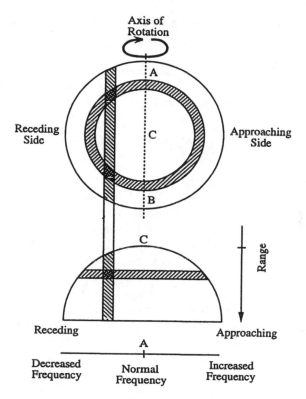

Fig. 1.1. Diagram showing contours of constant delay and Doppler for planetary imaging [10]

that enables high resolution and accuracy in either one-dimensional or two-dimensional imaging. Hence, high quality images showing much detail can be constructed.

As we stated previously, a binary phase code is a sequence of N pulses with a random phase of 0 or π associated with each pulse. The transmitted signal $s(t)$ is expressed by the following equation:

$$s(t) = \sum_{i=0}^{N-1} rect[(t - iT)/T]e^{j(2\pi f_c t + \phi_i)}, \qquad (3)$$

where f_c is the carrier frequency and ϕ_i is the random phase associated with each pulse. The function $rect[(t - iT)/T]$ is a pulse function of duration T which denotes the switching from pulse to pulse. If we omit the carrier frequency, the signal looks like an amplitude modulated waveform with random values of $+1$ and -1, as shown in Figure 1.2. The autocorrelation of such a sequence exhibits some desirable characteristics. For instance, the width of the autocorrelation's main lobe is equal to the duration of a single pulse. Thus, the range delay resolution is equal to the pulsewidth T. A direct result of this is that delay resolution is limited by the switching time from pulse to pulse that current technology allows. (The current switching time for the Arecibo systems is $\approx 100\mu s$.)

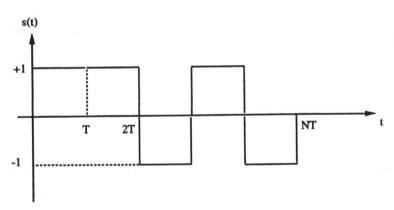

Fig. 1.2. Random binary phase code

The average sidelobe level of the autocorrelation for random binary phase codes is equal to $\sqrt{1/N}$.[6] Thus, the transmission of a longer sequence will yield an autocorrelation with a lower average sidelobe level. In addition, a longer sequence will have more resolution in Doppler. For our purposes, we

---

[6] This sidelobe level is for a normalized autocorrelation where the main lobe peaks at a level of 1.

select codes with very little tolerance for Doppler shifts [14] so that the main response measured in the correlator output is negligible for moving targets. For binary phase coding this tolerance for Doppler shifts defines a Doppler resolution, given by

$$\Delta f_d = 1/(NT). \tag{4}$$

1.5 Barker codes

Optimum binary phase codes, called Barker codes, are known for lengths $N = 2, 3, 4, 5, 7, 11$, and 13. The actual codes are shown in the table below. The autocorrelations for these codes have a uniform sidelobe structure, in which all sidelobes peak at a level $1/N$.

Code length N	Code elements
2	+−, ++
3	+ + −
4	+ + − +, + + + −
5	+ + + − +
7	+ + + − − + −
11	+ + + − − − + − − + −
13	+ + + + + − − + + − + − +

For the longest of these codes, the sidelobe level (SLL) is $1/N = 1/13 \approx 0.077$. Unfortunately, longer Barker codes do not seem to exist. This has been proven for *odd* N [29]. For *even* N, computer experiments have shown that no such codes exist for reasonably high N, and it appears highly unlikely that such codes will be found for larger, even N [4, 9, 28].

Incidentally, the situation with the codes for which $SLL \leq 1/N$ is not much better if, instead of allowing only the change in phase from 0 to π, we allow *arbitrary* phase changes every T seconds. In other words, if we consider sequences $s_1, \ldots, s_N$ of complex numbers for which $|s_i| = 1$. Such sequences with $SLL \leq 1/N$ are called *generalized Barker codes*. Generalized Barker codes were first found in [15] for all lengths $N \leq 13$. In [3], generalized Barker codes for $N = 14, \ldots, 19$ are given. No generalized Barker codes longer than 19 are known.

2. Formulation of the GA Problem

For the sake of better imaging, we set out to find binary phase codes whose autocorrelation's sidelobe level is as low as possible for $N > 13$. We call these codes *suboptimum* binary phase codes.

Long codes (large N), for which the autocorrelation's sidelobes will tend to be small (approximately $\sqrt{1/N}$), are desired. This will minimize the relative error of image reconstruction. As stated above, for $N \leq 13$, Barker codes which are optimal in the sense that all sidelobes peak uniformly at a level $1/N$, exist. For $N > 13$, such "optimal" signals are not known to exist, and therefore heuristic methods are used to find the next best signals.

In the search for suboptimum binary phase codes, there are increasingly large solution spaces to be explored. As the number of pulses increases, the number of combinations for these random binary sequences increases exponentially. Specifically, binary phase codes of length N pulses have a population size given by

$$n_{pop} = 2^N. \tag{5}$$

For small N, an exhaustive search, which looks at every code in the solution space, can be used with relative ease. For large N, an exhaustive search is not feasible. Therefore we have chosen to implement a genetic algorithm to search for suboptimum binary phase codes of large lengths from very large solution spaces.

In order to formulate the problem as a mathematical optimization problem, we select an objective function that combines or looks at all of the sidelobes. Each of the sidelobes contributes to the self noise of the resulting image; the larger the SLL, the worse the corresponding self noise. To a large extent, the image distortion is determined by the number of high sidelobes and the highest sidelobe level. In order to improve the image, it is natural to minimize the sidelobe level. Thus, we set out to minimize the function

$$J = \max(e_1, \ldots, e_{N-1}), \tag{6}$$

where e_i denotes the peak level (magnitude-wise) of each lobe of a normalized autocorrelation. Note that the main lobe, which is given by e_0, is not included in the minimization. In addition, we take advantage of the symmetry about zero delay of the autocorrelation to ease the amount of computations necessary. The objective function will therefore only look at one side of the autocorrelation ($e_1, \ldots, e_{N-1}$, not $e_{-(N-1)}, \ldots, e_{N-1}$).

The objective function in Equation 6 correctly describes the effect of the worst sidelobe e_i, but it does not take into consideration the influence of the other sidelobes. To take this influence into account, the overall influence of the sidelobes is minimized. This can be estimated as the sum of influences of different sidelobe levels e_i. This approximation corresponds to the fact that the influences are small, and therefore, their combined effects on the image can be neglected. If we denote the influence caused by the ith sidelobe level ($e_i \neq 0$) by $w(e_i)$, our approach leads to the minimization of the following objective function:

$$J = \sum_{i=1}^{N-1} w(e_i). \tag{7}$$

There are several reasonable choices for the function $w(e)$:

– For small e_i, a smooth function can be approximated by its first terms in the Taylor series: $w(e_i) = w(0) + w'(0)e_i + (1/2) \cdot w''(0)e_i^2 + \ldots$ For $e_i = 0$, we have $w = w(0)$. Since w attains its minimum when $e_i = 0$, we have $w'(0) = 0$. Therefore, the first non-trivial term is quadratic. Hence, we can use the approximate expression $w(e_i) = C_1 + C_2 e_i^2$, where C_1 and C_2 are arbitrary constants. This expression can be further simplified by realizing that we are not interested in the absolute value of $J = \sum w(e_i)$, but only in the codes that minimize it. Minimizing the sum $\sum C_2 e_i^2 = C_2 \sum e_i^2$ is equivalent to minimizing $\sum e_i^2$ [22]. So, we can take

$$w(e_i) = e_i^2. \tag{8}$$

This expression is known as the *square entropy measure* [11, 20].

– Another objective function can be derived from statistical arguments if we argue that e_i is proportional to the probability that a scatterer is located at τ_i. We may consider a method that enables us to choose the probability distribution in case several distributions are possible. This method chooses the probabilities $p_1, \ldots, p_n$ for which the entropy $S = -\sum p_i \cdot \log(p_i)$ takes the largest possible value. If we do not have *a priori* information on the distribution, then the maximum entropy S is achieved when all the values p_i are equal to each other. Vice versa, the minimum of the entropy is achieved when one of the values p_i is equal to 1 and the rest are 0. This minimum entropy result is similar to what we want to achieve for the values e_i. Hence, it is natural to minimize $J = \sum_{i=0}^{N-1} w(e_i)$, where

$$w(e_i) = -e_i \cdot \log(e_i). \tag{9}$$

This expression is the *standard entropy* measure [20].

– The function $w(p) = -p \cdot \log(p)$ used in the expression for the standard entropy is not the only function for which the minimum of $\sum w(p_i)$ is achieved when one of the p_i is equal to 1 and the rest are equal to 0. Other functions with the same property exist. In particular, we will consider a function of the form

$$w(e_i) = e_i \cdot \exp(-e_i). \tag{10}$$

This expression, which is known as *exponential entropy*, has been useful in a related radar image processing problem [11, 20].

Using these three objective functions, the best possible signals will be determined.

3. Application of Genetic Optimization

The goal is to find the best signal code $s_1, \ldots, s_N$ for which the objective function J takes the smallest possible value. Traditional numerical optimization methods are not applicable for two reasons:

- These techniques are designed for smooth well-behaved objective functions with few local extrema. In comparison, the dependence of J on s_i is highly irregular, as numerical experiments show that this dependence has many local extrema.
- These methods are designed to find values of s_i that can take all possible values from a certain interval. However, we are interested only in the values $s_i = \pm 1$.

Hence, non-traditional optimization techniques are needed. Of these techniques, *genetic algorithms* [7, 13] seem to be the most suitable. Indeed they are designed for situations in which every variable takes exactly one of two possible values.[7]

In our application, each individual is a binary sequence of length N that represents a candidate signal. For simplicity, 0 is interpretted as +, and 1 as −. Following this idea, the algorithm generates random candidates for the first generation. Once a generation is formed, individuals are selected for the next generation. A survival-of-the-fittest rule is applied to select parent codes where each individual k is selected with the probability $p_k = F_k / \sum F_j$ that is proportional to its fitness, where F is a measure of fitness. Strings are then recombined to form new codes, and mutations are applied with a low probability to finalize the new individual's code.

Recombination between two parent codes $s_i^{(1)}$ and $s_i^{(2)}$ is usually done by selecting a few random places (from 1 to N) that divide the code into several segments. In this chapter, a simple recombination strategy is applied, which corresponds to breaking at one random place. The crossover probability is set to 1.

The algorithm was tested with a probability of mutation of 0.01 per bit. Experimenting, we tried several other probabilities, for which the results were worse. For example, for 0.1, no convergence was attained; for 0.001, the convergence was much slower than for 0.01. We therefore maintained the mutation probability of 0.01.

4. Results

As a starting point, we tested the algorithm for a code length of 13 and used the fitness function given by Equation 6. We began with a population size of 10, and incremented the population size in steps of 20 up to 500 (i.e., 10, 30, 50, 70, ..., 500). We ran the genetic algorithm for twenty generations at 10,000 runs per population size. We also tested the algorithm for binary phase codes of length $N = 24$. For this length, we used *suboptimum* codes,

---

[7] To be more precise, genetic algorithms can also be applied to optimization with continuous variables, in which case, each continuous variable x must be represented as a sequence of discrete binary variables, e.g., as a sequence of variables that are described by bits in the binary representation of x.

that were previously discovered using an exhaustive search, to determine a convergence measure. The results for these test cases are given in [20]. The resulting *optimal* population size (in the sense that 90% convergence is attained) is approximately 250, for each case ($N = 13, 24$). This corresponds to the calculation of approximately 5,000 objective functions.

For the training case $N = 13$, we tested all objective functions mentioned above, namely the sidelobe minimization, standard entropy, exponential entropy, and square entropy measures. We found the Barker code 90% of the time after calculating 5,000 objective functions. This is slightly better than an exhaustive search of all 2^{13} possible binary sequences of size 13.

To check how well the algorithm would work for larger N, we applied it to the case $N = 24$. For this length code, we know the *suboptimum* codes have a maximum sidelobe level of $3/N$. Again, we calculated 5,000 objective functions to obtain 90% convergence. This is only a small fraction of the total number of calculations needed for an exhaustive search, which requires 16,777,216 objective function calculations.

Refer to Figures 4.1–4.4 for the autocorrelations of the signals found using the genetic search algorithm with a population size of 500. The resulting best signals are:

- For the worst-case criterion (Equation 6), the best signal has $J = 3/N$. As we mentioned above, this signal has already been determined via an exhaustive search. The corresponding code in hexadecimal notation is 047274. Translating hex into binary and binary into signs, we get

$$+ + + + \; + - + + \; + - - - \; + + - + \; + - - - \; + - + +$$

For this signal, $|e_i| \le 3/24$ for all $i \ne 0$.
- For the function $w(e_i) = e_i^2$, the best signal is 6F4BAE, i.e.,

$$+ - - + \; - - - - \; + - + + \; - + - - \; - + - + \; - - - +$$

For this signal, there are two sidelobes with $|e_i| = 5/24$ near the main lobe, two sidelobes at a level $3/24$, and for all the others, $|e_i| \le 2/24$.
- For the standard entropy function, the best signal is 6C4B55, i.e.,

$$+ - - + \; - - + + \; + - + + \; - + - - \; + - + - \; + - + -$$

For this signal, there are two very large sidelobes at a level $11/24$, few sidelobes with $|e_i| = 4/24$, and the rest have $|e_i| \le 2/24$.
- Finally, for the exponential entropy function, the best signal is 38156D, i.e.,

$$+ + - - \; - + + + \; + + + - \; + - + - \; + - - + \; - - + -$$

For this signal, the sidelobes are at a level of $4/24$ near the main lobe, while the others are $\le 3/24$.

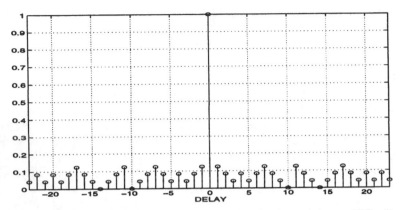

Fig. 4.1. Autocorrelation of binary phase code found using minimum SLL criterion. Code in hexadecimal = 047274

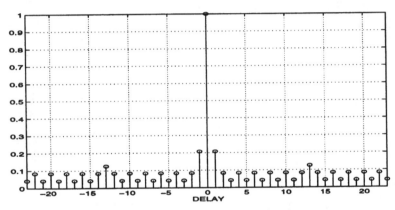

Fig. 4.2. Autocorrelation of binary phase code found using square entropy. Code in hexadecimal = $6F4BAE$

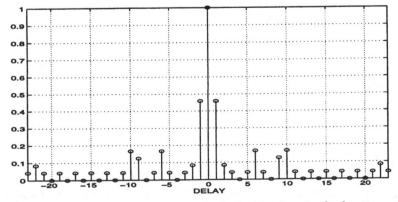

Fig. 4.3. Autocorrelation of binary phase code found using standard entropy. Code in hexadecimal = $6C4B55$

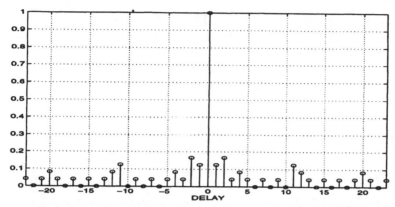

Fig. 4.4. Autocorrelation of binary phase code found using exponential entropy. Code in hexadecimal = 38156D

5. Conclusions and Future Work

A simple genetic algorithm was successfully implemented to search for optimum and suboptimum binary phase codes of a given length. The lengths $N = 13$ and 24 were used since the solutions for these lengths are already known. Four different fitness functions, which essentially characterize the sidelobe behavior of a code's autocorrelation, were tested to find the best fitness measure. For both lengths and for each of the objective functions, the genetic algorithm found the best code in more than 90% of the cases by testing 5,000 signals (20 generations of 250 individuals each). These results are very encouraging for the long length $N = 24$ codes, for which a very large solution space (2^{24} possible codes) exists. We believe that the GA can be used for even longer length binary phase codes for which the optimum/suboptimum are not known. Referring to the test results of Figures 4.1–4.4 for a population size of 500, the minimum sidelobe level criterion works the best out of the four fitness measures. All sidelobes are at the minimum level, and also appear to be somewhat uniform or evenly spread.

In this chapter, we formulated the criterion for choosing the best signal in terms of the autocorrelation, and used semi-heuristic arguments to reformulate and simplify this criterion in terms of a generalized entropy expression. This was based on the assumption that the image would only be corrupted by white noise. In reality, noise corruption is often better characterized as a fractal, random process [27]. In this case, we would need a different criterion, which may lead to different signals $s(t)$ as being optimal.

We are also interested in applying genetic algorithms to other waveform modulation techniques. In particular, random frequency hop coding is a good candidate. This technique, which is an alternative to binary phase coding, al-

lows higher resolution than is currently possible with binary phase codes.[8]
As is the case for binary phase codes, frequency hop codes are random waveforms, and therefore large solution spaces exist for different length codes.
Also, because of the similarity in signal processing for these two waveforms,
the same objective functions used above could be implemented.

Acknowledgments

Dr. Kreinovich was partially supported by NSF Grant No. EEC-9322370 and
by NASA Grant No. NAG 9-757. Dr. Flores was funded by NASA Contract No. NAS7-918, Task Order No. RE-252, under the development of the
Continuous Engineering Science and Technology Advancement (CUESTA)
for Undergraduate Minorities Program. The authors are thankful to D. Dasgupta and Z. Michalewicz, for their editorial comments.

References

1. R. S. Berkowitz, *Modern Radar*, John Wiley & Sons, N.Y., 1965.
2. A. B. Carlson, *Communication Systems, An Introduction to Signals and Noise
   in Electrical Communication*, McGraw-Hill, N.Y., 1986.
3. N. Chang and S. W. Golomb, "On n−Phase Barker Sequences", *IEEE Transactions on Information Theory*, 1994, Vol. 40, No. 4, pp. 1251–1253.
4. M. N. Cohen, P. E. Cohen, and M. Baden, "Biphase codes with minimum peak
   sidelobes", *IEEE National Radar Conference Proceedings*, 1990, pp. 62–66.
5. C. E. Cook and M. Bernfeld, *Radar Signals*, Academic Press, N.Y., 1967.
6. J. P. Costas, "A study of a class of detection waveforms having nearly ideal
   range-Doppler ambiguity properties", *Proceedings of the IEEE*, 1984, Vol. 72,
   No. 8, pp. 996–1009.
7. Yu. Davidor, *Genetic Algorithms and Robotics, A Heuristic Strategy for Optimization*, World Scientific, Singapore, 1991.
8. J. L. Eaves and E. K. Reedy (eds.), *Principles of Modern Radar*, Van Nostrand
   Reinhold Co., N.Y., 1987.
9. S. Eliahou, M. Kervaire, and B. Saffari, *A new restriction on the length of Golay
   complementary sequences*, Bellcore Tech. Mem. TM-ARH 012–829, October 24,
   1988.
10. J. V. Evans and T. Hagfors, *Radar Astronomy*, McGraw-Hill Book Company,
    N.Y., 1968.
11. B. C. Flores, A. Ugarte, and V. Kreinovich, "Choice of an entropy-like function
    for range-Doppler processing", *Proceedings of the SPIE/International Society
    for Optical Engineering, Vol. 1960, Automatic Object Recognition III*, 1993, pp.
    47–56.

---

[8] Recall that range delay resolution for binary phase codes is equal to the pulse
duration, and therefore is limited by the fastest switching time possible with
current technology. Range delay resolution is dependent on the modulation
bandwidth for frequency hop coding.

12. William A. Gardner, *Introduction to Random Processes with Applications to Signals and Systems*, McGraw-Hill, Inc., New York, NY, 1990.
13. D. E. Goldberg, *Genetic Algorithms in Search, Optimization, and Machine Learning*, Addison-Wesley, N.Y., 1989.
14. S. W. Golomb (ed.), *Digital Communications with Space Applications*, Prentice-Hall, Englewood Cliffs, NJ, 1964.
15. S. W. Golomb and R. A. Scholtz, "Generalized Barker sequences", *IEEE Trans. Inform. Theory*, 1965, Vol. IT-11, pp. 533–537.
16. J. R. Klauder, "The design of radar systems having both range resolution and high velocity resolution", *Bell Systems Technology Journal*, 1960, Vol. 39, pp. 809–819.
17. V. Kreinovich, C. Quintana, and O. Fuentes, "Genetic algorithms: what fitness scaling is optimal?" *Cybernetics and Systems: an International Journal*, 1993, Vol. 24, No. 1, pp. 9–26.
18. N. Levanon, "CW Alternatives to the Coherent Pulse Train–Signals and Processors", *IEEE Transactions on Aerospace and Electronic Systems*, 1993, Vol. 29, No. 1.
19. D. L. Mensa, *High Resolution Radar Cross-Section Imaging*, Atrech House, Norwood, MA, 1991.
20. J. L. Mora, B. C. Flores, and V. Kreinovich. "Suboptimum binary phase code search using a genetic algorithm", In: Satish D. Udpa and Hsui C. Han (eds.), *Advanced Microwave and Millimeter-Wave Detectors*, Proceedings of the SPIE/International Society for Optical Engineering, Vol. 2275, San Diego, CA, 1994, pp. 168–176.
21. F. E. Nathanson, *Radar Design Principles*, McGraw-Hill, N.Y., 1969.
22. H. T. Nguyen and V. Kreinovich, "On Re-Scaling In Fuzzy Control and Genetic Algorithms", *Proceedings of the 1996 IEEE International Conference on Fuzzy Systems, New Orleans, September 8–11, 1996* (to appear).
23. A. W. Rihaczek, *Principles of High-Resolution Radar*, McGraw-Hill, N.Y., 1969.
24. A. W. Rihaczek, "Radar waveform selection", *IEEE Transactions on Aerospace and Electronic Signals*, 1971, Vol. 7, No. 6, pp. 1078–1086.
25. M. I. Skolnik, *Introduction to Radar Systems*, Mc-Graw Hill, N.Y., 1980.
26. M. I. Skolnik (ed.), *Radar Handbook*, McGraw Hill, N.Y., 1990.
27. C. V. Stewart, B. Moghaddam, K. J. Hintz, and L. M. Novak, "Fractional Brownian motion models for synthetic aperture radar imagery scene segmentation", *Proceedings of the IEEE*, 1993, Vol. 81, No. 10, pp. 1511–1521.
28. R. J. Turyn, "On Barker codes of even length", *Proceedings of the IEEE*, 1963, Vol. 51, No. 9 (September), p. 1256.
29. R. J. Turyn and J. Storer, "On binary sequences", *Proc. Amer Math. Soc.*, 1961, Vol. 12, pp. 394–399.
30. D. R. Wehner, *High Resolution Radar*, Artech House, Norwood, MA, 1987.
31. K. M. Wong, Z. Q. Luo, and Q. Lin, "Design of optimal signals for the simultaneous estimation of time delay and Doppler shift", *IEEE Transactions on Signal Processing*, 1993, Vol. 41, No. 6, pp. 2141–2154.
32. P. M. Woodward, *Probability and Information Theory with Applications to Radar*, McGraw-Hill, N.Y., 1953.

Evolutionary Algorithms in Target Acquisition and Sensor Fusion

Steven P. Smith and Bertrand Daniel Dunay

System Dynamics International, St. Louis, MO 63026

Summary. In this paper, we present evolutionary algorithm techniques for signal processing. A proprietary evolutionary algorithm (EA) is applied to the task of discriminating stationary "targets" from "clutter" in both a radar and imaging sensor. Three applications are made of the EA. First, the EA is used to derive an algorithm for *radar* target detection including features and decision logics; second, the EA is used to derive an algorithm for an *imaging sensor* target detection; and third, the EA is used to derive a *fused detector*, which combines the information from both sensors to make a better detection decision. After detection, the EA may also be used to develop a fused classifier. The EA used can be categorized as a program induction algorithm. That is, the result of the evolutionary process is one or more computer programs that implements a detection or classification algorithm. A technique for *real time adaptation* of target acquisition algorithms is also discussed.

1. Introduction

It is often the case in signal processing that algorithms must be designed to filter or classify signals based on some available input data. Often, for these applications we have large amounts of data including examples of correctly filtered or classified cases, which have been collected under controlled conditions. It is the signal data which, in fact, dictates the design of the filter or classifier. This situation is perfectly suited for using Evolutionary Algorithms (EA) to discover the appropriate signal processing algorithm to perform the required filtering or classification. The example illustrated in this paper is a target detection problem using two sensors: a radar and an imaging sensor (see Figure 1.1).

The evolutionary algorithm used in this paper, named e, was developed at SDI and has been applied to a variety of target acquisition problems. The

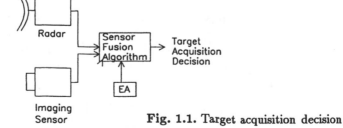

Fig. 1.1. Target acquisition decision

essential feature of e that makes it so useful for this type of problem is that it evolves procedural computer programs, which are provided as output in pseudo code or in compilable C, FORTRAN, or Basic. The advantage of e over traditional genetic algorithms is that there is no need to predetermine the form of the solution so that the genetic algorithm can evolve parameters of the predetermined form. Also, there is an advantage over what has come to be known as Genetic Programming [Koza, 1992], which produces LISP code, in that the procedural code is generally faster to execute and more readable than LISP.

The general scenario for which these algorithms are to be applied is acquisition of stationary ground targets. Radars used for this purpose are typically millimeter wavelength, frequency stepped, coherent, and fully polarimetric. For this paper, we are considering a real aperture radar such as might be mounted on a helicopter or ground vehicle. The imaging sensors could include FLIR (Forward-Looking Infrared Radar) or TV; however, the FLIR imagery is typically easier to use. For this paper, we have used simulated data for both the radar and imaging sensor.

2. System Block Diagram

Figure 2.1 is a system block diagram (simplified) of the target acquisition system we are considering. The two sensors are on the left. The radar is used to make the initial detections, and the imaging system is used to confirm these detections. This approach is consistent with the radar's capability to search very large areas in a very short time, which cannot be done with the imaging system. However, we expect that the imaging system can be used to eliminate radar false alarms because of the independent look provided by the imaging sensor. The radar sensor can be operated initially to pass almost all targets at the expense of a higher than acceptable false alarm rate. This insures that very few, if any, targets are not referred to the imaging sensor for evaluation. The fused detector, which uses information from both the radar and the imaging sensor, should be able to make better decisions than either sensor alone and is set to operate at the required false alarm rate.

2.1 Radar

The top line of the diagram represents the radar processing. Received radar returns are detected, mixed with signals from the coherent oscillator, and digitized to form in-phase (I) and quadrature (Q) values for each received pulse. Together the IQ data provides both amplitude and phase information on the returned pulse (Edde, 1993):

$$I = Acos\theta$$
$$Q = Asin\theta$$

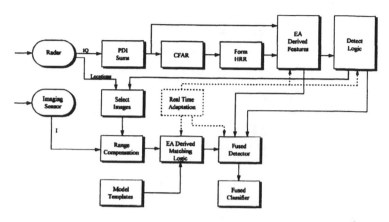

Fig. 2.1. System block diagram

where A is the amplitude of the signal and θ is the phase angle of the received signal ($\theta = 2\pi ft, f = frequency, t = time$). The received signal may be expressed as

$$I + jQ = A\cos + jA\sin = Aej$$

where $j = \sqrt{-1}$. Multiple pulses are transmitted during each dwell (time on target) at different frequency steps and different polarizations. In a linearly polarized radar, pulses are transmitted at both horizontal (H) polarizations and Vertical (V) polarizations and also received at both polarizations. This results in four types of polarization: HH, VV, HV, VH, where HH means transmit horizontal, receive horizontal, etc. For rotationally symmetric targets $HV = VH$, so that it is generally only necessary to have three channels. Each of these values are complex quantities of the form $HH = Ae^{j\theta} = I + jQ$. Thus, for each frequency step transmitted, HH, VV, and VH returns are collected. Before further processing, two of the three linear polarizations are converted to "odd" and "even" polarizations, so named because they conceptually represent the number of bounces off the scatterer taken by the return signal. For circular polarized signals (e.g. right circular polarization, R), each bounce will shift the return to the opposite polarization (e.g. left circular polarization, L). Thus, for an odd number of bounces an RL pulse (transmit R receive L) will be strong, and for an even number of bounces, an RR pulse will be strong. We can compute RR, RL, and LL from HH, VV, and VH as follows (Stutzman, 1993):

$$
\begin{aligned}
RR &= 0.5(HH - VV) - jVH \\
LL &= 0.5(HH - VV) + jVH \\
LR &= 0.5(HH - VV) = RL
\end{aligned}
$$

Now with these conversions, we define the three polarizations we will use to process the radar signal as:

$$odd = LR$$
$$even = 0.5(RR + LL)$$
$$cross = VH$$

The amplitude given by the IQ data for all frequency step pulses is summed for each polarization to get three Post Detection Integration (PDI) sums, one for each of HH, VV, and VH. The PDI sums are used in a constant false alarm (CFAR) stage to eliminate many clutter cells by simply comparing a radar cell's PDI sum to that of surrounding cells. For example, the even PDI sum of the cell under test (CUT) might be divided by the average of the odd PDI sums for surrounding cells (excluding those cells immediately adjacent to the CUT).

The next step in the processing is to form a High Range Resolution (HRR) (Wehner, 95) profile across the radar cell by performing a Fast Fourier Transform (FFT) (or inverse FFT) using the returns from the multiple frequency steps. This transforms the return pulses from the frequency domain to the time domain, which is also the "range" domain. The resolution of the HRR depends on the number of frequency stepped pulses transmitted. For our problem, we will resolve the low resolution range cell into 12 range bins. Figure 2.2 illustrates the coverage of the 12 HRR range bins within the range cell. Notice that although the HRR range bins give us a very fine resolution in the downrange direction (one meter in our data set), resolution is not improved across the dwell (the radar beam width). Since a target only occupies a small portion of the dwell width, depending on range, each range bin which falls on the target also includes a lot of clutter to the sides of the target.

To provide consistency in the presentation of the HRR profile to the signal processing algorithms, the HRR peak signal is centered (put in the sixth position) based on the even polarization HRR. Values are wrapped to the opposite ends, as required. Centering the HRR profiles (see Figure 2.3) insures that candidate target profiles are always located in the same portion of the HRR when presented to the EA.

The evolutionary algorithm is used to extract features from both the PDI sums and the HRR profiles. Next, these features are used in a decision logic to make preliminary target detection decisions. The decision logic may be as simple as thresholding a single feature or may be evolved using the EA and all of the features.

2.2 Imaging sensor

The imaging sensor provides a limited resolution image of the scene consisting of intensity values in an array indexed by x in the horizontal direction and y in the vertical direction, $I(x, y)$. The imaging sensor in our simulation is a

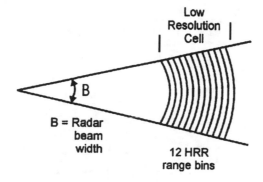

Fig. 2.2. Centering of HRR profiles

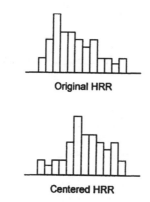

Fig. 2.3. Original and centered HRR

FLIR, (Forward Looking Infrared), which senses the radiant intensity emitted by the scene.

The imaging sensor is directed by the radar to image selected areas of the scene where the radar has detected candidate targets. As stated earlier, the candidate targets are chosen so as to have a very low probability of missing any real targets. After the scene is range compensated based on information from the radar, a template matching algorithm, as described in paragraph 6 below, is used to extract features from the scene. These features are used by the EA to derive higher level features which are subsequently used in the sensor fusion decision logic.

2.3 Fused detector and classifier

The fused detector and classifier are also based on EA derived logic. In this paper we will only present results for the detector, but the approach is very

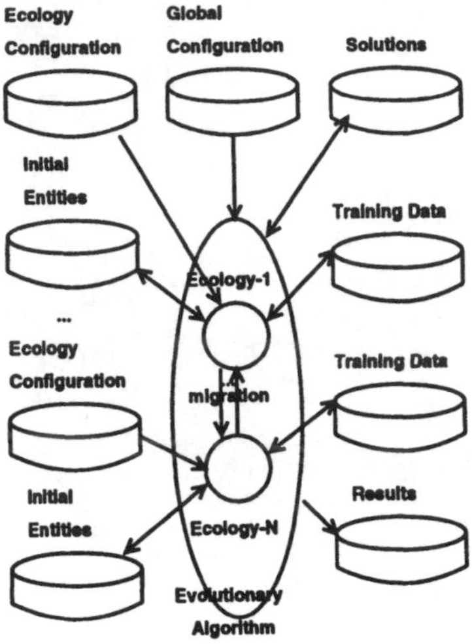

Fig. 2.4. e Evolutionary Algorithm

similar for the classifier. The fused detector uses feature level fusion. Five EA derived features from the radar and five EA derived features from the imaging sensor are provided in a training set together with the truth value (0 = clutter, 1 = target). The EA evolves the decision logic based on the features from both sensors.

3. Evolutionary Algorithm Description

The EA used is a C++ application called e. e combines a genetic algorithm [Holland, 1975] with techniques from genetic programming [Koza, 1992], hill climbing, heuristic search for coefficients [Press, 1994], and machine learning [Dunay, 1995]. The genetic algorithm uses traditional techniques for handling variable length genomes. e's genomes represent simple programs. The programming language is an imperative programming language which is a simple subset of C, but includes function calls, if ... then ... else, for loops, and a rich set of arithmetic operators. These programs can access the training data, global constants, their own local variables, and can evolve their own local constants.

e supports multiple ecosystems with multiple migration and intermarriage strategies. The topology of the ecosystems is programmable. Among the prebuilt topologies are queue, circular queue, and hierarchy. Migration occurs on demand from the destination ecosystem. The source ecosystem will send an individual based on the current migrant selection strategy. Intermarriage is accomplished by allowing mating between partners from different ecosystems. This permits the infusion of new genetic material from another ecosystem, but reduces the probability of dominance by the foreigner. A number of selection strategies are available for the choice of migrants and foreign marriage partners.

Reproduction within an ecosystem is done by allowing traditional crossover and X on linear chromosomes and subtree swapping at the arms of if ... then ... else statements.

When the GA discovers a new best program, the EA uses hill climbing to attempt to improve it. Hill climbing operates at the program element level and searches for a single point mutation which improves performance. Hill climbing continues on this new program until no further improvement is possible with a single point mutation. This newly discovered best individual is then the subject of a heuristic search to optimize all the local coefficients used in the program. Powell's Quadratically Convergent Method [Press, 1994] is used because of its ability to simultaneously optimize multiple coefficients.

One of the most powerful elements of e is its ability to leverage long term learning to produce better solutions to current problems. When a problem is solved, e may insert the new solution into the library.

When e is presented with a new problem, the library of sub-programs is automatically integrated into the search for solutions. The user may also insert programs into the library. This allows the user to integrate his/her insights into the problem with the efforts of e.

Solutions may appear in e's abbreviated syntax, or in the user's choice of C, FORTRAN, or Basic.

e is extensively configurable. All stochastic events (sexual reproduction, mutation, migration, inter-marriage) are linked to user configurable probabilities. Operations like hill climbing, migration, and intermarriage can be delayed. The creation of individual ecosystems can also be delayed to allow the development of supporting ecosystems. All operators and operands in the language can be configured to match a particular problem. The number of ecosystems and their topology can be chosen. The maximum number of instructions and the minimum and maximum size of local program storage can be configured.

e is implemented in C++. The training data class knows how to read its data from file and transparently cache requests for calculated data (max, min, average, etc.). The ecology configuration class also knows how to initialize itself from disk data and automatically resolves conflicting configuration parameters, or in extreme cases can shut down the EA because of bad configuration data. The EA class allocates the ecologies and mediates their requests for migrants, ecology configuration data, and training data. This allows the EA class to implement data sharing. If two ecosystems request the same training data or configuration data, the EA class will pass the pointers to the same configuration or training data classes.

4. Radar Features

There are a variety of radar features that might be formed for input to the evolutionary algorithm. However, we have chosen to take the most basic approach and let the EA evolve the higher order features.

Table 4.1 summarizes the basic features used. The PDI sums for the odd, even, and cross polarizations for the cell under test (CUT) are the first three features. The next three features are the average of the PDI sums for the cells surrounding the CUT, with guard cells excluded. This is illustrated in Figure 4.1.

Table 4.1. Basic radar features

Feature	PDI sum	PDI avg	HRR odd	HRR even	HRR cross
Number	3	3	12	12	12

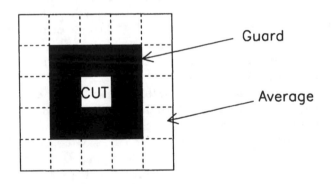

Fig. 4.1. PDI averages of surrounding cells

These average values allow the CUT to be compared to the surrounding area, most of which is presumably clutter. As seen from the figure, those cells nearest the CUT are excluded (they may contain the return from a portion of a target that straddles cells), and the 16 cells surrounding the CUT are averaged.

The three high range resolution profiles are the last three sets of features. As noted earlier, these are centered based on the even profile prior to presentation to the EA.

These radar features are formed for each low resolution radar cell and provided in the training set for the EA with the truth value for that cell. One low resolution radar cell corresponds to one row of data (truth value +

42 features) in the EA training set; and of course, once the EA has evolved the higher level features, these same basic 42 features are made available as input to the program evolved by the EA.

5. Imaging Features

Twelve features were formed for each image corresponding to twelve target templates (three targets times four poses for each target). Each feature was computed by finding the maximum value of the cross-correlation function, $R(m, n)$, over the image for a template. $R(m, n)$ is given by [Duda, 1973]:

$$R(m, n) = \sum \sum I(x, y) T(x - m, y - n)$$

where $I(x, y)$ is the image and $T(x - m, y - n)$ is the template of the target. The summation is performed over the template area, which is a 4x4 resolution cell area. $R(m, n)$ is further scaled by multiplying by 10^{-4} in order to match the magnitude of the radar features. The magnitude of the scale factor is arbitrary except that resulting values need to be of the same order of magnitude as the radar features to facilitate ease of use by the evolutionary algorithm. Figure 5.1 illustrates the manner in which the template is moved over the image to compute the cross-correlation function for each coordinate (m, n) in the image. The maximum cross-correlation value is saved as the feature associated with that template. For twelve templates we end up with twelve features.

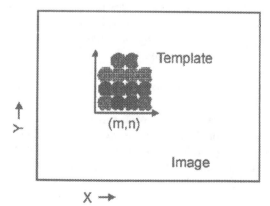

Fig. 5.1. Template and the image

The size of the image area to be searched with the template is equal to the dwell width horizontally and the elevation uncertainty vertically. This area will, therefore, contain the object detected by the radar. It may also contain other objects and/or targets that are not in the radar cell because they are located at longer or shorter ranges.

6. Fused Detector

The fused detector is simply an EA derived decision logic, which uses as input features from both the radar and imaging sensor. Five EA features were used from each sensor. With the total of ten features plus the truth value, the EA was run to produce the required decision logics.

7. Targets and Scenes

Scenes and targets were produced by a simulation and assigned intensity and radar cross-sections which provided a somewhat challenging detection problem. There are, of course, limitations to the realism of the simulated scenes, which make the problem "easier" than actual sensor data. Nevertheless, the simulation provides an excellent basis on which to demonstrate the techniques associated with the use of EAs in target acquisition.

7.1 Targets

Three target types were used, and each was allowed to have four poses (four aspects with respect to the sensor location). The twelve resulting target presentations are shown in image intensity format in Figure 7.1. Each row represents a different target type with front, side, rear, and opposite side poses. The first row is target type 1, second row target type 2, and third row target type 3. The four images within each row are from left to right, front pose, side pose, rear pose, other side pose. Each image resolution cell shown in the figure also corresponds to a radar scatterer, with a nominal radar cross-section for each of the three polarizations. One resolution cell in the image corresponds to approximately one meter diameter.

Targets were placed in the scene with random locations and orientations, and had noise randomly added to each intensity resolution cell and radar scatterer. In any target siting, the bottom two rows of the target could be masked by the terrain.

7.2 Image scene

Figure 7.2 shows an image which contains two targets. The brighter the image the greater the intensity. A type three target in an opposite side pose can clearly be seen in the upper left corner of the image with its lower left corner at coordinates $(0,4)$. (Note the index on the image starts with 0 in both

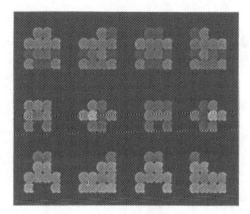

Fig. 7.1. Target templates

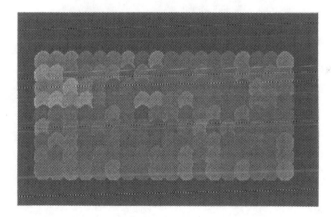

Fig. 7.2. Image scene with target

directions.) The second target, which is much more difficult to see, has its lower left corner at coordinates $(7, 2)$. It is a type one target in a rear pose. Figure 7.3 shows an image with no targets in the scene. The distribution of target resolution cell intensities as compared to scene intensities is shown in Figure 7.4.

7.3 Radar

Figure 7.5 is a B-Scope display for a portion of a radar scan. The B-Scope displays azimuth horizontally and range vertically. There are 10 azimuth dwells by 8 range dwells in Figure 7.5. Target locations are given in Table 7.1 (zero indexed). The intensity in the display is based on the even PDI sum for each cell. The brighter the spots represent the stronger the returns. We can see that some targets are clearly evident, but others are not.

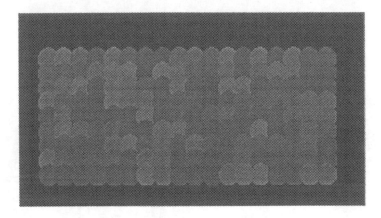

Fig. 7.3. Image scene without target

Target and Scene Intensities

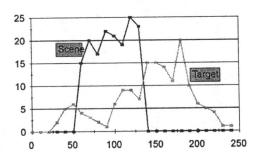

Fig. 7.4. Intensity histograms

Table 7.1. Target locations

Target Type	Dwell	Range
2	4	2
1	5	2
2	2	2
1	7	2
3	5	3
2	2	3
3	2	4
1	7	4

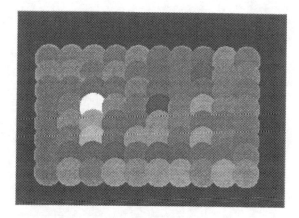

Fig. 7.5. Radar B-scope display

Figures 8.1, 8.2 and 8.3 display the HRR profiles for several clutter cells. The cross polarization is in front, then odd and even. Figures 8.4, 8.5, and 8.6 display the HRR profiles for several target examples. Figures 8.7, 8.8, and 8.9 are normalized histograms showing the distribution of PDI sums for cells with targets and cells without targets.

8. Evolutionary Algorithm Set-Up

Except for the actual training data, identical configurations were used to train e to distinguish target and clutter for both radar and image data. Five ecosystems were used. For purposes of migration and intermarriage, the ecosystems were arranged in a hierarchy that occasionally permitted the best program to migrate.

Each ecosystem consisted of approximately 70 entities. Each entity ranged in size from 6 to 20 instructions. Each instruction consisted of an operator/operand pair. The subset of operators applied to this problem is addition, subtraction, multiplication, division, load/store to three temporary variables, min function, max function, if...then...else control statements, and NOP (No Operation). Operands available included all the training data except the truth data, the average of a row of training data, the highest and the lowest value of training data, the constant "0", and the constant "−1". Hill climbing on instructions was enabled after 1000 reproductions. During hill climbing, instructions found not to contribute to fitness were removed.

Reproduction parameters were as follows. (Note that these operations are stochastic, and that the target rates are, therefore, only approximate.) Sexual reproduction was done 80% of the time, and mutation only was done 20% of the time. During sexual reproduction, 90% of the time instructions

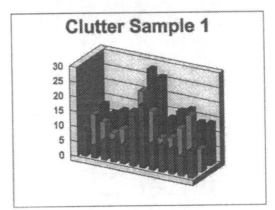

Fig. 8.1. Clutter sample 1

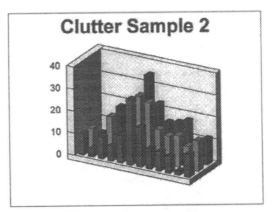

Fig. 8.2. Clutter sample 2

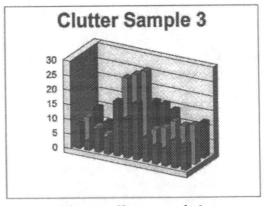

Fig. 8.3. Clutter sample 3

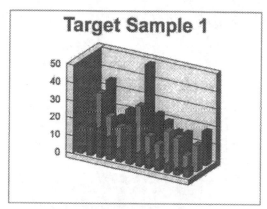

Fig. 8.4. Target sample 1

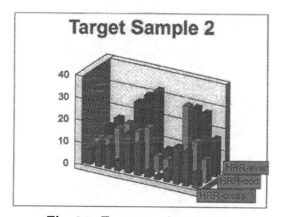

Fig. 8.5. Target sample 2

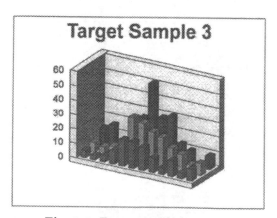

Fig. 8.6. Target sample 3

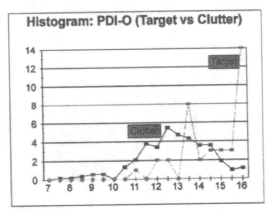

Fig. 8.7. Histogram PDI-O (Target vs. Clutter)

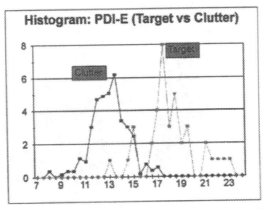

Fig. 8.8. Histogram PDI-E (Target vs. Clutter)

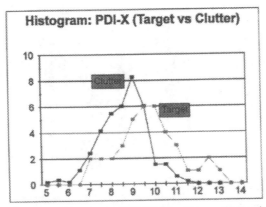

Fig. 8.9. Histogram PDI-X (Target vs. Clutter)

were mated, 10% of the time only coefficients (local constants) were mated. 3% of the child's instructions were mutated during birth. If both partners had "if.then.else" statements, then sexual reproduction was done via subtree swapping, otherwise crossover was done 90% of the time and X was done 10% of the time. The mutation rate would rise if no progress was observed for 100 reproductions. After 1000 reproductions, migration and intermarriage would be enabled. 1% of the time, a migrant was imported during reproduction. Also, 1% of the time, a foreign parent would participate in sexual reproduction, but would not actually be allowed to migrate.

Tournament selection was done with a tournament size of 4. Each program's score was calculated by:

$$score = err_0/n_0 + err_1/n_1$$

where err_0 and err_1 are the number of false alarms and missed targets respectively, and n_0 and n_1 are the number of clutter cells and targets in the training set respectively. No library learning was applied to this problem.

9. Results

Performance results for radar detection, imaging sensor detection, and fused detection are given below. Also, the programs resulting from each of the EA runs are given in the following paragraphs as C functions, which were provided as output from e.

9.1 Target acquisition performance

A test data set was generated using the same statistics as the original training set. This is representative of an operational situation in which the scene and targets presented to the sensors are similar to the data on which the algorithms are trained. Of course, this is not always the case. As targets or terrain or other target affecting conditions change, the performance of a specific algorithm can be expected to degrade. This leads to the requirement to adapt, which is discussed in paragraph 11 below.

Table 9.1 summarizes the performance results for the evolved detector for both the training and test data sets in terms of errors in detection. The results shown for the radar and imaging sensors by themselves are based on the "best" EA derived feature for each sensor.

Normally, target acquisition results would be stated in terms of probabilities of detection (Pd) and false alarm rate. Furthermore, a premium would be placed on a low false alarm rate. However, because we are solving a problem with simulated data, the Pd and false alarm values do not have an operationally significant meaning. The main purpose of this exercise was to demonstrate the basic approach.

Table 9.1. Target acquisition performance

Sensor	Percent Errors	
	Training Data	Test Data
Radar	2.5%	8.3%
Imaging	2.0%	8.0%
Fused	0.0%	4.2%

9.2 EA features for radar

Five EA features were evolved for the radar as shown below. The variables $X[1]$, $X[2]$, etc. refer to the input basic radar or image features in the order they are listed in the columns of the training file. The radar features are in the order listed in Table 4.1. The PDI sums and PDI averages are listed in the order of odd, even, and cross. So, for example, $X[1]$ = PDI sum odd, $X[2]$ = PDI sum even, etc. The first bin in the odd HRR profile is $X[7]$, and the maximum value in the odd HRR profile (due to centering) is $X[12]$.

```
float entity_0(float X[],int num)
{
float Accum,SIN,COS,LN,EXP,FUZZY,LAVG;
float TEMP1,TEMP2,TEMP3,CUBE,SQUARE;
short AR;
AR=1;
Accum=SIN=COS=LN=EXP=FUZZY=LAVG=0.0;
TEMP1=TEMP2=TEMP3=CUBE=SQUARE=0.0;
Accum = Accum - X[40];
Accum = Accum - X[36];
Accum = Accum + hi1(X,num);
Accum = Accum - X[27];
Accum = Accum - X[15];
Accum = dmin(Accum,LAVG);
Accum = Accum - X[15];
Accum = divide(Accum,X[5]);
Accum = Accum - X[4];
Accum = Accum + X[2];
return(Accum);
}
```

```
float entity_1(float X[],int num)
{
float Accum,SIN,COS,LN,EXP,FUZZY,LAVG;
float TEMP1,TEMP2,TEMP3,CUBE,SQUARE;
short AR;
AR=1;
Accum=SIN=COS=LN=EXP=FUZZY=LAVG=0.0;
TEMP1=TEMP2=TEMP3=CUBE=SQUARE=0.0;
Accum = Accum - X[17];
Accum = Accum + X[2];
Accum = dmax(Accum,X[31]);
Accum = Accum - X[27];
Accum = Accum - X[15];
Accum = Accum - X[15];
Accum = divide(Accum,X[5]);
Accum = Accum - X[4];
Accum = Accum + X[2];
return(Accum); }
```

```
float entity_2(float X[],int num)
{
float Accum,SIN,COS,LN,EXP,FUZZY,LAVG;
float TEMP1,TEMP2,TEMP3,CUBE,SQUARE;
short AR;
AR=1;
Accum=SIN=COS=LN=EXP=FUZZY=LAVG=0.0;
TEMP1=TEMP2=TEMP3=CUBE=SQUARE=0.0;
Accum = Accum - 0.521974;
Accum = Accum - X[40];
Accum = Accum - X[36];
Accum = Accum + X[24];
Accum = Accum + -1;
Accum = dmax(Accum,avg(X,num));
Accum = dmin(Accum,X[2]);
Accum = Accum - X[4];
return(Accum);
}
```

```
float entity_3(float X[],int num)
{
float Accum,SIN,COS,LN,EXP,FUZZY,LAVG;
float TEMP1,TEMP2,TEMP3,CUBE,SQUARE;
short AR;
AR=1;
Accum=SIN=COS=LN=EXP=FUZZY=LAVG=0.0;
TEMP1=TEMP2=TEMP3=CUBE=SQUARE=0.0;
Accum = X[2];
Accum = Accum - X[15];
TEMP3 = Accum;
Accum = dmax(Accum,X[7]);
Accum = Accum - X[15];
Accum = Accum - X[27];
Accum = Accum - X[15];
Accum = divide(Accum,X[5]);
Accum = Accum - X[4];
Accum = Accum + X[2];
return(Accum);
}
```

```
float entity_4(float X[],int num)
{
float Accum,SIN,COS,LN,EXP,FUZZY,LAVG;
float TEMP1,TEMP2,TEMP3,CUBE,SQUARE;
short AR;
AR=1;
Accum=SIN=COS=LN=EXP=FUZZY=LAVG=0.0;
TEMP1=TEMP2=TEMP3=CUBE=SQUARE=0.0;
Accum = dmax(Accum,X[42]);
Accum = dmax(Accum,X[20]);
Accum = divide(Accum,avg(X,num));
Accum = Accum + X[2];
Accum = Accum - X[4];
Accum = multiply(Accum,X[6]);
Accum = Accum - X[14];
Accum = Accum - X[28];
return(Accum);
}
```

9.3 EA features for imaging sensor

The twelve imaging sensor features which were input to the EA correspond
to the twelve target poses in order from left to right, top to bottom in Figure

7.1. These twelve features are the variables $X[1]$ through $X[12]$ in the five C functions below.

```
float entity_0(float X[],int num)
{
float Accum,SIN,COS,LN,EXP,FUZZY,LAVG;
float TEMP1,TEMP2,TEMP3,CUBE,SQUARE;
short AR;
AR=1;
Accum=SIN=COS=LN=EXP=FUZZY=LAVG=0.0;
TEMP1=TEMP2=TEMP3=CUBE=SQUARE=0.0;
Accum = hi1(X,num);
Accum = multiply(Accum,X[3]);
Accum = multiply(Accum,0.355817);
Accum = multiply(Accum,0.355817);
Accum = Accum - hi1(X,num);
Accum = Accum - X[4];
TEMP3 = Accum;
Accum = Accum + TEMP3;
Accum = Accum - X[2];
Accum = multiply(Accum,TEMP3);
Accum = Accum - X[11];
return(Accum);
}
```

```
float entity_1(float X[],int num)
{
float Accum,SIN,COS,LN,EXP,FUZZY,LAVG;
float TEMP1,TEMP2,TEMP3,CUBE,SQUARE;
short AR;
AR=1;
Accum=SIN=COS=LN=EXP=FUZZY=LAVG=0.0;
TEMP1=TEMP2=TEMP3=CUBE=SQUARE=0.0;
Accum = hi1(X,num);
Accum = multiply(Accum,X[3]);
Accum = multiply(Accum,0.318017);
Accum = multiply(Accum,0.318017);
Accum = Accum - hi1(X,num);
Accum = Accum - X[4];
return(Accum);
}
```

```
float entity_2(float X[],int num) {
float Accum,SIN,COS,LN,EXP,FUZZY,LAVG;
float TEMP1,TEMP2,TEMP3,CUBE,SQUARE;
short AR;
AR=1;
Accum=SIN=COS=LN=EXP=FUZZY=LAVG=0.0;
TEMP1=TEMP2=TEMP3=CUBE=SQUARE=0.0;
Accum = X[6];
Accum = Accum - 0.751934;
Accum = Accum - 0.751934;
Accum = divide(Accum,X[1]);
Accum = Accum - 0.696906;
return(Accum);
}
```

```
float entity_3(float X[],int num) {
float Accum,SIN,COS,LN,EXP,FUZZY,LAVG;
float TEMP1,TEMP2,TEMP3,CUBE,SQUARE;
short AR;
AR=1;
Accum=SIN=COS=LN=EXP=FUZZY=LAVG=0.0;
TEMP1=TEMP2=TEMP3=CUBE=SQUARE=0.0;
Accum = X[10];
Accum = multiply(Accum,X[6]);
Accum = multiply(Accum,0.346759);
Accum = multiply(Accum,0.346759);
Accum = Accum - hi1(X,num);
Accum = Accum - X[4];
return(Accum);
}
```

```
float entity_4(float X[],int num) {
float Accum,SIN,COS,LN,EXP,FUZZY,LAVG;
float TEMP1,TEMP2,TEMP3,CUBE,SQUARE;
short AR;
AR=1;
Accum=SIN=COS=LN=EXP=FUZZY=LAVG=0.0;
TEMP1=TEMP2=TEMP3=CUBE=SQUARE=0.0;
Accum = X[6];
Accum = multiply(Accum,0.751926);
TEMP3 = Accum;
Accum = Accum + TEMP3;
Accum = Accum - X[1];
Accum = multiply(Accum,X[7]);
Accum = Accum - hi2(X,num);
Accum = Accum - avg(X,num);
return(Accum);
}
```

9.4 Sensor fusion logic

The five EA derived radar features in 9.2 above and the five EA derived imaging features in 9.3 above were provided as input to still another EA run to develop the fused decision logic. In this input training set, $X[1]$ through $X[5]$ were the five radar features in order, and $X[6]$ through $X[10]$ were the five imaging sensor features in order.

Since training data was available for every range gate within a dwell for the radar and only on a per dwell basis (no range information) for the imaging sensor, the imaging feature data for each dwell was used multiple times for every range gate within a dwell.

The fusion decision logic evolved is given below.

10. Adaptive target acquisition

One of the additional advantages of using an EA to evolve features and decision logics is that the same approach can be used to adapt these features and logics in real-time. Of course, the speed with which this is done will depend on the on-board processing power available to the EA.

```
float entity(float X[],int num) {
float Accum,SIN,COS,LN,EXP,FUZZY,LAVG;
float TEMP1,TEMP2,TEMP3,CUBE,SQUARE;
short AR;
AR=1;
Accum=SIN=COS=LN=EXP=FUZZY=LAVG=0.0;
TEMP1=TEMP2=TEMP3=CUBE=SQUARE=0.0;
Accum = 0.702595;
TEMP2 = Accum;
TEMP1 = Accum;
Accum = divide(Accum,X[7]);
Accum = Accum + TEMP1;
Accum = dmin(Accum,X[6]);
Accum = dmin(Accum,LAVG);
Accum = Accum + X[1];
Accum = dmin(Accum,avg(X,num));
if (Accum <= 0.0) Accum = 0.0;else Accum = 1.0;
return(Accum);
}
```

The concept of using the EA for adaptation is shown at the top level in Figure 2.1 using dashed lines to indicate the adaptive control. The difference between training in the laboratory with "ground truth" data and adapting the features and logics in the operational environment lies in the source of the truth data to be used.

The key to making the adaptive scheme work is the following:

- First, the primary reason that the algorithm must be adapted due to differences in the clutter and the scene from those used for training data.
- Second, the targets are at least in principle knowable, and their characteristics can generally be measured in advance.
- Third, most of the radar returns are clutter and most of the scene is not targets. Therefore, it is not too difficult to obtain radar and image "clutter" in real-time and to be reasonably confident that it is clutter.

Figure 10.1 illustrates the adaptive training method for the radar. A similar approach is applicable to the imaging sensor. First, radar targets are stored in the computers aboard the operational vehicle. Second portions of the clutter which have very low probability of being targets are gathered in real-time. The clutter is used directly for the real-time training base, as well as for the imbedding of the stored targets in the clutter. The target IQ data is added coherently with the clutter IQ data to form the cell containing both target and clutter.

The EA runs in a background mode, continually evolving new and presumably better features and logics. Since the EA has a built-in scoring sys-

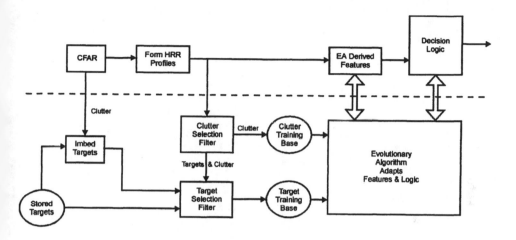

Fig. 10.1. Adaptive training method for the radar

tem, it can determine the best features and logics for the current terrain and operational situation, and use those in the on-line target acquisition process.

References

[Biermann, 1992] Biermann, A. W. (1992). Automatic Programming. Encyclopedia of Artificial Intelligence, 2nd ed. (ed S. Shapiro), Wiley NY, pp 65-83.

[Edde, 1993] Edde, B. (1993).   Radar: Principles, Technology, Applications. Prentice-Hall.

[Duda, 1973] Duda, R. O. and Hart, P. E. (1973). Pattern Classification and Scene Analysis. John Wiley & Sons.

[Dunay, 1995] Dunay, B. D. and Petry, F. E. (1995). Solving Complex Problems with Genetic Algorithms. Proceedings of the 6th International Conference on Genetic Algorithms, pp 264-270.

[Holland, 1975] Holland, J. H. (1975). Adaptation in Natural and Artificial Systems. MIT Press.

[Koza, 1992] Koza, J. R. (1992). Genetic Programming: On the Programming of Computers by Natural Selection. MIT Press.

[Petry, 1995] Petry, F. E. and Dunay, B. D. (1995). Automatic Programming and Program Maintenance with Genetic Programming. International Journal of Software Engineering and Knowledge Engineering.

[Press, 1994] Press, W. H., Teukolsky, S. A., Vetterling, W. T. and Flannery, B. P. (1994). Numerical Recipes in C. The Art of Scientific Computing, Cambridge University Press.

[Stutzman, 1993] Stutzman, W. L. (1993). Polarization in Electromagnetic Systems. Artech House.

[Wehner, 1995] Wehner, D. E. (1995). High Resolution Radar, Second Edition. Artech House.

Part V

Mechanical and Industrial Engineering

Strategies for the Integration
of Evolutionary/Adaptive Search
with the Engineering Design Process

I. C. Parmee

Plymouth Engineering Design Centre, University of Plymouth, Drakes Circus
Plymouth PL4 8AA, United Kingdom

Summary. The research concerns the development of evolutionary/adaptive search
strategies to enable their successful integration with the conceptual, embodiment
and detailed stages of the engineering design process. Global optimisation in relation
to engineering design is considered here in its broadest sense, i.e., as a complex, relatively continuous process that commences during the high risk stages of conceptual
design and progresses through the uncertainties of embodiment design to the more
deterministic, lower risk stages of detailed design. The objective during the early
stages is to identify optimal design direction (i.e., that direction that represents
best performance whilst best satisfying many qualitative and quantitative criteria
at least risk). During the more deterministic detailed design stages the emphasis
is upon minimisation of computational expense whilst identifying optimal design
solutions. Appropriate adaptive search integration involves the utilisation of design
models of varying detail commensurate with the degree of confidence in available
data and project specification. Results from the implementation of co-operative
search strategies also involving complementary soft computing techniques are presented and discussed. The development and integration of appropriate strategies
is illustrated with examples of real-world application from the mechanical, civil,
electronic, aerospace and power system engineering design domains.

1. Introduction

There are many examples of the application of evolutionary and adaptive
search (AS) algorithms to specific problems from the engineering design domain. However little research effort has been expended in moving from these
specific areas to investigate the utility of evolutionary/adaptive search within
the generic domain of the engineering design process as a whole. Research
within the Plymouth Engineering Design Centre (PEDC) at the University
of Plymouth has concentrated upon these generic aspects since its establishment in October 1991 [1]. The objective is to investigate, develop evolution-
ary/adaptive computing techniques and integrate them with each stage of
the engineering design process. Research concentrates upon those algorithms
which: are analogous to natural processes, incorporate co-operative elements
and display emergent properties.

In most cases the utility of certain characteristics of these techniques to
the various stages of design have been identified, reinforced and combined to
provide high-potential strategies that support the engineering designer during

conceptual, embodiment and detailed design. A high level of industrial collaboration with diverse sectors of the engineering industry provides complex design environments within which the potential capabilities of the various strategies can be tested. This problem-oriented approach to AS integration is essential due to the complex dynamics of the higher levels of the engineering design process in particular. A high level of interaction with designers and design teams is necessary to identify those areas where AS strategies would be of most benefit and to recognise specific problems relating to their successful integration.

Evolutionary and adaptive search techniques such as the genetic algorithm (GA) [2] and insect colony metaphors [3] are, in general, perceived to be powerful global optimisers that can negotiate highly complex, non-linear search domains described by some mathematical representation of an engineering system to provide globally optimal design solutions. A far broader view of global optimisation in engineering design describes a highly complex, largely continuous process which commences during the high-risk stages of conceptual design, continues through the uncertainties of preliminary/embodiment design to the more deterministic stages of detailed design. The main objective at the higher levels is to identify optimal design direction i.e., that direction which offers best satisfaction of both qualitative and quantitative criteria whilst representing least risk in terms of further development. The identification of optimal direction relies upon a highly interactive process involving computer-based search, engineering heuristics and design team decision-making. This iterative, cumulative process finally results in the identification of the globally optimal engineering solution relative to current knowledge, technology and financial considerations.

The following sections introduce a number of largely co-operative strategies that support the engineer at every level of engineering design by satisfactorily:

– providing concurrent multi-level processing of whole system design hierarchies,
– identifying high-performance regions of preliminary design spaces,
– providing relevant design information during the search process,
– improving the calibration of preliminary design software,
– accessing feasible regions of complex, heavily-constrained design spaces,
– manipulating computationally expensive analysis software.

The overall objective of these strategies is the establishment of highly interactive designer/AS environments. The resulting tools must be considered as powerful extensions of the design team stimulating innovative reasoning at the higher conceptual levels of the design process; providing diverse, high-performance solutions to support decision-making during preliminary design and acting as powerful global optimisers that can operate efficiently within highly complex domains during detailed design. Successful implementation at each level is illustrated and discussed and commonalties both in the design characteristics of each design stage and in the various search strategies are identified. This leads to the introduction of co-operative frameworks involving a number of search strategies/optimisation techniques operating concurrently within a single search environment.

2. Decomposition of Design Spaces During Conceptual/Embodiment Design

During the initial stages of a large bespoke engineering design project the engineer is faced with an extremely large space of possible design directions. Although the domain is likely to be well known there will be many discrete design options concerning large numbers of component subsets. If we take airframe design as an example, although the general requirements/functions are well-known and the likely possible components and basic structure to achieve these functions are apparent there remains a very large space of possible initial configurations from which the design process can progress. Discrete design options relating to various concepts (fixed vs variable geometry wing; short vs vertical take-off) and major design parameters (number and type of engines) will be combined with related continuous variable parameter sets relating to engine size, wing size and plan form, fuselage dimensions etc. Preliminary decisions regarding these parameters will depend upon specific operational concepts and performance requirements. The process is repeated at the large-scale component level i.e., having decided the number and power requirement of, say, the gas turbine engines their initial configuration must be determined at the whole-system level with design decisions relating to nacelle and annulus layout, overall cooling system design, mass throughput and pressure/temperature levels. These decisions will depend again upon performance requirements, working life, economic considerations etc.

Although the engineer is operating within a largely known design space this preliminary whole-system design stage is characterised by a high degree of uncertainty and associated risk. This uncertainty generally arises from low-confidence in available data and fuzziness within the initial specification, indeed the specification may well be evolving with the design as criteria develop from initial investigations. The operational concepts and performance requirements are likely to be vague for a significant period of time and this obviously has an effect on related major component design. However, the design process must progress despite this uncertainty and considerable expertise is thus required within the design team to avoid mistakes which, if not remedied in the short term, may prove impossible to rectify at a later date without incurring excessive expense or compromising overall performance to some extent.

The engineer will utilise a wide range of preliminary design models during this stage of design. These will be coarse representations of the engineering systems or, at their simplest, may merely consist of data correlations. Their complexity will be commensurate with the degree of confidence in available data and a current assessment of uncertainty in specification. A high degree of engineer interaction is essential to ensure appropriate interpretation of the results from such models. Budget and time constraints will govern the extent of search across the discrete/continuous hierarchy of design options leading to rapid concentration of design effort in a specific region. This decomposition of the design space will be based upon previous experience relating to similar design requirements. The possible design space is therefore immediately compromised in order to reduce parametric search and it is here that appropriate adaptive search strategies can contribute significantly by allowing the engineer to rapidly explore very large preliminary design spaces to identify regions of high-potential that may provide radical yet competitive design concepts/solutions within time and budget constraints. Such regions would not have been described within the compromised design space of a traditional approach.

3. Negotiating Whole System Design Hierarchies

During whole system design it is likely that a design hierarchy will contain both discrete variable parameters describing design options and continuous variable parameters describing dimensions and characteristics of related subset elements. It is difficult to achieve an efficient concurrent, multi-level search through this hierarchical structure due to the combination of discreteness and high dimensionality in terms of the continuous variables creating a very large, discontinuous space of possible design solutions. The tendency for premature convergence upon one of the discrete paths is considerable which frustrates the objective of providing the engineer with diverse high-performance solutions from different parts of the hierarchy.

The problems can be illustrated by the introduction of a large-scale civil engineering design environment. This involves a simple design hierarchy describing the major elements of a large-scale hydropower system. The problems associated with uncertainty and associated risk are prevalent during the early stages of a hydropower project. Typically the engineer is tasked with the identification of high-potential sites within some remote geographical region. Little data will initially be available regarding topography, geology, flow regimes, ease of access etc and this necessitates preliminary site investigations and the collation of information from existing ordnance survey maps, hydrological data and previous civils work (if any) within the area. Information will arrive daily and collation and analysis will allow a framework to be established to support a feasibility study. A concurrent economic evaluation will also be in progress to establish the competitiveness of hydropower when compared to other forms of generation. As the project progresses so the depth

of investigation increases and a firmer picture of the potential and characteristics of various sites is established. This can take a significant period of time
during which the preliminary design process must progress. A high degree
of estimation and assumption based upon available data must therefore take
place.

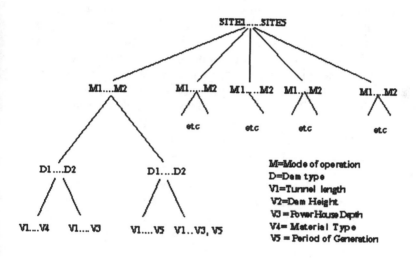

Fig. 3.1. Basic hydropower design ierarchy

The design hierarchy of Figure 3.1 represents practical design aspects of
some of the relevant major subsets of the overall hydropower system design.
Simple mathematical representations of the various subsets typical of the level
of representation utilised at this stage of design have been developed [4, 5].
Discrete design options relate to choice of site, mode of operation, dam type
and material selection whereas continuous variables describe pressure tunnel
length, dam height, powerhouse depth and period of generation. The overall
objective is to determine the site with the highest potential when combined
with the most appropriate overall system configuration to minimise unit cost
of power out. This unit cost is calculated as a ratio of the present worth of
the capital cost of the major system components and the present worth of
power out calculated over the projected working life of the system. The initial
requirement is to develop a global search algorithm that will achieve efficient
concurrent, multi-level processing of the various elements of this hierarchical
structure to identify a number of diverse high-performance design options.

4. The Development of Appropriate Evolutionary Regimes

Initial research involving the development of appropriate adaptive search strategies concentrated upon the integration of a structured genetic algorithm (stGA) [6] with the design models [4, 5]. Integration of a simple stGA, however, highlights the problems of premature convergence with the rapid identification of the optimal solution from the set of continuous variables along one particular discrete 'path' at the expense of search effort along the majority of the twenty possible paths as described by the three main discrete parameters (Site, Mode of Operation and Dam Type), i.e., very little search across the hierarchy takes place (Figure 4.1). The overall objective is to develop a search methodology that allows concurrent processing of far more complex hierarchies that will likely involve discrete branching below levels described by continuous variables. A high probability that the search algorithm is sampling each region of the hierarchy must therefore exist in order that a satisfactory degree of cover resulting in the identification of diverse high performance solutions can be achieved. Greater diversity of search can be achieved by assigning independent mutation probabilities to the high-level discrete parameter set (set A) and the continuous variable parameter set (set B). A mutation probability of 0.2 for discrete set A and 0.02 for continuous set B when combined with a non-elitist strategy produces the required diverse search with a much higher proportion of the design hierarchy being visited and a number of high-performance solutions located. A hybrid approach can be adopted in which the dual mutation regime with no elitism is implemented for the first fifty generations at which point single mutation plus elitism takes over for a further fifty generations. In this manner diversity of search can be maintained and better solutions can be achieved from several of the discrete configurations (Figure 4.2). However extremely complex stGA chromosomal representations result from an increase in the number of levels of the hierarchy thereby restricting development of the design model. Problems related to parameter redundancy within these chromosomal representations is also a cause for concern [7].

A simpler representation is achieved by the development of a strategy that ensures the avoidance of lethal parameter sets whilst allowing the information exchange evident during traditional crossover. Crossover is restricted to parameter strings that are identical in terms of the discrete parameters (set A). This allows an exchange of information relevant to a particular discrete design configuration and the subsequent evolution of that configuration in terms of the continuous variables (set B) whilst also allowing a simple chromosomal representation containing only the relevant parameters to that configuration. However this crossing of 'like' configurations in terms of the discrete parameters does not allow their perturbation and their subsequent improvement. This problem can be addressed by introducing the variable mutation probability approach described earlier i.e., introduce a high muta-

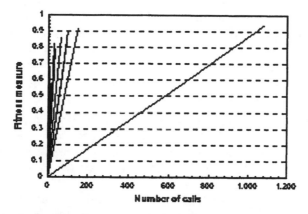

Each line represents one of the possible 20 paths created by the discrete parameter set A. The number of calls relates to the number of times each discrete set has been passed to the mathematical model. Fitness is relative to best fitness achieved during the experimentation.

Fig. 4.1. Simple stGA implementation

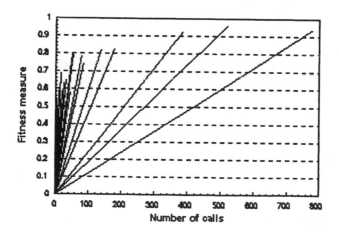

Fig. 4.2. Hybrid strategy

tion probability in discrete set A and a lower probability to continuous set B. The overall strategy would therefore involve two individual search agents, i.e., a simple hill climber manipulating the discrete set A and a genetic algorithm manipulating the continuous set B. Communication is inherent and exists at a lower level between the individual chromosomes of the continuous set B via crossover and at a higher level between the two search agents via the evaluation of each string and subsequent selection using roulette wheel. This can be improved by introducing lower level communication between the discrete sets (A) of each generation by introducing elements of the ant colony metaphor [3, 8] to the manipulation of the discrete set. Two particular Ant Colony 'operators' are of interest:

- **fitness proportionate distribution:** similar to fitness proportionate reproduction — in this case the number of software 'ants' (search agents) distributed down each trail is proportionate to the relative strength (i.e., fitness) of that trail. In this case each discrete path (set A) of the hierarchy represents a possible 'trail'.
- **evaporation:** if the strength of a particular trail does not improve over a preset number of cycles then that trail is evaporated and the released 'ant' resource is redistributed around the better trails.

The concepts supporting these two operations have been adapted slightly and integrated with the manipulation of the discrete parameter set. The resulting GAANT algorithm commences by randomly selecting a population of chromosomes each describing both the discrete and continuous variables. However although the continuous set (B) is allowed to evolve by generation using a standard GA, the discrete set (A) is fixed for a preset number (n) of generations. Crossover is only allowed between chromosomes describing the same system configuration in terms of set A. The average fitness of each chromosome is calculated over the preset number of generations and compared to the average fitness of the n-th generation. Fitness proportionate distribution and evaporation are applied to set A according to this relative fitness of each chromosome and the process continues in steps of n generations until satisfactory convergence upon a number of diverse solutions has been achieved. Preliminary results for this GAANT algorithm indicates that a more diverse search of the design hierarchy can be achieved which results in significantly better solutions from most of the discrete paths (Figure 4.3). A more comprehensive description of the dual-agent process can be found in [7] and [9] along with comparative results from the stGA, similarly a more extensive description of the stGA approach can be found in [4] and [5].

5. The Identification of High Performance Regions

Another evolution-based strategy for the automatic decomposition of preliminary design spaces involves the integration of a cluster-oriented GA (COGA) with models comprising primarily of continuous design variables [10]. This approach has been largely aimed at subset design although it is also relevant to the whole system stages of the previous section. The objective is the identification of high-performance regions of a design space as opposed to finding single local optimal solutions. The resolution of preliminary design models is

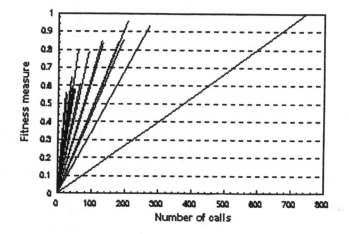

Fig. 4.3. GAANT strategy

such that there is a significant probability that extensive fine-grained search
will result in the identification of single solutions which will prove erroneous
when subjected to more in-depth analysis hence the concentration upon region identification. The extraction of relevant design information from these
regions can support the decision making process concerning the determination of optimal design direction. The engineer is presented with a selection
of regions each possessing differing design characteristics and each satisfying
to varying extent the set of qualitative criteria relating to specific aspects
of the design problem in addition to pre-defined quantitative criteria. The
philosophy is similar to that of the previous section in that it also allows the
engineer to rapidly negotiate the space of possible design solutions and to
identify regions of high-potential within budget and time constraints. This
results in the identification of regions containing competitive solutions that
may have been overlooked during the problem decomposition processes of
traditional heuristic design.

6. Cluster-oriented Genetic Algorithms

Initial research related to the identification of high-performance regions in
design domains described by continuous variables has been reported in the
literature [5, 12]. The work concentrated upon the establishment of variable
mutation regimes that encourage diversity during the early stages of a GA
search and encourage the formation of clusters of solutions in the better areas
of the overall design space. Variable mutation regimes form the basis for these
methods and data from selected populations is stored in a final clustering set.
A standard near neighbour clustering algorithm [12] which, if utilised in an

appropriate manner, requires little apriori knowledge of the design space, then identifies the naturally occurring clusters from the stored data.

With regard to extracting relevant design information from high-performance regions of a design space the initial clustering work was inspired by a desire to include solution sensitivity (i.e., the sensitivity of a design solution to mild perturbation of the variable parameters that describe it) [13] as part of the overall criteria from which the relative fitness can be determined. A perturbation-based sensitivity analysis of each solution of each generation is not feasible due to the number of calls to the design model that this would entail. However the preliminary identification of high-performance regions and the subsequent sensitivity analysis of the design solutions within those regions allows the relative robustness of each region to be assessed with an acceptable level of computational expense. Further search can then be continued in those robust, high-performance regions that satisfy some preset sensitivity criteria. The test function shown in Figure 6.1 has been developed to provide an experimental basis for the establishment of the techniques. The reasoning being that engineers would prefer robust solutions from region B to those from high-gradient region A where slight perturbation may lead to severe degradation of the design solution. A compromise between design performance and design sensitivity must therefore be made.

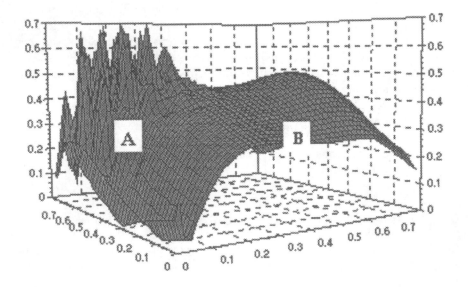

Fig. 6.1. Test Function 1 (TF1)

The basic strategy involves the initial introduction of a high mutation probability (circa 0.08) during the early generations of the search [7, 10]. This probability is then reduced linearly over a preset number of generations and the populations of pre-selected generations are extracted to form a final clustering set of solutions. The GA representation has the following characteristics:

- an initial structured population based upon a minimum fixed Hamming distance between each member of the population to ensure satisfactory cover of the search space at generation one [13],
- double-point reduced surrogate crossover to reduce incest and promote diversity [14],
- stochastic universal selection (SUS) [15].

An adaptive filter has been introduced which does not allow low performance solutions from the extracted populations to pass into the clustering set. The filter takes into account the relative fitness of the chromosomes of each of these populations in the following manner. The solutions from each of the selected generations are initially scaled and a threshold value (Rf) is introduced as shown in Figure 6.2. Chromosomes are either rejected or passed into the clustering set depending upon their scaled fitness in relation to the threshold value. The scaling allows the value of Rf to be preset or varied at each of the generations from which the populations are extracted and processed for inclusion in the final clustering set. This adaptive filter relies upon the relative fitness of the stored solutions of the extracted populations. This allows the technique to identify high-performance solutions and subsequently high-performance regions relative to the overall fitness landscape. The Rf threshold can be considered to be analogous to setting a lower bound on acceptable solution values. However this lower bound is adaptive in that it is dependant upon the value of known solutions that describe the surface topography to some extent the variation of Rf can therefore be utilised in an investigatory manner in initial experimental runs to gather information concerning the relative nature of differing regions of the design space.

A number of additional operators are introduced to improve region cover:

- a secondary local mutation regime in the later stages of the process say between generations fifteen and twenty-five. An additional probability for mutation is applied to the lower-order binary digits of the parameters only. This results in local perturbation around individuals that are, by this stage of the search, likely to be members of one of the high-performance regions,
- the elimination of solution duplication in the final clustering set and the introduction of a local perturbation around each of the remaining individuals prior to applying the clustering algorithm,
- the introduction of elitism in the later generations of the search process,
- any members of the populations of the intermediate generations which do not contribute to the final clustering set which have a fitness equal to or

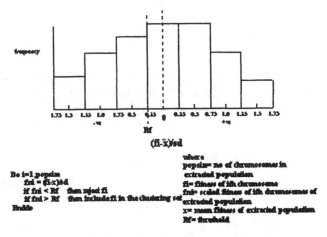

Fig. 6.2. Adaptive filter

greater than that related to the Rf value of the adaptive filter are stored and added to the final clustering set.

A more in-depth description of the COGA strategy outlined above and results from extensive experimentation can be found in [7] and [10]. Figures 6.3 and 6.4 illustrate the degree of cover of each of the regions of the test function of Figure 6.1 in and another from the test function set. Figure 6.5 shows, in two dimensions, results from the application of the strategy to a six dimensional gas turbine cooling hole geometry problem and corresponding information regarding region sensitivity. The objective function in this case is the minimisation of coolant flow rate.

Concurrent identification of all high-performance regions proves difficult as dimensionality increases. Strategies are now under development that will result in the rapid initial identification of the bounds of a region based upon a clustering of a few initial solutions. Fitness can then be suppressed within those bounds [16] allowing continuing search to identify further regions. The envisaged strategy involves a distributed environment manipulating search agents that are 'left behind' in the roughly identified regions to establish the region's real boundaries through local search based upon solution fitness relative to the Rf values from the adaptive filter. Some experimentation in this area is indicating potential but more research is required to establish a working strategy based upon the utilisation of independent search agents operating within a distributed, co-operative framework.

Solution sensitivity is just one example of the information concerning specific region characteristics that can be collated during the search. Coarse

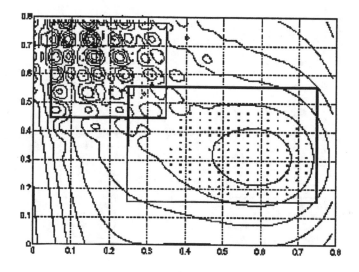

Fig. 6.3. Region cover – TF1

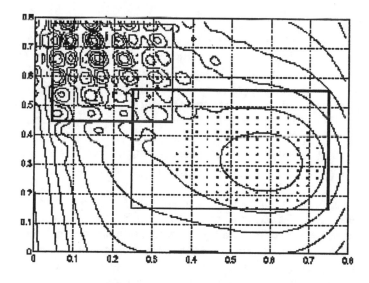

Fig. 6.4. Region cover – TF2

grained cover of each region can provide information relevant to the engineering designer in terms of qualitative criteria satisfaction in addition to design sensitivity. Information concerning the actual topographical nature of the region can also be extracted which may contribute to the choice of algorithm for further fine-grained search. The identification of diverse high

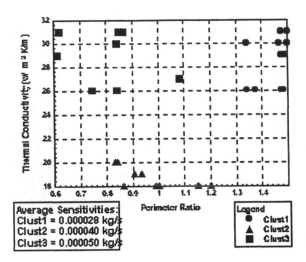

Fig. 6.5. Clustering in two dimensions with solution sensitivities

performance areas also presents the engineer with several design regions each with differing parameter characteristics. The engineer therefore has a choice of design paths each of which may offer a differing degree of multi-criteria satisfaction. Such criteria may be quantitative or qualitative in nature. Thus the engineer's intuition and specific knowledge of the design/manufacturing domain can be integrated with the search process by the subsequent selection of those regions warranting further investigation.

7. Extracting Design Information from High-performance Regions/Solutions

The COGA approach supports the reduction of the overall size of the initial search space by identifying high-performance regions and allowing the engineer to concentrate search in those regions which best satisfy both qualitative and quantitative design criteria. It can be seen therefore as a method that is providing support to the preliminary decision making processes. Further support can be provided by introducing an expert system which extracts relevant information during the search concerning characteristics of each solution or region of the design space. As high-performance solutions are identified, relevant information is dynamically extracted and stored [17]. This information may relate to such design aspects as parameter and solution sensitivity, extent of constraint violation and degree of qualitative criteria satisfaction. The overall objective is to provide as much information as possible during the search in order to enhance and subsequently utilise the engineer's design knowledge and intuition. The system provides an interface that allows an exchange of data between engineer and machine that results in an intelligent reduction of the overall design space.

In addition to the COGA approach, individual high-performance solutions can be identified using a variety of niching techniques although the single solutions should be treated with some caution due to the coarse design model representations. A recent in-house development of restricted tournament selection [18], adaptive restricted tournament selection (ARTS) [19] can identify individual optimum solutions. Qualitative evaluation of selected high-performance solutions is achieved via a fuzzy logic interface (Figure 7.1) using FuzzyCLIPS [20]. The overall effectiveness of each solution can then be assessed using the quantitative values returned from the GA search and the crisp values from the qualitative evaluations of the fuzzy interface. Three levels of knowledge representation are adopted:

- **Inter-variable Knowledge** — the importance of each variable parameter describing a solution is related to individual criteria such as manufacture-ability, choice of material etc. The values of parameters of relatively low importance to a particular criteria are compromised to some extent thereby allowing greater consideration for the values of more relevant parameters.
- **Intra-variable Knowledge** — each parameter has a range of values that can be considered to be the preferred range for a particular criteria i.e., minimum thickness of a section within appropriate stress considerations may be preferable in terms of weight criteria but not in terms of machining criteria. In such a case a medium range of thickness values may be preferred. Solutions with parameter values lying in the preferred ranges

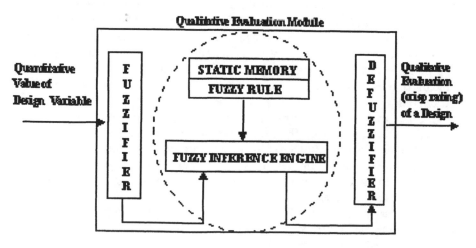

Fig. 7.1. Fuzzy logic interface

can be considered of a higher quality especially where those parameters
are of high importance in inter variable terms.
- **Heuristics** — both inter and intra variable aspects can be overruled in
specific cases where designer preference concerning a particular solution is
sufficiently well-defined.

The combination of the fuzzy preferences in terms of inter and intra variable
knowledge results in a crisp qualitative rating for each of the quantitative
high-performance solutions generated by ARTS or by COGA. This rating can
support the engineer in decisions concerning subsequent design direction.

8. Constraint Satisfaction and Constrained Optimisation

Evolutionary and adaptive algorithms can be applied to generic problems of
constraint satisfaction and constrained optimisation that are apparent to a
varying degree throughout the design process. Current research with Nuclear
Electric plc is investigating the integration of a genetic algorithm with thermal cycle control and design software [21]. The objective is to manipulate
system control variables whilst concurrently varying design parameters relating to system configuration in order to achieve optimum power output. Initial
work concentrated upon the automisation of the control variable optimisation
procedure. A typical control problem involves around twenty non-linear constraints and a refined linear programming technique was initially utilised to
locate a satisfactory solution. However this technique requires an initial feasible point. The problem therefore is the location of a feasible region within the
overall control search space. Traditionally a heuristic trial-and-error search
by an engineer with considerable experience of the software and of the domain would be utilised to locate a feasible starting point for the linear optimiser Such a search can take several days and is both tedious and expensive.
The involvement of the engineer also prohibited the overall integration of
an evolutionary-based search manipulating both design and control variable
parameters.

     In order to overcome this problem preliminary experimentation involved
the utilisation of a GA for the initial search for a feasible region. This has
resulted in the introduction of appropriate penalty functions to provide a
robust approach which will identify a 'good' feasible solution. The modified
linear programming technique is then utilised to improve upon this solution.
In this instance the diverse search characteristics of the GA are being utilised
to pre-process the problem. The GA is significantly reducing the size of the
search space by rapidly converging upon regions of feasibility. This initial success has reduced design time by 10-15% and translated a significant number
of man-hours to machine hours. Of equal importance however is the elimination of the engineer-based search for feasibility. This now allows the inclusion
of design variables relating to various elements of the thermal system and

the concurrent tuning of the design/control environment. Preliminary experimentation in this area is showing significant improvement in predicted power output.

The GA has been utilised in a similar manner in a project with British Aerospace plc [22]. The design domain involves preliminary air-frame design and definition of a flight trajectory for an air-launched winged rocket that will achieve orbit before returning to atmosphere for a conventional landing. The problem is extremely sensitive to a number of non-linear constraints relating to air-speed and angle of climb. Constraints also affect the physical parameters describing the airframe and engine configuration. The objective is to minimise the empty weight of the vehicle through a space of seven continuous variables and one discrete design variable. There was a degree of doubt as to whether a feasible solution to the problem is actually described by the system model supplied and preliminary experimentation utilising a GA with appropriate penalty functions could only achieve solutions exhibiting minimum constraint violation.

It was evident that a secondary search process was required. Investigation of the fitness landscape via the plotting of two-dimensional hyperplanes revealed a high degree of complexity and detail even when considering small neighbourhoods (0.01% of the parameter's range). This eliminates the utilisation of a hill climber due to problems of rapid convergence upon the many local optima. The ant colony model was therefore introduced to achieve a fine-grained search of the most promising regions surrounding the initial promising GA solutions. The inherent communication between the individual 'ants' eliminates the problems associated with premature convergence thereby allowing the search to progress across the complex landscape. This approach proved very successful and feasible solutions can now be identified from what appear to be very small disconnected regions of feasibility. Results from an initial GA search followed by the Ant Colony model are shown in Figures 8.1 and 8.2.

9. Software Calibration using Genetic Programming

Much has been said of the utilisation of basic preliminary design models during the early stages of the design process and the caution required when analysing results from them due to inherent system approximation. If the calibration of such models to empiric data or to results from a more in-depth analysis (FEA or CFD perhaps) could be improved then the element of associated risk would be correspondingly lessened. Current research is indicating that adaptive techniques and genetic programming (GP) [23] when utilised for system identification can improve the calibration of simple design software. This approach involves the investigation of the software coding to identify those areas where insufficient knowledge or the requirement of keeping computational expense to a minimum has resulted in unavoidable function

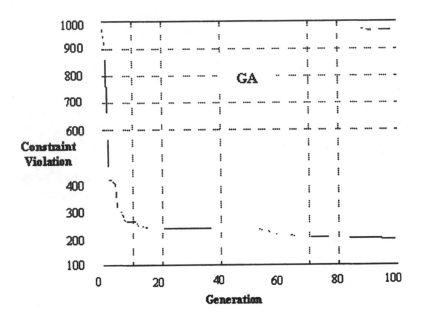

Fig. 8.1. GA search followed by the Ant Colony Model – part 1

approximation. A contributing factor may be the inclusion of empirically derived coefficients (ie discharge, drag etc). The objective is to evolve improved coding within these areas in order to achieve a better calibration with existing empiric data or results generated from a more in-depth, computationally expensive analysis.

The utilisation of a genetic algorithm to fit simple polynomials to various data sets soon reveals a requirement for apriori knowledge concerning the order and form of the equation. Genetic programming techniques however possess the ability to not only evolve the coefficients within a functional form but also the functional form itself. This is achieved by manipulating structures that are themselves general, hierarchical computer programs of dynamically varying size and shape. Although extensive comparisons can be made with other curve and surface fitting techniques it is felt that the real advantage of GP lies in this ability to identify function structure. In addition, the resulting expressions relate directly to the system which removes the 'black box' aspects of neural net system representation; an important factor when considering quality control and 'accountable' design software.

The GP manipulation of engineering physical relationships can provide impressive results. Recent research [24] has involved the GP generation of formula for pressure drop in turbulent pipe flow that compare favourably to Colebrook and White's formula whilst maintaining the explicit characteristics of Haaland's formula [25]. Other work has involved energy loss associated with

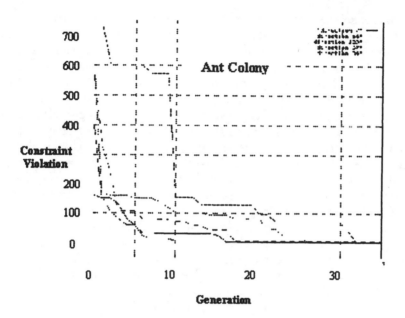

Fig. 8.2. GA search followed by the Ant Colony Model – part 2

sudden contraction or sudden expansion in incompressible pipeflow. The objective here is to evolve relationships dependant upon relative cross-sectional area and relative velocity that provide improved correlation to experimental results or results from a CFD representation when compared to standard pipeflow formulae. Results in both cases have been very encouraging and can be found in [24]. More complex pipeflow relationships are currently under investigation the overall objective being the improvement of preliminary design models again describing the cooling hole geometries of gas turbine engine blades. We are attempting here to reduce the risk associated with the use of preliminary design software without increasing computational expense. Such a reduction of risk will lead to the earlier implementation of the detailed design process and an overall reduction in design lead time.

10. Evolutionary and Adaptive Strategies for Detailed Design

As we progress to the detailed stages of the design process the problems facing the engineer change significantly. Progression to detailed design indicates that the subset designs achieved during the conceptual and embodiment stages can now be considered to be sufficiently low-risk to allow their detailed design using computationally expensive, fine-grained analysis techniques. The uncertainties of the higher levels of the design will now have largely disappeared and global optimisation of the various subset elements is necessary. Although several criteria may still be relevant they will be quantitative in nature and can be more easily accommodated within the fitness function. As uncertainty has largely been eliminated by a progressive refinement of design direction so the risks associated with the overall design task become less onerous. The requirements of the search/optimisation strategies also change considerably. The emphasis is upon the identification of the global optimum solution as opposed to the achievement of a number of high-performance designs. A degree of detailed analysis will already have been evident where it has been necessary to validate results obtained from preliminary design models in order to minimise risk and allow the design process to continue.

The major problem that now faces the successful integration of the adaptive techniques is the considerable computational expense associated with complex analysis such as finite element or computational fluid dynamics. It is essential that the number of calls to the system model is minimised if integration is to be considered practical. Research at the Centre has been addressing this problem by investigating high performance GA strategies and architectures based upon the co-evolution of differing representations of a design problem distributed across a number of separate 'evolution islands'. Appropriate communication between these co-evolving processes allows the rapid recognition of high potential design directions that not only minimise the number of generations required but also ensures that the major evolutionary effort utilises simple design representations thereby also minimising overall computational expense.

The Design Domain: The design test problem involves the identification of optimal design solutions for flat plates. The plate is represented in a grid type manner being divided into rectangular or square elements each with variable depth. Preliminary research has utilised a simple mathematical model based upon bending moment and complex stress analysis to ensure computational time is kept to a minimum thereby allowing extensive experimentation. The fitness of the design relates to the level of stress violation and the overall weight of the plate i.e. weight must be minimised within a given stress criteria. Theoretical direct stresses in both the X and Y planes are increased by a factor of 1.18 to allow to some extent for errors incurred through applying simple beam theory. Preliminary design solutions for the flat plate problem may be obtained with a relatively small number of design variables (15 to 50). However if we require a detailed design which utilises a number of support and load conditions we need to use in excess of 300 design variables (grid elements) to ensure accurate stress evaluation. Detailed design will also require the integration of finite element techniques to provide a sufficient depth of analysis.

Early work showed that a CHC genetic algorithm [26] performed significantly better than a standard GA. High-performance designs (relative to those achieved by heuristic industrial design methods) displaying very good signs of symmetry can be achieved. However plates containing in excess of 130 elements require considerably higher numbers of evaluations (circa 40000) to obtain solutions which meet the overall criteria. Although further improvement would be possible by system tuning it is considered unlikely that a sequential CHC approach would achieve satisfactory results within a search space described by 300 elements. In addition, 40000 calls to a finite element analysis for each of a number of support/load conditions for the plate is not feasible. However, the actual number of calls to the FE analysis can be significantly reduced by adopting an appropriate co-evolutionary approach.

The Injection Island Architecture: The "injection island" architecture (iiGA) [27] developed at Michigan State University involves the establishment of several islands of coevolution distributed around a number of processors. We are currently utilising the computationally inexpensive complex stress based design models integrated with the CHC GA, to allow extensive experimental work with the iiGA architecture. The objective is to minimise the number of calls required to the analysis tool whilst also achieving an optimal design. The flat plate is represented by a number of different resolution grids, each evolving as a separate process, the solutions of the coarse design representations are injected into the more detailed designs for fine grained refinement. Exchange of information is from low to high resolution after a set number of evaluations and also requires translation of the differing grids to maintain true representations (Figure 10.1). As the search space is proportionally smaller for lower resolution grids, fit solutions evolve rapidly. The best individuals may then be injected into higher resolution grids replacing the worst individuals in that subpopulation island. This allows search to occur in multiple encodings, each search taking place upon fitness landscapes of differing degrees of complexity.

Experimentation has shown that the average population fitness of the iiCHCGA using three subpopulation islands (consisting respectively of 25, 100 and 400 elements) of 20 chromosomes each results in an average 30% reduction in number of calls to the model to achieve identical fitness from a single population CHC GA of 60 chromosomes (400 elements). Preliminary results from this research are indicating a significant potential [28]. Further work will incorporate dynamic plate representation and combine complex stress models of varying resolution with more definitive finite element models within a single adaptive search framework comprising of several co-evolving islands. It is also intended to investigate the distribution of several evolutionary/adaptive algorithms plus more deterministic optimisation techniques around the islands and to allow their introduction/withdrawal depending upon relative performance. In this manner a dynamic system involving co-operating search agents concurrently negotiating fitness landscapes of varying complexity whilst exchanging relevant design information will be established. It is envisaged that this approach combined with the parallel processing of

Subpopulation 1   Subpopulation 2   Subpopulation 3

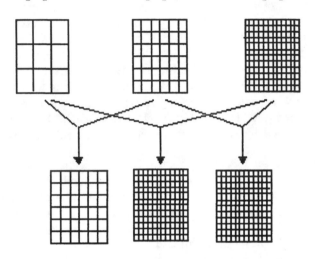

Fig. 10.1. Model representation for injection island

the FEA analysis will provide the required environment to allow feasible integration of the evolutionary/adaptive algorithms with detailed design. These strategies will be of particular benefit at those earlier stages of design where a degree of in-depth analysis is required to validate the results from preliminary design exploration.

11. Summary

Although only brief descriptions of various areas of the author's research within the PEDC have been presented it is hoped that it has been sufficient for the reader to perceive the potential of the evolutionary/adaptive algorithms at every level of the engineering design process and when applied to generic problem areas. Their utility as a support tool during the higher conceptual, whole system and preliminary design levels and, in the main, as apparent. a global optimiser during the more deterministic stages of detailed design is apparent. Most of the strategies described involve co-operative partnerships between both adaptive and more deterministic search and optimisation techniques and between co-evolving search processes. Such partnerships provide the engineer with the capability to rapidly survey the space of possible design solutions within project time constraints; access remote regions of feasibility in complex design spaces; concurrently evolve system control and hardware design environments and minimise computational expense during detailed design. The development and integration of complementary artificial intelligence and analysis techniques provides a means of extracting relevant qualitative and quantitative design information during the search process. This information enhances the engineer's knowledge of the design domain and stimulates innovation. There is a potential for the stimulation of creativity by allowing the recognition of commonalties between the characteristics of diverse design regions. This can lead to the development of new concepts and a redefinition of the design space.

The objective of reducing the risk associated with preliminary design models by improving system representation using GP and related techniques is to shorten lead times to the commencement of detailed design. However a great deal of overlap is evident between detailed and preliminary design in that in-depth analysis is regularly required to verify preliminary design direction. Improved preliminary design representations would be of great benefit during these verification procedures as would the co-evolution strategies of the last section. Preliminary work suggests that the development of co-evolution strategies combining design models of differing resolution and various adaptive and classical search and optimisation techniques within a single co-operative framework will greatly facilitate the integration of AS with the detailed design process.

There is much research and development required to achieve the overall successful integration of the various strategies with the engineering design process. This will involve not only the development of flexible search strategies and complementary techniques but also appropriate computing platforms that can best support these co-operative systems. It is essential that these strategies are seen as an extension of the design team. Human-centred aspects will play a major role in successful integration. There is no suggestion that these techniques will transform the design process into a computable function rather that they should be perceived as major participants within the design teams and the process as a whole.

Acknowledgements

The Plymouth Engineering Design Centre is one of six UK EPSRC Engineering Design Centres. This research is also supported by Rolls Royce plc, British Aerospace Military aircraft division,, Knight Piesold and Partners, Nuclear Electric and Redland. Technologies. The author thanks all of these organisations for their past and continuing support. Thanks also to the researchers of the Centre who have contributed so much to the projects described namely Harish Vekeria, George Bilchev, Rajkumar Roy, Andrew Watson, Kai Chen and Martin Beck.

References

1. Parmee I. C., Denham M. J. The Integration of Adaptive Search Techniques with Current Engineering Design Practice. Procs. of Adaptive Computing in Engineering Design and Control; University of Plymouth, UK; Sept. 1994; pp. 1-13.
2. Goldberg D. E., Genetic Algorithms in Search,Optimisation & Machine Learning. Addison - Wesley Publishing Co., Reading, Massachusetts, 1989.
3. A. Coloni, M. Dorigo, V. Maniezzo. An Investigation of Some Properties of the Ant Algorithm. Procs. PPSN'92, Elsevier Publishing, pp. 509-520.
4. Parmee I. C. Genetic Algorithms, Hydropower Systems and Design Hierarchies. Invited paper for Special Edition of Micro-Computers in Civil Engineering, to be published 1996.
5. Parmee I. C., Diverse Evolutionary Search for Preliminary Whole System Design. Procs. 4th International Conference on AI in Civil and Structural Engineering , Cambridge University, Civil-Comp Press, August 1995.
6. Dasgupta D., MacGregor D., A Structured Genetic Algorithm. Research Report IKBS-2-91, University of Strathclyde, UK, 1991.
7. Parmee I. C. The Maintenance of Search Diversity for Effective Design Space Decomposition using Cluster-oriented Genetic Algorithms (COGAs) and Multi-agent Strategies (GAANT). Procs. of Adaptive Computing in Engineering Design and Control; University of Plymouth, UK; March, 1996.
8. Bilchev G., Parmee I. C. The Ant Colony Algorithm for Searching Continuous Design Spaces. Evolutionary Computing, Lecture Notes in Computer Science 993, selected papers from AISB Workshop, Sheffield, UK, Springer Verlag, April 1995.
9. Parmee I. C. The Development of a Dual-Agent Search Strategy for Whole System Design Hierarchies. Procs. 4th International Conference on Parallel Problem Solving from Nature (PPSN IV), Berlin, September, 1996.
10. Parmee I. C. Cluster-oriented Genetic Algorithms (COGAs) for the Identification of high-performance Regions of Design Spaces. Procs 1st International Conference on Evolutionary Computation and Applications (EvCA'96), Moscow, June 1996.
11. Jarvis R. A. and Patrick, E. A. Clustering using a Similarity Measure Based on Shared Near Neighbours. IEEE Transactions on Computers, vol-22,no 11; 1973.
12. Parmee I. C., Johnson M., Burt S., Techniques to Aid Global search in Engineering Design. Procs Industrial and Engineering Applications of artificial Intelligence and Expert Systems; Austin, Texas, June 1994.
13. Reeves, C. R. Using Genetic Algorithms with Small Populations. Procs. Fifth International Conference on Genetic Algorithms, University of Illinois, Morgan-Kaufman, 1993.
14. Booker, L., Improving Search in Genetic Algorithms. In Genetic algorityhms and Simulated Annealing; L. Davis (ed.), Morgan-Kaufman, pp. 61-73, 1987.
15. Baker, J. E., Reducing Bias and Inefficiency in the Selection Algorithm. Proc International Conference on Genetic Algorithms 2, Lawrence Erlbaum Associates, pp. 14-21,1987.
16. Beasley D., Bull D. R., Martin R. R., A Sequential Niche Technique for Multimodal Function Optimisation. Journal of Evolutionary Computation 1 (2), MIT press, pp. 101-125, 1993.

17. Roy R., Parmee, I. C., Purchase, G. Integrating the Genetic Algorithm with the Preliminary Design of Gas Turbine Blade Cooling Systems. Procs. of Adaptive Computing in Engineering Design and Control; University of Plymouth, UK; March. 1996.
18. Harik G. Finding Multimodal Solutions Using Restricted Tournament Selection. Procs. 6th International Conference on Genetic Algorithms, Pittsburgh, 1995.
19. Roy R., Parmee I. C., Adaptive Restricted Tournament Selection for the Identification of Multiple Sub-optima in a Multimodal Function. Procs. AISB Workshop on Evolutionary Computing. Brighton, UK, 1996.
20. FuzzyClips Users Guide, 6.02A; 1994; Knowledge systems Laboratory, National Research Council, Canada.
21. Parmee I. C., Gane C., Donne M., Chen K. Genetic Strategies for the Design and Control of Thermal Systems. Procs. Fourth European Congress on Intelligent Techniques and Soft Computing; Aachen, September 1996.
22. G. Bilchev, I. C. Parmee. Constrained and Multi-modal Optimisation with an Ant Colony Search Model. Procs. of Adaptive Computing in Engineering Design and Control; University of Plymouth, UK; March, 1996.
23. Koza J., Genetic Programming. MIT Press Inc., 1992.
24. Watson A. H., Parmee I. C., Systems Identification Using Genetic Programming. Procs. of Adaptive Computing in Engineering Design and Control; University of Plymouth, UK; March, 1996.
25. Haaland S. E. Simple and Explicit Formulas for the Friction Factor in Turbulent Pipe Flow. Journal of Fluids Engineering 105, pp. 89-90, 1983.
26. L.J Eshelman. The CHC Adaptive Search Algorithm : How to Have Safe Search When Engaging in Nontraditional Genetic Recombination. In G.J.E Rawlins (editor), Foundations of Genetic Algorithms and Classifier Systems. Morgan Kaufmann, San Mateo, CA, 1991.
27. E.D Goodman, R.C Averill, W.F Punch, Y. Ding, B Mallot. Design of Special-Purpose Composite Material Plates Via Genetic Algorithms. Proc. of the Second Int. Conf. on Adaptive Computing in Engineering Design and Control, ed. I.C Parmee, Plymouth University, 1996.
28. Vekeria H., Parmee I. C. The Use of a Multi-level CHC GA for Structural Shape Optimisation. Procs. Fourth European Congress on Intelligent Techniques and Soft Computing; Aachen, September 1996.

Identification of Mechanical Inclusions

Marc Schoenauer, François Jouve, and Leila Kallel

CMAP – URA CNRS 756, Ecole Polytechnique, Palaiseau 91128, France

Summary. Evolutionary Algorithms provide a general approach to inverse problem solving: As optimization methods, they only require the computation of values of the function to optimize. Thus, the only prerequisite to efficiently handle inverse problems is a good numerical model of the direct problem, and a representation for potential solutions.

The identification of mechanical inclusion, even in the linear elasticity framework, is a difficult problem, theoretically ill-posed: Evolutionary Algorithms are in that context a good tentative choice for a robust numerical method, as standard deterministic algorithms have proven inaccurate and unstable. However, great attention must be given to the implementation. The representation, which determines the search space, is critical for a successful application of Evolutionary Algorithms to any problem. Two original representations are presented for the inclusion identification problem, together with the associated evolution operators (crossover and mutation). Both provide outstanding results on simple instances of the identification problem, including experimental robustness in presence of noise.

1. Introduction

Evolutionary Algorithms (EAs) are stochastic optimization methods that have been demonstrated useful to solve difficult, yet unsolved, optimization problems. Requiring no regularity of the objective function (or of the constraints), EAs are able to tackle optimization problems on different kinds of search spaces, such as continuous, discrete or mixed spaces, as well as spaces of graphs or lists. The only prerequisite are the definition of evolution operators such as crossover and mutation, satisfying as much as possible heuristically derived requirements. The two main drawbacks of EAs are first the large number of evaluation of the objective function they usually imply before eventually reaching a good, if not optimal, solution; and second, their stochastic aspect, weakening their robustness. Hence, EAs should be used with care, on problems beyond the reach of standard deterministic optimization methods.

In structural mechanics, the non-destructive identification of inclusions is such a difficult problem, resisting to-date numerical methods: in its simplest instance, a structure is known to be made of two different known materials, but their repartition in the structure is unknown. The available data consist of records of the mechanical behavior of the structure under known loadings. The goal is to find the geometrical repartition of both materials from these experimental data. In steel manufacturing plants, for instance, it is of vital importance to check if coal scories are included in steel parts, and if their repartition does not dangerously weaken the whole part. For a given repartition of both materials, the computation of the simulated mechanical behavior of the structure is straightforward (*e.g.* using any Finite Element package). The identification can then be viewed as an inverse problem.

This paper addresses this inverse problem using EAs. A possible objective function for such inverse problems is the difference between the simulated mechanical behavior of a tentative repartition of both materials and the actual experimental behavior. However, the main difficulty is to define the search space in which the EA will be defined. Considering past works on the Optimum Design problem, (a closely related problem, where the goal is to find a partition of a design domain into material and void), the straightforward representation is defined from a fixed mesh of the structure, leading to a fixed-length bitstring well-suited to Genetic Algorithms. However, this approach will not be considered here, as it makes the optimization problem intractable when the underlying mesh is refined. Instead, two non-standard representations (termed the *Voronoï representation* and the *H-representation*) are introduced, independent of any *a priori* discretization, but leading to variable-length "individuals". Hence, specific operators have to be designed and implemented.

Nevertheless, Evolutionary Algorithms working on these representations give outstanding results on the inclusion identification problem in the context of linear elasticity, outperforming previous deterministic numerical solutions. Experimental evidences appear that the quality of the results highly depends of the amount of experimental information the algorithm can rely upon. Furthermore, the robustness against noise in the experimental data is experimentally shown on simulated artificial noisy data.

The paper is organized the following way: In Section 2., a general framework to address inverse problems with EAs is introduced. The mechanical problem of inclusion identification is presented in details in Section 3. It is an instance of the general inverse problem of Section 2., and can thus be solved by Evolutionary Algorithms, once the search space has been skillfully designed: Two original representations for that problem are presented in Section 4., together with their specific evolution operators (crossover and mutations). The first results, using the Voronoï representation, are presented in Section 5., demonstrating outstanding performance on artificial instances of the inclusion identification problem. Comparative results for both representations are presented in Section 5.6, opening the discussion about the *a priori* choice of a representation for a given problem. This discussion leads to propose further directions of research, sketched in Section 6.

2. Evolutionary Inverse Problems Solving

This section presents a general framework to solve inverse problems by means of Evolutionary Algorithms, which will be further applied to the Mechanical problem introduced in Section 3.

2.1 Direct problems

Consider a process (*e.g.* a physical phenomenon) that produces output experimental results given some inputs (experimental conditions).

The successful resolution of the direct problem consists in building a simulated process able to accurately predict the experimental output of the physical phenomenon from the same experimental input data. Such simulation generally relies on a function (law, algorithm, command, ...) modeling the underlying physical behavior. Figure 2.1 gives a symbolic representation of a physical process together with its simulated model.

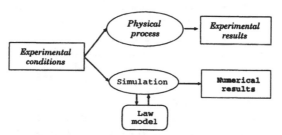

Fig. 2.1. Direct problem. Physical and simulated process. Simulation is successful (based on an accurate model) iff the error between experimental results and numerical results is small

2.2 Inverse problems

However, quite often, the model (law, command, ...) is not precisely known: The goal of inverse problems is to find a model such that the numerical simulations based on this model successfully approximates experimentations. The data of inverse problems are experimental conditions together with the corresponding actual experimental results.

Whenever a good simulation of a direct problem exists, Evolutionary Computation can be used to address the inverse problem. The fitness of a candidate solution can be computed as shown in Figure 2.2: the results of the numerical simulation performed using the individual at hand as the model to identify is compared to original experimental results, and the goal is to reach the smallest possible error.

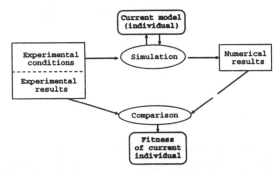

Fig. 2.2. Evolutionary approach for the inverse problem. The fitness of the individual at hand is achieved by comparing the actual experimental results with the numerical results obtained using the individual in place of the law model

2.3 The genotype space

The most critical step in evolutionary inverse problem solving is the choice of the representation, which defines the genotype space (*i.e.* the search space). In any case, the EA will find a good – if not the best – solution in the search space. Hence it seems that larger search spaces allow better solutions to be found. However, the larger is the search space, the more difficult is the optimization task, and a trade-off has to be found.

Yet another important issue in inverse problem solving is the *generalization capability* of the solution: How good is the resulting model when used with experimental conditions that are different from those used during the identification process? The usual answer in evolutionary inverse problem solving is to use, during the identification, many different experimental conditions, also termed *fitness cases*. The fitness is then the average of the error over all fitness cases. Needless to say, the total computational cost increases with the number of fitness cases.

3. The Mechanical Problem

This section gives a detailed presentation of the mechanical problem of inclusion identification, and states the simplification hypotheses made throughout this paper. The issues raised in Section 2. above will then be addressed on this instance of inverse problem.

3.1 Background

Consider an open bounded domain $\Omega \subset \mathbb{R}^N$ ($N = 2, 3$), with a smooth enough boundary $\partial\Omega$, filled with a linear elastic material. Under the hypothesis of small deformations (linear elasticity) the following equations hold for, respectively, the *strain* and the *stress* tensors:

$$\varepsilon(x) := \frac{1}{2}(\nabla u(x)\nabla u^T(x)), \quad \text{and} \quad \sigma(x) := A(x)\varepsilon(x), \qquad (3.1)$$

where $u(x)$ is the displacement field at point x and $A(x)$ is the elasticity tensor (a fourth order tensor) involved in the *Hooke's law* (3.1). A is supposed to be *inhomogeneous*, which means that its value depends on the point x. $A(x)$ is a positive definite tensor which satisfies some symmetry conditions.

When A is given, one can state two kinds of boundary problems, respectively of *Dirichlet* and *Neuman* type:

$$\begin{cases} \mathrm{div}\sigma = 0 & \text{in } \Omega, \\ u = u_0 & \text{on } \partial\Omega, \end{cases} \quad \text{and} \quad \begin{cases} \mathrm{div}\,\sigma = 0 & \text{in } \Omega, \\ \sigma.n = g_0 & \text{on } \partial\Omega, \end{cases}$$
$$(3.2)$$

where u_0 and g_0 are respectively a given displacement field and a given external force field on the boundary $\partial\Omega$.

It is well known (see *e.g.* [4]) that each of these problem has a unique solution (for the Neuman's problem, one has to impose an integral condition on g_0 to insure existence, and an integral condition on u to eliminate rigid displacements).

In the following, the *inverse problem* will be considered:

$$\text{find } A \text{ such that }, \forall i \in \{1, ..., n\}, \exists u, \quad \begin{cases} \mathrm{div}\,\sigma = 0 & \text{in } \Omega, \\ u = u_i & \text{on } \partial\Omega, \\ \sigma.n = g_i & \text{on } \partial\Omega, \\ \sigma = A\varepsilon, \end{cases} \quad (3.3)$$

where $(g_i)_{i=1,..,n}$ and $(u_i)_{i=1,..,n}$ are given.

Problem (3.3) is a discrete version of the "ideal" inverse problem:

$$\text{find } A, \text{ given application: } \Lambda_A : u_{|\partial\Omega} \longrightarrow \sigma_{|\partial\Omega}. \qquad (3.4)$$

The underlying physical problem is still lacking much more of known data than problem (3.3) since Λ_A is only known through a finite number of *experimental measures* performed at a finite number of points. Hence, the real identification problem treated by mechanical engineers can be stated as:

$$\text{find } A \text{ such that }, \forall 1 \le i \le n, \exists u, \quad \begin{cases} \mathrm{div}\,\sigma = 0 & \text{in } \Omega, \\ u(x^j) = u_i^j & \forall 1 \le j \le p, \\ \sigma(x^j).n = g_i^j & \forall 1 \le j \le p, \\ \sigma = A\varepsilon, \end{cases} \quad (3.5)$$

where information on the boundary is only known at a finite number of experimental points $(x^j)_{j=1,...,p}$ for a finite number (n) of experiments. In addition, these data may be known with a certain amount of experimental error or noise.

3.2 State of the art

If the aim is the numerical treatment of problem (3.4) (and *a fortiori* problem (3.3) or (3.5)) by "classical" (*i.e.* non-stochastic) methods, two theoretical points are crucial:

- existence and uniqueness of A as a function of Λ_A,
- continuity of the dependency of A with respect to Λ_A.

Existence is of course essential to the pertinence of the identification problem, but uniqueness and continuity are only needed to insure the reliability and the stability of deterministic numerical algorithms. On the other hand, EAs can deal with non-continuous functionals and non-unique solutions.

Problem (3.4) is the elastic equivalent of the so-called *tomography* problem where the elliptic operator is the conductivity operator ($\operatorname{div}(A\nabla u)$, u scalar field) instead of the elasticity one ($\operatorname{div}(A\varepsilon(u))$, u vector field).

The tomography problem as been widely studied. Under some hypothesis, existence and uniqueness have been proved. However, the continuity of the functional is only known in a weak sense, that cannot help for numerical simulations.

The elasticity problem (3.4) is more difficult. Existence and uniqueness have been proved for isotropic Hooke's laws, but there is no continuity result. For a comprehensive bibliographical discussion on this subject, see [6].

Numerical simulations by classical methods have shown that both tomography [19] and elastic identification problems [6] are ill-posed, and thus EAs are good tentative choice for a robust optimization method.

3.3 The direct problem

In this paper, attention has been focused on representation and on specific operators for EAs. To highlight specific problems involved in these algorithms, the mechanical problem (3.5) was restricted to a two-dimensional simpler class of problems:

Let A_1 and A_2 be two isotropic elasticity tensors, fully defined by Young's moduli E_1 and E_2 and Poisson ratios ν_1 and ν_2. The aim is to solve Problems (3.3) and (3.5), restricting allowable values of $A(x)$ to

$$A(x) = \begin{cases} A_1 & \text{if } \chi(x) = 0 \\ A_2 & \text{if } \chi(x) = 1 \end{cases} \tag{3.6}$$

where χ is a characteristic function defined on Ω: The target is hence a partition of Ω in two subsets, each made of a material with known characteristics.

These problems, although less general than (3.3) and (3.5) are still beyond the capabilities of deterministic algorithms (see [6]). Moreover, the general Problem (3.5) can be treated in the same way, as well as identification in non-linear elasticity (as discussed in Section 6.).

The direct elasticity problems have been solved by a classical finite element method (as described in [13]). All geometrical and mechanical details are specified in Section 5.

4. Representations for Mechanical Inclusions

This section introduces two non-standard representations for the problem described in Section 3. Both are variable-length representations, and use real numbers as main components. Hence, specific operators (*e.g.* crossover and mutation) have to be designed for each representation.

4.1 Prerequisites

A solution to the inclusion identification problem is a partition of the domain of the structure into two subsets, each subset representing one of the materials involved. Moreover, all connected components of any subset should have a non-void interior and a regular boundary.

A theoretical framework has been developed by Ghaddar & al. [10] in the context of Structural Optimum Design: The search space is restricted to partitions with polygonal boundaries. Theoretical results are proven, approximation spaces are introduced and corresponding approximation results are obtained. Though the objective function considered in this paper is quite different from the one in [10], the same search space will be used here.

However, a significant difference between the objective functions in [10] and the one to be used here is that the inclusion identification problem requires a Finite Element Analysis on the direct problem (see Section 3.3) to compute the fitness of a point of the search space (*i.e.* a given repartition of both materials), as introduced in the general framework of Section 2., and detailed in Section 5.2 It is well-known that meshing is a source of numerical errors [5]. Hence, for any Evolutionary Algorithm, using a fitness function based on the outputs of two Finite Element Analyses performed on different meshes is bound to failure, at least when the actual differences of behavior becomes smaller than the unavoidable numerical noise due to remeshing. The use of the same mesh for all Finite Element Analyses (at least inside the same generation) is thus mandatory to obtain significant results.

4.2 The bitstring representation

Once the decision to use a fixed mesh has been taken, and with even very little knowledge of EAs, the straightforward representation for a partition of the given domain is that of bitstrings: each element of the fixed mesh belongs to either one of the subsets of the partition, which can be symbolically labeled 0 or 1. The resulting representation is a bitstring – or, more precisely, a bitarray, as the pure bitstring point of view can be misleading, see [18]. Hence, almost all previous works using Genetic Algorithms in Optimum Design did use that representation [11, 3, 17].

However, the limits of this bitstring representation clearly appear when it comes to refining the mesh, in order either to get more accurate results or to solve 3-dimensional problems: this would imply huge bitstring, as the size of the bitstring is that of the underlying mesh. However, the size of the population should increase proportionally to that of the bitstring, according to both theoretical results [2] and empirical studies [27]. Moreover, more generations are also needed to reach convergence, and the resulting algorithm rapidly becomes intractable.

These considerations show the need for other representations, not relying on a given mesh – even if a fixed mesh is used during the computation of the fitness function. Two of such representations have been designed, and successfully used on the Optimum Design problem [23, 24].

4.3 The Voronoï representation

A possible way of representing partitions of a given domain comes from computational geometry, more precisely from the theory of Voronoï diagrams. The ideas of Voronoï diagrams are already well-known in the Finite Element community, as a powerful tool to generate good meshes [9]. However, the representation of partitions by Voronoï diagrams aiming at their evolutionary optimization seems to be original.

Voronoï diagrams: Consider a finite number of points $V_0, \ldots, V_N$ (the *Voronoï sites*) of a given subset of $\mathbb{R}^n$ (the design domain). To each site V_i is associated the set of all points of the design domain for which the closest Voronoï site is V_i, termed *Voronoï cell*. The *Voronoï diagram* is the partition of the design domain defined by the Voronoï cells. Each cell is a polyhedral subset of the design domain, and any partition of a domain of $\mathbb{R}^n$ into polyhedral subsets is the Voronoï diagram of at least one set of Voronoï sites (see [20, 1] for a detailed introduction to Voronoï diagrams).

The genotype: Consider now a (variable length) list of Voronoï sites, each site being labeled 0 or 1. The corresponding Voronoï diagram represents a partition of the design domain into two subsets, if each Voronoï cell is labeled as the associated site (here the Voronoï diagram is supposed regular, *i.e.* to each cell corresponds exactly one site). Example of Voronoï representations can be seen in Figure 4.1. The Voronoï sites are the dots in the center of the cells. Note that this representation does not depend in any way on the mesh that will be used to compute the mechanical behavior of the structure. Furthermore, Voronoï diagrams being defined in any dimension, the extension of this representation to $\mathbb{R}^3$ and $\mathbb{R}^n$ is straightforward.

An important remark is that this representation presents a high degree of *epistasis* (the influence of one Voronoï site on the mechanical structure is highly dependant on all neighbor sites). This will be discussed in more details in Section 6.

Decoding: Practically, and for the reasons stated in Section 4.1 the fitness of all structures will be evaluated using the same fixed mesh. A partition described by Voronoï sites is thus mapped on this fixed mesh: the subset an element belongs to is determined from the label of the Voronoï cell in which the center of gravity of that element lies in.

Crossover operator: The idea of the crossover operators is to exchange subsets of geometrically linked Voronoï sites. In this respect, it is similar to the specific bitarray crossover described in [18, 17]; moreover, this mechanism easily extends to any dimension [15]. Figure 4.1 demonstrates an application of this crossover operator.

Mutation operators: Different mutation operators have been designed. They are applied in turn, based on user-defined probabilities.

A first mutation operator modifies the coordinates of the Voronoï sites as in the now-standard Gaussian mutation for real-valued variables from Evolution Strategies [25] (i.e. by addition of a Gaussian random variable of mean 0 and user-defined standard deviation); another mutation randomly flips the boolean attribute of some sites; finally, dealing with variable-length representations, one has to include as mutation operators the random addition and destruction of some Voronoï sites.

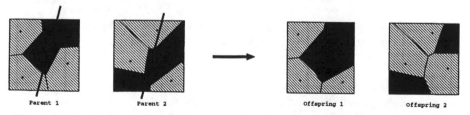

Parent 1   Parent 2   Offspring 1   Offspring 2

Fig. 4.1. The Voronoï representation crossover operator. A random line is drawn across both diagrams, and the sites on one side are exchanged

4.4 H-representation

Another representation for partitions is based on an old-time heuristic method in Topological Optimum Design (TOD): from the initial design domain, considered as plain material, remove material at locations where the mechanical stress is minimal, until the constraints are violated. However, the lack of backtracking makes this method useless in most TOD problems. Nevertheless, this idea gave birth to the "holes" representation [7], later termed H-representation.

The representation: The design domain is by default made of one material, and a (variable length) list of "holes" describes the repartition of the other material. These holes are elementary shapes taken from a library of possible simples shapes. Only rectangular holes are considered at the moment, though on-going work is concerned with other elementary holes (*e.g.* triangles, circles) [26, 16].

Example of structures described in the H-representation are presented in Figure 4.2. The rectangles are taken in a domain larger than the design domain, in order not to bias the boundary parts of the design domain toward the default value.

The H-representation, as the Voronoï representation, is independent from any mesh, and hence its complexity does not depend on any required accuracy for the simulation of the mechanical behavior of the structure. Its merits and limitations will be discussed in the light of the experimental results presented in next sections.

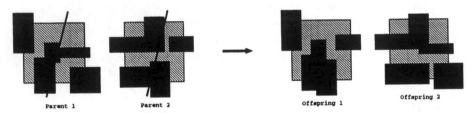

Parent 1   Parent 2   Offspring 1   Offspring 2

Fig. 4.2. The H-representation crossover operator. A random line is drawn across both structures, and the holes on one side are exchanged

Decoding: As for the Voronoï representation, the simulated behavior of the shapes is computed on a given fixed mesh, to limit the numerical noise due to re-meshing. The criterion to decide which subset an element does belong to, is based on whether its center of gravity belongs to a hole (in which case the whole element is void) or not.

Evolution operators: The evolution operators are quite similar to those of the Voronoï representation:

• crossover by geometrical (2D or 3D) exchange of holes (see Figure 4.2 for an example);

• mutation by Gaussian modification of the characteristics (coordinates of the center, width and length) of some holes;

• mutation by random addition or destruction of some holes;

5. Numerical Results

This section presents the very first experiments (to the best of our knowledge) on the inclusion identification problem using Evolutionary Algorithms.

5.1 Experimental settings

All numerical results for problem (3.3) have been obtained on a two-dimensional square structure fixed on its left-side, the forces being applied at points of the three other sides.

The aim is to identify the repartition of two materials into a given domain: a hard material ($E = 1.5$ and $\nu = 0.3$) and a soft material ($E = 1$ and $\nu = 0.3$). A fixed mesh of size 24×24 was used throughout the experiments.

The reference experimental loading cases considered to compute one fitness value (see Section 2.) were here actually computed from a known configuration of both materials in the structure. The optimal solution is thus known, which allows a better insight and understanding during the evolution of the population. Moreover, much flexibility was required during the tuning of the overall process, that actual experimental results could not have bought. Finally, considerations about the noise in the experimental reference recording also favor simulated results: there still is a bias (due to numerical error in the Finite Element Analysis), but, as this bias hopefully is the same during the fitness computation, it should not weaken the results of the evolutionary algorithm, as could unpredictable noise in actual measures. Of course, further work will have to consider actual experimental results to fully validate the approach.

5.2 The fitness functions

In real world situations, the design of the fitness function should take into account as many loading cases as available, in order to use as much information as possible about the behavior of the structure to be identified. However, the choice made in these preliminary experiments of having simulated "reference experiments" makes it possible to use as many loading cases as needed. In an attempt to have a sampling of the mechanical behavior of the structure as uniform as possible over the domain, 37 different loading cases were used: Each loading case consists in pulling with a given force at one point of the boundary of the structure, following the normal to the boundary of the unloaded structure. The 37 loading points are equidistributed on the free boundary of the square structure.

Another degree of freedom offered by the use of simulated experimental values for the reference displacements addresses the number of points where these reference displacements were available. In real world situations, some gauges are placed to the boundary of the structure, and only the displacements at those points is available. However, it seems clear that the more measure points, the easier the identification task. Hence, three different fitness functions have been used throughout the numerical experiments presented below, all using the 37 loading cases described above, but different in the number of measure points used to compute the error between the reference displacements (the so-called "experimental results") and the displacements of the partition at hand:

• The most informative fitness function, hereafter referred to as the *total fitness*, takes into account the displacements at *all nodes* of the mesh.

• An intermediate fitness function uses only the displacements of all nodes lying at the boundary fo the structure, and is termed the *boundary fitness*.

• The *real-world fitness* uses only 9 measure points equidistributed on the free boundary of the square structure, incorporating much less information to the fitness function, but resembling actual experimental conditions.

5.3 The Evolutionary Algorithm

The Evolutionary Algorithm used for all experiments presented in this paper uses a standard GA scheme: linear ranking proportional selection, crossover rate of 0.6, mutation rate of 0.2, all offspring replace all parents. The population size is set to 100, and at most 300 generations of the algorithms are allowed – but it stops whenever 50 generations are run without any improvement of the overall best fitness. Hence, between 10000 and 30000 Finite Element Analyses were performed for each run, requiring around 9h on a middle range HP workstation (715/75).

However, all experiments were run at least 10 times with independent initial populations, to avoid as much as possible stochastic bias (as clearly stated in [14], "You should never draw any conclusion of a single run of any Evolutionary Algorithm").

5.4 Results using the Voronoï representation

The very first experiments were performed using the Voronoï representation, described in Section 4.3

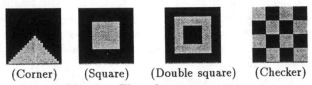

(Corner)   (Square)   (Double square)   (Checker)

Fig. 5.1. The reference structures

They were obtained on the reference partitions represented in Figure 5.1, where black areas represent the soft material and white areas the harder material (the examples range from the easiest to the most difficult).

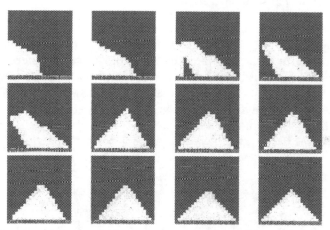

Fig. 5.2. A successful run on the corner problem, using boundary fitness (Perf). Plots of the best individual at different generations of the evolution

The first results on the "corner" example of Figure 5.1-a were astonishingly good: the exact solution was found 3 times out of 10 runs, in less than 100 generations when using the "total" fitness, and even once (in 250 generations) using the boundary fitness. Figure 5.2 shows an example of a successful run, where the best individual at different stages of evolution is plot together with the error, and the number of Voronoï sites it contains. The Voronoï sites represented on the figure by grey little dots.

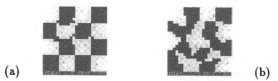

(a)   (b)

Fig. 5.3. The checker problem: (a): with total fitness. (b): with real fitness

On the "square" example of Figure 5.1-b, some difference began to appear between the different fitnesses, but this example is still an easy problem. This phenomenon was more and more visible as the difficulty of the problem increased. When it came to the "checker" example of Figure 5.1-d, the total fitness gave much better results than the real-world fitness, as can be seen in Figure 5.3-a and -b. However, the real-world fitness gave some interesting results, as can be seen on Figure 5.3-b: the actual values are clearly identified along the boundary, except along the fixed side of the boundary, where too little information is available.

5.5 Results in presence of noise

After these first satisfactory results on exact data, further validation of the proposed approach had to take into account possible errors and noise in the data. In order to test the sensibility of the algorithm to noise, artificial noise was purposely introduced in the reference "experimental" displacements. Figure 5.4 shows the results obtained on the (easy) "corner" example, with 2% and 5% noise (*i.e.* when all reference displacements were multiplied by a term $(1 + \varepsilon)$, ε being a random variable uniformly distributed in $[-0.02, 0.02]$ and $[-0.05, 0.05]$ respectively). The results are – of course ! – degraded, but they demonstrate a fairly good robustness, at least on this example.

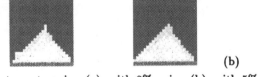

(a)   (b)

Fig. 5.4. Robustness to noise. (a): with 2% noise. (b): with 5% noise

5.6 Comparative results

Further experiments were run in order to compare both the Voronoï and the H-representations. In fact, three representations were tested on together on different problems: the Voronoï representation, the H-representation where rectangles represent the soft material (termed H-0), and the H-representation where rectangles represent the hard material (termed H-1).

Regarding the comparison between the two H-representation, their behavior was what could be expected, on reference structures like the "square" example (Figure 5.1-b): it is by far easier to find rectangles describing the inside square than rectangles approximating the outside of that square, and the H-1 representation consistently outperformed the H-0 representation. Moreover, this phenomenon increases when the size of the inside square decreases.

However, from the limited experiments performed so far, it seems that the Voronoï representation slightly outperforms the H-representations, in contradiction with the situation in the domain of Optimum Design [23]. Figure 5.5 shows an example of such a situation, on the problem of the "double square" of Figure 5.1-c with "real" fitness. The plots represent the average over 10 independent runs of the best fitness (*i.e.* the smallest error) along generations for all three representations. Note that the variance of the results for all representation was very small, and all but one runs using the

Voronoï representation reached smaller error than the best runs of using the H-representation. The best results for both the H-1 and the Voronoï representation are presented on Figure 5.6.

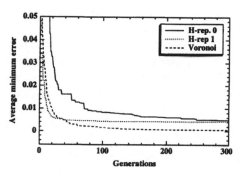

Fig. 5.5. Comparative results on the "double-square" problem with "real" fitness. The Voronoï representation slightly but consistently outperforms both H-representations

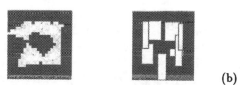

(a)   (b)

Fig. 5.6. Best results on the "double" example with real fitness for the Voronoï and the H-1 representation

6. Discussion and Further Work

Experimental comparisons of different representations encounter the difficulty of designing "fair" experiments: the mutation operators are not the same for all representations, forbidding any satisfactory way of comparing the mutation strengths. Moreover, experimental comparative results on one problem can hardly be generalized to too different problems.

Hence, it is essential to design a methodology, or at least some heuristics, to guide a future practitioner of evolutionary inclusion identification in his (or her) choice: Having so many possible representations for partitions os materials (the bitarray of section 4.2, the Voronoï representations, both H-0 and H-1 representations described in section 5.6) makes it more difficult to choose among them when facing a specific instance of a problem. Some promising directions are given in the literature.

The fitness variance theory of Radcliffe [22] studies the variance of the fitness as a function of the order of an extension of schemas called *formae* [21], and, simply put, shows that the complexity and difficulties of evolution increases with the average variance of the fitness as a function of the formae order. But if the formae and their order (or their precision) are well-defined on any binary representation, including the bit-array representation rapidly presented in Section 4.1, it is not straightforward to extend these definitions to variable length representations, as the ones presented in this paper.

Moreover, Radcliffe's fitness variance does not take into account the possible evolution operators. Further step in that direction would be to study the variance of the change of fitness with respect to a given evolution operator (*e.g.* the Gaussian mutation of Voronoï sites for different standard deviations), in the line of the work in [8].

The fitness distance correlation of Jones [12] studies the correlation between the distance to the optimal point and the fitness. Simply put again, the idea is that, the stronger this correlation, the narrower the peak the optimum belongs to. Conjectures based on this remark are experimentally confirmed in the bitstring frame. Nevertheless, the difficulty in variable-length representations is to define a distance which is meaningful for both the representation and the problem at hand. Preliminary work addressing this issue define a purely *genotypic* distance, based on partial matching of items of the variable-length lists representing the two individuals. The first – on-going – studies [16] demonstrate that the results of [12] seem to extend to the variable-length case: the correlation between the distance to optimum and the fitness is a good predictor of the performance of the representation on a given problem. Moreover, in the case where a good correlation exists, equivalent results are obtained when considering either the distance to the actual optimum or the distance to the best individual in the current sample: if this was not true, the method would be of poor practical interest, as the global optimum is usually unknown.

Another direction of research regards the link between the representation and the evolution scheme: as stated and partially demonstrated on the Optimum Design problem in [24], the higher degree of epistasis (*i.e.* interaction among genetic material of the same individual when going from the genotype to the actual mechanical structure) in the representation should favor top-down approaches, *e.g.* ES or EP schemes relying mostly upon mutation operator, rather than the GA bottom-up approach relying upon the gradual assembly of small building blocks to construct the a good solution. As opposed to the Optimum Design problem, the inclusion identification problem is tunable and hence allows precise experiments to be conducted: the optimum partition is known, and can be tailored at will: for instance, the respective amount of both material can be prescribed, as well as the number of connected components of both materials.

Further work will also consider the general problem (3.3) of Section 3. instead of the boolean simplification (3.6): the extension of both representations to handle real-valued labels instead of boolean labels is straightforward for the Voronoï representation (replace the boolean label of each site by a real number), and fairly simple to imagine for the H-representation (*e.g.* assign a real-valued label to each "hole", and, for each element, compute the mean value of the labels of all holes covering the center of the element).

7. Conclusion

This paper presents preliminary results using Evolutionary Computation on an important engineering problem: non destructive inclusion identification in Structural Mechanics. In spite of theoretical results on identifiability in the linear elastic case, the standard deterministic numerical methods have to face a ill-conditioned problem, and demonstrated to be both inaccurate and unstable. On the opposite, the evolutionary method demonstrates powerful on the simplified problem of linear elasticity involving two materials of known characteristics. These results required the design of non-standard representations together with *ad hoc* genetic operators. The main feature of these representation is their independence toward any discretization of the structure: hence the complexity of the algorithm itself, in terms of number of fitness evaluations, only depends on the problem at hand, regardless of the numerical method used to compute the fitness. Moreover, these representations can — and will — be extended easily to identify unknown materials in a given structure.

References

1. J.-D. Boissonnat and M. Yvinec. *Géométrie algorithmique*. Ediscience International, 1995.
2. R. Cerf. An asymptotic theory of genetic algorithms. In J.-M. Alliot, E. Lutton, E. Ronald, M. Schoenauer, and D. Snyers, editors, *Artificial Evolution*, volume 1063 of *LNCS*. Springer Verlag, 1996.
3. C. D. Chapman, K. Saitou, and M. J. Jakiela. Genetic algorithms as an approach to configuration and topology design. *Journal of Mechanical Design*, 116:1005–1012, 1994.
4. P. G. Ciarlet. *Mathematical Elasticity, Vol I : Three-Dimensional Elasticity*. North-Holland, Amsterdam, 1978.
5. P. G. Ciarlet. *The Finite Element Method for Elliptic Problems*. North-Holland, Amsterdam, 1988.
6. A. Constantinescu. *Sur l'identification des modules élastiques*. PhD thesis, Ecole Polytechnique, June 1994.
7. J. Dejonghe. Allègement de platines métalliques par algorithmes génétiques. Rapport de stage d'option B2 de l'Ecole Polytechnique. Palaiseau, Juin 1993.

8. D. B. Fogel. Phenotypes, genotypes and operators in evolutionary computation. In D. B. Fogel, editor, *Proceedings of the Second IEEE International Conference on Evolutionary Computation*. IEEE, 1995.
9. P.L. George. *Automatic mesh generation, application to Finite Element Methods*. Wiley & Sons, 1991.
10. C. Ghaddar, Y. Maday, and A. T. Patera. Analysis of a part design procedure. *Submitted to Nümerishe Mathematik*, 1995.
11. E. Jensen. *Topological Structural Design using Genetic Algorithms*. PhD thesis, Purdue University, November 1992.
12. T. Jones and S. Forrest. Fitness distance correlation as a measure of problem difficulty for genetic algorithms. In L. J. Eshelman, editor, *Proceedings of the 6th International Conference on Genetic Algorithms*, pages 184–192. Morgan Kaufmann, 1995.
13. F. Jouve. *Modélisation mathématique de l'œil en élasticité non-linéaire*, volume RMA 26. Masson Paris, 1993.
14. K. E. Kinnear Jr. A perspective on gp. In Jr K. E. Kinnear, editor, *Advances in Genetic Programming*, pages 3–19. MIT Press, Cambridge, MA, 1994.
15. A. B. Kahng and B. R. Moon. Toward more powerful recombinations. In L. J. Eshelman, editor, *Proceedings of the 6th International Conference on Genetic Algorithms*, pages 96–103. Morgan Kaufmann, 1995.
16. L. Kallel and M. Schoenauer. Fitness distance correlation for variable length representations. In preparation, 1996.
17. C. Kane. *Algorithmes génétiques et Optimisation topologique*. PhD thesis, Université de Paris VI, July 1996.
18. C. Kane and M. Schoenauer. Genetic operators for two-dimensional shape optimization. In J.-M. Alliot, E. Lutton, E. Ronald, M. Schoenauer, and D. Snyers, editors, *Artificial Evolution*, LNCS 1063. Springer-Verlag, Septembre 1995.
19. R. V. Kohn and A. McKenney. Numerical implementation of a variational method for electric impedance tomography. *Inverse Problems*, 6:389–414, 1990.
20. F. P. Preparata and M. I. Shamos. *Computational Geometry: an introduction*. Springer-Verlag, 1985.
21. N. J. Radcliffe. Equivalence class analysis of genetic algorithms. *Complex Systems*, 5:183–20, 1991.
22. N. J. Radcliffe and P. D. Surry. Fitness variance of formae and performance prediction. In D. Whitley and M. Vose, editors, *Foundations of Genetic Algorithms 3*, pages 51–72. Morgan Kaufmann, 1994.
23. M. Schoenauer. Representations for evolutionary optimization and identification in structural mechanics. In J. Périaux and G. Winter, editors, *Genetic Algorithms in Engineering and Computer Sciences*, 443–464. John Wiley, 1995.
24. M. Schoenauer. Shape representations and evolution schemes. In L. J. Fogel, P. J. Angeline, and T. Bäck, editors, *Proceedings of the 5th Annual Conference on Evolutionary Programming*. MIT Press, 1996.
25. H.-P. Schwefel. *Numerical Optimization of Computer Models*. John Wiley & Sons, New-York, 1981. 1995 – 2nd edition.
26. M. Seguin. Optimisation de formes par évolution artificielle. etude de deux représentations, Dec. 1995. Rapport de DEA d'Analyse Numérique de l'Université de Paris VI.
27. D. Thierens and D.E. Goldberg. Mixing in genetic algorithms. In S. Forrest, editor, *Proceedings of the 5th International Conference on Genetic Algorithms*, pages 38–55. Morgan Kaufmann, 1993.

GeneAS: A Robust Optimal Design Technique for Mechanical Component Design

Kalyanmoy Deb

Department of Mechanical Engineering, Indian Institute of Technology, Kanpur, UP 208 016, India

Summary. A robust optimal design algorithm for solving nonlinear engineering design optimization problems is presented. The algorithm works according to the principles of genetic algorithms (GAs). Since most engineering problems involve mixed variables (zero-one, discrete, continuous), a combination of binary GAs and real-coded GAs is used to allow a natural way of handling these mixed variables. The combined approach is called GeneAS to abbreviate **Genetic Adaptive Search**. The robustness and flexibility of the algorithm come from its restricted search to the permissible values of the variables. This also makes the search efficient by requiring a reduced search effort in converging to the optimum solution. The efficiency and ease of application of the proposed method are demonstrated by solving four mechanical component design problems borrowed from the optimization literature. The proposed technique is compared with traditional optimization methods. In all cases, the solutions obtained using GeneAS are better than those obtained with the traditional methods. These results show how GeneAS can be effectively used in other mechanical component design problems.

1. Introduction

In an engineering design problem including mechanical component design, the goal is either to minimize or to maximize a design objective and simultaneously satisfy a number of equality and inequality constraints. Representing the design variables as a vector of real numbers $x = \{x_1, \ldots, x_N\}$, an engineering design optimization problem is usually expressed in a nonlinear programming (NLP) problem. In general, there are J inequality constraints and K equality constraints. A generic NLP problem can be expressed as follows:

$$\text{Minimize} \quad f(x)$$

subject to

$$\left. \begin{array}{ll} g_j(x) \geq 0, & j = 1, 2, \ldots, J; \\ h_k(x) = 0, & k = 1, 2, \ldots, K; \\ x_i^{(L)} \leq x_i \leq x_i^{(U)}, & i = 1, 2, \ldots, N. \end{array} \right\} \quad (1.1)$$

Different engineering design problems constitute different NLP problems, because the objective function, constraints, and the design variables are usually of different types. When faced with different NLP problems involving various types of objective function, constraints, and variables, designers must rely on an optimization algorithm which is robust enough to handle a wide variety of such problems. In this paper, we call an optimization algorithm *robust* if it can handle different types of objective functions, constraints, and variables without a major change in the algorithm. Most traditional optimization algorithms exploit the structure of the objective function and constraints. This restricts them to be used in different problems, because those algorithms can only be used to solve a particular type of objective functions and constraints that fit into the purview of those algorithms. However, the algorithms based on the penalty functions do not exploit the functional form of the objective functions and constraints, thereby allowing them to used in many problems (Kannan and Kramer, 1993, Sandgren, 1988).

However, the traditional optimization algorithms are handicapped in terms of handling discrete or zero-one type of variables efficiently. Moreover, mechanical component design problems usually involve a combination of real and discrete variables. In some problems, the discrete variables can take values in steps of a fixed size, or they can take values in any arbitrary steps. Some discrete variables can only take one of two options, such as the material being steel or aluminum (a zero-one variable). Since most traditional optimization methods are designed to work with continuous variables only, in dealing with design problems having discrete variables, these methods do not work as well (Deb, 1995). However, in practice, discrete variables are handled by adding artificial constraints (Kannan and Kramer, 1993). In those methods, discrete variables are treated as continuous variables in the simulation of the optimization algorithm and the search algorithm discourages the solutions to converge to a non-feasible value of the variables by introducing artificial constraints (Deb, 1995; Kannan and Kramer, 1993; Reklaitis, Ravindran, and Ragsdell, 1983). This fix-up not only increases the complexity of the underlying problem, but the algorithm also spends a considerable amount of effort in evaluating non-feasible solutions.

In this paper, a robust optimization algorithm based on genetic algorithms (GAs) has been suggested to handle mixed variables and mixed functional form for objective functions and constraints. The normalized penalty function approach is used to handle constraints and objective function complexities, however a flexible coding scheme is used to allow mixed variables. The algorithm developed here (called GeneAS, pronounced as 'genius') is somewhat different from the conventional GA. Instead of usual practice of using binary strings, a mixed coding of binary strings and real numbers are allowed. In applying GA to engineering design problems, one difficulty is the use of binary coding of continuous variables to any arbitrary precision and other difficult arises in coding discrete variables having any arbitrary number of choices. Although a large precision in a continuous variable can be achieved by using a long binary string to code the variable, the computational effort needed to solve the problem increases with string length (Goldberg, Deb, and Clark, 1992). Recently, a study has shown that the GA operators can be suitably modified to work with continuous variables directly by retaining the same search power of GA (Deb and Agrawal, 1995). Based on these findings, in this article, we present a robust optimization algorithm which easily allows a mixed representation of different types of design variables.

In the remainder of the paper, we present the GeneAS technique and then demonstrate its power by presenting a number of case studies of mechanical component design.

2. Genetic Adaptive Search (GeneAS)

Genetic adaptive search (GeneAS) technique is a modification of the genetic algorithm (GA) search technique (Goldberg, 1989; Holland, 1975). In the following, we outline the differences of GeneAS from the traditional optimization algorithms.

- GeneAS allows a natural representation of the design variables.
- GeneAS works with more than one solutions in one iteration.
- GeneAS requires only objective function and constraint values.
- GeneAS uses probability to guide its search process.

We now discuss each of the above issues in details. In GeneAS, the design variables are represented by a vector. GeneAS allows any design variables— continuous, discrete, and zero-one—to be used in an optimization problem. The zero-one variables appear in design problems to represent the existence or nonexistence of certain features of the design. For example, the following solution represents a complete design of a cantilever beam having four design variables:

$$(1) \quad 10 \quad 23.457 \quad (1011)$$

The first variable can take only one of two values (either 1 or 0). The value 1 represents a circular cross-section for the cantilever beam and the value 0 represents a square cross-section of the cantilever beam. The second variable represents the diameter of the circular section if the first variable is a 1 or the side of the square if the first variable is a 0. This variable can take only one of a few pre-specified values in discrete form depending on the available sizes. Thus, the second variable is a discrete variable. The third variable represents the length of the cantilever beam, which can take any real value. Thus, it is a continuous variable. The fourth variable is a discrete variable representing the material of the cantilever beam. This material can take one of 16 pre-specified materials. A set of four binary digits (totalling 2^4 or 16 values) can be used to code the variable. The above solution represent the 12th (one more than the the the decoded value of the string (1011)) material in the specified list. Thus the above string represents a cantilever beam made of the 12th material from a prescribed list of 16 materials having a circular cross-section with a diameter 10 mm and having a length of 23.457 mm. With the above coding, any combination of cross sectional shape and size, material specifications, and length of the cantilever beam can be represented. This flexibility in the representation of a design solution is not possible with traditional optimization methods. This flexibility makes GeneAS robust to use in many engineering design problems. Moreover, each design variable is allowed to take only permissible values (for example, no hexagonal shape will be tried or no unavailable diameter of a circular cross-section will be chosen). Thus, it is likely that the computational time for searching the optimal solution is also substantially less.

Unlike most of the traditional optimization methods, GeneAS works with a number of solutions, instead of one solution. GeneAS begins with a set of random solutions and three genetic operators are applied on this population of solutions to create a new and hopefully improved population of solutions. The advantage of using a population of solutions instead of one solution is many. Firstly, it may allow GeneAS to overcome the convergence to a locally optimal solution and may help to converge to the globally optimal solution. Secondly, it helps GeneAS to implicitly process the search information parallely, giving its speed of convergence. Thirdly, if needed, GeneAS can be used to find multiple optimal solutions simultaneously in a population, a matter which has been exploited to solve multimodal and multiobjective engineering optimization problems (Srinivas and Deb, 1995; Deb and Kumar, 1996).

Figure 2.1 shows a pseudo code of GeneAS. After a population of solutions are randomly created, each solution is evaluated to find a *fitness* value. The evaluation process is completed in two steps. First, all design variables are extracted from a solution and the corresponding objective function value $f(x)$ is computed. Constraints are usually handled by penalizing the objective function value by a term related to the constraint violation. Thus, the fitness of a solution vector x is calculated as follows:

$$\mathcal{F}(x) = \bar{f}(x) + r \left(\sum_{j=1}^{J} \langle \bar{g}_j(x) \rangle^2 + \sum_{k-1}^{K} (\bar{h}_k(x))^2 \right), \qquad (2.1)$$

where $\langle \rangle$ is a bracket operator resulting the operand, if the operand is negative; and zero, otherwise. The penalty parameter r is usually kept constant. The constraints $\bar{g}_j(x)$ and $\bar{h}_k(x)$ are normalized constraints and $\bar{f}(x)$ is the normalized objective function. Since, all constraints and objective function are normalized, a reasonable fixed value of r (10^0 to 10^3) can be used. Thus, for minimization problems, the feasible solutions with smaller objective function value are good solutions with smaller fitness value. GeneAS can be used to solve both minimization and maximization problems by simply changing the sign of the penalty parameter r and by changing the relational operator used in the reproduction operator. Although there are many possible ways to handle constraints in GAs (Michalewicz, 1995), the bracket operator penalty term is a popular approach for handling inequality constraints (Deb, 1995).

After each solution in the initial population is evaluated to find a fitness value, the population is operated by the reproduction operator. Since the

```
/* Choose a coding for handling mixed variables */
for i = 1 to population_size
        old_solution[i] = random_solution;
        old_solution[i].fitness = fitness(old_solution[i]);
generation = 0;
repeat
        generation = generation + 1;
        for i = 1 to population_size step 2
                parent1 = reproduction(old_solution);
                parent2 = reproduction(old_solution);

                (child1, child2) = crossover(parent1, parent2);

                new_solution[i] = mutation(child1);
                new_solution[i+1] = mutation(child2);

                new_solution[i].fitness = fitness(new_solution[i]);
                new_solution[i+1].fitness = fitness(new_solution[i+1]);

        for i = 1 to population_size
                old_solution[i] = new_solution[i];
until (termination);
end.
```

Fig. 2.1. Pseudo-code of GeneAS

population is created at random, some good and some bad solutions (based on their fitness values) are expected. In the reproduction phase, good solutions in the population are duplicated and are used to replace bad solutions. In its simplest form (called the tournament selection), two solutions are picked at random and compared. For a minimization problem, the solution with the smaller fitness value is selected and kept in an intermediate population. For a maximization problem, the solution with a higher fitness value is selected. This operator is exactly the same as that used in the binary GAs.

Once the population is enriched with good solutions, a pair of solutions are chosen at random and crossed to produce two new children solutions. This is the crossover operation. This is where GeneAS differs to a great extent from the traditional GAs. The operator works as follows. If the variable is a binary string representing a discrete variable, a cross site is chosen at random along its length and some bits are exchanged between the strings. This is the same as the single-point crossover operator used in binary GAs. On the other hand, if the variable chosen to be crossed is a continuous variable, the parent values produce two children values based on a probability distribution. The derivation of obtaining this probability distribution is presented elsewhere (Deb and Agrawal, 1995). Here, we discuss the procedure of computing two children solutions $x_i^{(1,t+1)}$ and $x_i^{(2,t+1)}$ from two parent values $x_i^{(1,t)}$ and $x_i^{(2,t)}$. A non-dimensionalized parameter β is defined as follows:

$$\beta = \left| \frac{x_i^{(1,t+1)} - x_i^{(2,t+1)}}{x_i^{(1,t)} - x_i^{(2,t)}} \right|. \tag{2.2}$$

The children solutions are found by using the following polynomial probability distribution:

$$C(\beta) = \begin{cases} 0.5(n+1)\beta^n, & \text{if } 0 \le \beta \le 1; \\ 0.5(n+1)\frac{1}{\beta^{n+2}}, & \text{if } \beta > 1. \end{cases} \tag{2.3}$$

In order to create a child solution, a random number u between 0 and 1 is first chosen. Thereafter, from the above probability distribution function, the ordinate $\bar{\beta}$ is found so that the area under the probability curve from 0 to $\bar{\beta}$ is equal to the chosen random number u:

$$\int_0^{\bar{\beta}} C(\beta)d\beta = u. \tag{2.4}$$

After obtaining $\bar{\beta}$, the children solutions are calculated as follows:

$$x_i^{(1,t+1)} = 0.5\left[(1-\bar{\beta})x_i^{(1,t)} + (1+\bar{\beta})x_i^{(2,t)}\right], \tag{2.5}$$

$$x_i^{(2,t+1)} = 0.5\left[(1+\bar{\beta})x_i^{(1,t)} + (1-\bar{\beta})x_i^{(2,t)}\right]. \tag{2.6}$$

Note that two children solutions are symmetric about the parent solutions. This is used to avoid any bias towards any parent solution.

There is a specific reason for using the polynomial probability distribution described in Equation 2.3. The distribution matches closely with the underlying probability distribution of creating children solutions in the case of binary crossover operator. Figure 2.2 shows the above probability distribution with $n = 2$ for creating children solutions from two parent solutions ($x = 5$ and $x = 10$) in the real space. For example, if the chosen random number is $u = 0.25$, Equation 2.4 yields, $\bar{\beta} = 0.794$. Using this value of $\bar{\beta}$ and Equations 2.5 and 2.6, we calculate the children solutions as $x^{(1,t+1)} = 5.515$ and $x^{(2,t+1)} = 9.485$.

In the above probability expression, the distribution index n is any non-negative real number. A large value of n gives a higher probability for creating near parent values and a small value of n allows distant points to be selected as children solutions.

In some cases, the binary-coded strings pose a difficulty for handling discrete variables having any arbitrary search space, because in a binary-coded string of length k the exact number of choices for a variable must be 2^k. Often, this problem is solved by using an additional inequality constraint penalizing the extra choices. For example, if the diameter (d) of the cantilever

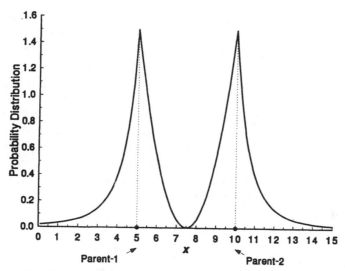

Fig. 2.2. Probability distribution for creating children solutions of continuous variables

beam discussed earlier can take any integer value between 10 to 100 mm, a minimum of seven-bit string coding can be used to code the variable in the range (10, 137) with 2^7 or 128 integer values. The string (0000000) would then represent the solution 10 mm and the string (1111111) would represent the solution 137 mm. Any other 126 integer values can be represented by a specific seven-bit string. Since the solutions having values greater than 100 mm are not desirable, a constraint $d \leq 100$ needs to be added to the list of other functional constraints. This difficulty of using an additional constraint can be eliminated by using a discrete version of the above probability distribution (equation 2.3) for the crossover operation.

After the reproduction and crossover operators are applied, a mutation operator is used sparingly with a small mutation probability p_m. If the variable selected for mutation is a binary string, a random bit is changed from 1 to 0 or vice versa and the solution is changed. For a continuous variable, the current value of the variable is changed to a neighboring value using a polynomial probability distribution similar to that used for the crossover operator. For discrete variables, a neighboring feasible value is chosen with a discrete probability distribution. Since, mutation probability is kept zero in this study, the details are omitted here. Interested readers may refer to an earlier study (Agrawal, 1995).

This completes one cycle of GeneAS. As can be seen from the pseudo-code of GeneAS (Figure 2.1), three operators mentioned above are iteratively applied to the current population of solutions until a termination criteria is satisfied. As seen from the above discussion that GeneAS is a direct optimization method requiring only the objective function and constraint values. Since no gradient or any other auxiliary information are needed, GeneAS can be applied to a wide variety of problems.

Like GAs, one distinct feature of GeneAS is that all its operators use probability measures. This is intentionally introduced to provide flexibility in its search. The stochasticity allows to have a global search of the variable space before hastily converging to any solution. Traditional optimization methods use deterministic rules for moving from one solution to another. In solving problems with multiple optimal solutions, a solution in the local basin will be modified by traditional methods only towards the locally optimal solution. Since fixed transition rules are not followed in GeneAS, the solutions, even in the local basin, are not always destined to move towards the locally optimal solution.

Binary-coded GAs have been largely successful in solving many engineering design problems including gas pipeline optimization (Goldberg, 1983), turbine blade optimization (Powell, Tang, and Skolnick, 1989), composite laminate stacking problem (Riche and Haftka, 1992), structural optimization (Hajela, 1990, Jenkins,1991), and others. Recently, the real-coded genetic algorithm with the above crossover operator has been successfully applied to solve truss-structure optimization problems (Chaturvedi, Deb, and Chakraborty, 1995). In this article, we have combined the two approaches to handle mixed variables and applied GeneAS to a number mechanical component design problems.

3. Mechanical Component Design

Three different problems are borrowed from the literature and solved using GeneAS. These problems were solved using different traditional optimization algorithm such as an augmented Lagrange multiplier method (Kannan and Kramer, 1993), the Branch and Bound method (Sandgren, 1988), and Hooke-Jeeves pattern search method (Deb, 1995). In the first method, Powell's zeroth-order method and conjugate gradient method (Reklaitis, Ravindran, and Ragsdell, 1983) are used alternatively as the unconstrained optimizer. The zero-one or discrete variables are handled by adding one additional equality constraint for each variable. This constraint is carefully crafted to penalize infeasible values of the variables. As the authors have mentioned, the added equality constraints are nondifferentiable and cause multiple artificial 'bumps' making the search space unnecessary complex and multimodal. The Branch and Bound method is a technique to handle discrete variables by assuming the variable to be continuous and then successively adding extra constraints to lead the algorithm to converge to a discrete solution (Deb, 1995; Reklaitis, Ravindran, and Ragsdell, 1983). Hooke and Jeeves pattern search method is a combination of exploratory searches and pattern moves applied on continuous variables.

In the following subsections, we compare the solution obtained using GeneAS with those from the above traditional methods. In all simulation runs of GeneAS, a crossover probability of 0.9 and a mutation probability of 0.0 are used, although a small non-zero mutation probability may be beneficial. A penalty parameter of $r = 100$ is used in all problems. A simulation run is terminated after a specific number of function evaluations have been computed.

3.1 Pressure vessel design

In this problem, a cylindrical pressure vessel with two hemispherical heads are designed for minimum cost of fabrication. The total cost comprises of the cost of the material and cost of forming and welding. Four variables are used. They are the thickness of pressure vessel T_s, thickness of head T_h, inner radius of the vessel R, and the length of the vessel without the heads L. Of the four variables, R and L are continuous variables and T_s and T_h are discrete variables with a constant increment of 0.0625 inch.

Denoting the variable vector $x = (T_s, T_h, R, L)$, the NLP problem can be written as follows (Kannan and Kramer, 1993):

Minimize $\quad f(x) = 0.6224 T_s T_h R + 1.7781 T_h R^2 + 3.1661 T_s^2 L + 19.84 T_s^2 R$
Subject to $\quad g_1(x) = T_s - 0.0193 R \geq 0,$
$\qquad\qquad g_2(x) = T_h - 0.00954 R \geq 0,$
$\qquad\qquad g_3(x) = \pi R^2 L + 1.333 \pi R^3 - 1,296,000 \geq 0,$
$\qquad\qquad g_4(x) = -L + 240 \geq 0,$

$$T_s \text{ and } T_h \text{ are integer multiples of } 0.0625,$$
$$R \text{ and } L \text{ are continuous.}$$

The first and second constraints limit the thickness of cylinder and head to lie below a factor of the cylinder diameter. The third constraint ensures that the enclosed volume is greater than a specified value. The fourth constraint limits the length of the cylinder to a prescribed value. Each constraint $g_j(x)$ is normalized so as to have more or less an equal weightage in the penalized function:

$$\bar{g}_1(x) = 1 - 0.0193 \frac{R}{T_s} \geq 0,$$

$$\bar{g}_2(x) = 1 - 0.00954 \frac{R}{T_h} \geq 0,$$

$$\bar{g}_3(x) = \frac{\pi R^2 L + 1.333 \pi R^3}{1,296,000} - 1 \geq 0,$$

$$\bar{g}_4(x) = 1 - \frac{L}{240} \geq 0.$$

The initial random population of 70 solutions are created in a wide range $0.0625 \leq T_s, T_h \leq 5$ (in steps of 0.0625) and $10 \leq R, L \leq 200$. However, the crossover and mutation operators are allowed to create any solution outside these bounds. The optimal solution obtained using GeneAS is presented in Table 3.1 and also compared with the solutions reported in Kannan and Kramer (1993) and Sandgren (1988). The table shows that GeneAS is able to find a solution with about 11% and 21% better in cost than the solution presented in Kannan and Kramer (1993) and Sandgren (1988), respectively.

Table 3.1. Optimal solution of the pressure vessel design

Design variables	Optimum solution (GeneAS)	(Kannan and Kramer)	(Sandgren)
T_s (T_s)	0.9375	1.125	1.125
T_h (T_h)	0.5000	0.625	0.625
R (R)	48.3290	58.291	47.700
L (L)	112.6790	43.690	117.701
$g_1(x)$	0.005	0.000	0.204
$g_2(x)$	0.039	0.069	0.170
$g_3(x)$	3652.898	21.216	$6.375(10^{06})$
$g_4(x)$	127.321	196.225	122.299
$f(x)$	6410.381	7198.200	8129.800

To investigate the robustness of the algorithm, a simulation of GeneAS is performed with an initial population away from the above optimal solution. The initial population is seeded in the range $2 \leq T_s, T_h \leq 5$ and $75 \leq R, L \leq 200$. The optimal feasible solution obtained is $T_s = 1.000, T_h = 0.500$, $R = 51.752$, and $L = 85.343$ with a cost of 6426.951, which is 0.25% worse than the optimal solution obtained earlier, but much better than the solutions presented in Kannan and Kramer (1993) or Sandgren (1988). This result demonstrates that even if the initial population is away from the desired optimal solution, GeneAS can find a solution close to it. This is because of the flexibility of its search operator, a matter which is unattainable with the binary GAs.

3.2 Belleville spring design

The objective is to design a Belleville spring having minimum weight and satisfying a number of constraints. A detailed description of the optimal design problem is presented in Siddall (1982). The problem has four design variables—external diameter of the spring (D_e), internal diameter of the spring (D_i), the thickness of the spring (t), and the height of the spring (h) as shown in Figure 3.1. We consider two different cases—one with all variables being continuous and the other with one variable being discrete.

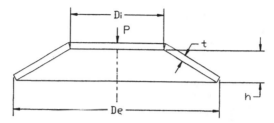

Fig. 3.1. A belleville spring

Seven different inequality constraints based on stress, deflection, and some physical limitations are used. The complete NLP problem is shown below:

Minimize $f(x) = 0.07075\pi(D_e^2 - D_i^2)t$

Subject to $g_1(x) = S - \frac{4E\delta_{max}}{(1-\mu^2)\alpha D_e^2}[\beta(h - \delta_{max}/2) + \gamma t] \geq 0,$

$g_2(x) = \frac{4E\delta}{(1-\mu^2)\alpha D_e^2}[(h - \delta/2)(h - \delta)t + t^3]\Big|_{\delta=\delta_{max}} - P_{max} \geq 0,$

$g_3(x) = \delta_l - \delta_{max} \geq 0,$

$g_4(x) = H - h - t \geq 0,$

$g_5(x) = D_{max} - D_e \geq 0,$

$g_6(x) = D_e - D_i \geq 0,$

$g_7(x) = 0.3 - \frac{h}{D_e - D_i} \geq 0.$

The parameters $P_{max} = 5,400$ lb and $\delta_{max} = 0.2$ inch are the desired maximum load and deflection, respectively. The parameters $S = 200$ kPsi, $E = 30(10^6)$ psi and $\mu = 0.3$ are the allowable strength, the modulus of elasticity and the Poisson's ratio for the material used. The parameter δ_l is the limiting value of the maximum deflection. The parameters $H = 2$ inch and $D_{max} = 12.01$ inch are the maximum limit on the overall height and outside diameter of the spring. Assuming $K = D_e/D_i$, we present the other parameters in the following:

$$\alpha = \frac{6}{\pi \ln K}\left(\frac{K-1}{K}\right)^2,$$

$$\beta = \frac{6}{\pi \ln K}\left(\frac{K-1}{\ln K} - 1\right),$$

$$\gamma = \frac{6}{\pi \ln K}\left(\frac{K-1}{2}\right).$$

The first constraint limits the maximum compressive stress developed in the spring to the allowable compressive strength (S) of the spring material (Shigley and Mischke, 1986). The second constraint limits the maximum deflection (P) of the spring to be at least equal to the desired maximum deflection (P_{max}). Although the ideal situation would be to have an equality constraint $P = P_{max}$, the optimal solution will achieve the equality together with the first constraint and the objective of minimizing the weight of the spring. In order to achieve the desired maximum deflection to be smaller than the height of the spring, the third constraint is added. In this constraint, a nonlinear functional form of the limiting deflection δ_l is assumed (Siddall, 1982), so that δ_l is always smaller than or equal to h. The fourth constraint takes care of the total height of the spring to be lower than the specified maximum limit (H). The fifth constraint limits the outside diameter of the spring to be at most equal to the maximum limit $(D_{\max})$. The sixth constraint allows the outside diameter of the spring to be larger than the inside diameter of the spring. The final constraint limits the slope of the conical portion of the spring to be at most equal to 0.3 (about 16.7 degrees).

At first, we consider that all four variables are continuous. We call this case as GeneAS-I. The optimization algorithm is used by normalizing the constraints as discussed in the previous example problem. The population is initialized in the following intervals:

$$0.01 \leq t \leq 0.6, \quad 0.05 \leq h \leq 0.5, \quad 5 \leq D_i \leq 15, \quad 5 \leq D_e \leq 15.$$

Using a population size of 100 and simulating GeneAS up to 100 generations, we obtain the optimal spring having 2.019 lb weight. The corresponding solution is shown in Table 3.2. The table shows that none of the constraints is

Table 3.2. Optimal solutions obtained using four different techniques

Design	Optimal solution			
variables	GeneAS-I	GeneAS-II	BGA	(Siddall)
t	0.205	0.210	0.213	0.204
h	0.201	0.204	0.201	0.200
D_i	9.534	9.268	8.966	10.030
D_e	11.627	11.499	11.252	12.010
g_1	58.866	1988.370	520.347	134.082
g_2	2.838	197.726	419.286	−12.537
g_3	0.001	0.004	0.001	0.000
g_4	1.594	1.586	1.586	1.596
g_5	0.383	0.511	0.758	0.000
g_6	2.094	2.230	2.286	1.980
g_7	0.204	0.208	0.212	0.199
Weight	2.019	2.162	2.190	1.980

violated. The load-deflection characteristics for the optimal spring is shown in Table 3.3. The table also shows that the maximum load required to achieve the maximum deflection is close to $P_{\max} = 5,400$lb.

In some cases, any real value of the thickness of the spring may not be allowed; instead the spring may be available only for certain thickness values. In GeneAS-II, we have declared the thickness variable as discrete (available

Table 3.3. Load-deflection characteristics of the optimum solutions obtained using different methods (Load in lb. and deflection in inch)

Deflection	0.02	0.04	0.06	0.08	0.10
Siddall	981.49	1823.17	2540.57	3149.22	3664.67
GeneAS-I	982.14	1825.07	2544.15	3154.78	3672.36
GeneAS-II	1003.77	1868.65	2609.73	3242.09	3780.81
BGA	1021.92	1905.14	2665.01	3316.86	3876.05
Deflection	0.12	0.14	0.16	0.18	0.20
Siddall	4102.44	4478.07	4807.10	5105.06	5387.48
GeneAS-I	4112.25	4489.86	4820.57	5119.77	5402.84
GeneAS-II	4240.98	4637.68	4986.00	5301.02	5597.82
BGA	4357.93	4777.83	5151.12	5493.13	5819.21

in steps of 0.01 inches) and kept the other three variables as continuous. The crossover operator with discrete probability distribution is used to search this variable. Using the same parameters, we obtain a solution with 2.162 lb, about 7% heavier than the solution obtained in GeneAS-I. The solution is shown in Table 3.2 and the load-deflection characteristic of this spring in shown in Table 3.3. In this case, the maximum load required to achieve the maximum deflection of 0.2 inch need not be exactly 5,400 lb, because of the discreteness in the search space. The latter table shows that this maximum deflection is somewhat larger than 5,400 lb.

Finally, we have applied binary GAs (BGA) to solve the same problem. Using 16 bits to code each variable and keeping all GA parameters the same as before, we obtain an optimum spring with 2.189 lb. The corresponding solution is shown in Table 3.2. As seen from the comparison that GeneAS-I and GeneAS-II have found better solutions than that obtained by the binary GA.

We have also compared these results with the solution reported in Siddall (1982). With our parameter setting[1] and using the optimal values of the variables reported in Siddall (1982), we observe that the second constraint is not satisfied (g_2 is negative). This is because the maximum load required to achieve the desired maximum deflection is smaller than the desired load of 5,400 lb (Table 3.3). Although this solution has a weight of 1.980 lb (better in weight than even GeneAS-I), the solution is not feasible according to our calculation. Thus, we can not compare the solutions of GeneAS with that reported in Siddall (1982).

---

[1] Sometimes, the accuracy of the solution depends on the machine precision and the exact value of the constant such as π etc.

3.3 Hydrostatic thrust bearing design

The objective of this problem is to minimize the power loss during the operation of a hydrostatic thrust bearing subject to satisfying a number of constraints (Figure 3.2). A hydrostatic thrust bearing has to withstand a specified load while providing an axial support.

In the optimal design problem, four variables are considered. They are bearing step radius R, recess radius R_0, oil viscosity μ, and the flow rate Q. All variables are assumed to take real values. The details of the problem formulation is presented in Siddall (1982). There are seven constraints associated with the minimum load carrying capacity, inlet oil pressure requirements, oil temperature rise, oil film thickness, and some physical constraints. The optimal design formulation in NLP form is given below:

$$\text{Minimize} \quad \frac{QP_0}{0.7} + E_f$$

$$\text{Subject to} \quad g_1(x) = \frac{\pi P_0}{2} \frac{R^2 - R_0^2}{\ln(R/R_0)} - W_s \geq 0,$$

$$g_2(x) = P_{\max} - P_0 \geq 0,$$

$$g_3(x) = \Delta T_{\max} - \Delta T \geq 0,$$

$$g_4(x) = h - h_{\min} \geq 0,$$

$$g_5(x) = R - R_0 \geq 0,$$

$$g_6(x) = 0.001 - \frac{\gamma}{gP_0}\left(\frac{Q}{2\pi Rh}\right)^2 \geq 0,$$

$$g_7(x) = 5000 - \frac{W}{\pi(R^2 - R_0^2)} \geq 0.$$

The first term in the objective function is the required pumping power, which depends on the inlet pressure, the flow rate and the pump efficiency. The inlet pressure P_0 is defined as follows:

$$P_0 = \frac{6\mu Q}{\pi h^3} \ln \frac{R}{R_0}. \tag{3.1}$$

This expression requires a parameter h, which we define later. The second term corresponds to the power loss due to friction, which can be calculated from the temperature rise ΔT in the oil:

$$E_f = 9336.0 Q\gamma C \Delta T, \tag{3.2}$$

where $\gamma = 0.0307$ lb/in^3 is the weight density of oil and $C = 0.5$ Btu/lb ^{0}F is the specific heat of oil. The temperature rise ΔT can be calculated by solving the following equation (Siddall, 1982):

$$n \log(560 + \Delta T/2) = \log\log(8.112(10^6)\mu + 0.8) - C_1,$$

where n and C_1 are two constants dependent on the grade of oil used. For SAE 20 grade oil, the values of n and C_1 are -3.55 and 10.04, respectively. After calculating E_f from the above equation, we can calculate the film thickness h from the following equation:

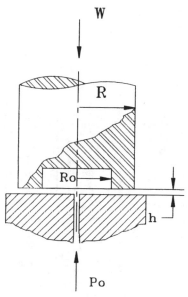

Fig. 3.2. The hydrostatic thrust bearing

$$h = \left(\frac{2\pi N}{60}\right)^2 \frac{2\pi\mu}{E_f}\left(\frac{R^4}{4} - \frac{R_0^4}{4}\right). \tag{3.3}$$

The parameters $W_s = 101,000$ lb, $P_{max} = 1000$ psi, $\Delta T_{max} = 50$ F, and $h_{min} = 0.001$ inch are the specified load the bearing is designed for, maximum inlet oil pressure, maximum oil temperature rise, and minimum oil film thickness, respectively. The parameter g is the acceleration due to gravity.

The first constraint restricts the weight carrying capacity of the bearing to be at least W_s. The second constraint limits the inlet oil pressure to be at most P_{max}. The third constraint restricts the oil temperature rise to ΔT_{max}, and the fourth constraint restricts the oil film thickness to be at least h_{min}. The fifth constraint allows the recess diameter to be smaller than the bearing step diameter. The sixth constraint ensures the laminar flow of oil and restricts the entrance and exit pressure loss of oil to 0.1% of the pressure drop P_0. The final constraint avoids surface contact between mating components by restricting the average pressure to 5000 psi (assumed to be an acceptable limit for avoiding surface damage (Siddall, 1982)).

GeneAS is used to solve this problem assuming that all four variables are continuous. The constraints are normalized as discussed in the first problem. A population size of 100 is used. The population is initialized in the following range:

$$1 \le R \le 16, \quad 1 \le R_0 \le 16, \quad 10^{-6} \le \mu \le 16(10^{-6}), \quad 1 \le Q \le 16.$$

Table 3.4. Optimal solution obtained using three different techniques

Design variables	Optimal solution		
	GeneAS	BGA	(Siddall)
R (in)	6.778	7.077	7.155
R_0 (in)	6.234	6.549	6.689
$\mu \times (10^{-6})$ (lb-s/in^2)	6.096	6.619	8.321
Q (in^3/s)	3.809	4.849	9.168
g_1	8299.233	1216.140	-11166.644
g_2	177.662	298.830	402.959
g_3	10.684	17.358	35.055
g_4	0.001	0.001	0.001
g_5	0.544	0.528	0.466
g_6	0.001	0.001	0.001
g_7	86.365	480.358	571.292
Power loss (ft-lb/s)	2161.6	2295.1	2288.0
Weight capacity (lb)	109,298	102,215	89,833
Film thk (in)	0.00165	0.00189	0.00254

The optimal solution is shown in Table 3.4. The optimal solution obtained using GeneAS corresponds to a power loss of 2,161.6 ft-lb/s (3.91 hp). When binary GAs are applied with 16 bit strings mapped in the above ranges, the optimal solution with a power loss of 2,295.1 ft-lb/s (4.15 hp) is obtained. The solution reported in Siddall (1982) obtained using Hooke and Jeeves pattern search method corresponds to a power loss of 2,288.0 ft-lb/s (4.14 hp). Moreover, the solution is found to be infeasible with our calculations, since the solution violates the first constraint as shown in Table 3.4. The table shows that the solution obtained by GeneAS is much better than that obtained by other two techniques and can withstand more weight with a smaller film thickness.

4. Conclusions

In this article, a new yet robust optimization algorithm is developed and used to solve a number of mechanical component design problems. The algorithm allows a natural coding of design variables by considering discrete and/or continuous variables. In the coding of GeneAS, the discrete variables are either coded in a binary string or used directly and the continuous variables are coded directly. In the search of discrete variable space coded in binary strings, the binary single-point crossover operator is used. However, in the case of continuous variables, the SBX operator developed in an earlier study (Deb and Agrawal, 1995) is used. When discrete variables are used directly, a discrete probability distribution is used in the crossover operator. The mutation operator is also modified in order to create feasible solutions.

The efficacy of GeneAS is demonstrated by solving three different mechanical component design problems, which were tried to solve using various traditional optimization algorithms. In all cases, it is observed that GeneAS has been able to find a better solution than the best available solutions. Moreover, the ease of accommodating different types of variables in a mixed-variable programming problem is also demonstrated. Most mechanical component design problems have mixed variables. The results of this study are encouraging and suggest the application of GeneAS in other mechanical component design problems.

It is worth mentioning here that some studies using this technique have been performed in solving multimodal and multiobjective design optimization problems (Deb and Kumar, 1996; Kumar, 1996). These extensions can also be used to solve mechanical component design problems where the objective is not to find only one optimal solution but to find multiple optimal solutions or a set of Pareto-optimal solutions.

Acknowledgments

The author appreciates the programming help provided by Mayank Goyal. This work is funded by the Department of Science and Technology, New Delhi under grant SR/SY/E-06/93.

References

Agrawal, R. B. (1995). Simulated binary crossover for real-coded genetic algorithms: Development and application in ambiguous shape modeling (Masters thesis). Kanpur: Department of Mechanical Engineering, Indian Institute of Technology, Kanpur, India.

Chaturvedi, D., Deb, K., and Chakrabarty, S. K. (1995). Structural optimization using real-coded genetic algorithms. In P. K. Roy and S. D. Mehta (Eds.), *Proceedings of the Symposium on Genetic Algorithms* Dehradun: B.S. Mahendra Pal Singh, 23-A New Connaught Place, Dehradun 248001, India (pp. 73–82).

Deb, K. (1995). *Optimization for engineering design: Algorithms and examples.* New Delhi: Prentice-Hall.

Deb, K. and Agrawal, R. (1995). Simulated binary crossover for continuous search space. *Complex Systems, 9* 115–148.

Deb, K. and Kumar, A. (1996). Real-coded genetic algorithms with simulated binary crossover: Studies on multimodal and multiobjective problems. (Report No. IITK/ME/SMD-96002). Kanpur: Department of Mechanical Engineering, Indian Institute of Technology, India.

Goldberg, D. E. (1983). Computer-aided gas pipeline operation using genetic algorithms and rule learning. *Dissertation Abstracts International, 44*, 10, 3174B. (University Microfilms No. 8402282)

Goldberg, D. E. (1989). *Genetic algorithms in search, optimization, and machine learning.* Reading: Addison-Wesley.

Goldberg, D. E., Deb, K., and Clark, J. H. (1992). Genetic algorithms, noise, and the sizing of populations. *Complex Systems, 6,* 333–362.

Riche, R. L. and Haftka, R. T. (1993). Optimization of laminate stacking sequence for buckling load maximization by genetic algorithm. Personal Communication.

Hajela, P. (1990). Genetic search: An approach to the nonconvex optimization problem. *AAAI Journal, 28,* pp. 1205–1210.

Holland, J. H. (1975). *Adaptation in Natural and Artificial Systems.* Ann Arbor: University of Michigan Press.

Jenkins, W. M. (1991). Towards structural optimization via the genetic algorithm. *Computers and Structures. 40,* 5, pp. 1321–1327.

Kannan, B. K. and Kramer, S. N. (1995). An augmented Lagrange multiplier based method for mixed integer discrete continuous optimization and its applications to mechanical design. *Journal of Mechanical Design, 116,* 405–411.

Kumar, A. (1996). Multimodal and multiobjective optimization using real-coded genetic algorithms. (Masters thesis). Kanpur: Department of Mechanical Engineering, Indian Institute of Technology, India.

Michalewicz, Z. (1995). Genetic algorithms, numerical optimization, and constraints. *In* L. Eshelman, (Ed.) *Proceedings of the Sixth International Conference on Genetic Algorithms* (pp. 151–158).

Powell, D. J., Tong, S. S., and Skolnick, M. M. (1989). EnGENEous domain independent, machine learning for design optimization. *In* J. D. Schaffer, (Ed.) *Proceedings of the Third International Conference on Genetic Algorithms* (pp. 151–159).

Reklaitis, G. V., Ravindran, A., and Ragsdell, K. M. (1983). *Engineering Optimization—Methods and Applications.* New York: Wiley.

Sandgren, E. (1988). Nonlinear integer and discrete programming in mechanical design. *Proceedings of the ASME Design Technology Conference,* Kissimee, FL, 95–105.

Shigley, J. E. and Mischke, C. R. (1986). *Standard handbook of machine design.* New York: McGraw-Hill.

Siddall, J. N. (1982). *Optimal engineering design.* New York: Marcel Dekker.

Srinivas, N. and Deb, K. (1995). Multiobjective function optimization using nondominated sorting genetic algorithms, *Evolutionary Computation, 2*(3), 221–248.

Genetic Algorithms for Optimal Cutting

Toshihiko Ono and Gen Watanabe

Dept. of Communication and Computer Eng., Fukuoka Institute of Technology, 3-30-1 Wajiro-higashi, Higashi-ku, Fukuoka 811-02, Japan

Summary. This chapter describes two kinds of the optimal cutting methods of material using genetic algorithms. The first is the optimal cutting of bars to satisfy two requirements of the minimum length of scraps produced and the balance among the numbers of finished products of each length to meet customer's order, when a series of raw bars are cut to various ordered length. The second is to determine the layout of various patterns to be cut on a sheet to make the required sheet length minimum. By combining layout determing algorithms(LDAs) with order-type genetic algorithms the required calculation time is considerably reduced. System configuration including the expression of genes, fitness function and how to integrate other proper systems into genetic algorithms, and the results of simulation studies are explained.

1. Introduction

This chapter deals with two methods for the optimal cutting of different shapes of material. The first is the optimal cutting of one-dimensional material. For instance a steel bar manufacturing process produces required number of bars(hereafter referred to as finished bars) for various lengths according to customer's requirements, by cutting a series of long raw bars. In this process, the cutting should be done to make losses minimal to achieve the highest productivity. Therefore, the optimal cutting method by GAs introduced here has to satisfy two requirements simultaneously : 1) the minimum length of total scrap produced to make waste minimal, 2) the number of finished bars for each length should conform to orders from customers not to produce unnecessary amounts of bars. To satisfy these two requirements, the optimal cutting method adopted here consists of two procedures: minimum scrap length cut and production balance cut, combined into one system. In the early stage of production, the combination of cutting lengths of a raw bar is determined mainly by making the length of each piece of scrap produced minimal and as the production proceeds, the ratio between the numbers of finished bars of each length gets more consideration.

The second optimization problem is how to cut a set of various two-dimensional patterns from a sheet, to make the required sheet length minimal. Concerning the two-dimensional optimal cutting, there are two kinds of problems : the optimal cut of rectangular patterns and that of free patterns. The optimal cutting of rectangular pieces from a sheet has been studied by many researchers[1, 2, 4, 13]. On the other hand, as to the optimal cutting of free patterns, which will be required in the cutting processes of clothes, metal sheets and the like, there have not been so many studies[7, 12], because of the difficulty of theoretical treatment. After much consideration, we have succeeded to solve it by GAs which incorporate special algorithms.

The given problem is originally a kind of two-dimensional search. As the space of two-dimensions is defined by the product of each dimension, it becomes quite large compared with that of one-dimension. Therefore, with the application of GAs to this problem, it is difficult to get proper results with a usually adopted tactic, in which genes represent the two-dimensional positions of patterns and the search is done two-dimensionally. In the method adopted here the problem is solved as a (one-dimensional) ordering by incorporating layout determining algorithms(LDAs) into GAs to reduce the dimension of search space. At the same time, LDAs are designed to get the solution as quickly as possible.

When applying GAs to various problems, it is important to consider how to represent the given problem with genes, how to calculate the fitness value and how to incorporate the GAs with other suitable systems. These are illustrated in our applications.

2. Optimal Cutting of Bars

The problem dealt with here is how to determine the optimal combination of lengths of finished bars in a raw bar, to attain the minimum amount of loss, when the finished bars of different length are produced by cutting raw bars of different length. Figure 2.1 shows an example of such production processes for steel bars, where a series of raw bars coming one after another are cut to finished bars of various lengths ordered by customers. Since the lengths of raw bars are different from each other due to the losses in the previous processes, the first problem to solve is how to determine the combination of cutting lengths in each raw bar so that the length of the scrap produced is minimal.

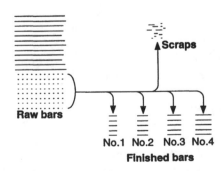

Fig. 2.1. Production of finished bars from raw bars

If the production proceeds with this method only, the imbalance will occur between the number of finished bars ordered by customers and that of those actually manufactured for each length and this will bring about the losses in production. Therefore, keeping balance among the numbers of the finished bars of various length at the end of production is the second requirement to satisfy. As the result, the optimal selection method of cutting length should satisfy these two requirements at the same time. Now, we will explain what system is adopted to solve these requirement by GAs [10, 11].

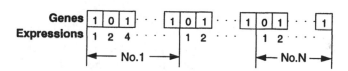

Fig. 2.2. Representation of genes

1) **Expression of genes** : As the purpose of the system is to get the combination of cutting lengths in a raw bar, genes are designed to express the number of finished bars to be cut from each raw bar. Figure 2.2 shows the composition of the genes, where N stands for the number of kinds of the finished bar lengths.

2) **Fitness function** : The fitness function consists of following two functions to satisfy the above two requirements :

1. Fitness function $F_1(k)$ which is used to make the length of scrap minimal.
2. Fitness function $F_2(k)$ which is introduced to keep balance among the numbers of finished bars for all lengths ordered.

These two functions are combined to make a fitness function $F(k)$ as follows :

$$F(k) = a(k)F_1(k) + b(k)F_2(k), \qquad (2.1)$$

where k stands for the k-th raw bar, and $a(k)$ and $b(k)$ are weighting coefficients explained later.

The function $F_1(k)$ is reduced as follows :

$$F_1(k) = f_1(\Delta L(k)); \qquad (2.2)$$

$$\Delta L(k) = L(k) - \sum_{i=1}^{N} L_i n_i(k), \qquad (2.3)$$

where $L(k)$ is the length of the k-th raw bar, N is the number of kinds of finished bar lengths, L_i is the i-th finished bar length and $n_i(k)$ the number of finished bars of i-th length cut from the k-th raw bar. Since $\Delta L(k)$ is the length of the scrap produced, it should be non-negative. To make the scrap length minimal, the function $f_1(x)$ is selected as a decreasing function whose value is non-negative in the operating range and maximum at $x = 0$. As a function to satisfy these requirements, $f_1(x) = \exp(-c_1 x^2)$ is adopted.

On the other hand, the function $F_2(k)$ expresses how well the number of finished bars for each length matches the number of those ordered by customers and is determined as follows :

$$F_2(k) = \sum_{i=1}^{N} f_2(\Delta r_i(k)); \tag{2.4}$$

$$\Delta r_i(k) = \frac{r_i}{\sum_{j=1}^{N} r_j} - \frac{p_i(k) + n_i(k)}{\sum_{j=1}^{N}(p_j(k) + n_j(k))}, \tag{2.5}$$

where r_i stands for the target value of production ratio for the i-th finished bars and p_i stands for the number of the i-th finished bars produced. Hence $\Delta r_i(k)$ becomes the deviation of production ratio from its object value. The function $f_2(*)$ is also to be a non-negative decreasing function to make the deviation of the ratios minimal and therefore the same function as $f_1(*)$ is applied.

Since the lengths of raw bars are assumed to be known only at cutting time, the combination of cutting lengths can't be determined before the time of cutting and therefore, it is imposible to take the lengths of succeeding raw bars into consideration. On the other hand the number of finished bars for each length should match the object value at the end of production. Considering these requirements, to attain the minimum length of scrap we adopted an inclining coefficient method (referred to as Inclining Weight Method), so that the combination of cutting lengths in each raw bar is determined, in the early stage of production mainly to make the scrap length minimal, in the final stage to keep the balance of products, and in the intermediate stage to partly satisfy both requirements. To realize this, the coefficient $a(k)$ is set large at the initial stage and is gradually decreased as the production proceeds. On the other hand, $b(k)$ is changed in the opposite way, that is, from small to large. The actual pattern of change is determined from the results of simulation as shown later.

3) Genetic operation : We adopted simple genetic algorithms(SGA), which consist of the following three operations:

1. Selection by roulette wheel method,
2. One point crossover,
3. Mutation with fixed probability.

Figure 2.3 shows the flow chart of simulation. The combination of cutting lengths in each raw bar is determined by GAs. Before transition to the next bar, the parameters such as p_i, $a(k)$ and $b(k)$ are reset to the new values calculated by the data collected. After the above procedure having been repeated to each raw bar, the final performance figures were obtained and evaluated.

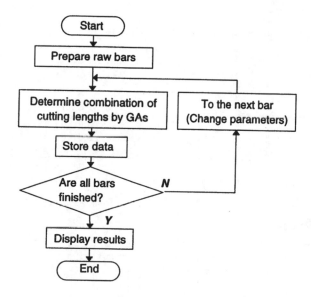

Fig. 2.3. Flow chart of simulation

2.1 Results of simulation

The following two kinds of simulation studies were conducted to investigate the optimizing characteristics of the proposed method, primarily to see how well the two requirements of both the minimum length of scrap and the balance among the number of finished bars are attained.

1. Optimization for each bar, in which only the minimum length of scrap for each bar is considered. Therefore, the coefficient $b(k)$ is set at zero.
2. Optimization as a whole, in which both the minimum scrap length and the balance of products are satisfied. The simulations are conducted under various production ratios.

Table 2.1 shows the specifications of raw and finished bars, As to the parameters related to GAs, the population is 50, the probability of crossover 0.1 and the probability of mutation 0.1.

Figure 2.4 shows the patterns of variation for the weighting coefficients $a(k)$ and $b(k)$ as to k, the number of raw bars processed.

Table 2.2 shows the results of the simulation for Case 1, in which only the optimization for each bar is considered. From the table we can see that the total length of scrap is kept to zero, and perfect optimization is attained. Both Table 2.3 and Table 2.4 show the results of optimization as a total system at various production ratios. In the both cases the production ratios attained by GAs are nearly equal to the taget values, while the total length of scrap is kept fairly small. Ideally this length is to be zero, but it is almost impossible to attain it, because it is difficult to achieve the optimization of

Table 2.1. Specifications of raw and finished bars

Raw bars	finished bars			
Length[m]	Length[m]	Target of ratio[%]		
		Case-1	Case-2	Case-3
123	11	-	25.0	33.3
131	13	-	25.0	33.3
137	17	-	25.0	16.7
145	21	-	25.0	16.7
151				

Table 2.2. Optimal cut for each bar

Length[m]		Each raw bar					Sum	Ratio[%]	
		123	131	137	145	151	687	Target	Result
Number	11m	5	0	2	2	1	10	-	23.3
of	13m	1	1	3	1	3	9	-	20.9
cuts	17m	2	2	2	4	1	11	-	25.6
	21m	1	4	2	2	4	13	-	30.2
L. of scrap[m]		0	0	0	0	0	0	-	100.0

Table 2.3. Optimal cut for all of bars : case-1

Length[m]		Each raw bar					Sum	Ratio[%]	
		123	131	137	145	151	687	Target	Result
Number	11m	3	1	2	3	2	11	25.0	25.0
of	13m	1	2	3	4	1	11	25.0	25.0
cuts	17m	2	3	2	1	3	11	25.0	25.0
	21m	2	2	2	2	3	11	25.0	25.0
L. of scrap[m]		1	1	0	1	2	5	100.0	100.0

Table 2.4. Optimal cut for all of bars : case-2

Length[m]		Each raw bar					Sum	Ratio[%]	
		123	131	137	145	151	687	Target	Result
Number	11m	3	3	2	5	4	17	33.3	35.4
of	13m	4	2	3	4	2	15	33.3	31.2
cuts	17m	1	3	2	1	1	8	16.7	16.7
	21m	1	1	2	1	3	8	16.7	16.7
L. of scrap[m]		0	0	0	0	1	1	100.0	100.0

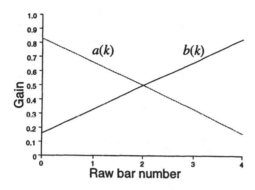

Fig. 2.4. Patterns of weighting coefficients $a(k)$ and $b(k)$

all the objectives at the same time and therefore some trade-off among the optimized values of those objectives is indispensable, in a multi-objective optimal problem such as this. We also conducted studies to realize a similar system using the neural networks of cross-connected Hopfield type [9]. The GAs method, however, proved to be much better than the neural networks. Especially, we had problems of a local minimum phenomenon at the neural networks, even though we added a proper annealing operation to get rid of this problem. From these experiences the GAs method seems suitable to the problems having many local minima such as those treated here.

3. Optimal Cutting of Two-dimensional Patterns

In this section, how to lay out a set of various two-dimensional patterns (polygons including concave ones) on a sheet to satisfy the following two requirements is discussed, when these patterns are produced by cutting the sheet.

1. To lay them out on the sheet without mutual overlapping,
2. To make the required sheet length minimal.

In applying GAs to this kind of problem, the normally adopted representation of genes will be the position of each pattern on a sheet. From the estimation of the size of search space, however, we have concluded that this representation method is impractical. In this method, since the genes will be a binary representation consisting of both the positions in width and those in length to all of the patterns, the length of the genes is $N(\log_2 W + \log_2 L) = N \log_2(WL)$ and thus the size of search space becomes $2^{N \log_2(WL)} = W^N L^N$, where N is the number of patterns, and W and L are the width and length of the sheet respectively.

On the other hand, if it is possible to solve the problem as an ordering one, the size of search space becomes $N!$, that is, the permutation of N.

Considering the amount of W, L, N , we can easily understand that a considerable reduction of search space can be obtained with this method. From these results, we decided to solve the problem by an ordering GAs, introducing layout determining algorithms(LDAs) to change a two-dimensional search into a one-dimensional (ordering) one. Figure 3.1 shows the flow chart of the systems.

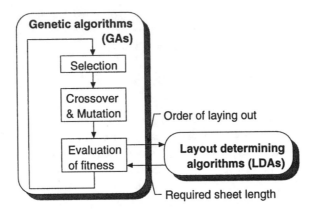

Fig. 3.1. Configuration of systems

The GAs and LDAs work sequentially, communicating each other. The GAs determine the order of patterns to arrange them on a sheet and send it to the LDAs, while the LDAs fix the position of each pattern according to the order given by the GAs so that each pattern on the sheet doesn't have an overlap and an extra opening. After the layout of all the patterns is fixed, the LDAs return the required sheet length to arrange the patterns to the GAs. With this length data, the GAs determine the order of pattern arrangement to make the object value(the required sheet length) minimal, by genetic operations such as selection, crossover and mutation.

As the applications of GAs to the ordering problem have been already studied in a traveling salesman problem [8], the results gained there were used in our systems.

For the convenience of explanations, the arrangement of polygon pieces is treated at first, and then the method will be expanded to free patterns later.

3.1 Layout determining algorithms (LDAs)

To determine the position of each pattern on a sheet without overlapping and extra openings between patterns according to the order given by the GAs, we have introduced layout determining algorithms. As shown in Figure 3.2, the allowable position for each pattern is searched on the sheet, each starting from the origin of the sheet and walking in the direction of width(Y-axis), while stepping in the longitudinal direction(X-axis). The allowable position being found, the pattern is set there and a marking is given to the sheet data to store the occupied area by the pattern. To minimize the required search time, the followings are considered.

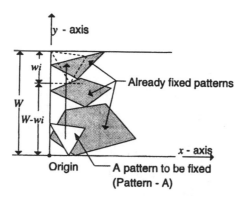

Fig. 3.2. Search of overlapping

Let pattern-A be the pattern to be searched for its position on the sheet. At first the check is done to see if there is an overlap between the pattern-A and those already fixed on the sheet. Even though this check is originally two-dimensional, it can be done under an one-dimensional basis or less, by checking overlapping on the boundary line and at the vertices of the pattern-A, as shown in Figure 3.3.

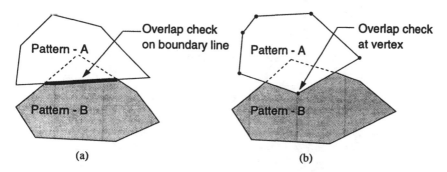

(a)   (b)

Fig. 3.3. Boundary line and vertices

There is no overlap, as long as no part of the boundary line is inside of already fixed patterns on the sheet. Therefore, by checking if the boundary line is inside the already fixed patterns, the necessary overlap-check can be made and time saving is attained because this checking is done along the boundary line of one-dimension. Moreover, if at least one vertex is within the fixed patterns, it means that there is an overlap. As the vertices are fewer than the points on the boundary line in general, the check at vertex prior to the same on boundary line contributes to another savings in calculation time.

As Figure 3.4 shows, the check on boundary line and the same at vertex are integrated in the LDAs and activate serially.

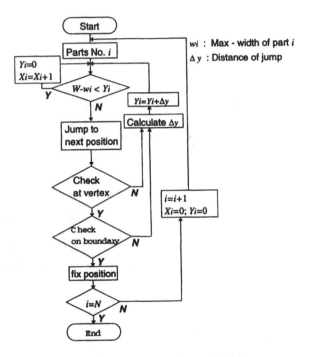

Fig. 3.4. Flow chart of LDAs

If one of them fails, the pattern-A jumps to the next search position by an amount equal to the distance calculated by the following method. After both of these checks have succeeded, the pattern-A is set there.

When an overlap is found, the distance of the jump to the next position is calculated, referring to the mutual position between the pattern-A and each of the overlapped fixed patterns on the sheet. As shown in Figure 3.5, the method is divided into two methods according to the mutual position. Taking the check at vertex as an example, these methods are explained.

At first the calculation method at one of the vertices is explained. As shown in (a)of Figure 3.5, to the vertex P in the upper side of the pattern-A, the necessary jump distance Δy of the pattern-A in the Y-axis direction to get out of the pattern-B consists of Δy_1 and Δy_2, because the portion of

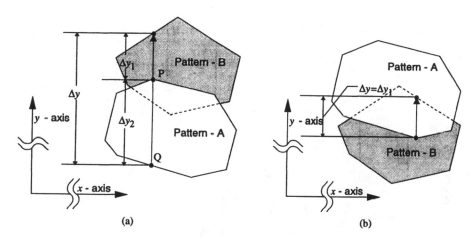

Fig. 3.5. Calculation of jump distance

the pattern-A between the vertices P and Q should be out of the pattern-B. The distance Δy_1 is obtained by scanning upward from the vertex P until the end of the fixed pattern-B. On the other hand, since Δy_2 is inherent to each vertex, it can be calculated beforehand and stored in memory. As for the vertex in the lower side, Δy_2 is zero as can be seen in (b) of Figure 3.5, hence only the calculation of Δy_1 is required. Therefore, the required jump distance is calculated as the maximum value of Δy for all the vertices of the pattern-A.

The calculation method of the jump distance for the boundary line is the same as the above, except for the vertices being replaced by the points on the boundary line.

The flow chart also shows the procedures at the near end of the sheet in the width side. In the scanning in the Y-direction, It is not necessary to scan up to the upper end of the sheet, but it is enough to scan for the range reduced by as much as the width of the pattern-A.

3.2 Genetic Algorithms (GAs)

As to the application of GAs to an ordering problem, various methods have been studied. Among them, we have used Genitor [3][8] as a software workbench of GAs. The genes are represented by the path representation, with the rank method for the selection. For the crossover, three methods have been compared among various methods applied to the path representation. As to the mutation, the method exchanging one random element of genes with another random one is added to Genitor.

3.3 Results of simulations

Among the various results obtained by simulation studies, two typical examples will be explained here. The first simulation is for the case where the optimal value is clear. As shown in Figure 3.6, the optimal arrangement of a set of 14 patterns has been obtained at the 147th generation with 50 individuals, starting from a random state.

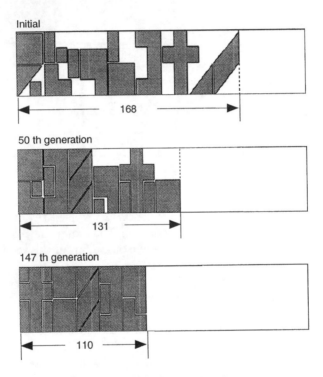

Fig. 3.6. Optimal arrangement of 14 patterns

Since one generation in Genitor corresponds to the calculation of each individual, the 147th generation in Genitor are equivalent to the third generation by the ordinary representation, considering the number of individuals being 50.

Figure 3.7 shows the comparison of the convergence characteristics for three crossover methods : CX(Cyclic crossover), PMX(Partially-mapped crossover) and OX(Order crossover)[8]. The CX method resulted in the best performance of all.

We have investigated the effect of the check at vertex on the overlap check. The calculation time needed to obtain the optimal state was 298.4 seconds without the check at vertex, while 34.4 seconds with it, on a workstation(Sun sparc classic), that is, the calculation time is reduced to 1/9 by the check at vertex.

The second simulation is for the optimal layout of 36 polygons as shown in Figure 3.8. With 50 individuals, the results that seem optimal are gained at the 497th generation (tenth generation by ordinary representation). Even though there is no theoretical method to confirm that the results gained here are optimal, the above result seems optimal, from the fact that the same results were obtained on various initial random states and from the results of the first simulation.

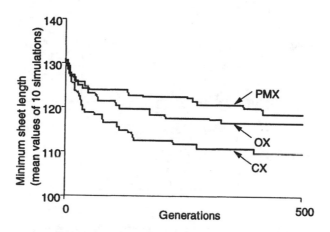

Fig. 3.7. Comparison of convergence characteristics

3.4 Extension to free patterns

We have explained the optimal layout of polygons(including concave one) so far. We can, however, easily extend this method to the free patterns of curved boundary line, by applying both the check-at-vertex method to proper number of the points selected from the points on the boundary line and the check-on-boundary method to the boundary line; those are the same as the methods explained. Figure 3.9 shows one of examples, for 22 free pattens, which was obtained with 20 individuals.

4. Conclusions

In this chapter we have explained two kinds of application of GAs to the optimal cutting method. One is the optimal cut of bars, which requires multi-objective optimization. By applying GAs in multiple stages with varying

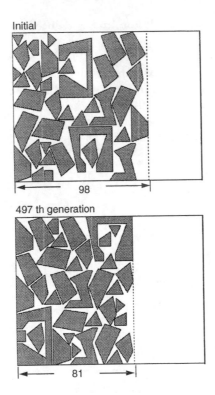

Fig. 3.8. Optimal arrangement of 36 patterns

weight coefficients two requirements of minimum scrap length and production balance are attained. The other is the optimal cut of two dimension free patterns from a sheet. By combining GAs with layout determining algorithms to solve a two-dimensional problem as an one-dimensional order one, the required calculation time is considerably reduced. Hopefully these results will be applied to not only these problems but similar ones as well.

References

1. Chauny, F. and Loulou, R. et al. : A Two-phase Heuristic for the Two-dimensional Cutting-stock Problem, Journal of the Operational Research Society, Vol.42, No.1, pp.39–47, 1991
2. Christofides, N. and Hadjiconstantinou, E. : An exact algorithm for orthogonal 2-D cutting problems using guillotine cuts, European Journal of Operational Research 83, pp.21–38, 1995
3. Davis, L. Ed. : Handbook of Genetic Algorithms, Van Nostrand Reinhold, 385p, 1991

Initial

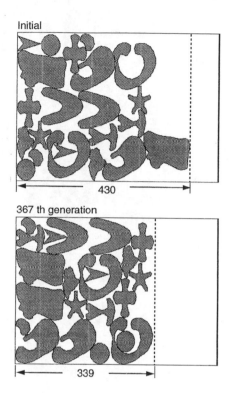

430

367 th generation

339

Fig. 3.9. Optimal arrangement of 22 patterns

4. Dyson, R. G. and Gregory, A. S. : The Cutting Stock Problem in the Flat Glass Industry, Operational Research Quarterly, Vol.25, No.1, pp.41–53, 1974
5. Goldberg, D.E. : Genetic Algorithms in Search, Optimization, and Machine Learning, Addison-Wesley Pub. Co., 412p, 1989
6. Holland, J.H. : Adaptation in Natural and Artificial Systems, The University of Michigan Press, 211p, 1975
7. Jakobs, S. : On genetic algorithms for the packing of polygons, Proc. of the KI-94 Workshop, pp.84–85, 1994
8. Michalewicz, Z. : Genetic Algorithms + Data Structures = Evolution Programs, Second, Extended Edition, Springer-Verlag, 340p, 1994
9. Ono, T. : Application of Neural Networks to Optimal Selection of Cutting Length of Bars (in Japanese), T.IEE Japan, Vol. 113-D, No.12, pp.1371–1377,1993
10. Ono, T. and Watanabe, G. : Application of Genetic Algorithms to Optimizing Problems(In Japanese), Language and Information Processing (Fukuoka Institute of Technology), Vol.6, pp.89-96, 1995
11. Ono, T. and Watanabe, G. : Application of Genetic Algorithms to Optimal Selection of Cutting Length of Bars, Proceedings of Fifth FIT-Ajou University Joint Seminar, pp.24-31, 1995
12. Petridis, V. and Kazarlis, S. : Varying Quality Function in Genetic Algorithms and The Cutting Problem, Proc. of the First IEEE Conf. on Evolutionary Computation, pp.166–169, 1994

13. Vasko, F. J. : A computational improvement to Wang's two-dimensional cutting stock algorithm, Computers ind. Engng, Vol.16, No.1, pp.109–115, 1989
14. Whitley, L.D. and Vose, M.D. Ed. : Foundations of Genetic Algorithms.3, Morgan Kaufmann Pub., 336p, 1995

Practical Issues and Recent Advances in Job- and Open-Shop Scheduling

Dave Corne[1] and Peter Ross[2]

[1] Department of Computer Science, University of Reading
[2] Department of Artificial Intelligence, University of Edinburgh

Summary. The chief advantage of evolutionary techniques in scheduling is the ability to provide good solutions to awkward problems with fast development time. There are still various issues and challenges that arise from the need to deal with the complexities of real-world problems rather than abstract 'academic' problems. This chapter reviews techniques and progress to date, and highlights what we believe to be important challenges and directions for future progress.

1. Introduction

Scheduling problems occur wherever there is a need to get a number of tasks done with limited resources. This very wide definition naturally covers many different kinds of problem, such as timetabling [7], the traveling salesperson problem [27], the vehicle routing problem [23], the job-shop scheduling problem [3], and many more. In this chapter, we will be mainly concerned with job-shop scheduling and related problems (JSSPs). This is a field which has received much attention from evolutionary algorithm (EA) researchers, and led to various successful applications to very complex, multi-constrained, multi-objective, and generally messy problems of commercial and industrial importance. We first present a short overview of EA research to date in this field. The overview highlights certain key points such as the representations and operator strategies used and overall EA design. We also discuss a list of interesting advances and the main issues which EA/scheduling researchers are currently facing. These are the need to deal with the complexity and stochastic nature of real-world problems, the recent success found in using approximate evaluation, and the existence of phase transitions in scheduling problem spaces.

2. Job-Shop Scheduling and Related Problems

Most EA/scheduling research concerns *job-shop* scheduling. At the heart of a job-shop scheduling problem are a number m of *machines*, and a list of j jobs which need to be performed. Each job involves a particular collection of tasks, and each task needs to be performed on a given machine for a given period of time. For example, a factory which produces wooden furniture may have a number of machines devoted to different tasks; eg, one for sanding the

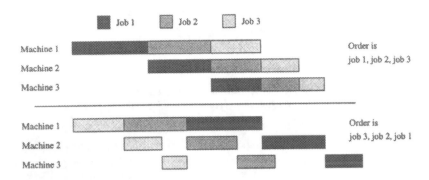

Fig. 2.1. Two schedules for a simple 3-job scheduling problem

surface of items after assembly, another for varnishing, another for quality inspection, and so on. A single job, perhaps corresponding to an order for a wardrobe, would usually require a visit to each machine for a specified (or estimated) period of time at each. In this example, it is also clear that the tasks within a job must often be done in a particular order.

Given a number of such jobs, the factory manager faces a job-shop scheduling problem. The problem is to get the jobs done in such a way as to satisfy the various constraints and optimise (minimise or maximise) some criterion. One such criterion might be to minimise the 'makespan', which is the time to complete all the jobs. Figure 2.1 illustrates an example schedule for a contrived three-job problem in our factory example. It shows two different possible schedules, of which the second has a longer makespan. If that was all that mattered, for example if makespan translated directly into overtime, then the first schedule would be preferable. However, if it was unwise to keep machines running continuously then the second might be preferable.

In practice, the makespan of a set of jobs is of very little interest. Often, new jobs must be added to the set in progress and existing ones get changed or suspended; job-processing tends to be a continual process. Jobs usually have agreed delivery dates and the manager may be more concerned to minimise the average lateness (tardiness) of jobs or to minimise the worst case of a job being late, or to try to achieve the delivery dates for certain jobs at the expense of others. We will return to this issue later on.

First, having introduced the general picture, we will quickly note the broad range of kinds of scheduling problem which typically occur. *Flow Shop Scheduling*, of which figure 2.1 was an illustration, occurs when tasks in each job 'flow' in the same direction. That is, there is a strict ordering in which tasks have to be performed (eg: first do the machine 1 task, then the machine 2 task, and so on) and this ordering is the same for each job. The term *Job-Shop Scheduling* tends involves a strict ordering on the tasks within each job, but the ordering may be different for different jobs. Next, a *General Job-Shop* problem is one which involves a partial ordering on tasks within a job.

For example, a car requiring an oil-change, new front seats, and valeting may have oil-change and valeting in any order, but the valeting will obviously need to wait until after the front seats have been replaced. Finally, in *Open-Shop Scheduling*, the tasks within each job may be performed in any order.

Beyond these standard classifications, many additional complications often arise. A typical example is *set-up time*. A machine or workstation may need to cool down, or be cleaned, for example, between tasks. Another complication is the issue of alternative *process plans* for a job. The given list of ordered or partially ordered tasks for a job may not be the only way to get the job done. For example, a given task may be capable of being performed on any of a selection of machines, perhaps with different processing times and/or different costs on each one. A further complication is that we must often reconsider the assumption (made so far) that everything about the problem is known at the start. But job details and other important scheduling data may constantly change as materials or resources are late to arrive, jobs are cancelled or altered, machines break down or operate below par, and so on. Such difficulties lead to what has become known as the *rescheduling* problem, which we will briefly look at later.

There are several other complications which constantly arise, and there are many more kinds of scheduling problem than can possibly be treated fairly in one survey. For example, one that is of considerable practical importance is that of kanban allocation. A production-line may be somehow divided into a number of consecutive sections, each with a fixed number of 'tickets', or kanbans, associated with it. A job must have a ticket to enter a section; if none are yet available it must wait until one becomes free when some job ahead of it leaves that section. This Japanese-inspired technique provides a good way to control the amount of work in progress automatically, and is important for just-in-time systems. A practical example of using an EA to segment a production line and allocate kanbans to sections can be found in [12].

3. An Overview of EA/Scheduling Research

Job-shop scheduling is known to be NP-hard [20]. It has been addressed by a great variety of researchers over the years, utilising a vast selection of heuristic methods. Good summaries appear in [1] and more recently, from an EA viewpoint, in [25], while a recent overview of applications appears in [4]. The story as regards EA-based approaches seems to start in 1985 when Davis [10] applied an EA to a simple JSSP, demonstrating the feasibility of the approach, but using a rather memory-intensive chromosome representation and an *ad-hoc* choice of operators. Skipping isolated examples of similar work we later find Nakano's efforts [31], which seems to be the first application of an EA to a set of benchmark JSSPs, hence facilitating the comparison of EA approaches with other methods.

Table 3.1. Published makespan benchmark results

Paper/Report	Algorithms	10x10 makespan	20x5 makespan
Balas 69	Branch & Bound	1177	1231
McMahon 75	Branch & Bound	972	1165
Baker 85	Branch & Bound	960	1303
ABZ 88	Shifting Bottleneck	930	1178
Carlier 89	Branch & Bound	930	1165
Nakano 91	EA	965	1215
Yamada & Nakano 92	EA	930	1184
Croce et al 92	EA	946	1178
Dorndorf & Pesh 93	EA	938	1178
Atlan et al 93	EA	943	1182
Fang 94	EA	939	1165
Mattfeld et al 94	EA	930	1165
Ono et al 96	EA	930	1165

Taking Nakano's bait, Fang *et al* [16] used a different chromosome representation and got better results overall on these JSSP benchmarks. Using the same problems, continued improved EA approaches emerged from [9], then [28], and more recently [32]. Table 3.1 gives a summary of such progress on an illustrative pair of well-studied benchmark JSSPs originally described in [19]. It includes results for both EA and non-EA based methods. As indicated, earlier work on these problems tended, naturally, to use classical operational research methods. These, particularly branch and bound search, enabled researchers to continually find better makespans, but tended to be highly computationally expensive. The later EA-based approaches similarly show a general improvement over the years although, just as with non-evolutionary methods, progress is generally sketchy until we arrive at Mattfeld et al's results [28] and Ono *et als*' results [32], achieving the known optima on both of the problems, hence competing with Carlier's branch & bound results, but with less computational cost.

It is instructive to view the research story so far in terms of the representation methods used, use of problem-specific knowledge, and other such issues. Nakano [31], for example, followed the stubborn tradition of using a binary chromosome representation. But this meant that standard crossover and mutation operations were likely to yield infeasible schedules needing repair. Fang *et al* [16] improved on this with a more natural representation, which yielded feasible schedules after standard genetic operations. Meanwhile, Bagchi *et al* [2] looked into the use of problem-specific knowledge, finding that the inclusion of such in the representation and operators generally helped, and more so the more such information was included.

This is not terribly surprising, either from the viewpoint of intuition or general EA folklore, and indeed stands for any general search method. The better EA results on the 10×10 and 20×5 Muth & Thompson benchmarks are a good illustration of this. In Yamada & Nakano's case, this achievement came through combining the EA with Giffler & Thompsons' active scheduling algorithm [22]. Ono *et al*'s also used Giffler's procedure, combining it with a carefully design 'job-based order crossover' operator (discussed later). In Mattfeld *et al*'s case, the achievement came via hybridisation with a local search heuristic, although the reported results only arrived when this in turn was set within a sophisticated population and recombination management strategy (see later).

Nevertheless, part of the challenge of scheduling is the existence of many awkward problems for which no good problem-specific heuristics exist. This situation tends to occur when the quality of a schedule is measured by criteria other than makespan alone. There are a great variety of other such criteria, mainly relating to jobs' due dates. Appropriate criteria may include *maximum tardiness* (Tmax), the tardiness of the job finished most past its due date, or *average tardiness* (Tave), the average tardiness of all jobs finished beyond their due date. Tables 3.2 and 3.3 respectively illustrate EA success [15] on a variety of benchmark job shop scheduling problems from [29], compared with a variety of well-known heuristic methods. We reproduce here the results on the larger job-shop problems, labeled LJB; a ** means that the EA found better results than the other rules, and * marks a known optimum. Similarly impressive EA results were found with the smaller job-shop benchmarks and also benchmark flow-shop scheduling problems [15].

Table 3.2. Fang's Maximum Tardiness results on Morton & Penticos' Job-Shop benchmarks compared with standard due-date related heuristic scheduling rules

Problem	SPT	EGD	EOD	EMOD	MST	S/OP	EA(ux)
LJB1	66	77	55	55	77	77	48**
LJB2	195	123	118	243	124	124	112**
LJB7	110*	179	161	110*	152	152	135
LJB9	684	240	219	688	214*	214*	323
LJB10	112	92	84*	104	107	107	94
LJB12	190	88	84*	189	85	85	120

In tables 3.2 and 3.3, the EA is compared with a range of heuristics more generally used when the criteria involved apply. These include shortest processing time (SPT), earliest global due date (EGD), earliest operational due date (EOD), earliest modified operational due date (EMOD), minimum slack time (MST), and minimal slack per operation (S/OP) [29]. Each such heuristic guides the choice of next task to schedule during the operation of a simple schedule-builder.

The EA used in these experiments used a fairly standard approach to scheduling (the 'indirect encoding' method, described in section 4.), without

Table 3.3. Fang's Average Tardiness results on Morton & Penticos' Job-Shop benchmarks compared with standard due-date related heuristic scheduling rules

Problem	EMOD	SPT	EGD	EOD	MST	S/OP	EA(UX)
LJB1	16	16	18	16	20	20	13**
LJB2	26	28	27	26	29	29	24**
LJB7	18	19	16	22	16*	17	17
LJB9	68*	74	119	109	111	111	84
LJB10	25	30	40	31	46	46	24**
LJB12	25	31	31	32	35	35	25*

extra sophistication, so the evident success of the EA is very encouraging. The implication is that awkward 'non-makespan' criteria for which it is difficult to design good problem-specific heuristics are best approached using an EA or similar. Much further evidence for this suggestion appears in [14]. In that article, results are presented comparing a simple EA with a similar range of heuristic methods on the same problem set from [29], and also with stochastic hillclimbing. Looking at a wide range of criteria, an EA clearly turns out to be best overall.

EAs have been found similarly successful in benchmark open-shop scheduling problems (OSSPs). This is a less well-studied area, but published results so far are very encouraging from the EA viewpoint. Table 3.4 summarises such results for a collection of OSSPs taken from [15], where the EA uses a method also described in [17]. The UB column gives the best-known result to date. The LB column gives a lower bound on the optimum, derived by very expensive computation and reported in [37]. The % column shows the percentage by which the EA result differs from the best known; a negative percentage indicates that the EA is better. Also, a * means that the EA equaled the best-known and a # means that the EA improved on it. For each size of problem (such as 7x7) there are 10 different benchmark problems, for which space constraints limit us to recording results on the first 5, labeled 7x7.0 to 7x7.4 and so on.

4. Applying EAs to Scheduling Problems

Direct Encodings Choice of chromosome representation for a schedule tend to lie on a continuum with *direct representations* at one end and *hybrid methods* at the other. In a direct representation, the chromosome essentially *is* the schedule. For example, the list of lists: { { 20 35 59 } 60 90 172 { 2 75 81 } } could represent a list of three jobs, in which the first, second, and third tasks of the first job start at times 20, 35, and 59 respectively, and so on for the other jobs. Combining this with given machine and task processing time information, this corresponds directly to a unique schedule. Such directness evidently makes it difficult to design operators which preserve schedule

Table 3.4. EA Results on Benchmark Open-Shop Scheduling Problems

SMALL					LARGE				
OSSP	UB	LB	EA	%	OSSP	UB	LB	EA	%
7x7.0	438	435	435#*	-0.7	10x10.0	645	637	641#	-0.6
7x7.1	449	443	446#	-0.7	10x10.1	588*	588	590	0.3
7x7.2	479	468	472#	-1.5	10x10.2	611	598	614	0.5
7x7.3	467	463	466#	-0.2	10x10.3	577*	577	577*	0
7x7.4	419	416	417#	-0.5	10x10.4	641	640	654	2.2
15x15.0	937*	937	937*	0	20x20.0	1155*	1155	1155*	0
15x15.1	918*	918	919	0.1	20x20.1	1244	1241	1252	0.6
15x15.2	871*	871	871*	0	20x20.2	1257*	1257	1257*	0
15x15.3	934*	934	934*	0	20x20.3	1248*	1248	1253	0.4
15x15.4	950	946	954	0.4	20x20.4	1256*	1256	1256*	0

feasibility, and hence direct representations require sophisticated and intelligent operators. One example of this is Bruns' work [5]. Another, although not strictly a direct representation, is Husbands' work on complex job-shop problems involving (among other complications) alternative process plans for each job [26, 25]. Husbands representation yields schedules with conflicting process plans for different jobs; these conflicts are dealt with by 'Arbitrators' (themselves evolved in step with schedules), which help to yield a workable schedule.

Permutation-Based Encodings A generally more favoured approach is the use of a hybrid method. Here, schedules are produced by a schedule-building algorithm, which receives its input from the chromosome. Typically, the schedule-builder builds a schedule by placing tasks in the schedule one at a time. It may use its own heuristics to decide where precisely to put the task. The decision of which task to schedule next, however, lies in the chromosome. For example, a 3-machine 2-job JSSP involves 6 tasks, which we can name $J1_1, J1_2, J1_3, J2_1, J2_2$, and $J2_3$, with the obvious meaning. A chromosome for such a problem within a hybrid approach might be a permutation, such as: { J1, J1, J2, J1, J2, J2 }. The first, second, and third instances of job Jk in the string respectively represent Jk_1, Jk_2, and Jk_3. This chromosome therefore says: "first schedule $J1_1$, then schedule $J1_2$, then schedule $J2_1, \ldots$ and so on. A schedule-building algorithm takes tasks in this specified order and schedules them one by one.

Several researchers favour this kind of technique since it allows the use of well-tried and developed operators for permutation encodings, such as Whitley's edge-recombination operator [39, 38]. Mainly developed with the traveling salesperson problem in mind, however, such operators can be rather below par when applied to scheduling [35], but researchers have lately experimented with more JSSP-oriented versions which have tended to yield successful results, such as Falkenauer & Bouffouix's Linear Order Crossover for JSSPs and OSSPs [13], and Ono et al's recent Job-based Order Crossover [32]. We

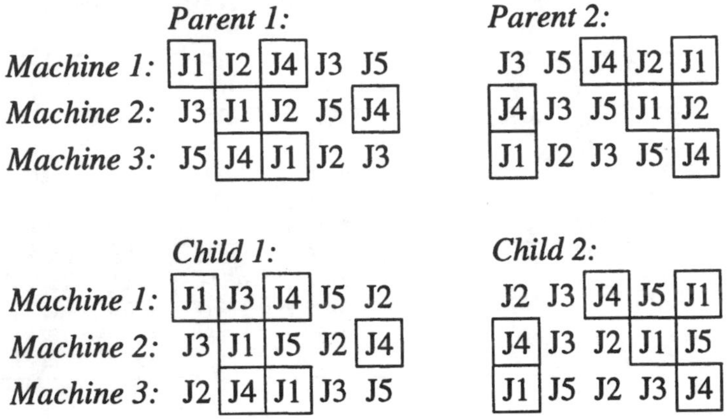

Parent 1:   *Parent 2:*

Machine 1: J1 | J2 | J4 | J3 J5   J3 J5 | J4 | J2 | J1

Machine 2: J3 | J1 | J2 J5 | J4   J4 | J3 J5 | J1 | J2

Machine 3: J5 | J4 | J1 | J2 J3   J1 | J2 J3 J5 | J4

Child 1:   *Child 2:*

Machine 1: J1 | J3 | J4 | J5 J2   J2 J3 | J4 | J5 | J1

Machine 2: J3 | J1 | J5 J2 | J4   J4 | J3 J2 | J1 | J5

Machine 3: J2 | J4 | J1 | J3 J5   J1 | J5 J2 J3 | J4

Fig. 4.1. Ono et al's Job-Based Order Crossover Technique

illustrate Ono *et al*'s method in figure 4.1. Here, PARENT1 and PARENT2 are shown in terms of the job sequence on each machine. There is a relatively direct way of translating this into a unique schedule. The figure illustrates a 5-job 3-machine example problem. A number of jobs are first chosen at random; in this case jobs 1 and 4. CHILD1 and CHILD2 are now partially built, CHILD1 contains the selected jobs in the same loci as they were in PARENT1, and similarly for CHILD2 and PARENT2, The remaining jobs forCHILD1 are then filled in, in the order in which they appear on each machine in PARENT2, and similarly for CHILD2 and PARENT1.

This operator produces feasible children from feasible parents; it is an obvious extension of Davis' order-based crossover as developed for the traveling salesperson problem, with the bonus of preserving some elements of the job sequence on each machine. This intuitively translates into possible building blocks for job-shop schedules. Child schedules may not be *active* however. meaning that it may be possible to perform certain tasks earlier without changing the sequence of tasks within a job. Ono *et al* hence use a variation on Giffler & Thompson's method [22] to turn such children into active schedules.

Indirect Encodings An alternative to permutation-based encodings with associated permutation-based operators are so-called 'indirect' encodings which may use standard EA operators such as uniform crossover [36]. Like permutation-based encodings, an indirect encoding translates into a task sequence which is then fed into a schedule-building algorithm. Fang *et al*'s JSSP work involved such an encoding [16]. A schedule was a list of numbers $abcdef...$, interpreted to mean: "schedule the next task of the ath unfinished job, schedule the next task of the bth unfinished job, ...", and so on. Chromosomes were free to be any k-ary string of length L, where L was the total number of tasks to schedule, and k was the number of jobs. A circular list of unfinished jobs was maintained, so that if at any point the remaining jobs were, say, $J1$, $J5$, and $J9$, a request to schedule the 8th unfinished job would lead to scheduling the next task of $J5$, since 8 mod 3 (remaining jobs) is 2, and $J5$ is the second in the list.

Such a representation raises the 'competing conventions' issue, because different strings can represent the same task sequence. Nevertheless, work with this method has been reasonably successful [16, 17]. When used in open-shop scheduling [17], a chromosome and its interpretation can be extended in the obvious way, with $AaBbCc$... meaning "schedule the Ath remaining task of the ath unfinished job, schedule the Bth remaining task of the bth unfinished job, ..." and so on. This is because "next task" is meaningful in the JSSP, where a job's tasks must be processed in a strict order, but this is not so in the OSSP.

So far, these indirectly encoded job-shop or open-shop scheduling sequences feed a fairly dumb schedule builder, which simply places a task in the earliest place it can go. The chromosome fully specifies which task to schedule next. A more successful approach involves using a cleverer schedule builder, using heuristics such as those surveyed and described by Morton & Pentico [29]. For example, a chromosome $abcdefg$.. might mean "heuristically choose a task from the ath unfinished job and schedule it, heuristically choose a task from the bth unfinished job and schedule it, ..." and so on. Much more successfully, however, Fang et al also used a representation in which $AaBbCc$... meant "from the Ath unfinished job, use heuristic a to decide which task to schedule next, from the Bth unfinished job, use heuristic b to decide which task to schedule next, ..." and so on. Thus the EA evolves the job-sequence and the choice of heuristic for each step. It is therefore not so susceptible to the specific weaknesses of any single heuristic. This is the method which led to a collection of 'new best' makespans for a variety of benchmark open-shop scheduling problems, as partly shown in table 3.4.

EA Design The design of the EA itself, ie: population size, population structure, control structure, etc ..., may be less important on the whole than the choice of representation and operators. Nevertheless, some impressive improvements can arise from careful and inspired design. Usually this happens to help improve results on a scheduling-based application, rather than be tailored to the problem itself. For example, Davidor et al [9] use the ECO grid-structured population framework [8] to produce improved results on Muth & Thompson benchmark JSSPs [19] than achieved in [31] and [40]. Another such example is Mattfeld et al's control of recombination in a grid-structured population by means of 'social-like behaviour' [28]. This is a sophisticated regime, whereby individual chromosomes are either 'co-operative' , 'critical' or 'conservative' according to their past recombination experience. In particular, Mattfeld et al report excellent results on the two commonly used Muth & Thompson benchmarks (see table 3.1) with this method, whereas, within a more standard EA their results were less impressive (best results respectively 934 and 1173 for the 10×10 and 20×5 benchmarks).

More traditional aspects of EA design, such as choosing the right population size, have also been looked at in a JSSP context. Nakano *et al* [30] used a statistical model of EA runs to determine the optimal population size under constant computation cost. Experiments using hard JSSPs yielded promise and plausibility for the model, although rather more work is required before an optimal population size can be predicted without the need for intensive initial experimentation.

5. Key Issues

In this section we will look at a small list of important issues which are challenging researchers in the field, progress in which will lead eventually to better EA (or hybrid) approaches, and better overall understanding.

5.1 Dealing with real-world problems

Two of the most important real-world issues are the frequent need to *reschedule* in the light of delays, breakdowns, etc ..., and the complexity of real-world schedule quality criteria. 'Academic schedulers' have been often criticised for their preoccupation with minimising makespan at the expense of dealing with more realistic criteria, and also for seemingly ignoring the rescheduling issue. These criticisms are being taken seriously on board by researchers in recent years, but there remains much work to do in this vein. We have already discussed 'non-makespan' criteria above, but the real problem with many real-world job shop situations is that they suffer from multiple conflicting criteria. For example, in unpublished work, the authors have studied a daily scheduling problem faced by an agricultural products firm which involves at least ten conflicting objectives. These include overtime, slack time, just-in-time criteria, lorry-usage, travel-time, and production-line balance. Tradeoff concerns between these criteria may change daily, so in addition to requiring a method which deals adequately with the overall problem, there needs to be a good way of adjusting the various tradeoff priorities with appropriate results. EA/scheduling literature is sparse in this area, not yet benefiting from the progress elsewhere in pareto-tradeoff based multiobjective approaches [24] or adaptive penalty methods [34, 11]. Further work in this vein is evidently needed, and would represent a major step towards the more general industrial applicability of EA/scheduling research.

Rescheduling is a similarly important issue; but one of the strengths of the EA approach to scheduling is that the same framework applies simply and easily to rescheduling. For example, an EA can used to develop a schedule based on initially available information; when information changes, the current schedule can be analysed to see how much is affected by the changes, and the EA can be applied only to the affected elements. We have conducted a range of experiments to demonstrate the feasibility of this. In essence, new or changed data leads to a new and typically simpler problem which can be solved by the same EA approach as used in the first place. A range of results in this vein appear in [16, 15]. In summary, there is relatively little EA work reported so far on the rescheduling issue, but the underlying problem appears amenable using standard EA/scheduling methods.

5.2 Approximate evaluation

Many scheduling systems obtain the fitness of a chromosome through some kind of simulation, and usually a simulator makes some assumptions about the true nature of the problem. For example, it may be assumed that machine downtime can be modeled by a random variable with some kind of Poisson distribution. It is unclear how much these assumptions affect the quality of results. In an unusual and detailed study of its kind, [33] use an EA to tackle the problem of scheduling the movement of pallets of beer from the production lines to lorries and rail goods vans in order to dispatch customers' orders in good time. Among many other issues studied they compared the use of a crude but fast simulation that assumed a load could be moved to its destination in an essentially constant time, with that of a much more detailed simulation which modeled the movement of fork-lift trucks and their interactions. When comparing different scheduling algorithms, only an EA seemed to be able to produce reasonably consistent high-quality results whichever simulator was used; and their results were appreciably better than those used in actual practice by the brewery. It remains to be seen whether this observation holds true in other practical applications of EAs to scheduling.

5.3 Problem difficulty

There is increasing interest in 'phase transitions' in the context of the difficulty of solving combinatorial optimisation problems. The difficulty experienced by a particular technique in solving such problems is usually illustrated by a graph of difficulty (eg: the percentage of trials which do not find an optimal solution) against some suitable measure of problem variation.

Often, such graphs are largely flat with a sharp peak. This is the phase transition. In graph 3-colouring, for example, the difficulty of finding a 3-colouring seems to increase very sharply as the density of the graph (number of edges in proportion to maximum possible number of edges) becomes around 15problems with few edges are easy, problems with more than 20% edge density are also easy, but, in between, the problems are generally very hard to solve. This feature of certain combinatorial problems has a number of practical consequences. Most importantly, it would be very useful to know if

No. of 'best' runs

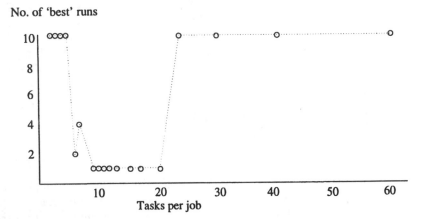

Fig. 5.1. Number of 'best' runs out of ten for 120-task JSSPs with varying numbers of task per job

and when real world problem instances may lie in a phase transition region. Knowing this would beneficially guide problem solving effort. Also, it can be argued that algorithm comparison studies should always incorporate tests on problems in phase-transition regions.

Job-shop scheduling is an important area of combinatorial optimisation as yet barely studied in terms of phase transitions. However, intuition from existing work suggests that these phenomena may well occur in this new context, and indeed some preliminary work bears out these intuitions [6]. Figure 5.1 summarises the results of some experiments on 120-task job-shop scheduling problems. The experiments involved 19 problems, each with a total of 120 tasks, but varying in terms of numbers of jobs and machines from a 2-job 60-machine problem to a 60-job 2-machine problem.

Makespan results were recorded for ten runs on each problem. Where m_p is the best makespan found during the ten runs on problem p, figure 5.1 plots the number of times a schedule with makespan m_p was found against the number tasks per job ratio of problem p. We rely on the intuition that the higher this number is, the easier the underlying problem. The graph shows three clear phases. When there are relatively few tasks per job, the problems appear easy. This also seems true when there are relatively many tasks per job, but inbetween there is a wide 'phase' in which there was great variation in EA performance in different trials, suggesting that these problems were much harder.

To some extent, we can expect this: with a constant number of tasks, relatively few or relatively many tasks per job both signal relatively little interaction between tasks. Few per job gives great leeway for tasks within a job to move around in relation to each other, while many per job (and hence fewer jobs) indicates little interaction between tasks from different jobs. What seems interesting and surprising is the apparent sharpness of the transitions between easy and hard problems. Much more work is needed here, of course, but initial investigations suggest such work is merited.

6. Discussion

Current EA/scheduling faces many important challenges when applied to real-world job shop situations. Chiefly, these are the ability to deal well with multiple conflicting objectives, the need for speed, and the need to cope with continual rescheduling.

Progress so far is strong and exciting. In the first place, mounting evidence and folklore point to the general conclusion that EAs are a highly appropriate and successful technology for scheduling applications. Secondly, progress in the 'key challenge' areas is rapid. Rana et al's 'approximate evaluation' work [33], for example, highlights great promise for a little-studied but important corner of the field. Continued work in that vein should lead to much speedier EA solutions to a wider range of scheduling problem without unduly compromising schedule quality. Fang's preliminary investigations on rescheduling [15] again point to potential industrial merit, while preliminary scheduling phase-transition studies suggest further potential mileage in the context of understanding the difficulty of scheduling problems.

In conclusion, this brief survey and discussion has tried to capture the flavour, strengths, and direction of EA/scheduling research. Building on early seminal work [10, 2, 31], research has progressed rapidly to a point at which EA/scheduling methods are now increasingly used in real job-shop environments, generally using variations on the methods outlined in section 4., for example [18, 21, 12]. Many challenges remain, however, in order to widen EA/scheduling usage and increase its general success, As we have tried to indicate, though, these challenges are being promisingly addressed, and the future of EA/scheduling research looks bright.

References

1. D. Applegate and W. Cook. A computational study of the job-shop scheduling problem. *ORSA Journal on Computing*, 3(2):149–156, Spring 1991.
2. Sugato Bagchi, Serdar Uckun, Yutaka Miyabe, and Kazuhiko Kawamura. Exploring problem-specific recombination operators for job shop scheduling. In R.K. Belew and L.B. Booker, editors, *Proceedings of the Fourth International Conference on Genetic Algorithms*, pages 10–17. San Mateo: Morgan Kaufmann, 1991.
3. K. Baker. *Introduction to Sequencing and Scheduling*. John Wiley and Sons, Inc., 1974.

4. J. Biethahn and V. Nissen. *Evolutionary Algorithms in Management Applications*. Springer-Verlag, Berlin, 1995.
5. Ralf Bruns. Direct chromosome representation and advanced genetic operators for production scheduling. In Stephanie Forrest, editor, *Proceedings of the Fifth International Conference on Genetic Algorithms*, pages 352–359. San Mateo: Morgan Kaufmann, 1993.
6. D. Corne. Investigating phase transitions in job-shop scheduling. Technical report, University of reading Department of Computer Science, 1996.
7. D. Corne, P. Ross, and H-L. Fang. Evolving timetables. In L. Chambers, editor, *Practical Handbook of Genetic Algorithms: Applications, Volume 1*, chapter 8, pages 219–276. CRC Press, 1995.
8. Yuval Davidor. A naturally occurring niche & species phenomenon: The model and first results. In R.K. Belew and L.B. Booker, editors, *Proceedings of the Fourth International Conference on Genetic Algorithms*, pages 257–263. San Mateo: Morgan Kaufmann, 1991.
9. Yuval Davidor, Takeshi Yamada, and Ryohei Nakano. The ECOlogical framework II: Improving GA performance at virtually zero cost. In Stephanie Forrest, editor, *Proceedings of the Fifth International Conference on Genetic Algorithms*, pages 171–176. San Mateo: Morgan Kaufmann, 1993.
10. L. Davis. Job shop scheduling with genetic algorithms. In J. J. Grefenstette, editor, *Proceedings of the International Conference on Genetic Algorithms and their Applications*, pages 136–140. San Mateo: Morgan Kaufmann, 1985.
11. A. Eiben and Zs. Ruttkay. Self-adaptivity for constraint satisfaction: Learning penalty functions. In *ICEC'96: Proceedings of the 1996 IEEE International Conference on Evolutionary Computation*, pages 258–261. IEEE, 1996.
12. M. Ettl and M. Schwehm. Determining the optimal network partition and kanban allocation in jit production lines. In J. Biethahn and V. Nissen, editors, *Evolutionary Algorithms in Management Applications*, pages 139–152. Springer-Verlag, Berlin, 1995.
13. E. Falkenauer and S. Bouffouix. A genetic algorithm for job shop. In *Proceedings 1991 IEEE International Conference on Robotics and Automation*, volume 1, pages 824–829, Sacramento, CA, April, 9-11 1991. IEEE Computer Society Press, Los Alamitos, CA. Cat. No. 91CH2969-4.
14. H-L. Fang, D. Corne, and P. Ross. A genetic algorithm for job-shop problems with various schedule quality criteria. In T. Fogarty, editor, *Evolutionary Computing: Proceedings of the 1996 AISB Workshop*, volume Lecture Notes in Computer Science. Springer, 1996. to appear.
15. Hsiao-Lan Fang. *Genetic Algorithms in Timetabling and Scheduling*. PhD thesis, Department of Artificial Intelligence, University of Edinburgh, 1994.
16. Hsiao-Lan Fang, Peter Ross, and Dave Corne. A promising Genetic Algorithm approach to job-shop scheduling, rescheduling, and open-shop scheduling problems. In S. Forrest, editor, *Proceedings of the Fifth International Conference on Genetic Algorithms*, pages 375–382. San Mateo: Morgan Kaufmann, 1993.
17. Hsiao-Lan Fang, Peter Ross, and Dave Corne. A promising hybrid GA/heuristic approach for open-shop scheduling problems. In A. Cohn, editor, *Proceedings of the 11th European Conference on Artificial Intelligence*, pages 590–594. John Wiley & Sons, Ltd., 1994.
18. B. Filipic. A genetic algorithm applied to resource management in production systems. In J. Biethahn and V. Nissen, editors, *Evolutionary Algorithms in Management Applications*, pages 10–111. Springer-Verlag, Berlin, 1995.
19. H. Fisher and G. L. Thompson. Probabilistic learning combinations of local job-shop scheduling rules. In J. F. Muth and G. L. Thompson, editors, *Industrial Scheduling*, pages 225–251. Prentice Hall, Englewood Cliffs, New Jersey, 1963.

20. Michael R. Garey and David S. Johnson. *Computers and Intractability: a Guide to the Theory of NP-Completeness.* Freeman, 1979.
21. I. Gerdes. Application of genetic algorithms for solving problems related to free routing for aircraft. In J. Biethahn and V. Nissen, editors, *Evolutionary Algorithms in Management Applications*, pages 328–340. Springer-Verlag, Berlin, 1995.
22. B. Giffler and G. L. Thompson. Algorithms for solving production scheduling problems. *Operations Research*, 8(4):487–503, 1960.
23. B. Golden and A. Assad, editors. *Vehicle Routing: Methods and Studies*, number 16 in Studies in Management Science and Systems. Elsevier, 1991.
24. Jeffrey Horn and Nicholas Nafpliotis. Multiobjective optimisation using the niched pareto genetic algorithm. Technical Report 93005, Illinois Genetic Algorithms Laboratory (IlliGAL), July 1993.
25. P. Husbands. Genetic algorithms for scheduling. *AISB Quarterly*, Autumn (89): 38–45, 1994. ISSN 0268-4179.
26. P. Husbands and F. Mill. Simulated co-evolution as the mechanism for emergent planning and scheduling. In *Proceedings of the Fourth International Conference on Genetic Algorithms*, pages 264–270. San Mateo: Morgan Kaufmann, 1991.
27. E. Lawler, J. Lenstra, A. Rinnooy Kan, and D. Shmyos. *The Travelling Salesman Problem: A Guided Tour of Combinatorial Optimization.* Wiley-Interscience, 1985.
28. D. C. Mattfeld, H. Kopfer, and C. Bierwirth. Control of parallel population dynamics by social-like behavior of ga-individuals. In Y. Davidor, H-P. Schwefel, and R. Manner, editors, *Parallel Problem-Solving from Nature – PPSN III*, number 866 in Lecture Notes in Computer Science, pages 16–25. Springer-Verlag, 1994.
29. T.E. Morton and D.W. Pentico. *Heuristic Scheduling Systems.* John Wiley, 1993.
30. R. Nakano, Y. Davidor, and T. Yamada. Optimal population size under constant computation cost. In Y. Davidor, H-P. Schwefel, and B. Manner, editors, *Parallel Problem Solving from Nature - PPSN III*, number 866 in Lecture Notes in Computer Science, pages 130–148. Springer-Verlag, 1994.
31. Ryohei Nakano. Conventional genetic algorithms for job shop problems. In R.K. Belew and L.B. Booker, editors, *Proceedings of the Fourth International Conference on Genetic Algorithms*, pages 474–479. San Mateo: Morgan Kaufmann, 1991.
32. Ono. A genetic algorithm for job-shop scheduling problems using job-based order crossover. In *ICEC'96: Proceedings of the 1996 IEEE International Conference on Evolutionary Computation*, pages 547–552. IEEE, 1996.
33. S. Rana, A.E. Howe, and D. Whitley. Comparing heuristic, evolutionary and local search approaches to scheduling. In *Proceedings of Third International Conference on AI Planning Systems (AIPS-96) (to appear)*, May 1996.
34. Alice E. Smith and David M. Tate. Genetic optimisation using a penalty function. In S. Forrest, editor, *Proceedings of the Fifth International Conference on Genetic Algorithms*, pages 499–503. San Mateo: Morgan Kaufmann, 1993.
35. T. Starkweather, S. McDaniel, K. Mathias, D. Whitley, and C. Whitley. A comparison of genetic sequencing operators. In R.K. Belew and L.B. Booker, editors, *Proceedings of the Fourth International Conference on Genetic Algorithms*, pages 69–76. San Mateo: Morgan Kaufmann, 1991.
36. G. Syswerda. Uniform crossover in genetic algorithms. In J. D. Schaffer, editor, *Proceedings of the Third International Conference on Genetic Algorithms and their Applications*, pages 2–9. San Mateo: Morgan Kaufmann, 1989.

37. E. Taillard. Benchmarks for basic scheduling problems. *European Journal of Operations Research*, 64:278–285, 1993.
38. D. Whitley, T. Starkweather, and D. Shaner. Travelling salesman and sequence scheduling: Quality solutions using genetic edge recombination. In L. Davis, editor, *Handbook of Genetic Algorithms*, pages 350–372. New York: Van Nostrand Reinhold, 1991.
39. Darrell Whitley, Timothy Starkweather, and D'Ann Fuquay. Scheduling problems and travelling salesmen: The genetic edge recombination operator. In J. D. Schaffer, editor, *Proceedings of the Third International Conference on Genetic Algorithms*, pages 133–140. San Mateo: Morgan Kaufmann, 1989.
40. Takeshi Yamada and Ryohei Nakano. A genetic algorithm application to large-scale job-shop problems. In R. Manner and B. Manderick, editors, *Parallel Problem Solving from Nature II*, pages 281–290. Elsevier Science Publisher B.V., 1992.

The Key Steps to Achieve Mass Customization

Bill Fulkerson

Deere & Company, Moline IL 61265-8098

Summary. The successful manufacturing operation must focus outside of the walls of the manufacturing facility to accommodate a global supply chain, disaggregation of enterprise production resources, and customer focused product distribution. I present some examples of current and anticipated methods of operation and discuss the use evolutionary computing to enable their implementation.

1. Introduction

The world industrial economies have turned from a state of product scarcity to one of product abundance. In response to this change, industry has shifted its emphasis from the supply-side (raw materials, inbound logistics, and production processes) to the demand side (outbound logistics, marketing, and sales). As companies gather, organize, select, synthesize, and distribute information in the marketspace (abstract information space) while managing raw and manufactured goods in the marketplace (real resource conversion and market space), they have the opportunity to 'sense and respond' to customers desires rather than simply make and sell products and services [1]. According to Ken Press, senior fellow with the Agility Forum, Lehigh University [2]:

> "In the near future, simply having good product won't be good enough for a manufacturer. It will be the process used to deliver that product to market that will matter to the consumer. People want solutions now, not just products. Companies will begin to view their customers as subscribers, not just one-time buyers."

"Giving customers what they want — when they want it" has become a prerequisite for the success in many businesses.

This change in emphasis from supply to demand thinking has caused dramatic changes in business systems. The existing value chain of supply and demand processes have been reengineered to obtain efficiencies. Manufacturing methods have turned from mass production of standard products, through continuous process improvement, production of unique, quality products with methods of mass customization. In mass production, products are produced with preconfigured, scale-efficient work processes incorporating formal processes and work rules to take the economic advantage of largely homogeneous product or service. It creates value through meeting the needs of a large population of customers with a standard product. Continuous improvement adds adaptability to overcome the limitations of standardized products and services to compete on the basis of product-service quality. It exploits the dual environment of continuous process innovation and stable process efficiency to turn worker knowledge, teamwork, and increasing level of quality into market advantage. Mass customization incorporates a dynamic network of modular work to translate the needs of the marketplace into cost-efficient, individually tailored product-service solutions. It creates unique value at an affordable price for each customer by turning efficient production, custom product configuration, and precision made products into market advantage [3].

Enterprise Resource Planning (ERP) systems which integrate both supply and demand aspects have been created to enable these newer production methods. In practice, these systems are hierarchical, time lagged, batch production systems with an architecture reminiscent of the MRP II systems from which they have emerged. Their limitations include:

- production plan derived from a contrived master schedule,
- customer order identity lost within batches,
- time delays due to aggregation into economical production lots,
- fixed part routes to control machine utilization.

To reduce the effect of these limitations upon continuous improvement and mass customization methods, production lots have become smaller and smaller until lots of size one are possible.

A second (virtual) value chain now links the previous value chain and the customer through information technology. The supply emphasis and the need to service the marketspace (of information) has moved ERP systems closer to the customer as well. Previously, electronic data interchange (EDI) linked supplier and producer with real time transactions. Presently, customer orders are placed on the producer (in real time) by a sales force equipped with order systems that:

- automatically configure product options to match the customer's functional needs,
- accept orders on an available-to-promise (ATP) basis by reserving production capacity and the requisite resources,
- sequence product for efficient production while respecting the previously committed delivery date.

The time compression created by EDI, management of both committed and uncommitted production capacity, and optimal sequencing of assembly line production based upon constraints enables an agile response to customer demand. The sensitivity of assumptions in these systems requires the constant monitoring and evaluation. The balancing of local, enterprise, and global objectives requires use of optimization techniques to improve upon feasible solutions. Evolutionary computation has been effectively used for both of these tasks.

This remainder of this paper presents potential applications application of evolutionary computation in discrete manufacturing. There are few documented commercial applications of these methods. In fact, the topic of product configuration and order management have only recently appeared in the industrial press. John Layden has written insightful article, "A Rapidly Changing Landscape: MES, MRP, and scheduling" [4], in which he prescribes simulation rather than evolutionary computation to improve production. The following discussion relates applications in assembly line sequencing, decentralized control of assembly line support, and the supply chain.

2. Genetic Algorithms

The mission of this facility was to produce preconfigured customer orders for seeding implements while retaining the agility and economy provided by autonomous modular production and supply that has been the hallmark of John Deere for decades. The solution was to establish a schedule-centric mode of operation with products (model numbers) sequenced daily on the assembly line. The supporting components for assembly are either supplied by contiguous production cells or purchased with delivery at the point of use.

The planters consist of a rectangular frame with an even number of row units mounted at fixed intervals. The frames are welded at one of three frame stations, painted, and launched down the assembly line interspersed with purchased painted frames. Workers assemble the frames and components produced in adjacent focused factories at work stations along the line. Other workers assemble frames and purchased components that are delivered for use at point of delivery. Finished planters are shipped to dealers on trucks staged at the end of the line.

Efficient sequencing of the preconfigured product orders can be the difference between meeting or missing order delivery dates. An efficient product sequence must balance several competing objectives simultaneously. It optimizes production performance (worker productivity, operational efficiency, product quality and order cycle time) while controlling cost (among contributing production modules and suppliers). The production sequence must meet order due date while accommodating known manufacturing constraints. These constraints can take two forms: strong constraints (illegal and prohibited from occurring) and weak constraints (legal but penalized in proportion to cost). Non-manufacturing constraints such as customer service policies, market planning goals, order fill priorities, and product distribution strategies are implicit in the problem statement. Constraint categories represent types of production bottlenecks. For example, the frame build rate is a major constraint in determining the build sequence. Assume the assembly line throughput rate is R units per hour and the build rate for Model A and Model C frames is longer than R. Specifically, the build rate for a model A frame is 1.5R units per hour and for a model C frame is 4R. Let us examine some possible sub-sequences where A and C denote the respective model and x denotes unspecified models other than A or C. The sub-sequence {xxxxAAxxxx} is impossible for model A. It represents a strong back-to-back constraint. Likewise, the sub-sequence {CxxxxCxxxx} is barely possible for model B. This possibility would be represented as a weak apart constraint with a large penalty. The product sequencing system incorporates commercial software, OptiFlex, that incorporates a proprietary genetic algorithm for automatic sequence generation and an intelligent GUI to enable manual sequencing and sequence repair. An efficient constraint computation engine enables OptiFlex to compute a production sequence with no hard constraint violations while it minimizes the sum of the weak constraint penalty scores. OptiFlex operates on a PC under Microsoft Windows with a relational database that stores constraints, dealer orders, product sequences, and operational data.

3. Agents

Evolutionary computation also includes computation performed by autonomous software agents. Autonomous differ from the more conventional intelligent agent that are indistinguishable from a virus (except for their intent) as they wonder through cyberspace. Intelligent agents function in as personal assistant that can gradually be tuned to assist better by (1) observing the user, (2) accepting positive and negative, feedback from the user (3) accepting instructions from the user or (4) interrogating other agents for information.

Autonomous agents perform general, independent tasks and assume a persona (or learn) through exchanging messages with their environment and/or other agents. When present in large numbers they self-organize to form bottom-up systems called swarms. "[The emergent behavior of swarms of agents] is often more robust, flexible, and fault tolerant than programmed, top-down organized complexity. This is the case because none of the components is really in charge of producing this complexity. None of the components is more critical than another one. When one of them breaks down, the system demonstrates a graceful degradation of performance. Because all of the components interact in parallel, the system is also able to adapt more quickly to environmental changes. Often the system explores multiple solutions in parallel, so that as soon as certain variables change, the system is capable of changing to an alternative way of doing things" [5]. The multiple parallel solutions available in swarm intelligence can be readily adapted to evolutionary computation.

3.1 Intelligent agents

Although limited in functionality and in the capability to learn, intelligent agents can be used to advantage in manufacturing. Paint color and component options are used to differentiate standard products to achieve some measure of mass customization. However, the savings available by painting long sequences of common colors often violates other resource constraints in prior assembly steps.

Intelligent agents can operate paint operations as self-scheduling (or dispatching) systems which accept any color sequence stream without regard to sequencing [6]. Truck bodies enter the paint shop at a rate of one per minute in random color sequence and are immediately dispatched to one of ten paint modules to be painted one of twelve base coat colors, sealed with a clear coat, and inspected within a three minute cycle. The process of choosing the appropriate paint booth is controlled by agents that negotiate with each other by messages. After the "dispatcher" posts the next job, the active "modules" bid on the job. The "dispatcher" acknowledges all bids before assigning the job to the winning "module" that must accept the job assignment. The "module" agent bids follow these rules: (1) Paint the Same Color (if possible); (2) Paint a New Color (based upon chemistry/cost/time); (3) Paint any Color (if not busy). The bids are made based upon length of travel time, cost to change paint to the posted color, and queue length at the respective booth.

Intelligent agents and dispatch rules can be applied to shopfloor control to enable self-scheduling of parts and machines [7]. They enable dynamic part routings that occur spontaneously in response to local conditions (such as manpower availability, machine capacity, and inventory levels) rather than previously prescribed plans. Depleted inventory levels initiate a customer trigger that signals all available producers and suppliers. Local rules identify the current best producer and supplier from among the candidates. These dynamic process routings enable worker autonomy, eliminate grouping multiple customer orders for the same part number into a single batch, and reduce the need for hierarchical checks and controls [8]. In the abstract, a manufacturing facility reduces to a network of intelligent nodes which can receive, consume, convert, and ship and that contract with each other to accomplish these required tasks. Nodes assume the dual roles of supplier and customer — customer to upstream nodes and supplier to downstream nodes through a common messaging scheme which enables an information exchange between the pair. This approach readily generalizes from cells to factories or virtual factories. For example, prototype system to demonstrate an autonomous agent based factory scheduler is being developed at the Rock Island Arsenal.[1]

Dispatch rules limit the effect of system bottlenecks by effective (but non-optimal) local heuristics. Their performance is directly proportional to the amount of redundant resources available in the system. In the paint example, there must be sufficient paint booths and inventory queues to eliminate paint color as an assembly constraint. In the shopfloor example, there must be

---

[1] This project is part of ARPA contract F33615-95-C-5524, officially titled "Large Scale System Simulation and Resource Scheduling Based on Autonomous Agents", under BAA 94-31, awarded to Intelligent Automation, Inc. of Rockville, MD, Principal Investigator: Leonard Haynes.

redundant machines or multi-function machines to enable adaptive efficient response to local conditions.

3.2 Intelligent agents—applications

Intelligent agent applications are still in the prototype stage. Autonomous Agents for Rock Island Arsenal (AARIA) is an industrial-strength agent-based shop-floor scheduling and control system prototype, sponsored by the ARPA Agile program.[2] The agents exchange messages (transactions) as part of a contract negotiation which incorporate the true cost of actions and decisions. The best (quickest and cheapest) routings are dynamically developed for the product parts, assemblies, and components.

In the AARIA[3] architecture, fine grained agents represent the essential resources in the factory such as people, machines, gauges, and tools while course grained agents oversee the order fulfillment process. The role of the inventory management agent (called a part broker) is to minimize inventory holding costs and to shorten lead-times by managing material handling resources. A part broker charges its customers for time a part spends in inventory. The architecture allows the part broker to learn order frequency distributions which it uses to build inventory in advance of anticipated orders if such a decision is cost justified. Resource agents (called resource brokers) learn the value of un-committed capacity at different demand levels and auction the uncommitted capacity when it is available. Orders can be fulfilled on a make-to-order basis or a part broker may elect to fulfilled the order from stock. The AARIA system agents enable a direct quote to a customer price which reflects delivery date and quantity. When the customer order is accepted, production capacity is reserved to ensure a reliable delivery date. When conditions change, future commitments must not violate past commitments.

Agent technology has come of age in the 1990s [9] and it will play a central role in the development of complex distributed systems, networked information systems, and computer interfaces during the twenty-first century. The computer power to enable agent based modeling of the enterprise is readily available and economical. However, there is a need for an efficient computation language to exploit this power. The Swarm Simulation System[4] is a general-purpose framework for simulating concurrent, distributed artificial worlds. It has been developed by the Artificial Life Group, the Santa Fe Institute, to provide a general architecture for simulation in a wide variety of disciplines. The generality of Swarm applications (ranging from physics to

[2] This project is part of ARPA contract F33615-95-C-5524, officially titled "Large Scale System Simulation and Resource Scheduling Based on Autonomous Agents", under BAA 94-31, awarded to Intelligent Automation, Inc. of Rockville, MD, Principal Investigator: Leonard Haynes.

[3] Dr. Albert D. Baker, Assistant Professor, Electrical & Computer Engineering and Computer Science Department, 834A Rhodes Hall - Mail Location 30.

[4] See URL http://www.santafe.edu/projects/swarm/

biology and to economics) requires that its software components address an unusually wide range requirements. This generality applies both to the structure of objects or agents belonging to a simulated world and to the actions which these objects perform.

Programmed using an object-oriented approach (coded in the Objective C Language and compiled by the GNU C compiler.), the agent is a data structure or object. Messages from other agents or system control objects trigger local changes within either single or multiple message recipients (agents). An Agent consists of [10]:

1. A data structure containing internal state variables local to the agent. This data structure enables an individual-based modeling approach in which past experience affect the next action through the internal state variables, e.g., a bill-of-materials or operational procedures for a process step.
2. A step function to specify action for an active agent prior to receiving a message, e.g., time-of-day warm-up or backup procedures.
3. Multiple step functions to specify action when triggered by messages from other agents or system control objects, e.g., orders, tool breakage, unavailable materials.

The recursive management of time supports a spectrum of time and synchrony management. It can range from a strict, lock-step synchrony managed by a single schedule object to a loose asynchrony of multiple schedule objects with effective parallelism. Initially, Swarm applications operating on sequential processor(s) are comparable to conventional discrete-event simulation system. However, the abstract representation of time represented in Swarm Schedulers enables efficient simulations on either serial or parallel computer systems. This parallelism will enable multiple concurrent simulations of a scenario that will enable reflective evaluation of competing alternatives. Swarm s architecture of modular components enables the use of reusable, machine-independent code organized into software object libraries.

4. Summary

Decentralized control and emergent behaviors available from evolutionary computation offer the promise of adaptive solutions to organizational and systems design that can adapt quickly to customer driven requirements. "Previous visions, efforts, and results of integration worldwide have yet to sufficiently consider the customer's evolving needs. Solutions emphasize the achievement of synergism across enterprise functions given *requirements fixed at a particular point in time*. [However] the basic competitive strategy of a corporation increasingly requires that the enterprise be able to respond rapidly to market conditions with a high degree of product differentiation and value-based customer services. These requirements cause *rapid changes* in the enterprise which *can no longer be dealt with by using managerial savvy alone* [11]. An adaptive enterprise must:

- effect shorter cycle times,
- produce lower volumes of identical items,
- produce a variety of products on an assembly line,
- adapt to change in product design,
- design facilities to provide flexibility and adaptability,
- increase product customization,
- provide parallel business functions and processes,
- closely couple vendors and customers.

Decentralized control and emergent behavior can offer insight that transcends managerial savvy! Effective management of information, which is essential to achievement of these goals, requires radical change rather than incremental improvement.

References

1. J. Rayport and J. Sviokla, "Exploiting the Value Chain", *Harvard Business Review*, November-December 1995, pages 75–85.
2. David Blanchar, editor, "Mass Customization Replacing Mass Production", *Intelligent Manufacturing*, June 1995.
3. B. Victor, A. Boynton, J. Daniels, M. Shank, "Back to Work!: The Right Path to Process Transformation". Unpublished Working Paper, June 1995.
4. J. Layden, "A Rapidly Changing Landscape: MES, MRP, and Scheduling", *Manufacturing Systems* — Supplement, March, 1996, pages 10a–18a.
5. Pattie Maes, "Modeling Adaptive Autonomous Agents", *Artificial Life*, Vol.1, No 1/2, page 141.
6. "Worried about PLC Programming Productivity? Try Another Flavor", *Advanced Manufacturing Research Control Strategy Report*, Advanced Manufacturing Research, Boston MA, Vol.1 No.6, June 1992.
7. "Agent Architecture Distributes Decisions for The Agile Manufacturer: Re-engineering at AlliedSignal Safety Restraint Systems", *Advanced Manufacturing Research CIM Strategies*, Advanced Manufacturing Research, Boston MA, June 1994, Vol.11, No.6.
8. Gilbert Staffend, "A Programmable Enterprise: The Ultimate in Reengineering for Agile Manufacturing, *ISADS-International Symposium on Autonomous Distributed Systems*, IEEE Press, April 1995.
9. The Third International Workshop on Agent Theories, Architectures, and Languages, Budapest, Hungary, August 12–13, 1996.
10. Fru-Ren Lin, "Reeingineering the Order Fulfillment Process in Supply Chain Networks: A Multi-Agent Information Systems Approach". An unpublished Dissertation Proposal, The University of Illinois, Urbana-Champaign, January 1996.
11. Cheng Hsu, Lester Gerhardt, David Spooner, and Alan Rubenstein, "Adaptive Integrated Manufacturing Enterprises: Information Technology for the Next Decade", *IEEE Transactions on Systems, Man, and Cybernetics*, Vol.24, No.5, May 1994, page 828.

Springer
and the
environment

At Springer we firmly believe that an international science publisher has a special obligation to the environment, and our corporate policies consistently reflect this conviction.

We also expect our business partners – paper mills, printers, packaging manufacturers, etc. – to commit themselves to using materials and production processes that do not harm the environment. The paper in this book is made from low- or no-chlorine pulp and is acid free, in conformance with international standards for paper permanency.

Springer

Druck: STRAUSS OFFSETDRUCK, MÖRLENBACH
Verarbeitung: SCHÄFFER, GRÜNSTADT